Beyond Octonion Cosmology II: Origin of the Quantum; A New Generalized Field Theory (GiFT); A Proof of the Spectrum of Universes; Atoms in Higher Universes

Stephen Blaha Ph. D.
Blaha Research

Pingree-Hill Publishing
MMXXI

ISBN: 978-1-7372640-7-1

Rev. 00/00/01 October 24, 2021

To Margaret

Some Other Books by Stephen Blaha

All the Megaverse! Starships Exploring the Endless Universes of the Cosmos using the Baryonic Force (Blaha Research, Auburn, NH, 2014)

SuperCivilizations: Civilizations as Superorganisms (McMann-Fisher Publishing, Auburn, NH, 2010)

All the Universe! Faster Than Light Tachyon Quark Starships & Particle Accelerators with the LHC as a Prototype Starship Drive Scientific Edition (Pingree-Hill Publishing, Auburn, NH, 2011).

Unification of God Theory and Unified SuperStandard Model THIRD EDITION (Pingree Hill Publishing, Auburn, NH, 2018).

The Exact QED Calculation of the Fine Structure Constant Implies ALL 4D Universes have the Same Physics/Life Prospects (Pingree Hill Publishing, Auburn, NH, 2019).

Unified SuperStandard Theory and the SuperUniverse Model: The Foundation of Science (Pingree Hill Publishing, Auburn, NH, 2018).

Quaternion Unified SuperStandard Theory (The QUeST) and Megaverse Octonion SuperStandard Theory (MOST) (Pingree Hill Publishing, Auburn, NH, 2020).

Unified SuperStandard Theories for Quaternion Universes & The Octonion Megaverse (Pingree Hill Publishing, Auburn, NH, 2020).

The Essence of Eternity: Quaternion & Octonion SuperStandard Theories (Pingree Hill Publishing, Auburn, NH, 2020).

A Very Conscious Universe (Pingree Hill Publishing, Auburn, NH, 2020).

From Octonion Cosmology to the Unified SuperStandard Theory of Particles (Pingree Hill Publishing, Auburn, NH, 2020).

Beyond Octonion Cosmology (Pingree Hill Publishing, Auburn, NH, 2021).

Available on Amazon.com, bn.com Amazon.co.uk and other international web sites as well as at better bookstores (through Ingram Distributors).

CONTENTS

FIGURES and TABLES

Introduction

Condensed Matter Physics has been blessed with a very wide range of phenomena that has attracted broad continuing interest. Elementary Particle Physics has been moribund for several decades—mostly due to the absence of major new experimental results at higher energies since the 1970s. This book, as well as earlier books, attempts to broaden theoretical efforts with new concepts and ideas due to Octonion Cosmology.

To that end it develops a deeper basis for Quantum Theory. The new basis is required by the higher dimension universes that appear in Octonion Cosmology ranging up to 20 space-time dimensions. As a result Quantum Theory requires a new formulation of Field Theory: the author's solution is a Generalized Field Theory (GiFT), which leads ultimately to Quantum Mechanics in a series of stages with each dependent on the previous:

Origin: Generalized Field Theory (GiFT)
to
Quantum Field Theory
to
Quantum Mechanics

Quantum Mechanics issues are resolved by Quantum Field Theory or ultimately by GiFT.

The book develops an Extended Quantum Mechanics that combines quantum and classical mechanics with each being a limiting case of the extended theory. Extended Quantum Mechanics is the Quantum Mechanical equivalent of PseudoQuantum Field Theory, a part of GiFT.

GiFT is based on Two-Tier Quantum Theory and PseudoQuantum Field Theory, both of which the author developed some time ago. They resolve divergence problems, and also non-static coordinate system quantization issues, that appear in higher dimension universes. Four dimension renormalization methods do not work in higher dimensions. Particle quantization has difficulties in non-static coordinate systems with no Killing vector. GiFT resolves those difficulties. GiFT contains a local SU(1, 1) group structure that gives a new interpretation to creation/annihilation operators in field theory.

The extended Quantum Mechanics version of GiFT is CQ Mechanics – a Quantum Mechanics that also embodies a classical mechanics limit. It appears in appendix C.

The Octonion Universes Spectrum is closely related to spinor arrays. We have shown spinor arrays have a similar block structure to hypercomplex numbers, which, in turn, are closely related to the structure of internal symmetry groups. In this book we derive the Octonion Cosmology Spectrum of universes partly based on spinor array features.

Lastly, QED-like interactions in higher dimension universes are considered including the EM potential and Bohr-like atomic energy levels. It suggests our kind of Life is unlikely in higher dimension universes.

1. Higher Dimensions Quantum Theory

We are familiar with Quantum Theory, as it is usually developed, in our universe. In universes with higher space-time dimensions there is a need for a more expansive Quantum Field Theory. We presented this theory in earlier books on Two-Tier Quantum Theory[1] and PseudoQuantum Field Theory.[2] It resolved all divergences in ElectroWeak and Strong interaction theories. It also eliminated divergences in other types of Quantum Field Theories including theories with higher order derivatives and four fermion interactions.

In this book we focus on the higher space-time dimension spaces (universes) of Octonion Cosmology.[3] In these universes we find Two-Tier Quantum Theory is *absolutely* necessary to avoid divergences in perturbation theory calculations. The universes of Octonion Cosmology have Feynman propagators and Perturbation Theory terms with integrations of the form:

$$\int d^n k\ f(k) \tag{1.1}$$

where n = 4, 8, 10, 12, 14, 16, and 18 (and also a less interesting case n = 2.)

In eight space-time dimensions the second order single fermion loop vacuum polarization is sextic divergent: $\int d^8k/k^2 \sim k^6$. Two-Tier Quantum Theory eliminates divergences in Perturbation Theory with exponentiated Gaussian quadratic terms in momenta of the form:

$$\exp(-ak^2) \tag{1.2}$$

where a is a constant. Two-Tier Quantum Theory is required for Perturbation Theory calculations in higher dimensions.

PseudoQuantum Field Theory is also needed in Octonion Cosmology. It is needed for proper quantization in arbitrary coordinate systems that might be relevant in higher dimension Octonion Cosmology spaces. For example, consider coordinate systems for non-static space-times where no time-like Killing vector exists.[4] It also enables canonical higher order derivative theories for quark confinement[5] and other

---

[1] See Blaha (2005a) *Quantum Theory of the Third Kind*, which is reprinted later in this volume in Appendix B.

[2] Blaha (2018e) and earlier books.

[3] Appendix A contains a detailed outline of Octonion Cosmology to which the reader is directed.

[4] B. DeWitt, Phys. Rep. **19**, 295 (1975) and references therein. S. Blaha, "New Framework for Gauge Field Theories", IL Nuovo Cimento **49A**, 113 (1979). See Appendix E.

[5] See Appendix F.

purposes. And it dovetails with Two-Tier Coordinates to "dress" bare fermion and boson particles.

1.1 Reasons For Two-Tier Quantum Theory

Originally Two-Tier coordinates were developed by this author to remove infinities that appear in perturbation theory calculations. We showed that the quantum smeared coordinates of Two-Tier Quantum Field Theory succeeded in removing all ultra-violet infinities in perturbation theory including the fermion triangle infinities.

Remarkably the high precision, "low" energy[6] *predictions of QED remained true in Two-Tier QED and thus remained consistent with experiment to a hitherto unsurpassed level of accuracy.* "Low" energy predictions in other quantum field theories also remained unchanged. At high energies, Two-Tier perturbation theory results are finite and consequently all ultra-violet infinities, to any order in perturbation theory, in *any number of space-time dimensions* were eliminated.

In addition to removing perturbation theory infinities, Two-Tier coordinates enable us to define finite theories of Quantum Gravity and 'non-renormalizable' quantum field theories based on polynomial Lagrangians, to tame vacuum fluctuations, to eliminate infinities associated with the Big Bang, and possibly to generate the explosive growth of the universe in its role as a type of Dark Energy.[7]

1.2 Two-Tier Features in 4-Dimensional Space-Time

Two-Tier Quantum Field Theory[8] is based on a new method[9] in the Calculus of Variations that uses two 'layers' of fields to introduce quantum coordinates. We shall consider this technique for the specific case of a massless vector field $Y^\mu(y)$ where the index μ ranges from 0 through 3. It is.analogous to the electromagnetic field.

The X^μ coordinate system, where it appears, has a c-number real part and a q-number imaginary part. Thus particle fields which are normally defined on real four-dimensional real space-time will now be defined on a "slightly" complex four-dimensional space-time:

$$X^\mu(y) = y^\mu + i\, Y^\mu(y)/M_c^2 \tag{1.3}$$

where M_c is an extremely large mass of the order of the Planck mass or larger.

The $Y^\mu(y)$ field is a function of the space y coordinates. The real part of the space-time dimensions will be taken to be the space of real-valued y coordinates.[10]

The imaginary part of space-time coordinates is the massless $Y^\mu(y)$ vector quantum field that is suppressed by the very large mass scale. The effects of Quantum Dimensions only become appreciable in quantum field theory at energies of the order of

---

[6] Relative to a mass scale that was perhaps of the order of the Planck mass.

[7] See Blaha (2017b) and earlier books for details. This section is basically a summary of some features.

[8] See Blaha (2005a), and Blaha (2002), for discussions of this new method to eliminate infinities in quantum field theory calculations. See Appendix B.

[9] See Blaha (2005a)..

[10] In a deeper theory the real part might also be a quantum field that undergoes a condensation to generate c-number coordinates. We will not consider this possibility in this book.

M_c. At these energies exponential Gaussian factors in each particle (and ghost) propagator are generated by the Quantum Dimensions and serve to make *all* perturbation theory calculations ultra-violet finite – including calculations in Quantum Gravity. Later we will see that the Two-Tier formalism may be extended directly to the universes of Octonion Cosmology with similar results – finiteness in Perturbation Theory.

The Two-Tier formalism introduces a new form of interaction that does not have the form of the simple polynomial interactions that have hitherto dominated quantum field theories. This form of interaction takes place via the composition of quantum fields and can be called a *Dimensional Interaction* or an *Interdimensional Interaction* since it affects particle behavior through Quantum Dimensions.

The basic *ansatz* of the Two-Tier formalism is to replace every appearance of a coordinate x in a quantum field with the variable

$$x^\mu \rightarrow X^\mu = (y^0, \mathbf{y} + \mathbf{Y}(y^0, \mathbf{y})/M_c^2) \tag{1.4}$$

where $\mathbf{Y}(y^0, \mathbf{y})$ is the spatial part of a free massless vector field with features that are identical to the free QED field in the Radiation gauge.

Then one finds that the momentum space free field Feynman propagators G(k) of all particles acquires a Gaussian factor exp(h(k)):

$$G(k) \rightarrow G(k)\exp(h(k)) \tag{1.5}$$

so that all perturbation theory diagrams are finite. The result is finite perturbative results for all calculations to any order in perturbation theory. Blaha (2005a) shows that Two-Tier theories are finite, Poincare covariant, and unitary. (See Blaha (2005a) or chapter 5 in Appendix B, for a complete discussion.)

1.3 Two-Tier Quantum Coordinates Formalism

In this section we will introduce the basic Two-Tier formalism. Taking the Lagrangian described in Blaha (2005a):[11]

$$\mathscr{L}(y) = \mathscr{L}_F(X^\mu(y))J + \mathscr{L}_C(X^\mu(y), \partial X^\mu(y)/\partial y^\nu, y) \tag{1.6}$$

where

$$X^\mu(y) = y^\mu + i\,Y^\mu(y)/M_c^2 \tag{1.7}$$

with M_c being a large mass scale, $Y_\mu(y)$ a vector quantum field, and where J is the absolute value of the Jacobian of the transformation from X to y coordinates:

$$J = |\partial(X)/\partial(y)|$$

The Lagrangian term $\mathscr{L}_C$ is

[11] Eq. 7.1 in Appendix B..

$$\mathscr{L}_C = +¼\, M_c{}^4 F^{\mu\nu} F_{\mu\nu}$$

with

$$F_{\mu\nu} = \partial X_\mu/\partial y^\nu - \partial X_\nu/\partial y^\mu \qquad (1.8)$$
$$\equiv i\,(\partial Y_\mu/\partial y^\nu - \partial Y_\nu/\partial y^\mu)/M_c{}^2$$

The Lagrangian term $\mathscr{L}_F$ ($X^\mu(y)$) contains the terms for scalar, fermion and other gauge terms in general. The sign in $\mathscr{L}_C$ is not negative – contrary to the conventional electromagnetic Lagrangian. The reason for this difference is that the quantum field part of X^μ is imaginary. Thus $\mathscr{L}_C$ ends up having the correct sign after taking account of the factor of i in the field strength $F_{\mu\nu}$.

Defining

$$F_{Y\mu\nu} = (\partial Y_\mu/\partial y^\nu - \partial Y_\nu/\partial y^\mu)$$

we see the Lagrangian assumes the form of the conventional electromagnetic Lagrangian:

$$\mathscr{L}_C = -¼\, F_Y{}^{\mu\nu} F_{Y\mu\nu}$$

The action of this theory has the form

$$I = \int d^4y\, \mathscr{L}(y)$$

Since $X^\mu(y)$ has an imaginary part there would appear to be an issue with the conservation of probability (unitarity). *We show unitarity is not a problem later in this chapter in section 1.8.*

1.4 Y^μ Gauge

The gauge invariance of the Lagrangian allows us to choose a convenient gauge. The gauge invariance of the full Lagrangian:

$$\mathscr{L}_S = L_F(\phi(X), \partial\phi/\partial X^\mu)\, J + \mathscr{L}_C(X^\mu(y), \partial X^\mu(y)/\partial y^\nu)$$

is based on the standard gauge invariance of $\mathscr{L}_C$, and the gauge invariance of $J\mathscr{L}_F$ in the form of translational invariance

$$X^\mu(y) \rightarrow X^\mu(y) + \delta X^\mu(y)$$

for the special case of a translation of X with the form of a gauge transformation:

$$\delta X^\mu(y) = \partial\Lambda(y)/\partial y_\mu$$

In this case we find

$$\int d^4y\ \Lambda(y)\ \partial\,[\,J\,\partial/\partial X^\mu\ \mathscr{T}_{F\mu\nu}\,]/\partial y_\nu = 0 \tag{1.9}$$

after a partial integration. Thus we have the differential conservation law:

$$\partial\,[\,J\,\partial\mathscr{T}_{F\mu\nu}/\partial X^\mu]/\partial y_\nu = 0$$

since $\Lambda(y)$ is arbitrary. This conservation law is trivially obeyed:

$$\partial\mathscr{T}_{F\mu\nu}/\partial X^\mu = 0 \tag{1.10}$$

Thus translational invariance in the $\mathscr{L}_F$ sector together with standard gauge invariance in the $\mathscr{L}_C$ sector automatically guarantees Y field gauge invariance of the total Lagrangian. We use the separate invariance of each term of

$$L = \int d^4y\ [\mathscr{L}_F\, J + \mathscr{L}_C\,] = \int d^4X\ \mathscr{L}_F + \int d^4y\ \mathscr{L}_C = L_F + L_C$$

under a constant translation $X^\mu \rightarrow X^\mu + \delta X^\mu$ where δX^μ is constant. Then we consider a position dependent translation/gauge transformation, which taken together with the above equation, establishes the invariance under the position dependent translation/gauge transformation.

An alternate approach that leads to the same result is to start with the particle part of the Lagrangian $\mathscr{L}_F$ rewritten to be invariant under general coordinate transformations, as it must, when we generalize to include General Relativity. Since position dependent translations are a form of general coordinate transformation the full theory must be invariant under position dependent translations due to invariance under general coordinate transformations.

Having established invariance under gauge transformations we now choose to use the most convenient gauge – the radiation gauge[12]:

$$\partial Y^i/\partial y^i = 0 \tag{1.11}$$

where i = 1, 2, 3. In the absence of external sources, we set

$$Y^0 = 0$$

since Y^0 does not have a canonically conjugate momentum. A conventional treatment leads to the equal time commutation relations:

---

[12] It is also possible to quantize using an indefinite metric that preserves manifest Lorentz covariance as was done by Gupta and Bleuler for the electromagnetic field. We will use the Gupta-Bleuler approach later to establish covariance under special relativity later. Now we opt for manifest positivity and use the radiation gauge.

$$[Y^\mu(\mathbf{y}, y^0), Y^\nu(\mathbf{y}', y^0)] = [\pi^\mu(\mathbf{y}, y^0), \pi^\nu(\mathbf{y}', y^0)] = 0 \tag{1.12}$$
$$[\pi_j(\mathbf{y}, y^0), Y_k(\mathbf{y}', y^0)] = -i\,\delta^{tr}{}_{jk}(\mathbf{y} - \mathbf{y}')$$

where (Note the locations of the j indexes above introduce a minus sign.)

$$\pi^k = \partial\mathscr{L}_C/\partial Y_k'$$
$$\pi^0 = 0$$

$$\delta^{tr}{}_{jk}(\mathbf{y} - \mathbf{y}') = \int d^3k\, e^{i\,\mathbf{k}\cdot(\mathbf{y} - \mathbf{y}')}(\delta_{jk} - k_j k_k/\mathbf{k}^2)/(2\pi)^3 \tag{1.13}$$
$$Y_k' = \partial Y_k/\partial y^0$$

The Radiation gauge reveals the two degrees of freedom that are present in the vector potential. The Fourier expansion of the vector potential is:

$$Y^i(y) = \int d^3k\, N_0(k) \sum_{\lambda=1}^{2} \varepsilon^i(k, \lambda)[a(k,\lambda)\, e^{-ik\cdot y} + a^\dagger(k,\lambda)\, e^{ik\cdot y}] \tag{1.14}$$

where

$$N_0(k) = [(2\pi)^3 2\omega_k]^{-\frac{1}{2}}$$

and (since m = 0)

$$\omega_k = (\mathbf{k}^2)^{\frac{1}{2}} = k^0$$

with $\vec{\epsilon}(k, \lambda)$ being the polarization unit vectors for $\lambda = 1,2$ and $k^\mu k_\mu = 0$.

The further development of Two-Tier theory is described in Part 3 of Blaha (2005a) which is reprinted in Appendix B.

1.5 Two-Tier Uncertainty Principle

The Uncertainty Relation for Quantum Mechanics is based on the coordinate-momentum commutator. Similarly, in defining Quantum Coordinates we have established a commutation relation based on Y^μ. Its conjugate momentum is

$$P^\mu(y) = i\pi_Y{}^\mu(y)/M_c{}^2 \tag{1.15}$$

where

$$\pi_Y{}^\mu(y) = -\, dY^\mu/(y)dy^0$$

In the Radiation gauge (eq. 1.11) we see

$$\pi_Y{}^0(y) = 0 \tag{1.16}$$

and

$$[X^0, P^0] = [y^0, p^0] = 0 \tag{1.17}$$

The non-zero equal time commutation relation expressing a quantum Uncertainty Relation is

$$[P^i(\mathbf{y}, y^0), Y^k(\mathbf{y}', y^0)] = -[\pi_Y{}^i(y), Y^k(y),]/M_c{}^4 \quad (1.18)$$

$$= i\delta^{trik}(\mathbf{y} - \mathbf{y}')/M_c{}^4$$

using eq. 1.12.

At low energy the impact of the Two-Tier Uncertainty Relation is diminished (more or less eliminated) by the factor $M_c{}^4$. Then Two-Tier Quantum Field Theory becomes ordinary Quantum Field Theory with the same results in Perturbation Theory. This limit is described in detail in Blaha (2005a), which appears in in Appendix B.

In the "low" energy limit the conventional Heisenberg Uncertainty Condition becomes evident as shown by Heitler (1954) and others. *Conventional Quantum Mechanics is a result of Second Quantization.*

1.6 Quantum Mechanics vs. Quantum Field Theory vs. Two-Tier Quantum Theory

Historically, Quantum Mechanics predated Quantum Field Theory, which, in turn, predates the Two-Tier Quantum Theory developed by the author in the early 2000's.

Logically, Two-Tier Quantum Theory, which embodies an Uncertainty Relation at ultrahigh energies, is the fundamental Quantum Theory. It is the predecessor of Quantum Field Theory, which only embodies the Heisenberg Uncertainty Relation. Quantum Field Theory is flawed by high energy divergences just as atomic physics, before the Bohr atom and Quantum Mechanics, was flawed by infinities in hydrogen atom models.

Quantum Field Theory (a "low" energy theory) implies Quantum Mechanics[13] as shown in Heitler[14] as well as elsewhere.

Quantum Mechanics is often treated as a complete, self-contained theory. On occasion paradoxes and ambiguities are found in Quantum Mechanics studies. They may be resolved within Quantum Mechanics. Any, that are not so resolved, should be considered within the framework of Quantum Field Theory, which is the ultimate forum for all quantum phenomena at "low" energy.

Note the trend of quantum theories, from the earliest at "low" energy theory, Quantum Mechanics, to the highest energy quantum theory: Two-Tier Quantum Theory,

Two-Tier Quantum Field Theory is a welcome extension of Quantum Theory to the deepest levels of Quantum Theory.

1.7 Two-Tier Perturbation Theory

The form of Two-Tier Perturbation Theory is similar to the Perturbation Theory of conventional Quantum Field Theory. It is described in *Quantum Theory of the Third*

[13] Quantum Mechanics is often treated as an independent theory. This practice may be valid mathematically but it is not valid physically. For example, many phenomena in atomic theory require explanation in terms of Quantum Field Theory.

[14] Heitler (1954).

Kind, which is in Appendix B in chapters 5 and 6. The reduction development for the U-matrix and the S-matrix are presented there.

1.8 Two-Tier Unitarity

The unitarity of Two-Tier Quantum Field Theory can be viewed in the cases of "low" energy phenomena and "high" energy phenomena with the scale mass of M_c separating the cases.

In the case of "low" energy phenomena the Two-Tier S-matrix gives results identical to conventional S-matrix results. Thus there is no conflict with unitarity for Two-Tier S-matrix results at "low" energy.

At "high" energy there are two potential issues: asymptotic states containing Y quanta; and conservation of probability. The first issue is resolved in the chapter 6 discussion in Appendix B.

The second issue leads to a requirement to renormalize S-marix probability amplitudes such that their absolute values squared sum to unity – the unitarity condition. If we let S-matrix elements have the form S_{fi} where i represents an initial state and f represents one of its final states, then

$$\sum_n S_{nf}{}^* S_{ni} = \delta_{fi} \tag{1.19}$$

expresses the conventional unitarity condition. In Two-Tier Quantum Field Theory the unitarity condition for energies much less than M_c is the same as eq. 1.19 since perturbation theory results are the same as conventional Quantum Field Theory. The sum over intermediate states n is restricted to states whose total energy is less than M_c by energy conservation.

For initial states i with energy E_i of the order of or greater than M_c, the intermediate state total energy and the final state total energy are also E_i by energy conservation. In this case the sum in eq. 1.19 *may* be

$$\sum_n S_{nf}{}^* S_{ni} = g_i \delta_{fi} \tag{1.20}$$

where g_i is a constant term dependent on the state i. In this circumstance unitarity may be recovered by redefining S_{ni} and S_{nf} as

$$S_{ni}{}' = S_{ni}/\sqrt{g_i} \tag{1.21}$$
$$S_{nf}{}' = S_{nf}/\sqrt{g_f} \tag{1.22}$$

Then the *normalized* Two-Tier unitarity condition can be expressed as

$$\sum_n S_{nf}{}'^* S_{ni}{}' = \delta_{fi} \tag{1.23}$$

with the understanding that $g_i = 1$ for initial states with energies much less than M_c. Thus Two-Tier Quantum Field Theory satisfies an enhanced unitarity condition.

1.9 Two-Tier Quantum Space Theory

Quantum Space Theory is the theory of particles containing spaces internally. It was recently developed by the author.[15] This theory can directly put in the form of Two-Tier Quantum Space Theory by following the procedure described in section 1.2. The structure of the Octonion Spaces Spectrum (Appendix A has a summary) is not changed by this formulation.

1.10 CQ Mechanics – A Union of Classical and Quantum Mechanics

The author developed a larger theory, CQ Mechanics, containing both Quantum Mechanics and Classical Mechanics that is described in Blaha (2016f). In this theory a phenomenon can be described as classical in one limit and as quantum in another limit. The description is determined by an angle that specifies one, or the other, limiting case. Intermediate cases combining both quantum and classical are also present. The theory appears here as Appendix C.

This theory can bridge the classical and quantum regimes. We discuss some applications in Appendix C: a generalized Feynman path integral formalism, a generalized Schrödinger equation, a generalized Boltzmann equation, the Fokker-Planck equation, a generalized approach to quantum and classical chaos, and to quantum entanglement as well as semi-quantum entanglement. Our formalism applies to Quantum Mechanics as well as the path integrals, the Fokker-Planck equation and the Boltzmann equation.

This "Mechanics" theory has an analogue at the field theoretic level, PseudoQuantum Field Theory that is briefly described next.

1.11 PseudoQuantum Field Theory

PseudoQuantum Field Theory[16] was developed by the author in the 1970's and presented in a series of articles including the papers below that appear in Appendices E and F :

S. Blaha, "The Local Definition of Asymptotic Particle States", IL Nuovo Cimento **49A**, 35 (1979). It describes the PseudoQuantization of boson and fermion field theories for use in the quantization of fields in universes and the Megaverse.

S. Blaha, "New Framework for Gauge Field Theories", IL Nuovo Cimento **49A**, 113 (1979). It describes the PseudoQuantization of gauge field theories for the purposes of defining higher derivative field theories and for use in the quantization of fields in universes and the Megaverse.

PseudoQuantum Field Theory (and its Quantum Mechanics analogue CQ Mechanics[17]) originated in the need to second quantize in unusual coordinate systems,

---

[15] Blaha (2021f) and (2021g).

[16] Appendix D describes PseudoQuantum Theory features in some detail.

[17] See Appendix C for details, which contains Blaha (2016f). CQ Mechanics encompasses both classical mechanics and quantum mechanics, and provides a method of rotating between them. It has applications to transitions between Quantum/Semi-Classical Entanglement, and Quantum/Classical Path Integrals, and Quantum/Classical Chaos.

and in curved space-time coordinate systems. It is also very relevant for the canonical formulation of higher derivative field theories for quark confinement and other applications.

PseudoQuantum Field Theory is formulated by duplicating all fields, both fermion and boson fields, in a "normal" Lagrangian theory. Scalar field theory provides a simple example that illustrates the PseudoQuantum Field Theory procedure.

We duplicate a scalar field generating two scalar fields: $\varphi_1(x)$ and $\varphi_2(x)$. We choose $\varphi_1(x)$ to have a zero equal time commutator with $d\varphi_1(x)/dx^0$ and $\varphi_2(x)$ to have a conventional equal time commutator with $d\varphi_2(x)/dx^0$. Conceptually $\varphi_1(x)$ is a "classical" field and $\varphi_2(x)$ is a quantum field. A Lagrangian that implements these choices of commutation relations is:

$$\mathcal{L} = \partial^\mu \varphi_1(x)\partial_\mu\varphi_2(x) - ½\, \partial^\mu \varphi_1(x)\partial_\mu\varphi_1(x) - m_2^2\, \varphi_1(x)\varphi_2(x) + ½\, m_1^2\, \varphi_1(x)^2 \quad (1.24)$$

$$(\square + m_2^2)\varphi_1(x) = 0 \quad (1.25)$$

$$(\square + m_2^2)\varphi_2(x) - (\square + m_1^2)\varphi_1(x) \quad = 0 \quad (1.26)$$

The canonical momenta are

$$\pi_1 = d\varphi_2(x)/dt - d\varphi_1(x)/dt \quad (1.27)$$
$$\pi_2 = d\varphi_1(x)/dt \quad (1.28)$$

and the equal time commutation relations are

$$[\varphi_i(x), \pi_j(y)] = i\delta_{ij}\delta(\mathbf{x} - \mathbf{y}) \quad (1.29)$$
$$[\varphi_i(x), \varphi_j(y)] = [\pi_i(x), \pi_j(y)] = 0 \quad (1.30)$$

implying

$$[\varphi_1(x), d\varphi_1(y)/dt] = 0 \quad (1.31)$$
$$[\varphi_2(x)\, , d\varphi_2(y)/dt] = i\delta(\mathbf{x} - \mathbf{y}) \quad (1.32)$$
$$[\varphi_1(x)\, , d\varphi_2(y)/dt] = i\delta(\mathbf{x} - \mathbf{y}) \quad (1.33)$$

1.11.1 Color Confinement

Appendix F contains papers on a non-Abelian gauge field theory that gives explicit color confinement using a PseudoQuantum formulation that leads to a confining r-potential.

If we set $m_2 = 0$ in the above example, we see a fourth order field equation results

$$\square^2\varphi_2(x) \;= 0 \quad (1.34)$$

Color confinement results in the non-Abelian gauge field case in a similar manner.

1.11.2 PseudoQuantum Quantization for Non-Static Coordinate Systems

The PseudoQuantum formalism that was developed by the author for "normal" and non-static coordinate systems in the 1970s appears in papers reprinted in Appendix E.

1.11.3 Advantages of PseudoQuantum Quantization

In this section we point out some of its advantages in a variety of field theory contexts that are relevant for Octonion Cosmology and Quantum Field Theory in general. Appendix D describes PseudoQuantum features in more detail.

Some advantages of PseudoQuantum Field Theory are:

1. Quantization in any coordinate system in flat or curved space-times with an invariant definition of asymptotic particle states. *This is especially important for the higher dimension spaces of Octonion Cosmology.* An n particle asymptotic state in one coordinate system is a unitarily equivalent n particle asymptotic state in any other coordinate system. Therefore particle number is invariant under change of coordinate system. This is important for the Unified SuperStandard Theory in curved space-times. It is also important for quantization in higher dimensional spaces such as those of Octonion Cosmology. The method was developed in the late 1970's by the author to provide a quantization procedure which supports a unique particle interpretation of states in arbitrary non-static space-times where no global time-like coordinate (Killing vector) exists. PseudoQuantum Field Theory which we developed in a series of books[18] also can be formulated in the Octonion Spectrum of spaces. For example, we can use it to implement the Higgs Mechanism to generate particle masses and symmetry breaking.

2. PseudoQuantum Field Theory enables one to define Higgs particle dynamics in such a way that a non-zero vacuum expectation value cleanly separates from the quantum field part of the Higgs fields. This technique can be used in symmetry breaking mechanisms, mass generation, and possible generation of coupling constants as vacuum expectation values.

3. It supports the canonical definition of higher derivative field theories through the use of the Ostrogradski bootstrap. We have used it to construct a fourth order theory of the Strong interaction that has color confinement and a linear r potential. The potential part of this theory was used by the Cornell group to calculate the Charmonium spectrum. (See Blaha (2017b) for details.)

1.12 Combination of Two-Tier and PseudoQuantum Formalisms

These formalisms can be directly combined with no issues. See chapters 2 and 3.

---

[18] See Blaha (2017b) for the discussion of the PseudoQuantum field theory formalism for Higgs particles in our Extended Standard Model. See chapter 20 of Blaha (2017b), and earlier books, for a more detailed view than that presented here.

1.13 A Model Illustrating Scalar Field Quantization Using X^μ

We begin by considering the case of a scalar quantum field theory. We assume a real underlying y subspace. Since X^μ is a set of coordinates, we choose to define a scalar field ϕ as a function of X^μ, which, in turn, is a function of the y^ν coordinates. We will provisionally second quantize ϕ treating X^μ as c-number coordinates using a conventional approach.[19]

We assume a Lagrangian, with the momentum conjugate to ϕ:

$$\pi_\phi = \partial L_F / \partial \phi' \equiv \partial L_F / \partial(\partial\phi/\partial X^0) \tag{1.35}$$

Following the canonical quantization procedure, π and ϕ become Hermitian operators with equal time ($X^0 = X^{0\prime}$) commutation rules:

$$\begin{aligned} [\phi(X), \phi(X')] &= [\pi_\phi(X), \pi_\phi(X')] = 0 \\ [\pi_\phi(X), \phi(X')] &= -i\,\delta^3(\mathbf{X} - \mathbf{X}') \end{aligned} \tag{1.36}$$

The standard Fourier expansion of the solution to the Klein-Gordon equation is:

$$\phi(X) = \int d^3p\, N_m(p)\, [a(p)\, e^{-ip\cdot X} + a^\dagger(p)\, e^{ip\cdot X}] \tag{1.36a}$$
$$= \int d^{3k}p\, N_m(p)\, [a(p) \exp(-ip\cdot(y + Y/M_c^2)) + + a^\dagger(p) \exp(ip\cdot(y + Y/M_c^2))] \tag{1.36b}$$

where

$$N_m(p) = [(2\pi)^3 2\omega_p]^{-1/2}$$

and

$$\omega_p = (\mathbf{p}^2 + m^2)^{1/2}$$

The commutation relations of the Fourier coefficient operators are:

$$\begin{aligned} [a(p), a^\dagger(p')] &= \delta^3(\mathbf{p} - \mathbf{p}') \\ [a^\dagger(p), a^\dagger(p')] &= [a(p), a(p')] = 0 \end{aligned}$$

The reader will recognize the quantization procedure is formally identical to the standard canonical quantization procedure of a free scalar quantum field.

In the case of spin ½, spin 1 and spin 2 fields the standard quantization procedure *in terms of the X coordinate system* can also be followed in a way similar to the procedure in standard texts.

---

[19] Some texts are: Bogoliubov, N. N., Shirkov, D. V., *Introduction to the Theory of Quantized Fields* (Wiley-Interscience Publishers Inc., New York, 1959); Bjorken, J. D., Drell, S. D., *Relativistic Quantum Fields* (McGraw-Hill, New York, 1965); Huang, K., *Quarks, Leptons & Gauge Fields Second Edition* (World Scientific, River Edge, NJ, 1992); Kaku, M., *Quantum Field Theory* (Oxford University Press, New York, 1993); Weinberg, S., *The Quantum Theory of Fields* (Cambridge University Press, New York, 1995).

1.14 Scalar Feynman Propagators

The momentum space free field Feynman propagators G…(k) of all particles and ghosts in all Two-Tier Quantum Field Theories acquires a Gaussian factor exp(h(k)):

$$G\ldots(k) \rightarrow G\ldots(k)\ \exp(h(k))$$

so that all perturbation theory diagrams are finite. The result is a finite perturbative result in all calculations to any order in perturbation theory. Blaha (2005a) shows that Two-Tier theories are finite, Poincare covariant, and unitary.

An example of the Two-Tier effect on propagators is the case of the Two-Tier photon propagator. The Two-Tier photon propagator[20] is:

$$iD_F^{TT}(y_1 - y_2)_{\mu\nu} = -i \int \frac{d^4p\ e^{-ip\cdot z}\ g_{\mu\nu}\ R(\mathbf{p}, z)}{(2\pi)^4\ (p^2 + i\varepsilon)} \qquad (1.37)$$

(since the imaginary parts can be taken to be zero: $y_{1i}^{\mu} - y_{2i}^{\mu} = 0$) where

$$z^{\mu} = y_{1r}^{\mu} - y_{2r}^{\mu}$$

$$R(\mathbf{p}, \mathbf{z}) = \exp[-p^i p^j \Delta_{Tij}(z)/M_c^4]$$

$$= \exp\{ -\mathbf{p}^2[A(v) + B(v)\cos^2\theta] / [4\pi^2 M_c^4 |\mathbf{z}|^2$$

with i, j = 1, 2, 3, and with $\Delta_{Tij}(z)$ being the commutator of the positive frequency part $Y^{+}_{k}(y)$ and the negative frequency part $Y^{-}_{k}(y)$ of $Y_k(y)$:

$$\Delta_{Tij}(z) = [Y^{+}_{j}(y_{1r}), Y^{-}_{k}(y_{2r})] = \int d^3k\ e^{ik\cdot(y_{1r} - y_{2r})}\ (\delta_{jk} - k_j k_k/\mathbf{k}^2)/[(2\pi)^3 2\omega_k] \qquad (1.38)$$

and

$$v = |z^0|/|\mathbf{z}|$$
$$A(v) = (1 - v^2)^{-1} + .5v\ \ln[(v - 1)/(v + 1)]$$
$$B(v) = v^2(1 - v^2)^{-1} - 1.5v\ \ln[(v - 1)/(v + 1)]$$
$$\mathbf{p}\cdot\mathbf{z} = |\mathbf{p}|\ |\mathbf{z}|\ \cos\theta$$

with $|\mathbf{p}|$ denoting the length of a spatial vector $\mathbf{p}$, $|\mathbf{z}|$ denoting the length of a spatial vector $\mathbf{z}$, and with $|z^0|$ being the absolute value of z^0.

The Gaussian factors R(**p**, z) which appear in all Two-Tier propagators damp the large momentum behavior of all perturbation theory integrals producing a completely finite perturbation theory and yet give the usual results of perturbation theory at energies that are small compared to the mass scale M_c.

---

[20] Blaha (2005a).

1.15 String-like Substructure of the Theory

Two-Tier Quantum field Theory endows each particle with an extended structure that resembles the extended structure seen in bosonic string and Superstring theories. For example, Bailin (1994) use the operator[21]

$$V_\Lambda(k) = \int d^2\sigma \sqrt{-h}\, W_\Lambda(\tau, \sigma)\, e^{-ik.X}$$

where X^μ is a quantized Fourier expansion of the string fields (see eq. 7.22 of Bailin (1994)).

We note our X^μ coordinate-field has two transverse degrees of freedom due to gauge invariance, which also invites comparison to the bosonic string. A point of difference is that we have a well-defined quantum field theoretic formulation in conventional space-time that has the Standard Model as its "large distance" behavior thus introducing a note of reality that is not apparent in Superstring theories. We see that the interacting quantum field theories based on this approach also have good, finite, short distance behavior just as string theories.

The scalar, and other particles', Feynman propagators can be viewed as describing the propagation of a particle cloaked (accompanied) by a cloud of Y particles (which generates the $R(\mathbf{p}, y_1 - y_2)$ factor in the above propagator). If we examine the Fourier transform of R(p, z) we see:

$$(2\pi)^4 R(\mathbf{p}, q) = \int d^4z\, e^{iq.z} R(\mathbf{p}, z) = \int d^4z\, e^{iq.z} \exp[-p^i p^j \Delta_{Tij}(z)/M_c^4] \tag{1.39}$$

and we find

$$R(\mathbf{p},q) = \sum_{n=0}^{\infty} [i(2\pi M_c)^4]^{-n} (n!)^{-1} \prod_{j=1}^{n} [\int d^4k_j\, \theta(k_j^0)(\mathbf{p}^2 - (\mathbf{p}\cdot\mathbf{k}_j)^2/\mathbf{k}_j^2)/(k_j^2 + i\varepsilon)]\, \delta^4(q - \Sigma\, k_r)$$

which can be interpreted as a "cloud" of Y particles dressing the "bare" particle propagator. (The apparent divergences for R(p, q) are an artifact of the expansion and the subsequent Fourier transformation. They are not present in the $R(\mathbf{p}, y_1 - y_2)$ factor in the propagator. See Fig. 1.1 for the Feynman diagram of the Two-Tier 'cloaked' propagator as compared to the normal scalar particle Feynman propagator. The Two-Tier Feynman propagator is basically a conventional scalar propagator that is modified by coherent Y particle emission.[22]

[21] D. Bailin and A. Love, *Supersymmetric Gauge Field Theory and String Theory* (Institute of Physics Publishing, Philadelphia, PA, 1994) page 272.
[22] T. W. B. Kibble, Phys. Rev. **173**, 1527 (1968) and references therein. In particular see p. 1532 of Kibble's paper.

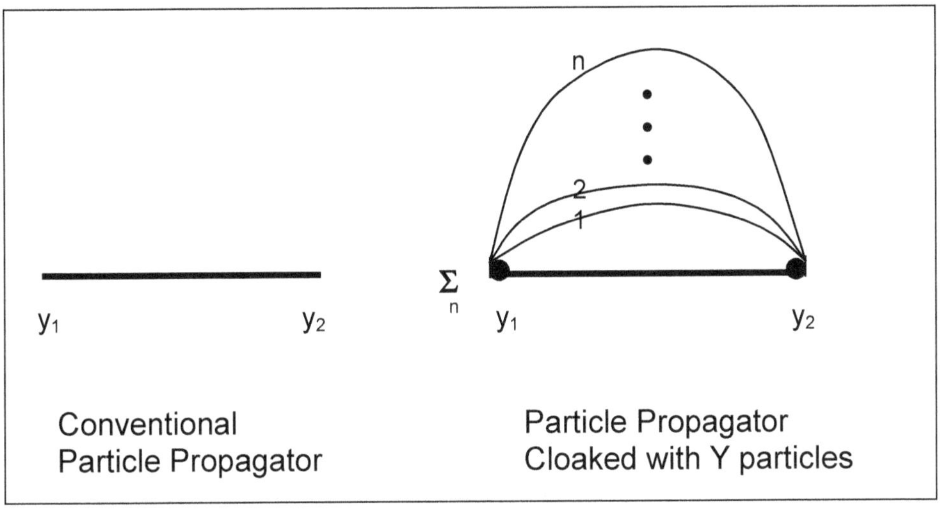

Figure 1.1. Feynman diagram for conventional and the nth diagram of a cloaked Two-Tier propagator.

We note that R(p, q) satisfies the convolution theorem:

$$\int d^4k\, R(\mathbf{p}, k)\, R(\mathbf{p}, q - k) = [R(\mathbf{p}, q)]^2$$

or

$$(2\pi)^4 \int d^4z\, e^{iq\cdot z}\, R(\mathbf{p}, z)\, R(\mathbf{p}, z) = [\int d^4z\, e^{iq\cdot z}\, R(\mathbf{p}, z)\,]^2 \qquad (1.40)$$

The proof follows from the Binomial theorem.

2. The Basis of Quantum Theory

Quantum Field Theory embodies Quantum Mechanics within it. The origin of Quantum Field Theory is somewhat unclear. It embodies the particulate nature of matter and energy in a relatively simple manner compared to other formalisms that might be developed. But the "quantum" aspect of Quantum Field Theory is somewhat elusive.

Some time ago this author saw a similarity between coordinates and particles. In some sense they both represented a cornerstone of Reality. Coordinates only make sense if matter/energy exists to demarcate coordinate points in space. On the other hand particles can only exist within a space. Thus they complement each other.

Octonion Cosmology has recognized this relation. Every one of its ten spaces has the total number of coordinates (space-time and internal symmetry coordinates) equaling the total number of fundamental fermions (normal and Dark, leptonic and quark) in the space.[23]

So there is a "sameness" in the set of fundamental fermions, the critical components for specifying coordinate location, and the set of coordinates (dimensions) of a space.

With this perspective in mind we now turn to establish a qualitative connection between Two-Tier Theory and PseudoQuantum Theory. As noted earlier conventional Quantum Field Theory can be directly extended to include both without any conflict. Two-Tier Theory eliminates high energy divergences that are present in conventional Quantum Field Theory. PseudoQuantum Theory eliminates the issues of particle number ambiguity in General Relativity when one transforms between arbitrary coordinate systems.

Two-Tier coordinates have the form:

$$X^{\mu}(y) = y^{\mu} + iY^{\mu}(y)$$

$X^{\mu}(y)$ has a classical c-number part y^{μ} and a quantum q-number part $Y^{\mu}(y)$.

Similarly a PseudoQuantum implementation has a c-number part and a q-number part. For example, in chapter 1 we considered a scalar particle model that was implemented with two fields $\varphi_1(x)$ and $\varphi_2(x)$. One field, φ_1, can be viewed as a "classical c-number" field. The other field, φ_2, can be viewed as a quantum q-number field.

The similarity between the coordinates and fields suggests the possibility of a deeper connection. We propose a new PseudoQuantum formulation in chapter 3 that unites it with Two-Tier Theory.

---

[23] The equivalence suggests QUeST space has the .right Dark sector fermion spectrum

3. Generalized Field Theory - GiFT

We propose a new formulation of Quantum Field Theory that we call Generalized Field Theory (GiFT) that combines PseudoQuantum Field Theory with Two-Tier Theory. The new theory originates in the author's 1979 paper, S. Blaha, "The Local Definition of Asymptotic Particle States", Il Nuovo Cimento **49A**, 35 (1979), which is reproduced in Appendix E. The purpose of the paper was to extend Quantum Field Theory in such a way as to allow particle states defined in a "conventional" coordinate system to be consistently defined in non-static space-times where no global time-like coordinate Killing vector exists. It appears this goal can be best achieved by associating each particle in a quantum field theory model with two quantum fields. This extension, which we call a PseudoQuantum formulation, must be accompanied by changes in the model's Lagrangian.

3.1 PseudoQuantum Formalism

We illustrate the procedure by considering the case of a scalar particle in four dimensions[24] paralleling the above mentioned paper (in Appendix E). We associate it with two scalar fields: $\varphi_1(x)$ and $\varphi_2(x)$. We choose $\varphi_1(x)$ to have a zero equal time commutator with $d\varphi_1(x)/dx^0$ and $\varphi_2(x)$ to have a conventional equal time commutator with $d\varphi_2(x)/dx^0$. Conceptually $\varphi_1(x)$ is a "classical" field and $\varphi_2(x)$ is a quantum field. A Lagrangian that implements these choices of commutation relations is:

$$\mathcal{L} = \partial^\mu \varphi_1(x)\partial_\mu\varphi_2(x) - ½\, \partial^\mu \varphi_1(x)\partial_\mu\varphi_1(x) - m^2\, \varphi_1(x)\varphi_2(x) + ½\, m^2\, \varphi_1(x)^2 \tag{3.1}$$

$$(\Box + m^2)\varphi_1(x) = 0 \tag{3.2}$$

$$(\Box + m^2)\varphi_2(x) - (\Box + m^2)\varphi_1(x) \quad = 0 \tag{3.3}$$

The canonical momenta are

$$\pi_1 = d\varphi_2(x)/dt - d\varphi_1(x)/dt \tag{3.4}$$
$$\pi_2 = d\varphi_1(x)/dt \tag{3.5}$$

and the equal time commutation relations are

$$[\varphi_i(x), \pi_j(y)] = i\delta_{ij}\delta(\mathbf{x} - \mathbf{y}) \tag{3.6}$$
$$[\varphi_i(x), \varphi_j(y)] = [\pi_i(x), \pi_j(y)] = 0 \tag{3.7}$$

for i, j = 1, 2 implying

$$[\varphi_1(x), d\varphi_1(y)/dt] = 0 \tag{3.8}$$

---

[24] The case of a space with a higher dimension spacetime is completely analogous.

$$[\varphi_2(x), d\varphi_2(y)/dt] = i\delta^3(\mathbf{x} - \mathbf{y}) \tag{3.9}$$
$$[\varphi_1(x), d\varphi_2(y)/dt] = i\delta^3(\mathbf{x} - \mathbf{y}) \tag{3.10}$$

The most general mode expansion of the fields is

$$\varphi_1(x) = \int d^3k\, [(c_{11}a_{1k} + c_{12}a_{2k})f_k(x) + (c'_{11}a^\dagger_{1k} + c'_{12}a^\dagger_{2k})f_k^*(x)] \tag{3.11}$$

$$\varphi_2(y) = \int d^3k\, [(c_{21}a_{1k} + c_{22}a_{2k})f_k(x) + (c'_{21}a^\dagger_{1k} + c'_{22}a^\dagger_{2k})f_k^*(x)] \tag{3.12}$$

with

$$f_k(x) = (2\pi)^{-3/2}\,(2\omega_k)^{-1/2}\exp(-ik\cdot x) \tag{3.13}$$

where the c_{ij} are all constants.

The creation/annihilation operators have the commutation relations:

$$[a_{ik'}, a_{jk}] = 0 \tag{3.14}$$
$$[a^\dagger_{ik'}, a^\dagger_{jk}] = 0$$

$$[a_{ik}, a^\dagger_{jk'}] = (1 - \delta_{ij})\delta^3(\mathbf{k} - \mathbf{k'}) \tag{3.15}$$

with i, j = 1, 2. Two related vacuums are defined: $|0>_1$ and $|0>_2$ by

$$a_{1k}|0>_2 = a^\dagger_{1k}|0>_2 = 0 \tag{3.16}$$
$$a_{2k}|0>_1 = a^\dagger_{2k}|0>_1 = 0$$

and

$$a_{ik}|0>_i \neq 0 \qquad a^\dagger_{ik}|0>_i \neq 0 \tag{3.17}$$

for i = 1, 2 and all k.

We define a Bogoliubov transformation (after Blaha) with

$$a_{ik}(\lambda_1, \lambda_2) \equiv B_{\lambda_1\lambda_2}a_{ik}B^{-1}_{\lambda_1\lambda_2} = \exp(i\lambda_1)\cosh(\lambda_2)a_{ik} + \exp(-i\lambda_1)\sinh(\lambda_2)a^\dagger_{ik} \tag{3.18}$$

where λ_1 and λ_2 are functions of k. The B transformation has the form:

$$B(\lambda_1(k), \lambda_2(k)) = B_{\lambda_1\lambda_2} = \exp[2i\int d^3k\, \lambda_1(k)\, \Gamma_{3k}]\exp[2i\int d^3k\, \lambda_2(k)\, \Gamma_{2k}] \tag{3.19}$$

where the Hermitian operators Γ_{ik} are

$$\Gamma_{3k} = (a^\dagger_{2k}a_{1k} + a_{2k}a^\dagger_{1k})/2 \tag{3.20}$$
$$\Gamma_{2k} = i(a^\dagger_{2k}a^\dagger_{1k} - a_{2k}a_{1k})/2 \tag{3.21}$$

We also define

$$\Gamma_{1k} = -(a^\dagger_{2k}a^\dagger_{1k} + a_{2k}a_{1k})/2 \tag{3.22}$$

Together these three operators satisfy SU(1, 1) algebra commutation relations:

$$[\Gamma_{1k}, \Gamma_{2k'}] = -i\delta_{kk'}\Gamma_{3k} \qquad [\Gamma_{2k}, \Gamma_{3k'}] = i\delta_{kk'}\Gamma_{1k} \qquad [\Gamma_{3k}, \Gamma_{1k'}] = i\delta_{kk'}\Gamma_{2k} \qquad (3.23)$$

The properties of this algebra are detailed in S. Blaha, "The Local Definition of Asymptotic Particle States", IL Nuovo Cimento **49A**, 35 (1979), which appears in Appendix E.

The equal-time commutation relations and the self-adjointness of φ_2 and the Hamiltonian H imply the fields in eq. 3.11 and 3.12 (after some algebra) have the parameters[25]

$$\begin{aligned} c_{11} &= \cos(\theta_1 - \theta_2)/\sin\theta_1 \\ c_{12} &= \sin(\theta_1 - \theta_2)/\sin\theta_1 \\ c'_{11} &= \cos(\theta_1 - \theta_2)/\cos\theta_1 \\ c'_{12} &= -\sin(\theta_1 - \theta_2)/\cos\theta_1 \end{aligned} \qquad (3.24)$$

$$\begin{aligned} c_{21} &= \cos\theta_2 \\ c_{22} &= \sin\theta_2 \\ c'_{21} &= \sin\theta_2 \\ c'_{22} &= \cos\theta_2 \end{aligned}$$

where $\theta_1(k)$ and $\theta_2(k)$ are functions of k that are not directly related to the Bogoliubov transformation of eq. 3.19.

3.2 Generalized Field Theory (GiFT)

The PseudoQuantum formalism can be extended to encompass a very general form of Quantum Field Theory. This generalization, Generalized Field Theory (GiFT), combines the PseudoQuantum formalism described above and in 1970s papers by the author printed in Appendices E and F.

The generalization is quite direct: The parameter $\lambda_1(k)$ is made a function of the $Y^\mu(y)$ field appearing in the expressions for Quantum Coordinates of Two-Tier Quantum Theory such as eq. 1.3. Thus $B(\lambda_1(k), \lambda_2(k))$ in eq. 3.19 generalizes to

$$B(\lambda_1(k) - k_\mu Y^\mu(y)/M_c^2, \lambda_2(k) = \exp[2i\int d^3k\, (\lambda_1(k) - k_\mu Y^\mu(y)/M_c^2)\, \Gamma_{3k}]\, \exp[2i\int d^3k\, \lambda_2(k)\, \Gamma_{2k}] \qquad (3.25)$$

Consequently the extended Bogoliubov transformation becomes[26]

$$\begin{aligned} a_{ik}(\lambda_1, \lambda_2) \equiv B_{\lambda_1\lambda_2} a_{ik} B^{-1}{}_{\lambda_1\lambda_2} = \exp(i\lambda_1 - ik_\mu Y^\mu(y)/M_c^2)\cosh(\lambda_2)a_{ik} + \\ + \exp(-i\lambda_1 + ik_\mu Y^\mu(y)/M_c^2)\sinh(\lambda_2)a^\dagger{}_{ik} \end{aligned} \qquad (3.26)$$

---

[25] See eqs. 41 and 42 of S. Blaha, Il Nuovo Cimento **49A**, 35 (1979) in Appendix E.

[26] $a_{ik}(\lambda_1, \lambda_2)$ displays a k parameter subscript only to indicate the implicit dependence of λ_1 and λ_2, and the term $k_\mu Y^\mu(y)$ on k.

If we set $\lambda_1 = \lambda_2 = 0$, then the Two-Tier form of a scalar particle Fourier expression emerges. See eq. 1.36b. The Bogoliubov transformation in this case transforms between coordinate systems with well-defined Killing vectors (time-like coordinates).

For the general case of eq. 3.25 one has the general case of Two-Tier coordinates and the PseudoQuantum formalism for transformations to non-static coordinate systems. GiFT unifies Two-Tier coordinates and the PseudoQuantum formalism.

The GiFT formalism can also be applied to other types of particles: fermions, gauge vector bosons, and gravitons. See Appendices E and F for papers on these cases.

GiFT is appropriate for the higher space-time dimension spaces of Octonion Cosmology. It enables all perturbation theory computations to be finite, and is suitable for all coordinate systems: static and non-static.

3.3 A Gauge Formalism for GiFT

The Bogoliubov transformations for GiFT can be generalized to a local gauge transformation theory. Eqs. 3.25 and 3.26 suggest a gauge theory with three fields:

$$B_i(y)$$

for i = 1, 2, 3. The local gauge transformation generalizing eq. 3.25 is

$$B(y) = \exp[i\ \Sigma_i \int d^3k\ B_i(k, y)\Gamma_{ik}] \tag{3.27}$$

3.4 GiFT Application to Fermions and Bosons

GiFT provides a *complete* formalism for field theories: quantum theories and classical theories, normal particle theories and space particle theories, and fermions and bosons.

The author developed most forms of the PseudoQuantum sector in published papers the 1970s. Most of these papers are presented in Appendices E and F. They show the PseudoQuantum formalism for fermions and bosons.

The Two-Tier formalism was developed in the early 2000s and described in the book, *Quantum Theory of the Third Kind: A New Type of Divergence-free Quantum Field Theory Supporting a Unified Standard Model of Elementary Particles and Quantum Gravity based on a New Method in the Calculus of Variation* This book is in Appendix B.

Some of the relevant 1970s papers for the PseudoQuantum formalism are:

<u>**Appendix E:**</u>

3.4.1 PseudoQuantum Formalism and Bogoliubov SU(1,1) Transformations

S. Blaha, "The Local Definition of Asymptotic Particle States", IL Nuovo Cimento **49A**, 35 (1979). The formalism and application to scalar fields and fermion fields.

3.4.2 Classical Field Theory

S. Blaha, "Embedding Classical Fields in Quantum Field Theories", Phys. Rev. **D17**, 994 (1978). Classical fields are handled by a sector in the PseudoQuantum formalism. Thus the prefix "Pseudo".

Appendix F:

3.4.3 Quantum Quark Confining Field Theory

S. Blaha, Phys. Rev. **D10**, 4268 (1974). "Towards a Field Theory of Hadron Binding."

S. Blaha, Phys. Rev. **D11**, 2921 (1974). "Secondquantized Non-Abelian Field Theory for Hadrons with Quark Confinement and Scaling Deep-Inelastic Structure Functions." NonAbelian PseudoQuantum gauge vector fields, and PseudoQuantum quark fermions.

S. Blaha, "New Framework for Gauge Field Theories", IL Nuovo Cimento **49A**, 113 (1979). Contains Strong Interaction model with PseudoQuantum gauge vector (Yang Mills) fields, PseudoQuantum fermion fields, and Gravitation with PseudoQuantum graviton fields. A Strong Interaction model for quark (color) confinement.

3.4.4 Space Particle Two-Tier Formalism

See chapter 4.

3.4.5 Space Particle PseudoQuantum Formalism

See chapter 5.

3.5 Purpose of PseudoQuantum Formulation

The PseudoQuantum-Two-Tier formalism enables Quantum Field Theory to be generalized to other universes with higher space-time dimensions such as those of Octonion Cosmology without infinities appearing in Perturbation Theory. Universes can have general quantum field theory (GiFT) formulations in non-static coordinate systems.

3.6 GiFT Procedure

The generation of a GiFT formulation can happen using the following steps:

1. Take the conventional formulation of a Quantum Field Theory model and modify all coordinates and derivatives in the manner of the Two-Tier Theory described earlier.
2. Transform the theory to non-static coordinates if that is desired.
3. Proceed to do Perturbation Theory and other calculations.

Note Two-Tier Theory directly generalizes to higher dimension coordinate systems. See chapter 4.

3.7 GiFT at Low Energy and in Static Coordinate Systems

Having developed a GiFT formulated theory one may ask how it relates to conventional Quantum Field Theory. From the considerations presented in this volume it is clear that a GiFT formulation and its results become identical to conventional

Quantum Field Theory for small momenta less than M_c and in static space-times with a Killing vector.

3.8 Conclusion

GiFT remedies major problems that appear in conventional Quantum Field Theory.

4. Proof of the Octonion Cosmology Space Spectrum

In developing the theory of Octonion Cosmology we assumed that it must be based on probability using definitions of probability amplitudes. To do otherwise, by assuming a classical deterministic theory, would inexorably lead to universes where the hydrogen atom (and other atoms) would be unstable and immediately collapse. Thus quantum theory is necessary to have universes with stable matter.

As a result the crucial feature is a quantum theory in a space[27] where a satisfactory norm exists with which to determine probabilities. This requirement is the primary reason, and the basis for the proof of the Cayley number spectrum of the spaces of Octonion Cosmology.

In our past efforts the author based the Octonion Cosmology spectrum of spaces on:

1. A hypercomplex sequence of Cayley numbers.
2. A map from spinor arrays to arrays of hypercomplex numbers
3. The generation of each space from particle-antiparticle annihilation in a parent space.

In this chapter we derive the spectrum of Octonion Cosmology spaces from the existence and properties of norms in the hypercomplex numbers spaces generated by the Cayley-Dickson construction. We will show that the hypercomplex number series corresponds to spaces, which are generated by fermion-antifermion annihilation.

Every Cayley-Dickson number n corresponds to a space of dimension $d_c = 2^n$. It defines a vector space. Each vector v in that space has a conjugate v* such that its norm is $(vv^*)^{1/2}$. Functions of the space's vectors may be defined that also have conjugates and thus a norm. (Note: For Cayley numbers a and b, the conjugate of ab is (ab)* = b*a*.)

A second space may be defined with dimension $d_d = d_c^2 = 2^{2n}$. This second space corresponds to Cayley-Dickson number 2n. It also is a normed vector space with functions that have norms. The 2nd space is isomorphic to the set of spinor square arrays whose columns and rows number 2^n.

Due to the existence of a norm, Quantum theory[28] is supported in each of the 2nd set of spaces. The 2nd space for each Cayley-Dickson number n corresponds to a universe of dimension d_d. The spectrum of universe spaces follow a Cayley-Dickson number sequence. The total dimension of each universe is set by the Cayley-Dickson number squared d_d.. See Fig. 4.1. The order of the sequence is determined by the set of fermion-antifermion annihilations that generate spaces.

---

[27] Normed spaces form a subset of the set of metric spaces. Metric spaces are a subset of the set of topological spaces. The norm condition is satisfied by the spaces generated by hypercomplex numbers.
[28] A norm is a necessary condition for a space to support Quantum Theory.

At this point we have a set of universes, each of which supports Quantum theory. The universes in the set are each uniquely associated with a Cayley-Dickson number.

4.1 Relation of the Universes by Fermion-Antifermion Annihilation

The set of universes are joined by assumption 3 above, which is required in addition to the association of each space with a Cayley-Dickson number.

We have noted previously that the Cayley-Dickson arrays of dimension d_d correspond to (square) fermion spinor arrays. Every fermion spinor array has a Cayley-Dickson number specifying the number of its rows and columns.

We unite the set of universes by requiring the creation of each (child) universe (space) to be the result of fermion-antifermion annihilation[29] in the (parent) universe above it.[30]

The link between a child universe of space-time dimension q and its parent universe of space-time dimension r sets the space-time to be *at minimum*

$$r - q = 2 \tag{4.1}$$

If r – q = 1, then the spinor arrays of both parent and child would have the same number of rows and columns since one universe would have an even space-time dimension and the other an odd space-time dimension.[31] This choice is ruled out to avoid duplicate numbering of Cayley-Dickson numbered universes. Thus eq. 4.1 is the minimal choice of difference between the parent and child space-time dimensions. Each universe has its unique Cayley-Dickson number. The set of universes begins with the n = 1 universe with space-time dimension 0 and proceeds upward in space-time dimension in units of 2. *These units of 2 cause the spinor array rows and columns to ascend through the Cayley-Dickson sequence of hypercomplex numbers. So the Cayley-Dickson sequence of numbers corresponds to the spectrum sequence of numbers.*

We assume the universe of CayleyDickson number n = 10 is generated by the spinor array of a point universe with space-time dimension 20.[32]

The Quantum norm requirement, universe generation by fermion-antifermion annihilation and the choice of space-time dimensions of parent and child differing by 2 gives the universe spectrum of Fig. 4.2. The form of the Octonion Cosmology universe spectrum is thus determined.

---

[29] The fermion and antifermion each contain an internal space. They are space particles. Their annihilation can generate a scalar or vector space particle with the internal spaces of the fermions combined.

[30] A child space dimension array (column 4 in Fig. 4.2) has the same form as the parent spinor arrays (column 6 in Fig. 4.2 below) of the annihilating fermions.

[31] The spinor array of a universe with *even* spacetime dimension r has a row/column size of $2^{r/2}$. The spinor array of a universe with *odd* spacetime dimension r has a row/column size of $2^{r(r-1)/2}$. Thus, for example, a space of 4 space-time dimensions has a spinor array of the same size as that of a space with spacetime dimension 5.

[32] See Blaha (2021g).

Cayley-Dickson Number n	1st Dimension Space $d_c = 2^n$	2nd Dimension Space $d_d = 2^{2n}$	Space-time Dimensions q	Parent Space-time Dimension
10	1024	1024 · 1024	18	20
9	512	512 · 512	16	18
8	256	256 · 256	14	16
7	128	128 · 128	12	14
6	64	64 · 64	10	12
5	32	32 · 32	8	10
4	16	16 · 16	6	8
3	8	8 · 8	4	4
2	4	4 · 4	2	2
1	2	2 · 2	0	0

Figure 4.1. Spaces with Cayley-Dickson numbers 1 through 10 with their dimensions indicated.

THE TEN UNIVERSE SPECTRUM

Octonion Space Number o_s	Cayley-Dickson Number n	Cayley Number Dimension d_c	Dimension Array Total d_d	Space-time-Dimension q	Fermion Spinor Array Total d_s	Cayley Number "Name"
0	10	1024	1024 × 1024	18	512 × 512	Complex Octonion Octonion Octonion
1	9	512	512 × 512	16	256 × 256	Octonion Octonion Octonion
2	8	256	256 × 256	14	128 × 128	Quaternion Octonion Octonion
3	7	128	128 × 128	12	64 × 64	Complex Octonion Octonion
4	6	64	64 × 64	10	32 × 32	Octonion Octonion
5	5	32	32 × 32	8	16 × 16	Quaternion Octonion
6	**4**	**16**	**16 × 16**	**6**[33]	**8 × 8**	**Complex Octonion**
7	**3**	8	8 × 8	4	4 × 4	Octonion
8	**2**	4	4 × 4	2	2 × 2	Quaternion
9	**1**	2	2 × 2	0	1 × 1	Complex

Figure 4.2. The Octonion Cosmology ten space spectrum. The space for our universe, number 6, (Cayley number 4) is in bold type.

[33] The space-time dimensions become 4 as in the Unified SuperStandard Theory (UST) of our universe through either transfer of dimensions to internal symmetries or by the compactification of two dimensions.

5. Two-Tier Quantum Field Theory in Higher Dimensions

This chapter illustrates the use of Two-Tier Theory to calculate the Feynman propagator for a vector field. The Two-Tier formalism extends directly to higher dimension space-times. We defer use of the PseudoQuantum formalism to later chapters.

5.1 Two-Tier Features in D-Dimensional Space-Time (such as the Megaverse)

Since a field, quantized in D-dimensional conventional coordinates (D > 4), would lead to divergences in perturbation theory calculations, we can use D-dimensional Two-Tier coordinates to avoid them in perturbation theory:[34]

$$X^i(y) = y^i + i\, Y_u{}^i(y)/M_u{}^{D/2} \tag{5.1}$$

where $X_u{}^i(y)$ for i = 0, 1, ..., D - 1 is a D-dimensional free gauge field and M_u is a mass of the order of the Planck mass or greater in the universe's space-time. The $Y_u{}^i(y)$ term adds a quantum field to the D coordinates making them a set of quantum coordinates. Quantum coordinate derivatives are defined by

$$\partial_i = \partial/\partial X^i(y) = \partial/\partial(y^i - Y_u{}^i(y)/M_u{}^{D/2}) \tag{5.2}$$

The use of these coordinates to quantize particle fields leads to a completely finite perturbation theory. We applied them in Blaha (2017b) to create a finite fundamental theory of mater. We applied them to fields in the Megaverse[35] to achieve a finite theory of Megaverse dynamics for elementary particles and universe particles.

The second quantization of a vector gauge field $V^i(y)$ is analogous to the second quantization of the electromagnetic field. The Lagrangian density terms for the free $V^i(X(y))$ fields is

$$\mathscr{L}_{Vu} = -\tfrac{1}{4}\, F_{Vu}{}^{ij}(X(y))F_{Vuij}(X(y)) \tag{5.3}$$

The Lagrangian is

$$L_{Vu} = \int d^Dy\, \mathscr{L}_{Vu}(X(y))$$

with

$$F_{Vuij} = \partial V_i(X(y))/\partial X^j(y) - \partial V_j(X(y))/\partial X^i(y)$$

[34] We assume one time dimension and D – 1 spatial dimensions.

[35] Blaha (2017c).

where the values of i and j range from 0 to D – 1 in this section.

The equal time commutation relations, using the 0th coordinate as the time coordinate, are specified in the usual way:

$$[V^i(X(\mathbf{y}, y^0)), V^j(X(\mathbf{y}', y^0))] = [\pi^i(X(\mathbf{y}, y^0)), \pi^j(X(\mathbf{y}', y^0))] = 0$$
$$[\pi_j(X(\mathbf{y}, y^0)), V_k(X(\mathbf{y}', y^0))] = -i\,\delta^{(D-1)tr}{}_{jk}(X(\mathbf{y},0) - X(\mathbf{y}',0))$$

where

$$\pi_u{}^k = \partial\mathscr{L}_{Vu}(V(X(y)))/\partial V_k{}'(X(y))$$
$$\pi_u{}^0 = 0$$

for k = 1, … , (D – 1), and

$$\delta^{(D-1)tr}{}_{jk}(\mathbf{y} - \mathbf{y}') = \int d^{(D-1)}k\; e^{i\,\mathbf{k}\cdot(X(\mathbf{y},0) - X(\mathbf{y}',0))}\,(\delta_{jk} - k_j k_k/\mathbf{k}^2)/(2\pi)^{D-1} \qquad (5.4)$$
$$V_k{}'(X(y)) = \partial V_k(X(y))/\partial y^0$$

for j, k = 1, 2, … , (D – 1).

If we choose the Radiation gauge for $V_k(X(y))$:

$$V^D(X(y)) = 0$$
$$\partial V^j(X(y))/\partial X^j(y) = 0 \qquad (5.5)$$

for j = 1, 2, … , (D – 1), then (D – 2) degrees of freedom (polarizations) are present in the vector potential.[36] The Fourier expansion of the vector potential $V^i(X(y))$ is:

$$V^i(X(y)) = \int d^{(D-1)}k\; N_{0V}(k) \sum_{\lambda=1}^{D-2} \varepsilon^i(k, \lambda)[a_V(k,\lambda) :e^{-ik.X(y)}: + a_V{}^\dagger(k,\lambda) :e^{ik.X(y)}:] \qquad (5.6)$$

for i = 1, … , (D – 2) where

$$N_{0V}(k) = [(2\pi)^{(D-1)} 2\omega_k]^{-1/2}$$

and (since the field is massless)

$$k^0 = \omega_k = (\mathbf{k}^2)^{1/2}$$

where k^D is the energy, and where the $\varepsilon^i(k, \lambda)$ are the polarization unit vectors for $\lambda = 1, \ldots, (D-2)$ and $k^\mu k_\mu = k^{D\,2} - \mathbf{k}^2 = 0$.

The commutation relations of the Fourier coefficient operators are:

$$[a_V(k,\lambda), a_V{}^\dagger(k',\lambda')] = \delta_{\lambda\lambda'}\delta^{D-1}(\mathbf{k} - \mathbf{k}')$$
$$[a_V{}^\dagger(k,\lambda), a_V{}^\dagger(k',\lambda')] = [a_V(k,\lambda), a_V(k',\lambda')] = 0$$

and the polarization vectors satisfy

---

[36] Note we use the Radiation gauge for $Y^\mu(y)$ also.

$$\sum_{\lambda=1}^{D-2} \varepsilon_i(k, \lambda)\varepsilon_j(k, \lambda) = (\delta_{ij} - k_i k_j/\mathbf{k}^2)$$

The V^μ Feynman propagator is

$$iD_F^{trTT}(y_1 - y_2)_{jk} = <0|T(V_j(X(y_1))V_k(X(y_2)))|0> \tag{5.7}$$

$$= -\, ig_{jk} \int \frac{d^Dk\, e^{-ik.(y_1 - y_2)}\, R(\mathbf{k}, y_1 - y_2)}{(2\pi)^D\, (k^2 + i\varepsilon)}$$

where g_{jk} is the D-dimensional Lorentz metric and where $R(\mathbf{k}, y_1 - y_2)$ is given by

$$\begin{aligned} R(\mathbf{k}, y_1 - y_2) &= \exp[-k^i k^j \Delta_{Tij}(y_1 - y_2)/M_u^D] \\ &= \exp\{-k^2[A(v) + B(v)\cos^2\theta] / [(2\pi)^{D-2}M_u^4 z^2]\} \end{aligned}$$

where k^2 is *the sum of the squares of the D – 1 spatial components* with

$$\begin{aligned} z^\mu &= y_1^\mu - y_2^\mu \\ z &= |\mathbf{z}| = |\mathbf{y_1} - \mathbf{y_2}| \\ k &= |\mathbf{k}| \\ v &= |z^0|/z \\ A(v) &= (1 - v^2)^{-1} + .5v\ \ln[(v-1)/(v+1)] \\ B(v) &= v^2(1 - v^2)^{-1} - 1.5v\ \ln[(v-1)/(v+1)] \end{aligned}$$

$$\mathbf{k}\cdot\mathbf{z} = kz\cos\theta$$

and $|\mathbf{k}|$ denoting the length of a spatial (D – 1)-vector $\mathbf{k}$ while $|z^0|$ is the absolute value of z^0.

As the above equations indicate, the Gaussian damping factor R(k, z) for *all* large spatial momentum k^j is the same for both the positive and negative frequency parts of the (Two Tier) V Feynman propagator. We are assuming the spatial momentum is real-valued in this discussion. It is also important to note that R(k, z) does not depend on k^0 (in the V and Y_u Radiation gauges) and thus the integration over k^0 proceeds in the usual way to produce time-ordered positive and negative frequency parts.

The Gaussian exponential factor in *all* spatial coordinates causes the Feynman propagator to be finite and, together with the Gaussian factor in universe particle propagators, causes all perturbation theory calculations when interactions are introduced to be finite as we have seen in Blaha (2017b).

For small momentum much less than M_u then $R(\mathbf{k}, y_1 - y_2) \to 1$ and the Feynman propagator is the "normal" propagator of conventional D-dimensional quantum field theory. For large momentum the corresponding potential approaches r^{D-3}

in contrast to the electromagnetic Coulomb potential r^{-1}. The V potential is highly non-singular at large energies.

Thus using Two-Tier Quantum Field Theory we can perform perturbation theory calculations that always yield a finite result.[37] This is not true if conventional Quantum Field is used.[38]

The Gaussian exponential damping factor appears in all propagators when the Two-Tier formalism is applied to all particle fields in a Lagrangian. Thus all perturbation theory calculations are finite in all orders.

---

[37] In particular, the fermion triangle divergence (anomaly) does not occur in our Two Tier Quantum Field Theory of the fermion sector. Thus there is no requirement for axion-like particles in the Megaverse (or in universes) although the possible existence of this type of particle is not ruled out.

[38] Blaha (2005a) and appendix B below provides a complete discussion of Two-Tier Quantum Field Theory.

6. Space Particle Two-Tier and PseudoQuantum Formalism

In this chapter we list the Two-Tier and PseudoQuantum free field space wave functions and Lagrangians for scalar, fermion, gauge vector boson and graviton *space* particles. Details of their features may be found in Blaha (2021g). The Two-Tier and PseudoQuantum formulations for normal fermion and boson particles are considered in the references in Appendices E and F.

6.1 Two-Tier and PseudoQuantum Free Fermion Space Particle Wave Function and Lagrangian

Following S. Blaha, Il Nuovo Cimento **49A**, 35 (1979) as we did in previous chapters we define the Two-Tier, PseudoQuantum Lagrangian for the spin ½ space particle interacting with a scalar space particle to be:

$$\mathcal{L} = \overline{\psi}_2(i\partial_\mu\gamma^\mu - m)\psi_1 + \overline{\psi}_1(i\partial_\mu\gamma^\mu - m)\psi_2 - \overline{\psi}_1(i\partial_\mu\gamma^\mu - m)\psi_1 + \mathcal{L}_\Phi(\psi_2,\Phi_2) \tag{6.1}$$

where Φ_2 is a scalar space field with a pair of spinor indices specified later, where the fields are all functions of $X^\mu = y^\mu + iY^\mu(y)$, and where $\partial_\mu = \partial/\partial X^\mu$.

From the development[39] in S. Blaha, Il Nuovo Cimento **49A**, 35 (1979) we find the free space fermion wave function for an r space-time dimension space fermion (with a q-dimension inner space-time) has the form[40]

$$\Psi_{ia}(r, q, X) = (\text{const}) \int d^{r-1}p(p^0)^{-½}\{e^{-ip\cdot X}\, U_{a\beta}(r, q, p)(c_{i1\beta}\, b_{1\beta}(p) + c_{i2\beta}\, b_{2\beta}(p)) +$$

$$+ \; e^{ip\cdot X}\, V_{a\beta}(r, q, p)\, (c'_{i1\beta}\, d^\dagger{}_{1\beta}(p) + c'_{i2\beta}\, d^\dagger{}_{2\beta}(p))\,\} \tag{6.2}$$

Its Hermitian conjugate is

$$\Psi^\dagger{}_{ia}(r, q, X) = (\text{const}) \int d^{r-1}p(p^0)^{-½}\{e^{ip\cdot X}\, U^\dagger{}_{a\beta}(r, q, p)(c_{i1\beta}\, b^\dagger{}_{1\beta}(p) + c_{i2\beta}\, b^\dagger{}_{2\beta}(p)) +$$

$$+ \; e^{-ip\cdot X}\, V^\dagger{}_{a\beta}(r, q, p)\, (c'_{i1\beta}\, d_{1\beta}(p) + c'_{i2\beta}\, d_{2\beta}(p))\,\} \tag{6.3}$$

where a and β are spinor indices.

The spinor arrays U and V have the form:[41]

---

[39] Eqs. 61- 68.

[40] This form makes β into the spinor index. It implements the close connection of the space feature and the spinors. Alternately an internal symmetry index could be introduced (as done in the scalar case later) for the internal symmetry described in chapter 4. The sets of b_β and of d_β operators each constitute a fundamental q-number representation of $U(2^{r/2})$.

$$U(r, q, p, m) = ((m + p^0)/2m)^{½}\gamma^0 S(p) = (\not{p} + m)/2m \quad (6.4)$$

$$V(r, q, p, m) = ((m - p^0)/2m)^{½}\gamma^0 S'(p) = (m - \not{p})/2m \quad (6.5)$$

One may visualize the form of the spinor matrix U as a matrix with $2^{r/2}$ spinor columns, in which each column is a spinor with $2^{r/2}$ components.[42] (Fig. 1.3) We view U as a matrix in spinor space. See chapter 5 of Blaha (2021g)..

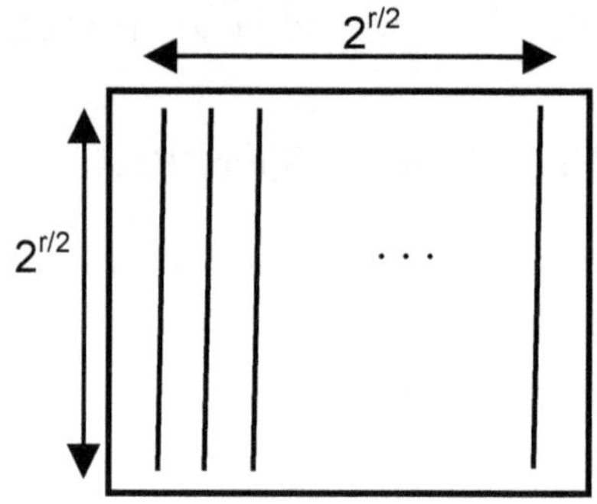

Figure 6.1. The U array with $2^{r/2}$ columns and $2^{r/2}$ rows. The V array has the same form.

The creation/annihilation operators anti-commute:

$$\{b_{i\beta}(p), b_{j\beta'}(p')\} = 0 \quad (6.6)$$
$$\{b^{\dagger}_{i\beta}(p), b^{\dagger}_{j\beta'}(p')\} = 0$$
$$\{b_{i\beta}(p), b^{\dagger}_{j\beta'}(p')\} = (1 - \delta_{ij})\delta_{\beta\beta'}\, \delta^{r-1}(\mathbf{p} - \mathbf{p'})$$

$$\{d_{i\beta}(p), d_{j\beta'}(p')\} = 0$$
$$\{d^{\dagger}_{i\beta}(p), d^{\dagger}_{j\beta'}(p') = 0$$
$$\{d_{i\beta}(p), d^{\dagger}_{j\beta'}(p')\} = (1 - \delta_{ij})\delta_{\beta\beta'}\, \delta^{r-1}(\mathbf{p} - \mathbf{p'})$$

for i, j = 1, 2.

The equal-time commutation relations and the self-adjointness of the Hamiltonian H imply the fields in eq. 6.2 and 6.3 (after some algebra) have the parameters[43]

$$c_{11\beta} = \cos(\theta_{1\beta} - \theta_{2\beta})/\sin\theta_{1\beta} \quad (6.7)$$
$$c_{12\beta} = \sin(\theta_{1\beta} - \theta_{2\beta})/\sin\theta_{1\beta}$$
$$c'_{11\beta} = \cos(\theta_{1\beta} - \theta_{2\beta})/\cos\theta_{1\beta}$$

---

[41] Blaha(2021g). Eqs. 1.32 and 1.33. The origin and properties of these spinor arrays are discussed in detail there.
[42] r is the spacetime dimension.
[43] See eqs. 41 and 42 of S. Blaha, Il Nuovo Cimento **49A**, 35 (1979) in Appendix E.

$$\begin{aligned} c'_{12\beta} &= -\sin(\theta_{1\beta} - \theta_{2\beta})/\cos\theta_{1\beta} \\ c_{21\beta} &= \sin\theta_{2\beta} \\ c_{22\beta} &= \cos\theta_{2\beta} \\ c'_{21\beta} &= \cos\theta_{2\beta} \\ c'_{22\beta} &= \sin\theta_{2\beta} \end{aligned}$$

where θ_1 and θ_2 are functions of k and the spinor index β. These constants must depend on β since the creation/annihilation operators are indexed by β and form a $U(2^{r/2})$ spinor group representation.

6.2 Two-Tier and PseudoQuantum Free Scalar Space Particle Wave Function and Lagrangian

Using eq. 6.1, we define the free scalar space particle Lagrangian terms as:

$$\mathscr{L}_\Phi(\psi_2,\Phi_2) = -g'\bar{\psi}\,\Phi_2\psi + \bar{\Phi}_1(i\partial_\mu\gamma^\mu - m')\,\Phi_2 - \tfrac{1}{2}\,\bar{\Phi}_1(i\partial_\mu\gamma^\mu - m')\,\Phi_1 \qquad (6.8)$$

In order to have a spinor array within a scalar space particle we require the free Lorentz *scalar* space particle Dirac equation to satisfy:[44]

$$(i\partial_\mu\gamma^\mu - m')\,\Phi(r, q, X) = 0 \qquad (6.9)$$

with $\partial_\mu = \partial/\partial X^\mu$. It implies the usual scalar Klein-Gordon equation

$$(\partial_\mu\partial^\mu + m'^2)\,\Phi(r, q, X) = 0 \qquad (6.10)$$

We choose the Two-Tier, PseudoQuantum, free scalar space particle wave function with second rank spinor indices for an r space-time dimension space boson (with a q-dimension inner space-time) to be

$$\begin{aligned} \Phi_{iab}(r, q, X) = \int d^{r-1}p(2\pi)^{-(r-1)}(m'/p^0)^{1/2}\,\{&\exp^{-ip\cdot X}\,U_{a\beta}(r, q, p, m')a'_{i\beta b}(p) + \\ &+ \exp^{ip\cdot X}\,V_{a\beta}(r, q, p, m')c''_{i\beta b}{}^{\dagger}(p)\} \end{aligned} \qquad (6.11)$$

where, due to a PseudoQuantum transformation, have the form

$$\begin{aligned} a'_{i\beta b}(p) &= c_{i1\beta}\,a_{1\beta}(p) + c_{i2\beta}\,a_{2\beta}(p) \\ c''_{i\beta b}{}^{\dagger}(p) &= c'_{i1\beta}\,c^{\dagger}_{1\beta}(p) + c'_{i2\beta}\,c^{\dagger}_{2\beta}(p) \end{aligned} \qquad (6.12a)$$

where a, b, and β are spinor internal symmetry indices. Its Hermitian conjugate is

[44] The Dirac equation appears necessary for scalar space particles. Together with the requirement that the scalar wave function have a pair of spino indices the form of the wave function is determined.

$$\Phi^{\dagger}{}_{iab}(r, q, X) = \int d^{r-1}p(2\pi)^{-(r-1)}(m'/p^0)^{1/2} \{\exp^{ip\cdot X}U^{\dagger}{}_{a\beta}(r, q, p, m')\, a'_{i\beta b}{}^{\dagger}(p) + $$
$$+ \exp^{-ip\cdot X} V^{\dagger}{}_{a\beta}(r, q, p, m')\, c''_{i\beta b}(p) \} \quad (6.12)$$

The transformation constants of the creation/annihilation operators in eq. 6.12a are

$$\begin{aligned} c_{11\beta} &= \cos(\theta_{1\beta} - \theta_{2\beta})/\sin \theta_{1\beta} \\ c_{12\beta} &= \sin(\theta_{1\beta} - \theta_{2\beta})/\sin \theta_{1\beta} \\ c'_{11\beta} &= \cos(\theta_{1\beta} - \theta_{2\beta})/\cos \theta_{1\beta} \\ c'_{12\beta} &= - \sin(\theta_{1\beta} - \theta_{2\beta})/\cos \theta_{1\beta} \end{aligned} \quad (6.13)$$

$$\begin{aligned} c_{21\beta} &= \cos \theta_{2\beta} \\ c_{22\beta} &= \sin \theta_{2\beta} \\ c'_{21\beta} &= \sin \theta_{2\beta} \\ c'_{22\beta} &= \cos \theta_{2\beta} \end{aligned}$$

The constant m' is the scalar particle mass as well as the energy of the created space, and U and V are the same as in fermion case. The spinor arrays, U and V, have $2^{r/2}$ columns and $2^{r/2}$ rows. The spinor arrays U and V have their size determined by r, which we call the *parent* space space-time dimension. The space-time dimension q is the *child* space space-time dimension. Note Φ is a second rank spinor. The sets of $a_{\beta b}$ and of $c_{\beta b}$ operators each constitute a q-number representation of $U(2^{r/2})$.

The creation/annihilation operators satisfy the commutation relations:

$$\begin{aligned} [a_{j\beta b}(p), a^{\dagger}{}_{j'\beta' b'}(p')] &= (1 - \delta_{jj'})\delta_{\beta\beta'}\, \delta_{b\, b'}\, \delta^{r-1}(\mathbf{p} - \mathbf{p'}) \quad (6.14) \\ [a_{j\beta b}(p), a_{j'\beta' b'}(p')] &= 0 \\ [a^{\dagger}{}_{j\beta b}(p), a^{\dagger}{}_{j'\beta' b'}(p')] &= 0 \\ [c_{j\beta b}(p), c^{\dagger}{}_{j'\beta' b'}(p')] &= (1 - \delta_{jj'})\delta_{\beta\beta'}\, \delta_{b\, b'}\, \delta^{r-1}(\mathbf{p} - \mathbf{p'}) \quad (6.15) \\ [c_{j\beta b}(p), c_{j'\beta' b'}(p')] &= 0 \\ [c^{\dagger}{}_{j\beta b}(p), c^{\dagger}{}_{j'\beta' b'}(p')] &= 0 \\ [c_{j\beta b}(p), a^{\dagger}{}_{j'\beta' b'}(p')] &= 0 \\ [c_{j\beta b}(p), a_{j'\beta' b'}(p')] &= 0 \\ [c^{\dagger}{}_{j\beta b}(p), a^{\dagger}{}_{j'\beta' b'}(p')] &= 0 \end{aligned}$$

The $a_{jk}(p)$ and $a^{\dagger}{}_{j'k'}(p')$ commute with $c_{jk}(p)$ and $c^{\dagger}{}_{j'k'}(p')$. There are two sets: 2^r operators $a_{jk}(p)$ and 2^r operators c_{jk}. Each set furnishes a representation of $U(2^{r/2})$.

6.3 Two-Tier and PseudoQuantum Free Vector Space Particle Wave Function and Lagrangian

In this section we define the wave functions of a free, vector space particle. We use T matrices as the generators of the $U(2^{r/2})$ Dirac-Good group,[45] with $A_{k\mu}$ as the gauge vector fields of that group, and g' is its coupling constant. The $U(2^{r/2})$ group will be seen to be a local Yang-mills gauge group. The group acts on spinors. Therefore T has spinor indices. T also consists of 2^r generators: the r Dirac matrices and $2^r - 1$ other generators. T has the form T^i_{ab} for $i = 1, \ldots, 2^r$ and for $a, b = 1, \ldots, 2^{r/2}$.

The vector field in the radiation gauge satisfies

$$(\partial_\mu \partial^\mu + \xi^2)\, A^i_k(r, q, X) = 0 \tag{6.16}$$

for k = 1, 2, for spatial coordinates $i = 1, \ldots, r - 1$, where ξ is a small mass that we take to zero at the end of calculations (as done in QED infrared studies.) In order to introduce the U and V spinor arrays seen in previous chapters we require

$$(i\partial_\mu \gamma^\mu - \xi)\, A_k{}^i(r, q, X) = 0 \tag{6.17}$$

Note eq. 6.16 follows from eq. 6.17.

Emulating the scalar wave function of section 6.2, we choose the vector space particle wave function for an r space-time dimension space gauge boson (with a q-dimension inner space-time) in the radiation gauge to be[46]

$$\mathbf{A}^i_{kab}(r, q, X) = \int d^{r-1}p(2\pi)^{-(r-1)}(2p^0)^{-1/2}\, \Sigma_{\lambda\alpha\beta}\varepsilon(p,\lambda)T^i_{a\alpha}\{\exp^{-ip\cdot X}\, U_{\alpha\beta}(r, q, p, \xi)\, w'_{k\beta b}(p, \lambda) +$$
$$+ \exp^{ip\cdot X}\, V_{\alpha\beta}(r, q, p, \xi)\, z'_{k\beta b}{}^\dagger(p,\lambda)\} \tag{6.18}$$

for k = 1, 2 where ξ is a mass that is set to zero after perturbation theory calculations; where $\boldsymbol{\varepsilon}$ is the polarization unit vector; w and z are creation/annihilation operators with spinor indices—each taking $2^{r/2}$ values, λ takes $r - 2$ values, and U and V are similar to U and V in the fermion case. $\mathbf{A}^i_{ab}(r, q, X)$ is a vector of the spatial components of the vector field in the radiation gauge. The creation/annihilation operators under a PseudoQuantum transformation have the form:

$$\begin{aligned} w'_{k\beta b}(p, \lambda) &= c_{k1\beta}\, w_{1\beta}(p, \lambda) + c_{k2\beta}\, w_{2\beta}(p, \lambda) \\ z'_{k\beta b}{}^\dagger(p, \lambda)) &= c'_{k1\beta}\, z^\dagger{}_{1\beta}(p, \lambda) + c'_{k2\beta}\, z^\dagger{}_{2\beta}(p, \lambda) \end{aligned} \tag{6.19}$$

where

$$\begin{aligned} c_{11\beta} &= \cos(\theta_{1\beta} - \theta_{2\beta})/\sin\theta_{1\beta} \\ c_{12\beta} &= \sin(\theta_{1\beta} - \theta_{2\beta})/\sin\theta_{1\beta} \\ c'_{11\beta} &= \cos(\theta_{1\beta} - \theta_{2\beta})/\cos\theta_{1\beta} \end{aligned} \tag{6.20}$$

---

[45] See Blaha (2021g).

[46] With repeated indices summed.

$$c'_{12\beta} = - \sin(\theta_{1\beta} - \theta_{2\beta})/\cos\theta_{1\beta}$$
$$c_{21\beta} = \cos\theta_{2\beta}$$
$$c_{22\beta} = \sin\theta_{2\beta}$$
$$c'_{21\beta} = \sin\theta_{2\beta}$$
$$c'_{22\beta} = \cos\theta_{2\beta}$$

The commutation relations of the creation/annihilation operators are

$$[w_{j\beta b}(p, \lambda), w^{\dagger}_{j'\beta' b'}(p', \lambda')] = (1 - \delta_{jj'})\delta_{\beta\beta'}\delta_{bb'}\delta_{\lambda\lambda'}\delta^{r-1}(\mathbf{p} - \mathbf{p}') \tag{6.21}$$
$$[w_{j\beta b}(p, \lambda), w_{j'\beta' b'}(p' \lambda')] = 0$$
$$[w^{\dagger}_{j\beta b}(p, \lambda), w^{\dagger}_{j'\beta' b'}(p' \lambda')] = 0$$
$$[z_{j\beta b}(p, \lambda), z^{\dagger}_{j'\beta' b'}(p' \lambda')] = (1 - \delta_{jj'})\delta_{\beta\beta'}\delta_{bb'}\delta_{\lambda\lambda'}\delta^{r-1}(\mathbf{p} - \mathbf{p}')$$
$$[z_{j\beta b}(p, \lambda), z_{j'\beta' b'}(p' \lambda')] = 0$$
$$[z^{\dagger}_{j\beta b}(p, \lambda), z^{\dagger}_{j'\beta' b'}(p' \lambda')] = 0$$
$$[z_{j\beta b}(p, \lambda), w^{\dagger}_{j'\beta' b'}(p' \lambda')] = 0$$
$$[z_{j\beta b}(p, \lambda), w_{j'\beta' b'}(p' \lambda')] = 0$$
$$[z^{\dagger}_{j\beta b}(p, \lambda), w^{\dagger}_{j'\beta' b'}(p' \lambda')] = 0$$

for j, j' = 1, 2.

The spinor arrays, U and V, have $2^{r/2}$ columns and $2^{r/2}$ rows. The spinor arrays U and V size is determined by r, the *parent* space space-time dimension. The space-time dimension q is the *child* space space-time dimension. Since U and V are not orthogonal for zero mass, we have introduced a "fictitious" mass ξ which is taken to zero at the end of calculations.

The Hermitian conjugate is

$$\mathbf{A}^{i}_{kab}(r, q, X)^{\dagger} = \int d^{r-1}p(2\pi)^{-(r-1)}(2p^0)^{-½}\Sigma_{\lambda\alpha\beta}\varepsilon(p,\lambda)T^{i}_{a\alpha}\{\exp^{ip\cdot X}\, U_{\alpha\beta}(r, q, p, \xi)^{\dagger} w'_{k\beta b}{}^{\dagger}(p, \lambda) +$$
$$+ \exp^{-ip\cdot X}\, V_{\alpha\beta}(r, q, p, \xi)^{\dagger} z'_{k\beta b}(p,\lambda)\} \tag{6.22}$$

6.4 Two-Tier and PseudoQuantum Free Graviton Space Particle Wave Function and Lagrangian

Our consideration of the fermion, scalar boson, and vector boson space fields leads to the possibility of graviton space fields. In a weak field approximation we define a metric tensor with two spinor indices a and b:

$$g_{ab}{}^{\mu\nu}(r, q, X) \cong \eta_{ab}{}^{\mu\nu}(r, q, X) + h_{ab}{}^{\mu\nu}(r, q, X) \tag{6.23}$$

Following the conventional approach we arrive at the dynamic equation:

$$-\tfrac{1}{2}\,\square(h_{ab}{}^{\mu\nu} - \tfrac{1}{2}\, h\, \eta_{ab}{}^{\mu\nu}) = -16\pi G T_{ab}{}^{\mu\nu} \tag{6.24}$$

based on the

$$R_{\mu\nu ab} - \tfrac{1}{2} R_{a\beta} g_{\mu\nu\beta b} = 8\pi G T_{\mu\nu ab}$$

The spin 2 $h_{ab}{}^{\mu\nu}(r, q, x)$ field can be PseudoQuantum expanded to

$$h^{\mu\nu}{}_{kab}(r, q, X) = \int d^{r-1}p(2\pi)^{-(r-1)}(2p^0)^{-\frac{1}{2}} \Sigma_{\lambda,\beta}\, \varepsilon^{\mu\nu}(p,\lambda)\{\exp^{-ip\cdot X}\, U_{a\beta}(r, q, p, \xi)\, w'_{k\beta b}(p, \lambda) +$$

$$+ \exp^{ip\cdot X}\, V_{a\beta}(r, q, p, \xi)\, z'_{k\beta b}{}^{\dagger}(p,\lambda)\} \tag{6.25}$$

with quantities defined similarly to the previous.
The creation/annihilation operators under a PseudoQuantum transformation have the form:

$$\begin{aligned} w'_{k\beta b}(p, \lambda) &= c_{k1\beta}\, w_{1\beta}(p, \lambda) + c_{k2\beta}\, w_{2\beta}(p, \lambda) \\ z'_{k\beta b}{}^{\dagger}(p, \lambda)) &= c'_{k1\beta}\, z^{\dagger}{}_{1\beta}(p, \lambda) + c'_{k2\beta}\, z^{\dagger}{}_{2\beta}(p, \lambda) \end{aligned} \tag{6.26}$$

where

$$\begin{aligned} c_{11\beta} &= \cos(\theta_{1\beta} - \theta_{2\beta})/\sin\theta_{1\beta} \\ c_{12\beta} &= \sin(\theta_{1\beta} - \theta_{2\beta})/\sin\theta_{1\beta} \\ c'_{11\beta} &= \cos(\theta_{1\beta} - \theta_{2\beta})/\cos\theta_{1\beta} \\ c'_{12\beta} &= -\sin(\theta_{1\beta} - \theta_{2\beta})/\cos\theta_{1\beta} \\ c_{21\beta} &= \cos\theta_{2\beta} \\ c_{22\beta} &= \sin\theta_{2\beta} \\ c'_{21\beta} &= \sin\theta_{2\beta} \\ c'_{22\beta} &= \cos\theta_{2\beta} \end{aligned} \tag{6.27}$$

The commutation relations of the creation/annihilation operators are

$$\begin{aligned} [w_{j\beta b}(p, \lambda), w^{\dagger}{}_{j'\beta'b'}(p', \lambda')] &= (1 - \delta_{jj'})\delta_{\beta\beta'}\,\delta_{b b'}\,\delta_{\lambda\lambda'}\,\delta^{r-1}(\mathbf{p} - \mathbf{p'}) \\ [w_{j\beta b}(p, \lambda), w_{j'\beta'b'}(p' \lambda')] &= 0 \\ [w^{\dagger}{}_{j\beta b}(p, \lambda), w^{\dagger}{}_{j'\beta'b'}(p' \lambda')] &= 0 \\ [z_{j\beta b}(p, \lambda), z^{\dagger}{}_{j'\beta'b'}(p' \lambda')] &= (1 - \delta_{jj'})\delta_{\beta\beta'}\,\delta_{b b'}\,\delta_{\lambda\lambda'}\delta^{r-1}(\mathbf{p} - \mathbf{p'}) \\ [z_{j\beta b}(p, \lambda), z_{j'\beta'b'}(p' \lambda')] &= 0 \\ [z^{\dagger}{}_{j\beta b}(p, \lambda), z^{\dagger}{}_{j'\beta'b'}(p' \lambda')] &= 0 \\ [z_{j\beta b}(p, \lambda), w^{\dagger}{}_{j'\beta'b'}(p' \lambda')] &= 0 \\ [z_{j\beta b}(p, \lambda), w_{j'\beta'b'}(p' \lambda')] &= 0 \\ [z^{\dagger}{}_{j\beta b}(p, \lambda), w^{\dagger}{}_{j'\beta'b'}(p' \lambda')] &= 0 \end{aligned} \tag{6.28}$$

for j, j' = 1, 2.

7. Some Higher Space-time Dimensions Results

7.1 Two-Tier Perturbation Theory

Two-Tier Perturbation theory is described in some detail in chapters 5 and 6 of *Quantum Theory of the Third Kind* appearing in Appendix B. The discussion includes the U matrix and the S-matrix as well as asymptotic states.

7.2 Two-Tier Quantum Electrodynamics

Quantum Electrodynamics is also covered in detail in chapter 7 in *Quantum Theory of the Third Kind* appearing in Appendix B.

7.3 Four Dimension Two-Tier Coulomb Interaction vs. Conventional Coulomb Potential

Chapter 7 in Appendix B describes the Coulomb interaction. The below material and figures are abstracted from Appendix B.

The familiar Coulomb potential in four dimensions is (for two particles of opposite unit electric charge):

$$V_{\text{Coulomb}} = -\,a/\,|\mathbf{r}| \qquad (7.1)$$

The Two-Tier QED Coulomb potential in four space-time dimensions is:

$$V_{\text{Two-TierCoul}} = -\,a\Phi(M_c^2\pi\,|\mathbf{r}|^2)/\,|\mathbf{r}| \qquad (7.2)$$

where $\Phi(x)$ is the error function.[47] At small distances ($\pi r^2 \ll M_c^{-2}$)

$$V_{\text{Two-TierCoul}} \rightarrow -\,2a\surd\pi\; M_c^2\,|\mathbf{r}| \qquad (7.3)$$

Is a linear potential, and at large distances ($\pi r^2 \gg M_c^{-2}$) a Coulomb potential:

$$V_{\text{Two-TierCoul}} \rightarrow V_{\text{Coulomb}} \qquad (7.4)$$

The Two-Tier Coulomb potential has a minimum at

$$M_c^2\pi\,|\mathbf{r}|^2 = 1 \qquad (7.5)$$

The Two-Tier potential is equivalent to the conventional potential except for ultra-short distances. Thus atoms will effectively experience only the conventional potential.

---

[47] W. Magnus and F. Oberhettinger, *Formulas and Theorems for the Special Functions of Mathematical Physics* (Chelsea Publishing Co., New York, 1949) page 96.

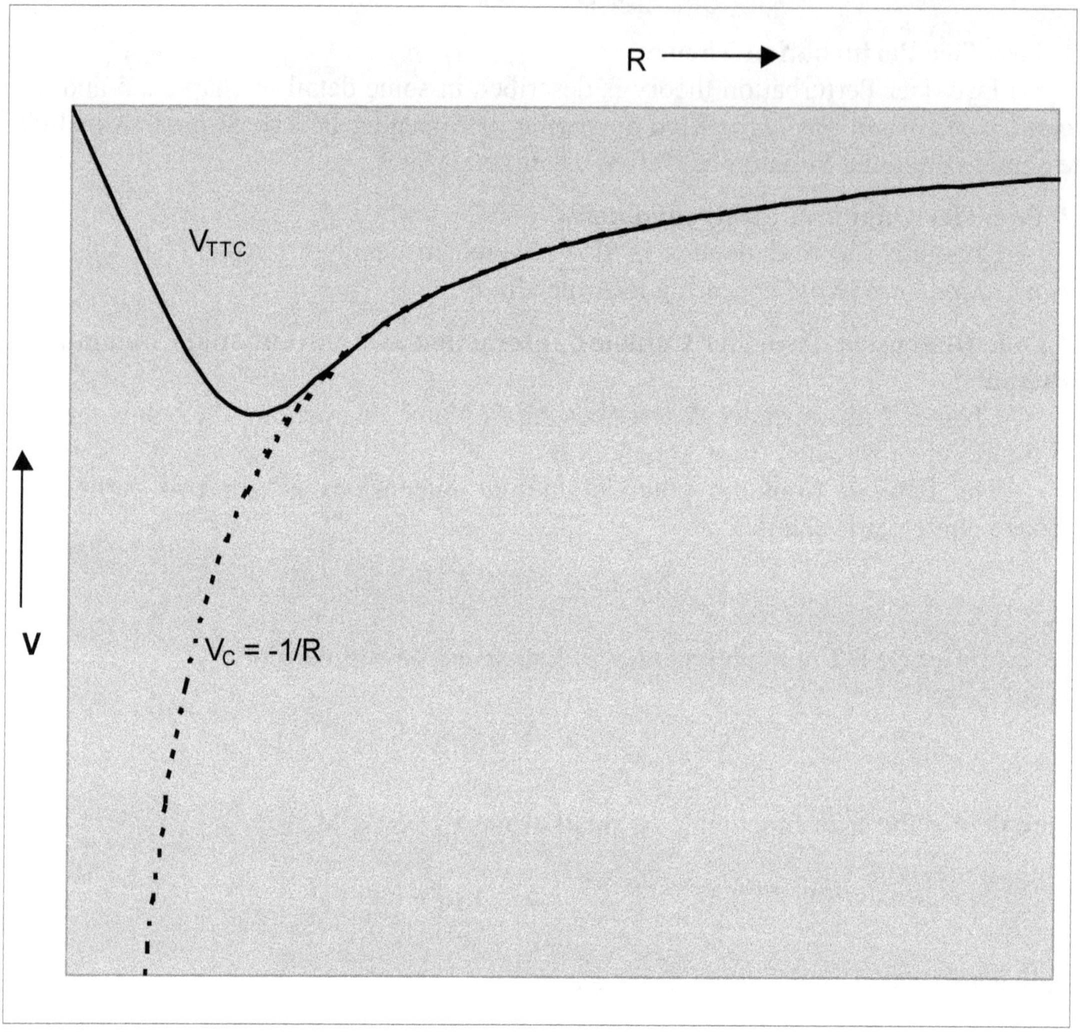

Figure 7.1. Plot of the form of the 4 dimension Two-Tier Coulomb "attractive" potential between particles of opposite unit charge divided by αM_c: $V_{TTC} = V_{Two\text{-}TierCoul}/(\alpha M_c)$. V_{TTC} is dimensionless. The dotted line is the conventional Coulomb attractive potential divided by αM_c: $V_c = V_{Coulomb}/(\alpha M_c) = 1/R$. Note the Two-Tier Coulomb force between particles of opposite charge is repulsive at short distance.

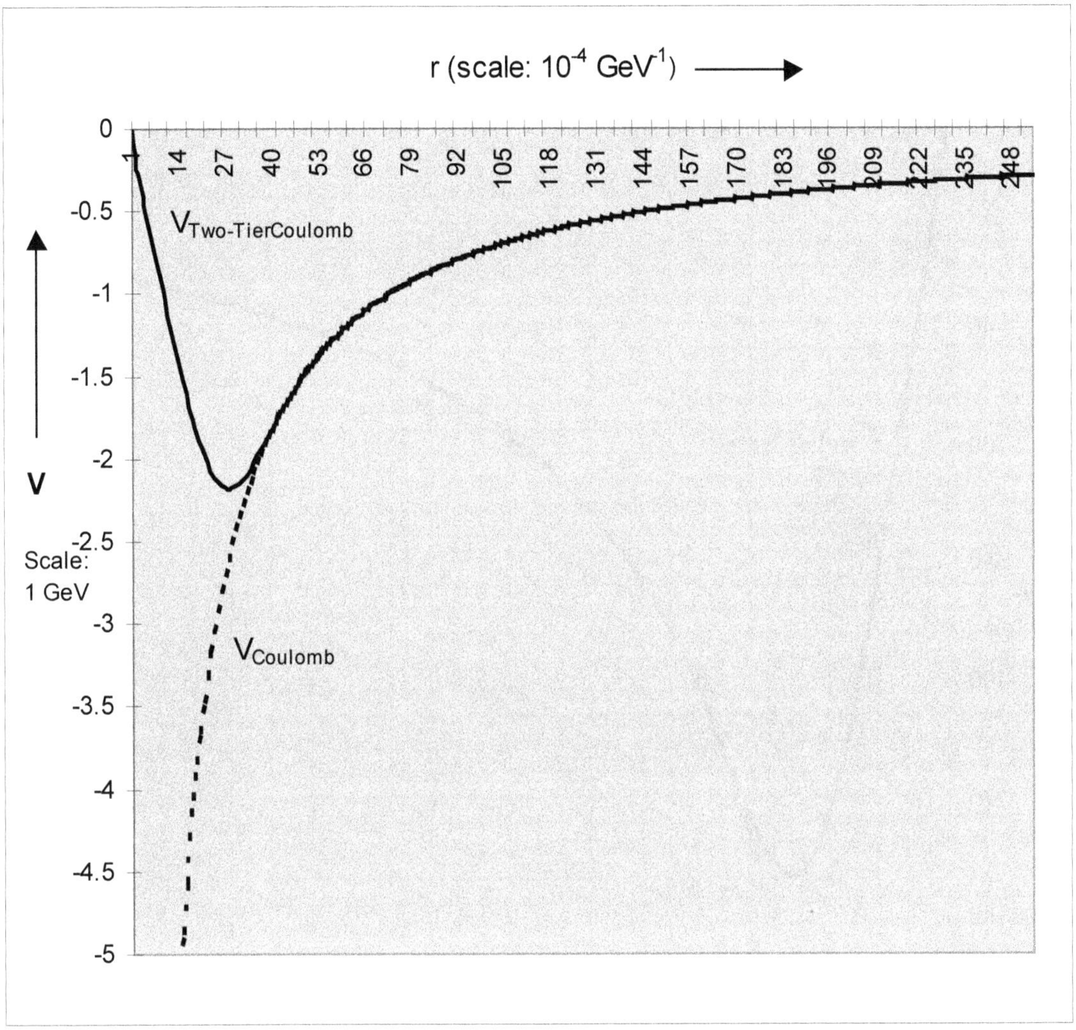

Figure 7.2. Two-Tier Coulomb Potential compared to conventional Coulomb potential for M_c = 200 GeV/c^2. Radial distance is measured in units of 10^{-4} GeV^{-1}. The potential energy for two opposite unit charges is measured in GeV units.

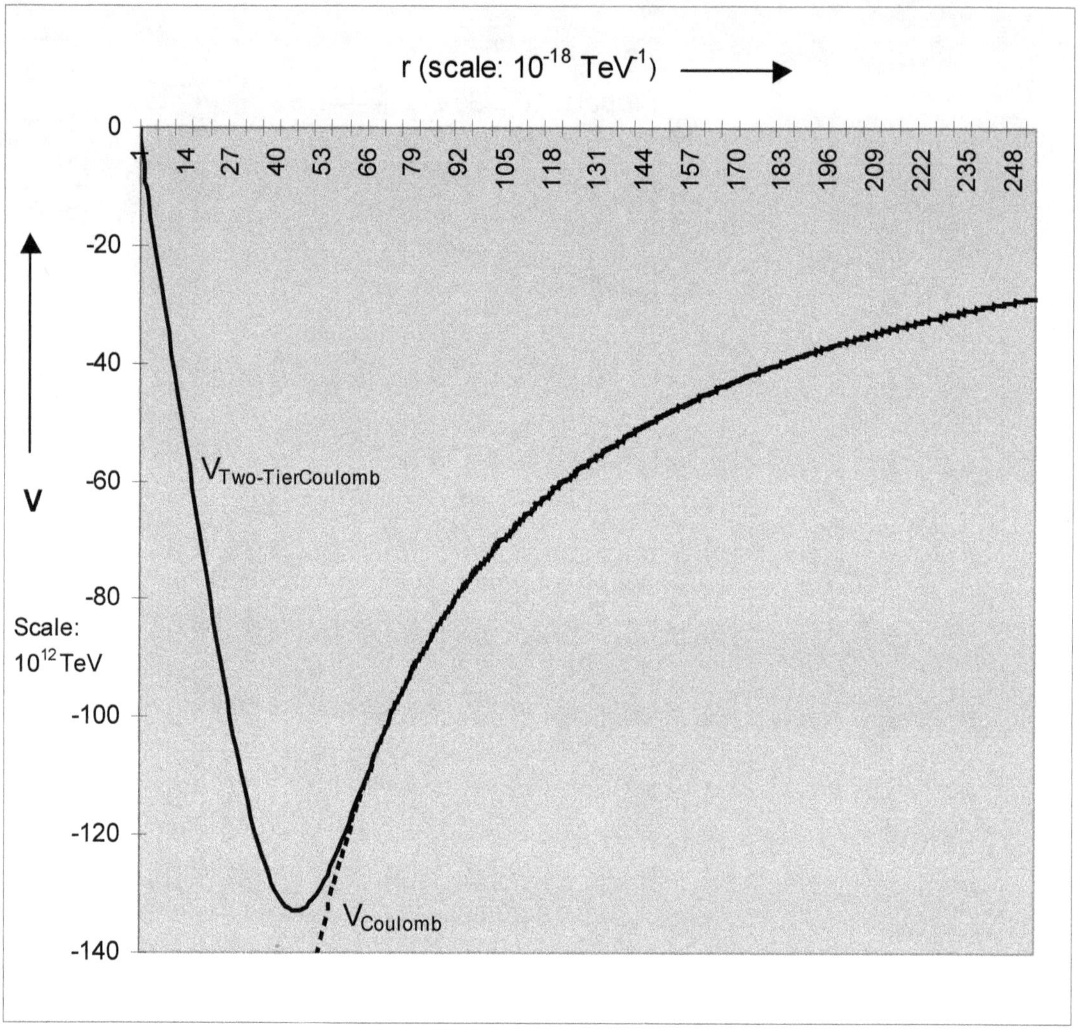

Figure 7.3. Two-Tier Coulomb Potential compared to conventional Coulomb potential for $M_c = M_{planck} = 1.22 \times 10^{19}$ GeV/c^2. Radial distance is in units of 10^{-18} TeV^{-1}. The potential energy for two opposite unit charges is measured in units of 10^{12} TeV.

7.4 Higher Space-time Dimension Coulomb Interaction

The Coulomb interaction is of great importance in Atomic Physics. In this section we calculate the form of the Coulomb potential in higher space-time dimension universes. We assume they have an equivalent to our four dimension Coulomb potential. The Two-Tier Coulomb potential in higher dimension space-time dimension q is:

$$V_{TT}(y_1 - y_2) = e^2 \int \frac{d^q p \; e^{-ip\cdot(y_1 - y_2)} R(\mathbf{p}, y_1 - y_2)}{(2\pi)^r \mathbf{p}^2} \tag{7.6}$$

We approximate VTT with the conventional potential with a view towards determining an approximate potential for atomic energy levels in higher dimensions.

$$V_{TT}(y_1 - y_2) \cong V(y_1 - y_2) = e^2 \int \frac{d^q p \; e^{-ip\cdot(y_1 - y_2)}}{(2\pi)^q \mathbf{p}^2} \tag{7.7}$$

Performing energy integration and hyperspherical angular integrations we obtain

$$V(y_1 - y_2)\, \delta(y_1{}^0 - y_2{}^0) = (\text{const}) \int dp\, p^{q-2}\, e^{-ipr}\, \delta(y_1{}^0 - y_2{}^0) \tag{7.8}$$

$$= (\text{const})(1/r^{q-3})\, \delta(y_1{}^0 - y_2{}^0) \tag{7.9}$$

where $r = \surd(\mathbf{y}_1 - \mathbf{y}_2)^2$ is the radial distance. Then

$$V(r) = (\text{const})/r^{q-3} = b/r^{q-3} \tag{7.10}$$

7.5 Higher Space-time Dimension Bohr-like Atom

We now obtain an estimate of the atomic energy levels in higher dimension space-times. To that end we now make the reasonable assumption of a higher space-time dimension Bohr-like atom consisting of an 'electron' circling a nucleus in a *non-relativistic* approximation in a circular orbit of radius a. The centripetal acceleration equals the Coulomb force:

$$mv^2/a = (q - 3)b/a^{q-4} \tag{7.11}$$

Thus the (non-relativistic) kinetic energy of the electron is

$$T = ½\, mv^2 = ½\, (q - 3)b/a^{q-3} \tag{7.12}$$

The stationary states of the Bohr atom have definite values of the angular momentum:

$$p_\varphi = mva_n = n\hbar \tag{7.13}$$

for the nth orbit. Thus

$$v = n\hbar/(ma_n)$$

Substituting in eq. 7.12 we find

$$a_n^{q-5} = (q-3)bm\,(1/n\hbar)^2 \quad (7.14)$$

and

$$a_n = [(q-3)bm/(n\hbar)^2]^{1/(q-5)}$$

The total energy is

$$E_n = T + V = \tfrac{1}{2}(q-3)b/a_n^{q-3} - b/a_n^{q-3}$$
$$= -[\tfrac{1}{2}(q-3) - 1]b/a_n^{q-3}$$

or

$$E_n = -\tfrac{1}{2}(q-5)b/[(q-3)bm/(n\hbar)^2]^{(q-3)/(q-5)} \quad (7.15)$$

For four dimension space-time the conventional Bohr energy levels emerge

$$q = 4: \qquad E_n = -\tfrac{1}{2}\, mb^2/(n^2\hbar^2)$$

For eight dimension space-time—Megaverse space-time with Cayley-Dickson number 5— the Bohr energy levels are:

$$q = 8 \qquad E_n = -1.5b/[5bm/(n^2\hbar^2)]^{5/3}$$

Note the inverse dependence of the energy on the coupling constant b and the mass m:

$$E_n \sim b^{-2/3}m^{-5/3}$$

The n dependence is also unusual

$$E_n \sim n^{10/3}$$

As the coupling increases the energy levels decrease. And as the electron mass increases the energy levels also decrease.

For the largest universe in the octonion space spectrum, where the space-time dimension q = 18, the energy levels are:

$$q = 18 \qquad E_n = -(13/2)b[n^2\hbar^2/(15bm]^{15/13} \quad (7.16)$$

The case of q = 3 is interesting. For q = 3 the potential V is a constant, and no binding results and

$$E_n = +b \text{ for all n.}$$

As $q \to \infty$

$$E_n \to -0.5\, n^2\hbar^2/m \quad (7.17)$$

with the energy levels rising as n^2 and an inverse m dependence.

The above forms of the hydrogen atom energies in higher dimensions suggest that life, if it exists, will be very different in the higher universes.

Appendix A. The 10 Spaces of Octonion Cosmology

This chapter outlines the theory of Octonion Cosmology. It is based on the author's books: *Quantum Space Theory With Application to Octonion Cosmology & Possibly To Fermionic Condensed Matter* and *Beyond Octonion Cosmology* as well as previous books by the author. The discussion will be in a narrative form to make it accessible to a wider range of readers.

Subsequent chapters describe new features of Natural Philosophy that emerge from the consideration of Octonion Cosmology.

1.1 Fundamental Required Features

Certain features are necessary for a satisfactory modern theory of fundamental Physics. These features include:

1. Space-times within which dynamic processes can take place.
2. A satisfactory Quantum formalism.
3. Matter.
4. Energy.
5. A space of internal symmetries for SU(3) and so on.

In addition to these necessary requirements are a viable Quantum Field Theory, and a satisfactory Quantum Space Theory[48] of fermion and boson quantum fields containing subspaces, and describing the generation and features of these spaces.

1.2 The Primary Space of Octonion Cosmology

Octonion Cosmology begins with the definition of a primary space,[49] which we will call space 0, from which the other spaces evolve. Space 0 has a twenty dimension space-time (nineteen spatial dimensions and one time dimension.) It contains $2^{22} - 20 =$ 4,194,284 additional dimensions for internal symmetries such as SU(2)⊗U(1) and SU(3) symmetries.

A space is a universe. Each universe has a space-time with dimensions ranging from zero to eighteen. In addition it has dimensions for irreducible representations of the internal symmetry groups that it supports. Examples of groups are SU(3) – the Strong interactions and SU(2)⊗U(1) – the ElectroWeak interactions. We will use the terms space and universe interchangeably.

---

[48] See Blaha (2021g).

[49] The space corresponds to Cayley number 2048. See Fig. 1.1 or Blaha (2021c) for details

We view space 0 as initially having a certain size in both space and time. In that space-time a fermion has a *spinor array* with row and column widths of 1024×1024 and thus 1,048,576 entries. We assume a fermion(s) exists in the 20 space-time dimension space 0.

We now contract the space-time to a 20 dimension point. The fermion spinor array size (rows and columns) remains the same and generates a spinor array. A *spinor space* with 1,048,576 dimensions is thus defined.[50] This space is Cayley-Dickson number 10 in Fig. 1.1. This space is a universe.

1.3 Space 10

A fermion of the 20 space-time dimension space 0 contains a spinor space (space 10) populated with matter and energy due to its mass. The spinor space, with its elementary particles and energies, is the space[51] of a higher dimension universe. It has an 18 dimension space-time. The choice of 18 dimensions is motivated by the goal of having a Cayley-Dickson spectrum of spaces.

In space 10 a fermion-antifermion annihilation can take place. It can be visualized with the following Feynman-like diagram:[52]

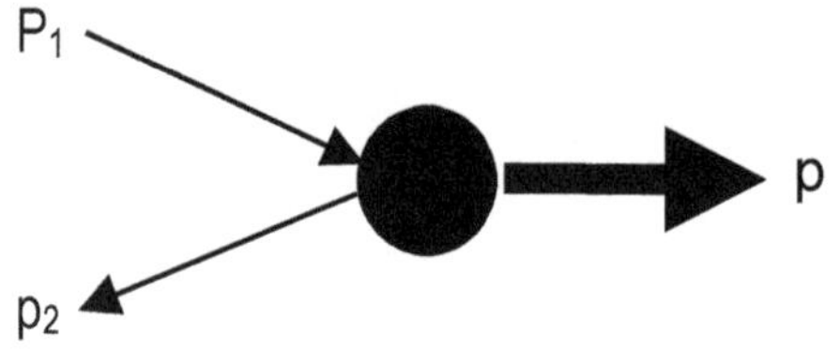

where p_1 and p_2 are the momentum of the fermion and antifermion respectively and p is the momentum of the produced space-containing particle. The produced (scalar) space particle, which exists in space 10,[53] contains a spinor array/space that is the space of another higher dimension universe of the type of space 9 in Fig. 1.1.

Space 9 with its mass and energy is created using the mass and energy from the annihilation in space 10. It has 512×512 = 262,144 dimensions, of which 16 are space-time dimensions.

This space-time is the first of the series of nine spaces (universes) generated through fermion-antifermion annihilations. They produce a Cayley-Dickson number based spectrum of spaces. See Figs. 1.1 and 1.2 below.

1.4 The Further Generation of Spaces

The spectrum of spaces (universes) is generated through a series of fermion-antifermion annihilations in space after space. The Cayley-Dickson pattern of spaces emerges from the choice of the number of space-time dimensions in each space. We

---

[50] The transformation of a fermion spinor array to a dimension array is described in detail in the case of fermion annihilation in chapter 6 of Blaha (2021g).

[51] See Fig. 1.1.

[52] We assume off shell fermions.

[53] Note the produced particle has a mass and momentum in space 10.

descend from space to space by decreasing the space-time dimension by two at each space.[54] The below Fig. A shows the general pattern of nesting of space instances.[55]

The result of the sequence of "annihilations" is the spectrum of Fig. 1.1 which corresponds to the Cayley-Dickson numbered spaces from 10 through 1 with space-time dimensions from 18 through 0.

In this iterative process there can be any number of subspace instances generated from a space instance by fermion-antifermion annihilations. See Fig. 1.5.

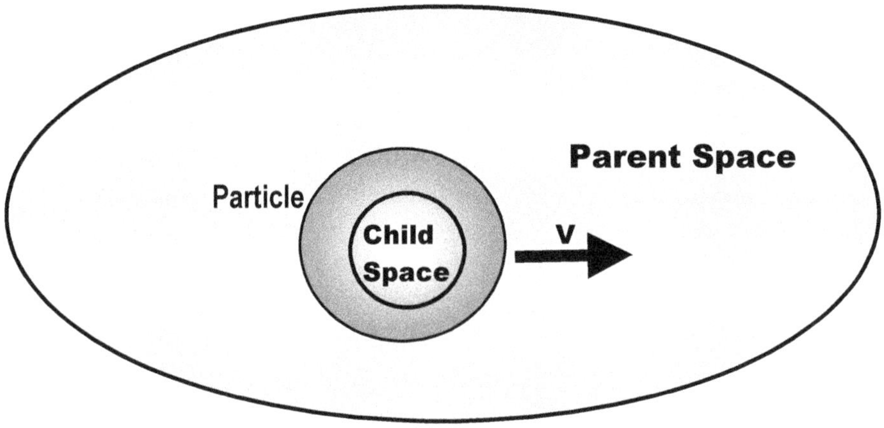

Figure A. The general pattern of a (parent) space containing a particle of velocity **V** that has an internal space instance (child) of lower dimension and lower Cayley-Dickson number. The space is itself within an instance of a higher space. The spectrum's spaces form a nested sequence. Each space contains the next lower space within it. The instances of spaces also form a nested sequence as shown in Fig. 1.2 below.

1.5 Comments

One might expect that generating the sequence of spaces would take an extraordinarily long time. However time, as we know it, only exists in our space 4 universe. Time is not a problem.

The octonion spaces have many dimensions beyond their space-time dimensions. These additional dimensions furnish the fundamental representations of the internal symmetry groups in the space. We show this for space 4 in Fig. 1.3. Fig. 1.4 shows the *form* of the fermion part of the instance of universe (described by our NEWQUeST theory.) The fermions of our universe have this form.

An interesting feature is the creation of space instances leads to a Big Bang for each instance—justifying the Big Bang for our universe for, perhaps, the first time.[56]

---

[54] An alternate ten space spectrum can be obtained by a diferent set of space-time dimensions. See Fig. 8.2 in Blaha (2021g).

[55] The definition of an instance appears below in the "Point of Clarification."

[5656] See Blaha (2021d) for the author's model for universe expansion and the Hubble Constant.

Point of Clarification: When we say a space (universe) is generated by a fermion-antifermion annihilation we mean that an *instance* of the space is generated. A space is defined independently of its contents. A space plus its contents is called an *instance* of the space. Annihilations generate instances. The source of the matter and energy of an instance is the mass-energy of the annihilating particles. We sometimes say space to refer to a space instance to avoid cumbersome text.

1.6 The Golden Triangle of Mass, Energy and Space

Examining the Octonion Cosmology spectrum and its origin in fermion-antifermion annihilations, we see a "golden" triangle of transformations between mass, energy, and space. Mass can be transformed into energy and vice versa.

Mass and/or energy can also be transformed into a space and vice versa. The creation of a space (instance) is the result of a mass and energy transformation. A space instance generates mass and energy transformations as it evolves. A space instance also generates instances of particles and of subspaces through fermion-antifermion annihilations within it.

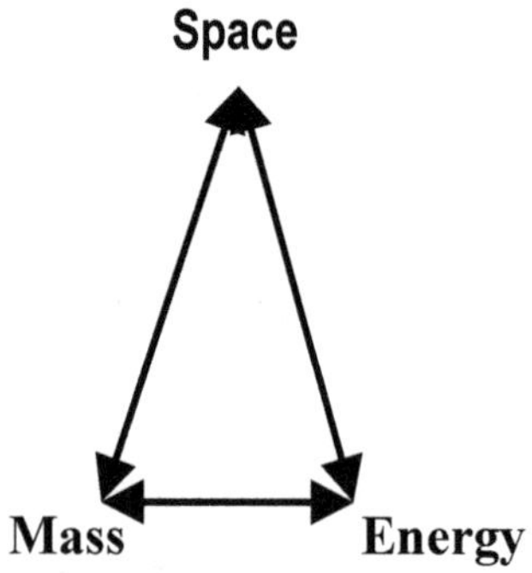

Figure B. The golden triangle of mass, energy, and space.

A remarkable reflection of the connection between mass and space is the equality of the number of fundamental fermion types in each of the ten spaces and the dimension of the space. In Cayley-Dickson number 4 space, which is that of our universe, we have found 256 fundamental fermion types matching the 16×16 = 256 dimensions of the space.

1.6 Octonion Cosmology Connection to Elementary Particle Theory

The Octonion Cosmology with its ten spaces originated in the author's Unified SuperStandard Theory (UST) that was developed over a number of years, and described in:

Blaha, 2018e, *Unification of God Theory and Unified SuperStandard Model THIRD EDITION*

Blaha, 2020a, *Quaternion Unified SuperStandard Theory (The QUeST) and Megaverse Octonion SuperStandard Theory (MOST)*

Blaha, 2020c, *Unified SuperStandard Theories for Quaternion Universes & The Octonion Megaverse*

Blaha, 2021c, *Beyond Octonion Cosmology*

Blaha, 2021d, *Universes are Particles*

Blaha, 2021e, *Octonion-like dna-based life, Universe expansion is decay, Emerging New Physics*

Blaha, 2021f, *The Science of Creation New Quantum Field Theory of Spaces*

In the Fall of 2019 the author considered the possibility that the UST was based on a deeper theory and proceeded to consider theories based on hypercomplex numbers such as quaternions and octonions. *Thus Octonion Cosmology did not "emerge out of thin air."* It provided a deeper base for the UST.

Remarkably the author found that the hypercomplex theories, QUeST and NEWQUeST, had the same internal symmetry groups, and fundamental fermions, as UST. In addition a hypercomplex theory NEWUTMOST furnished an acceptable theory of the Megaverse (Multiverse). Over the following two years 2020 and 2021 Octonion Cosmology emerged with its ten spaces.

The history of this effort, which has some relevance to Natural Philosophy, appears below. It is extracted from *Quantum Space Theory With Application to Octonion Cosmology Beyond Octonion Cosmology*. All **references below in this section 1.6 are to the contents of that book.**

EXTRACT FROM CHAPTERS 11 & 15 of *Quantum Space Theory*

The [UST and NEWUST] theories originated in the past twenty years from the Standard Model of Particles with SU(2)⊗U(1)⊗SU(3), and Two-Tier Quantum and PseudoQuantum Field Theory. Noting the presence of conserved particle numbers, and the presence of at least three fermion generations, we introduced the U(4) Generation Groups and the U(4) Layer Groups together with four layers of four generation fermions. The "Normal" fermions had a corresponding set of four layers of four generations of "Dark" fermions. The result was the Unified SuperStandard Theory (UST) symmetry:

$$\{[SU(2)\otimes U(1)\otimes SU(3)]^2\otimes U(4)^4\}^4$$

with an additional Strong Interaction U(1) group (analogous to that of ElectroWeak theory) found to be needed. Space-time had four dimensions.

In the Fall, 2019 the author discovered that an octonion-based theory that he constructed and named QUeST had the same internal symmetries as UST with the addition of $U(1)^8$. QUeST's internal symmetry, which could be based on a 16 × 16 dimension array, was

$$[SU(2)\otimes U(1)\otimes SU(3)\otimes U(1)]^8\otimes U(4)^{16}$$

The addition of $U(1)^8$ indicates that the Strong Interactions in the theory are broken Strong SU(4).

During 2020 the author developed an octonion space spectrum for both universes and Megaverses, and other spaces. The spectrum was shown to arise from a generation mechanism whereby fermion-antifermion annihilation in a higher space produced an instance of a lower space. A critical part of the derivation of the octonion spectrum was the realization that even space-time dimension spinor arrays are composed of Cayley number rows and columns. Spinor arrays of annihilating fermion-antifermion pairs were shown to generate the dimension arrays of subspace instances.

Analyzing the spinor arrays the author noted that the dimension array could be viewed as composed of 64 dimension blocks, which were further subdivided into 16 dimension subblocks.

The subblock structuring, using the known contents of the Standard Model plus Generation and Layer groups, gives the dimension array structure containing 4 × 4 subblocks in Figs. 11.1 and 11.2.

Thus there was a *most* satisfactory match between UST and QUeST with the only significant difference being the space-time: four octonion (complex quaternion) coordinates for QUeST and four real space-time coordinates for UST. This difference was resolved in NEWUST and NEWQUeST. See the below Connection groups [discussion].

The form of the square spinor arrays generated by fermion-antifermion annihilation gives 64 dimension blocks and 16 dimension blocks as well as 32 dimension composite blocks that are evidenced in the NEWQUeST (and NEWUTMOST) fermion spectrums and internal symmetry group structure.

Since we see only real dimensions in Reality, we transferred 28 QUeST dimensions from space-time to $U(2)^7$ internal symmetry dimensions. The set of internal symmetries was increased by $U(2)^7$, which we call Connection Groups. Each Connection group specifies interactions between corresponding fermions (e with e, q with q, and so on) in separate layers and between Normal and Dark fermions. The connections between the various blocks of fermions are shown in Fig. 11.3. *We implement the very practical rule that all blocks must be connected by interactions or they would not be of physical interest. A totally isolated block effectively does not exist physically (except possibly for gravitation effects).*

The interactions of the Connection groups must be very weak and/or their gauge bosons must be very massive.

The addition of the Connection Groups and the reduction of space-time dimensions accordingly results in NEWQUeST and NEWUST as summarized below. See appendix 11-A for a discussion of NEWQUeST groups.

Note: the Generation, Layer, and Connection groups are all badly broken. Their vector bosons must be very massive since they have not been detected in experiments.

11.1.1 Internal Symmetries

The groups are ElectroWeak SU(2)⊗U(1), Strong SU(3), Generation Group U(4), Layer Group U(4), and U(2) and U(4) Connection groups obtained by transfer from space-time coordinates (See Blaha 2012c). The SU(3)⊗U(1) symmetry may be a broken SU(4) symmetry in eqs. 11.1 – 11.4. The internal symmetries for the theories are:

UST

$$[SU(2)\otimes U(1)\otimes SU(3)]^8 \otimes U(4)^{16} \quad (11.1)$$

QUeST

$$[SU(2)\otimes U(1)\otimes SU(3)\otimes U(1)]^8 \otimes U(4)^{16} \quad (11.2)$$

NEWQUeST

$$[SU(2)\otimes U(1)\otimes SU(3)\otimes U(1)]^8 \otimes U(4)^{16} \otimes U(2)^7 \quad (11.3)$$

The only change is in internal symmetries: Twenty-eight real dimensions transferred from space-time coordinates to Connection group $U(2)^7$ internal symmetry.

NEWUST

$$[SU(2)\otimes U(1)\otimes SU(3)\otimes U(1)]^8 \otimes U(4)^{16} \otimes U(2)^7 \quad (11.4)$$

The only change in internal symmetries: Twenty-eight real dimensions added for $U(2)^7$ Connection group internal symmetry.

11.1.2 Space-Time Coordinates

UST

Four real space-time coordinates.

QUeST

Four octonion (complex quaternion) coordinates.

NEWUST

Four real space-time coordinates. No change from UST space-time.

NEWQUeST

Four real space-time coordinates. The six coordinates in the n = 4 octonion space (Figs. 1.5 and 1.6) were lowered to four space-time coordinates with two coordinates transferred to Connection groups.

The only change is in space-time coordinates: Fourteen dimensions transferred from QUeST space-time coordinates to Connection group $U(2)^7$ internal symmetry.

11.2 Fundamental Fermion Spectrum

There are 256 fundamental fermions[57] in NEWQUeST and NEWUST. Conceptually their structure can be viewed as an extrapolation of the known three generations of The Standard Model. For reasons given previously and in Appendix 11-A a fourth generation was indicated and a corresponding Dark sector of similar structure was added. In addition, because of the need for Layer groups (Appendix 11-A), the overall structure consisted of four copies of this layer.

Correspondingly, each layer also has its own set of internal symmetry gauge groups[58] to limit mixing between the layers to Layer group interactions and Connection group interactions.

Fig. 11.4 shows the structure of the NEWQUeST/NEWUST fermions. The blocks are 4 × 4 reflecting the origin of the NEWQUeST/NEWUST space (universe) instance from Megaverse fermion-antifermion annihilation. The spinor analysis of their spinor arrays yields a 16 dimension-based block structure. The 64 dimension fermion layers reflect the 64 dimension structuring of the Megaverse obtained from its creation by fermion-antifermion creation in the Maxiverse.

11.3 Total Dimensions

The total of internal symmetry and space-time dimensions is 256 in all four theories listed above. It is based on the 16 × 16 dimension array of the Cayley number n = 4 space of the Octonion spectrum.

11.4 Pattern of Internal Symmetries

The NEWQUeST dimension array for internal symmetries is subdivided into four layers of 56 dimensions—just as in NEWUST (and UST). Fig. 11.1 displays the layers using SU(4) in place of SU(3)⊗U(1). Each layer has a block of 28 dimensions for Normal and 28 dimensions for Dark sectors. There are also seven U(2) Connection groups[59] plus four real-valued space-time coordinates bringing the NEWQUeST total to 256 dimensions.= 4*56 + 28 + 4 = 256 dimensions. The Connection groups are shown in Fig. 11.2 (and section 11-A.4), which is a revision of the pattern shown in Blaha (2021c).

[57] See Fig. 9.2b for the fermion spectrum determined by a spinor array.
[58] See Fig. 9.2c for the pattern of symmetry groups.
[59] See Appendix 11-A.

Layers	NORMAL 4	4	DARK 4	4
4	SU(2)⊗U(1)⊗SU(3)⊗U(1) 4 Space-time Dimensions	Generation + Layer Groups	SU(2)⊗U(1)⊗SU(3)⊗U(1) 4 Space-time Dimensions	Generation + Layer Groups
4	SU(2)⊗U(1)⊗SU(3)⊗U(1) 4 Space-time Dimensions	Generation + Layer Groups	SU(2)⊗U(1)⊗SU(3)⊗U(1) 4 Space-time Dimensions	Generation + Layer Groups
4	SU(2)⊗U(1)⊗SU(3)⊗U(1) 4 Space-time Dimensions	Generation + Layer Groups	SU(2)⊗U(1)⊗SU(3)⊗U(1) 4 Space-time Dimensions	Generation + Layer Groups
4	SU(2)⊗U(1)⊗SU(3)⊗U(1) 4 Space-time Dimensions	Generation + Layer Groups	SU(2)⊗U(1)⊗SU(3)⊗U(1) 4 Space-time Dimensions	Generation + Layer Groups

Figure 11.2.. Four layers of Internal Symmetry groups in NEWQUeST (omitting Connection Groups). The groups in each layer are independent of those in other layers. The groups in each block of each layer are independent of those in the other blocks. Each block contains 16 dimensions. The block dimensions furnish fundamental representations for the groups listed. The entire set of blocks contains 256 dimensions. Each layer contains 56 internal symmetry dimensions. The first two columns are for the "Normal" sector. The last two columns are for the "Dark" sector (although most of the Normal sector is Dark observationally at present.) This figure also holds for UST with the addition of U(1) groups. The eight sets of 4 real dimension space-times combine to give a 4 real dimension space-time and seven U(2) Connection groups.

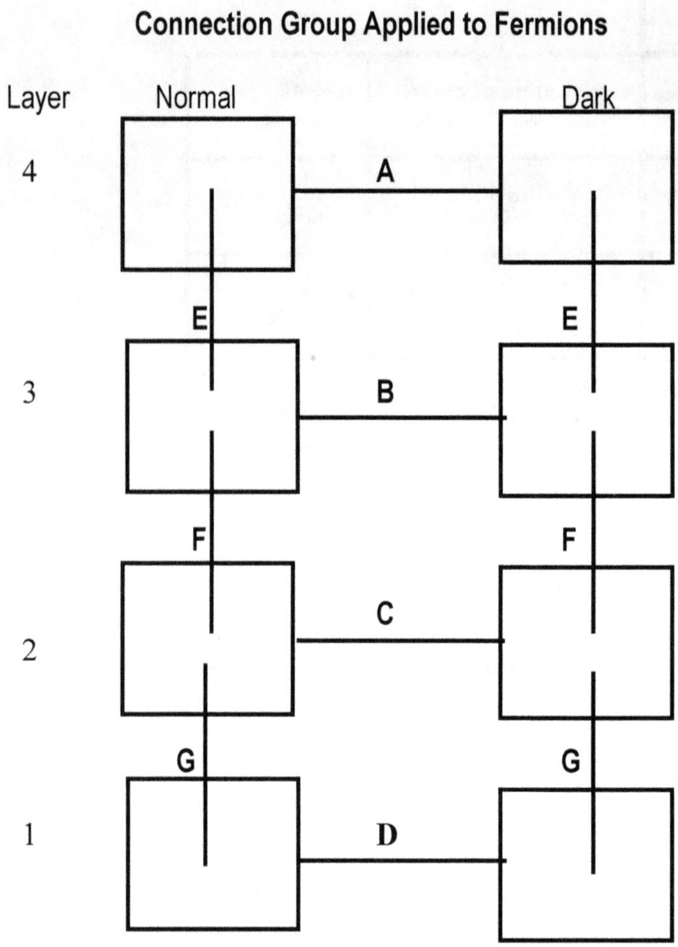

Figure 11.3. The seven U(2) Connection groups (shown as 10 lines) between the eight NEWQUeST/NEWUST blocks. Connection groups are obtained by transfering 28 dimensions from QUeST space-time to internal symmetries with the consequent reduction of the space-time from four octonion (complex quaternion) coordinates to four real coordinates. The Connection groups generate rotations and interactions between corresponding fermions and vector bosons of each pair of blocks. See Appendix 11-A. The Normal and Dark sector U(2) vertical connections above (E, F, G) represent the same U(2) groups.

Chapter 15. Possible Relation to SuperString Theories, Composition Algebras

Seven of the ten spaces of Octonion Cosmology appear to correspond to SuperString Theories:

Cayley-Dickson Number n	Space-time Dimension	Spinor Size	Cayley Number = Supercharge	Composition Algebra	SuperString Theory
4	6[60]	8	16	Complex Octonion	Type I
5	8	16	32	Quaternion Octonion	Type 2
6	10	32	64	Octonion Octonion	Heterotic?
7	12	64	128	Complex Octonion Octonion	?
8	14	128	256	Quaternion Octonion Octonion	?
9	16	256	512	Octonion Octonion Octonion	?
10	18	512	1024	Complex Octonion Octonion Octonion	?

The higher Cayley number spaces have yet to be matched to SuperString Theories. ***Their discovery would be of great interest!*** *Then a* possible match between Octonion Cosmology and SuperString theories may emerge.

The ? indicate additional theories that are suggested by the order of type I and type II supercharges, which are Cayley numbers. See Fig. 1.1.

Note the type I and type II SuperString Theories are assigned to different Octonion Cosmology spaces.

The comparison of Octonion Cosmology, with its series of steps connecting to NEWQUeST and the Unified SuperStandard Theory (NEWUST), with the goal of SuperString Theories to connect to Elementary Particles suggests direct steps from SuperString Theories to Particle Theory are likely to be difficult. The possibility of SuperString Theories as a basis of Octonion Cosmology is evident.

END OF EXTRACT FROM CHAPTERS 11 & 15 of *Quantum Space Theory*

[60] Possibly 4 with two compacted dimensions.

THE TEN OCTONION SPACES SPECTRUM

Octonion Space Number o_s	Cayley-Dickson Number n	Cayley Number d_c	Dimension Array Total d_d	Space-time-Dimension r	Fermion Spinor Array Total d_s
0	10	1024	1024×1024	18	512×512
1	9	512	512×512	16	256×256
2	8	256	256×256	14	128×128
3	7	128	128×128	12	64×64
4	6	64	64×64	10	32×32
5	5	32	32×32	8	16×16
6	**4**	**16**	**16×16**	**6**[61]	**8×8**
7	**3**	8	8×8	4	4×4
8	**2**	4	4×4	2	2×2
9	**1**	2	2×2	0	1×1

Figure 1.1. The Octonion Cosmology ten space spectrum. The space for our universe, number 6, (Cayley-Dickson number 4) is in bold type.

\

[61] The space-time dimensions become 4 as in the Unified SuperStandard Theory (UST) through either transfer of dimensions to internal symmetries or by the compactification of two dimensions. As a result the pattern of fermion-antifermion annihilations producing spaces 7, 8, and 9 is disrupted.

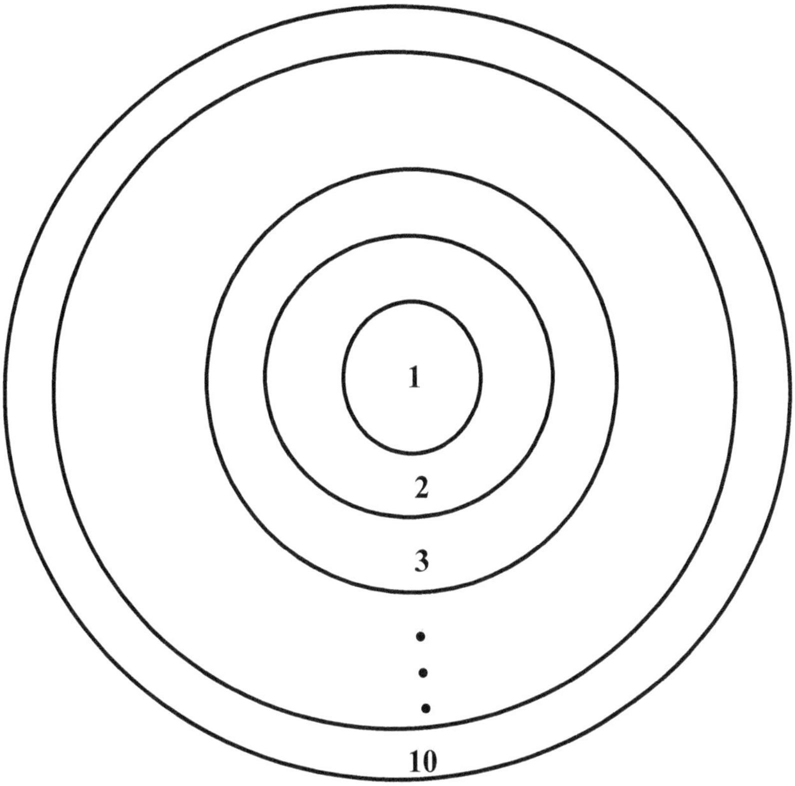

Figure 1.2. Numbered form of the ten space instances of Octonion Cosmology. They are generated by fermion-antifermion annhilations. The outermost space instance (Cayley-Dickson space 10) is generated first from the 20 space-time dimension space. The Space 9 instance is generated within space 10 and so on.

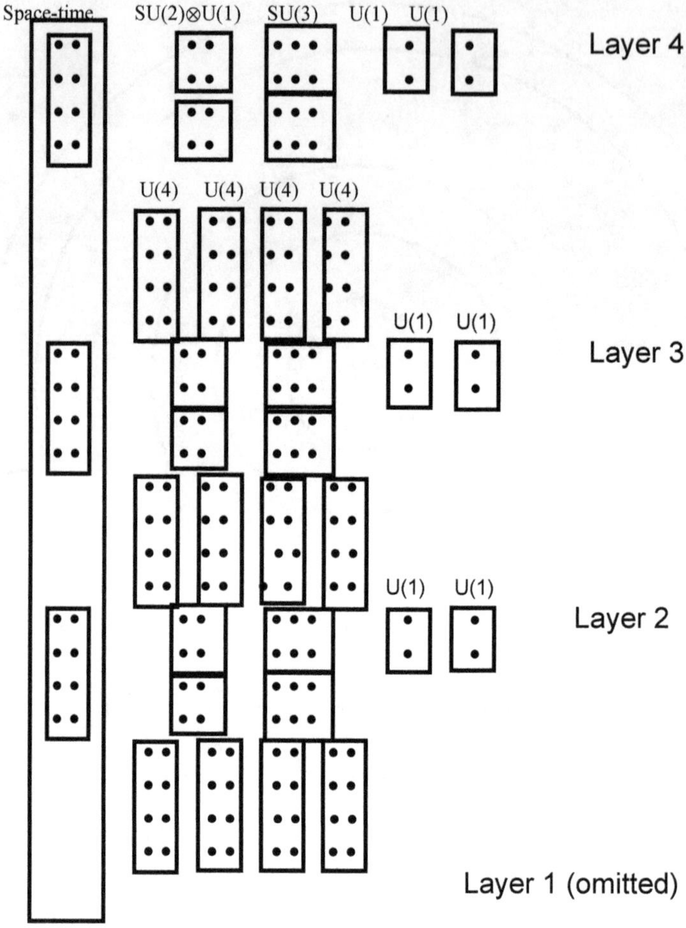

Figure 1.3. Three of the four layers of QUeST internal symmetry groups (and space-time) for Cayley-Dickson space 4. Layer 1 which has an identical form was omitted due to "page space" limitations. Note the left column of blocks combine to specify a 4 dimension octonion space-time. Note each layer has 64 dimensions.

The Fermion Periodic Table

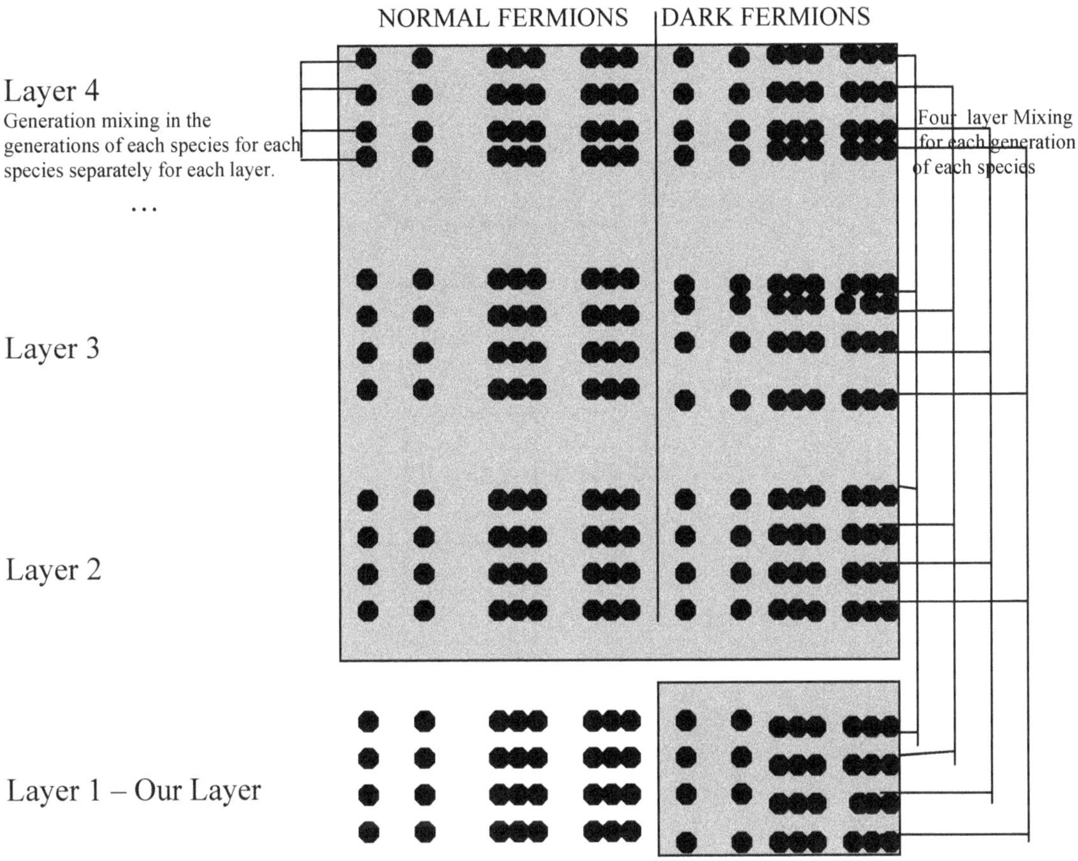

Figure 1.4. Fermion particle spectrum and partial examples of the pattern of mass mixing of the Generation group and of the Layer group. Unshaded parts are the known fermions with an additional, as yet not found, 4th generation. The lines on the left side (only shown for one layer) display the Generation mixing within each layer. The Generation mixing occurs within each layer using a separate Generation group for each layer. The lines on the right side show Layer group mixing (for Dark matter) with the mixing among all four layers for each of the four generations individually. There are four Layer groups for Normal matter and four Layer groups for Dark matter.. There are 256 fundamental fermions. QUeST and UST have the same fermion spectrum.

A SMALL COSMOS

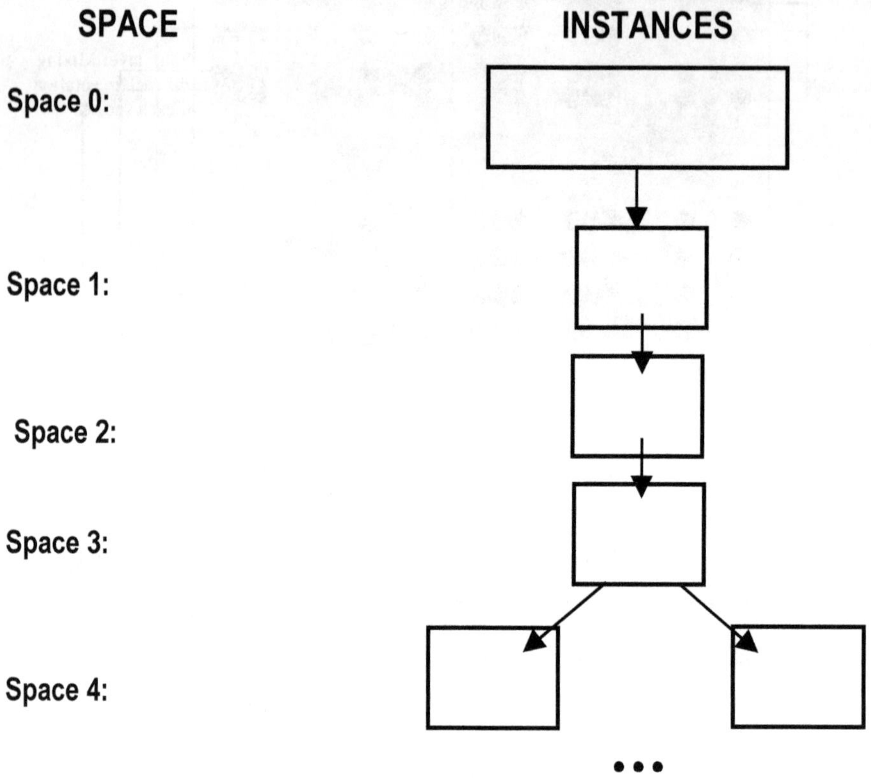

Figure 1.5. A hierarchy of space instances one “sibling” and three “cousin” universes. *The entire hierarchy resides in space 0.*

Appendix B. Quantum Theory of the Third Kind

Quantum Theory of the Third Kind

A New Type of Divergence-free Quantum Field Theory Supporting a Unified Standard Model of Elementary Particles and Quantum Gravity based on a New Method in the Calculus of Variations

Stephen Blaha, Ph.D.*

Pingree-Hill Publishing

* sblaha777@yahoo.com

Pingree-Hill Publishing
P. O. Box 368
Auburn, NH 03032 USA

Or email to: baliltd@compuserve.com

ISBN: 0-9746958-3-1

This book is printed on acid free paper.

1. Quantum Dimensions vs. Classical Dimensions

All beginnings are obscure.
H. Weyl – Space, Time, Matter

Beyond 4-Dimensional Space-time

There have been countless attempts since the 1920's to use additional dimensions beyond the known four dimensions of space and time to explicate and unify the fundamental forces of nature. The most noteworthy *recent* attempts along these lines have been Superstring theories and Technicolor theories.

In the opinion of this author the efforts in these directions are not justified by the results. The physics that these theories attempt to describe is simpler than the formulation of the theories with much fewer particles and interactions. Assigning high masses to undiscovered particles and placing undiscovered forces in the high energy regime beyond the limits of accelerators does not seem to be a satisfactory approach. Extrapolations of theories in the past, without the guidance and confirmation of experiment, have usually not been successful.

Therefore it seems reasonable to develop a deeper, sounder formulation of the Standard Model *as it is now* since it was developed through a close interplay of theory and experiment. It remains the preeminent theory of elementary particles—actually the *only* experimentally acceptable theory of elementary particles.

This book attempts to establish a deeper framework for the Standard Model that enables it to be combined with quantum gravity to form a divergence-free unified quantum field theory of nature.

It is clear from the existence of internal symmetries in the Standard Model that something is "going on" inside particles which is outside the framework of normal space and time. Otherwise, there would be no internal symmetries.

Therefore it is reasonable to consider the possibility of extra dimensions beyond normal space-time. The open question is how these extra dimensions enter into physical theory. Superstring theory appears to go too far in terms of the numbers of dimensions and the particles that it requires – not to mention – the complexity of its mathematics, appears to preclude all but the simplest calculations. Technicolor and extended Technicolor theories also introduce substantial additional complexities in order to explicate the pattern of symmetries of the Standard Model. We face the question of whether the cure is worse than the disease in these approaches.

Thus we ask if a more tractable theory is possible. Our first requirement is that it would improve the Standard Model by taming the divergences in quantum field theory so that the Standard Model can be unified with Quantum Gravity. This author has suggested[62] an alternate form of quantum field theory, Two-Tier quantum field theory that is ultra-violet divergence-free to all orders for theories of the type of the Standard Model, and for Quantum Gravity. This theory not only resolves divergence issues in the Standard Model and Quantum Gravity but also eliminates the singularities at the point of the Big Bang in Cosmology.[63]

This theory is *not* based on extra dimensions that become compacted as in Superstring theories. Rather it postulates that extra dimensions are directly generated by a free quantum field and thus constitute *quantum dimensions*™. Quantum dimensions are fluctuating quantum degrees of freedom that make each elementary particle a "fuzzy ball" that partly exists in imaginary space as well as real space.

Ideally, in this author's view, we would start with a concept of a pre-dimensional entity – an entity that is perhaps initially formless which evolves, perhaps through a form of self-organization, to develop dimensions, energy and quantum particles. Physics is far from developing such a grand scheme and must, at present, content itself with assuming the existence of dimensions, a Lagrangian of some sort, and quantum dynamical entities (particles). Thus we will make assumptions as to the number and nature of additional dimensions.

Quantum Dimensions vs. Classical Dimensions

We are all familiar with the concept of dimension. Euclid gave a geometrical definition of dimensions and a procedure for determining the number of dimensions: move in a straight line for a distance; then make a 90° turn; again move in a straight line in the new direction for a while; then make a 90° turn such that the new direction is perpendicular to the previous two directions of motion; continue this procedure until it is no longer possible to make a 90° turn to move in a new direction. This process establishes the direction of dimensions and their number in flat space.

From a Cartesian point of view the number of dimensions can initially be simply viewed as the number of independent coordinate axes. Each axis is broken into intervals in such a way as to allow us to specify a position in space with an ordered set of numbers. We can then define functions that depend on these ordered sets of numbers – the coordinates of a point. Such a function can then have a range of values as the coordinates change from point to point in space. We can denote this function as $f(x_1, x_2, x_3, \ldots x_n)$ if the space has n dimensions.

The range of the dimensions in a flat Cartesian space is usually from $-\infty$ to $+\infty$. One can also specify cyclic or compact dimensions that "form a circle" – a coordinate can range in value say from 0 to 2π. The point at 2π can then be made to coincide with the point at 0 so that one can view the dimension as a circle.

Another approach that was first introduced by Blaha (2003) is to use a new type of coordinate – a quantum coordinate. To understand this concept we imagine a 3-

62 See Blaha (2003) – the first edition of this book.
63 See Blaha (2004).

dimensional space of the normal sort with coordinates: x, y, z and values ranging from $-\infty$ to $+\infty$. Now suppose, for example, we introduce a sine function:

$$G(x, y) = \sin(x + y) \tag{1.1}$$

and require

$$z = G(x, y) \tag{1.2}$$

Then a function of the three coordinates of space f(x, y, z) becomes

$$f(x, y, z) = f(x, y, G(x, y)) \tag{1.3}$$

on the surface defined by eq. 1.2. We see f still has values at each point in 3-dimensional space. However, since eq. 1.2.2 defines a 2-dimensional surface within the 3-space the expression for f in eq. 1.3 is properly viewed as defined on that surface.

Now if we replace G(x, y) with a second quantized field Q(x, y) in eq. 1.3 then we obtain a qualitatively new entity:

$$f(x, y, Q(x, y)) \tag{1.4}$$

is now an operator expression. (We will assume that infinities and other issues are not present or resolvable.) Eq. 1.4 in itself does not have a value. It obtains a value when evaluated for quantum states |q>

$$q(x, y) = \langle q | f(x, y, Q(x, y)) | q \rangle \tag{1.5}$$

Thus the replacement of the z coordinate in f(x, y, z) with a field operator gives us a qualitatively new entity with well-defined values only when evaluated between states.

In a sense f(x, y, Q(x, y)) is dependent on three unknown quantities – the values of x and y, and the expectation value of f(x, y, Q(x, y)) between (as yet unidentified) quantum states. Thus we can regard f(x, y, Q(x, y)) as depending on two coordinates x and y, and on a quantum coordinate™ whose value is an undetermined quantum fluctuation quantity that only becomes known upon taking an expectation value.

We thus have developed a new form of dimension that is quite properly called a quantum dimension with values that are not c-numbers but in fact are q-numbers (determined only by taking expectation values between quantum states).

Quantum Theory of the Third Kind

First there was quantum mechanics – developed essentially in the period from 1914 - 1926. Quantum Mechanics postulated that position and momentum were non-commuting operators and so the position and momentum of a particle could not be measured simultaneously to arbitrary accuracy. Instead they satisfied the Heisenberg Uncertainty Principle:

$$\Delta x \Delta p \geq \hbar \qquad (1.6)$$

Then quantum field theory (second quantization) was developed roughly in the period from 1935 – 1960. Quantum theory postulated that any quantum field and its conjugate momentum operator satisfied an uncertainty principle:

$$\Delta\phi\Delta p_\phi \geq c_\phi\hbar \qquad (1.7)$$

for any quantum field ϕ and its conjugate momentum p_ϕ where c_ϕ is a factor dependent on the nature of the quantum field ϕ. As a result the arguments of the field operators ϕ - the position coordinates – were treated as ordinary parameters – c-numbers. Eq. 1.6 and eq. 1.7 were shown[64] to both be required for a consistent quantum theory. The standard example considered the effect of using a quantum electromagnetic field to measure the position and location of a particle.

We will suggest that there is a further quantum formulation beyond first and second quantization in which quantum fields are functions of quantum coordinates. This type of *quantum theory of the third kind* resolves the divergence problems that have plagued second quantization.

Loosely speaking by making the coordinate arguments of quantum fields "fuzzy" we avoid the infinities associated with evaluating products of quantum field operators at precisely the same point. We will proceed to consider quantum field operators that are functions of quantum coordinates in the remainder of this book.

Quantum Coordinates in 4-dimensional Space-time

Now that we see the nature of a quantum dimension we will turn to developing the form of space-time for a universe with both ordinary space-time dimensions and additional dimensions to resolve divergence problems. In Blaha (2003) and Blaha (2004) we were concerned with resolving the divergences of the Standard Model quantum field theory and the divergences of Quantum Gravity as well as the singularity issue at the point of the Big Bang in the Standard Model of Cosmology. We pointed out all these issues could be resolved by using complex quantum coordinates

$$X_\mu(y) = y_\mu + i\, Y_\mu(y)/M_c^{\,2} \qquad (1.8)$$

where $Y^\mu(y)$ is a real quantum field with properties identical to the free electromagnetic quantum field of Quantum Electrodynamics, y^μ is a Minkowski space-time 4-vector, and M_c is a large mass that is presumably of the order of or equal to the Planck mass.

All particle (boson and fermion) quantum fields were defined as q-number functions of the quantum coordinates $Y^\mu(y)$ and not of y^μ directly. Using a new method in the Calculus of Variations (Appendix A) that we called the "composition of extrema"

---

[64] Heitler (1954) pp. 79-86.

we developed quantum field theories – called *Two-Tier quantum field theories* – for particles of spin 0, ½, 1, and 2.

For example, the Two-Tier Lagrangian for a free scalar particle is

$$I = \int \mathscr{L}_s \, d^4y \tag{1.9}$$

with

$$\mathscr{L}_s = J \, \mathscr{L}_F(\phi(X), \partial\phi/\partial X^\mu) + \mathscr{L}_C(X^\mu(y), \partial X^\mu(y)/\partial y^\nu) \tag{1.10}$$

where ϕ is a scalar field, J is the Jacobian for the transformation from $X^\mu(y)$ coordinates to y^μ coordinates, and

$$\mathscr{L}_F = ½ \, [(\partial\phi/\partial X^\nu)^2 - m^2\phi^2] \tag{1.11}$$

If we define $X^\mu(y)$ using eq. 1.8 then

$$\mathscr{L}_C = -¼ \, F_Y^{\mu\nu} F_{Y\mu\nu} \tag{1.12}$$

where

$$F_{Y\mu\nu} = \partial Y_\mu/\partial y^\nu - \partial Y_\nu/\partial y^\mu \tag{1.13}$$

Upon variation in ϕ with y^μ (and thus $Y^\mu(y)$) held fixed (see Appendix A for a discussion of composition of extrema in the Calculus of Variations) we find

$$\partial\mathscr{L}/\partial\phi - \partial/\partial X^\mu \, [\partial\mathscr{L}/\partial(\partial\phi/\partial X^\mu)] = 0 \tag{1.14}$$

which gives us the Klein-Gordon field equation for ϕ

$$(\Box + m^2) \, \phi(X) = 0 \tag{1.15}$$

where

$$\Box = \partial/\partial X^\nu \, \partial/\partial X_\nu \tag{1.16}$$

A Fourier representation of the solution of eq. 1.15 is:

$$\phi(X) = \int dp \, \delta(p^2 - m^2)\theta(p^0) \, [a(p){:}e^{-ip\cdot X}{:} + a^\dagger(p){:}e^{ip\cdot X}{:}] \tag{1.17}$$

where a(p) is a function of p, † indicates Hermitian conjugation, and : : indicate normal ordering of the q-number expression in $X^\mu(y)$.

The manifold defined by $X^{\mu}(y)$ is found by variation of the Lagrangian with respect to $Y^{\mu}(y)$. In brief it yields field equations (and gauge invariance) that are identical to the case of the electromagnetic field. This scalar particle case and other cases are considered in detail in the following chapters. So we will defer further consideration of the scalar particle case until then.

The following points were proved in Blaha (2003) and Blaha (2004):

1. Free field theories created with the Two-Tier formulation (including quantum gravity) have propagators that are the same as those in the corresponding conventional quantum field theory except that the Fourier representation of each particle propagator contains a Gaussian momentum factor. This Gaussian factor eliminates all ultra-violet divergences in perturbative quantum field theories such as the Standard Model and Quantum Gravity. The gauge invariance of $Y^{\mu}(y)$ was required in order to have well-behaved Gaussian factors throughout the momentum region. Blaha (2003).

2. The resultant theories were proven to satisfy unitarity. Blaha (2003).

3. The resultant theories were proven to be Lorentz invariant in any gauge of $Y^{\mu}(y)$. Blaha (2004).

Thus this new approach to quantum field theory (Two-Tier quantum field theory), when applied to the Standard Model and Quantum Gravity, results in divergence-free quantum field theories enabling us to create a unified, divergence-free quantum field theory of all the forces of nature. (The question of creating this type of quantum field theory in curved space-time was successfully addressed in Blaha (2004).)

The nature of the point of the Big Bang in Cosmology has been an ongoing issue. When Two-Tier quantum field theory is applied to the Big Bang we find that it suggests the universe is very dense region of finite size and temperature at the time of the Big Bang with an inhomogeneous generalized Robertson-Walker metric. The Einstein equations of this metric are separable: one equation is dominated by classical physics; the other equation is quantum in nature. The universe is shown to rapidly change into a standard Robertson-Walker universe with the usual scale factor. Thus the Standard Model of Cosmology emerges shortly after the Big Bang but the singularity of the Standard Cosmological Model at the Big Bang is eliminated Blaha (2004).

Thus the Two-Tier quantum field theoretic approach eliminates the divergence problems of the Standard Model, Quantum Gravity and the Standard Cosmological Model. In addition it justifies treating the Big Bang epoch of the universe as having an ultradense, quasi-free energy density in the form of a perfect fluid.[65] This result follows from the factorization of the scale factor of the generalized Robertson-Walker metric into a factor that is dominated by the macroscopic energy density and is thus primarily classical in nature; and another factor embodying quantum effects that is independent of the energy density.

Features of Two-Tier Quantum field Theories

The currently known theories of fundamental interactions and elementary particles fall into two broad categories: conventional quantum field theories and string-based theories that are united with supersymmetry to form superstring theories. Many physicists believe that only theories of these types can meet the reasonable physical requirements of Lorentz invariance and unitarity (positive probabilities summing to unity) while accounting for spin and internal quantum numbers.

It appears that our new form of elementary particle theory is viable. In this type of theory quantum fields are functions of q-number coordinates with imaginary quantum dimensions. We will investigate a unified quantum field theory of this type. We will create a unified theory of the Standard Model and Quantum Gravity. The unified theory has the following features:

1. Consistency with the current Standard Model in all respects at current energies up to current maximal energies of several TeV.

2. Consistency with the classical Theory of Gravity. Classical gravity is the "large distance", "low energy" limit of the theory.

3. General relativistic covariance of the field equations, Lorentz invariance of the S matrix, and unitarity as required in physically acceptable theories.

4. The unified theory is divergence-free – it contains no infinities.

5. All interactions are modified by a short distance, high-energy, substructure that begins at some energy presumably much above currently accessible energies. The energy scale could be set by the Planck scale (10^{19} GeV) or could be a much lower energy such as 10^5 GeV.

6. It allows modifications of the Standard Model such as further unification through broken higher symmetries without losing features 1 – 5 above.

7. It predicts a number of new phenomena at extremely short distances such as ultra-relativistic dilepton resonances consisting of two leptons of the same charge. An example of this type of bound state is a bound resonance consisting of two electrons.

---

[65] A recent experiment in which gold nucei collided at high energy confirms that the collision region briefly contained a perfect fluid consisting of a quark-gluon plasma. This result is consistent with our Two-Tier Standard Model and the Two-Tier cosmological model developed in Blaha (2004).

These dilepton states could be created at ultrahigh energies by penetrating the repulsive Coulomb barrier. Inside the barrier near r = 0 the modified Coulomb potential is a linear potential. A dilepton bound state is highly unstable with a large decay rate due to quantum tunneling through the Coulomb barrier. Other exotic resonances are also possible.

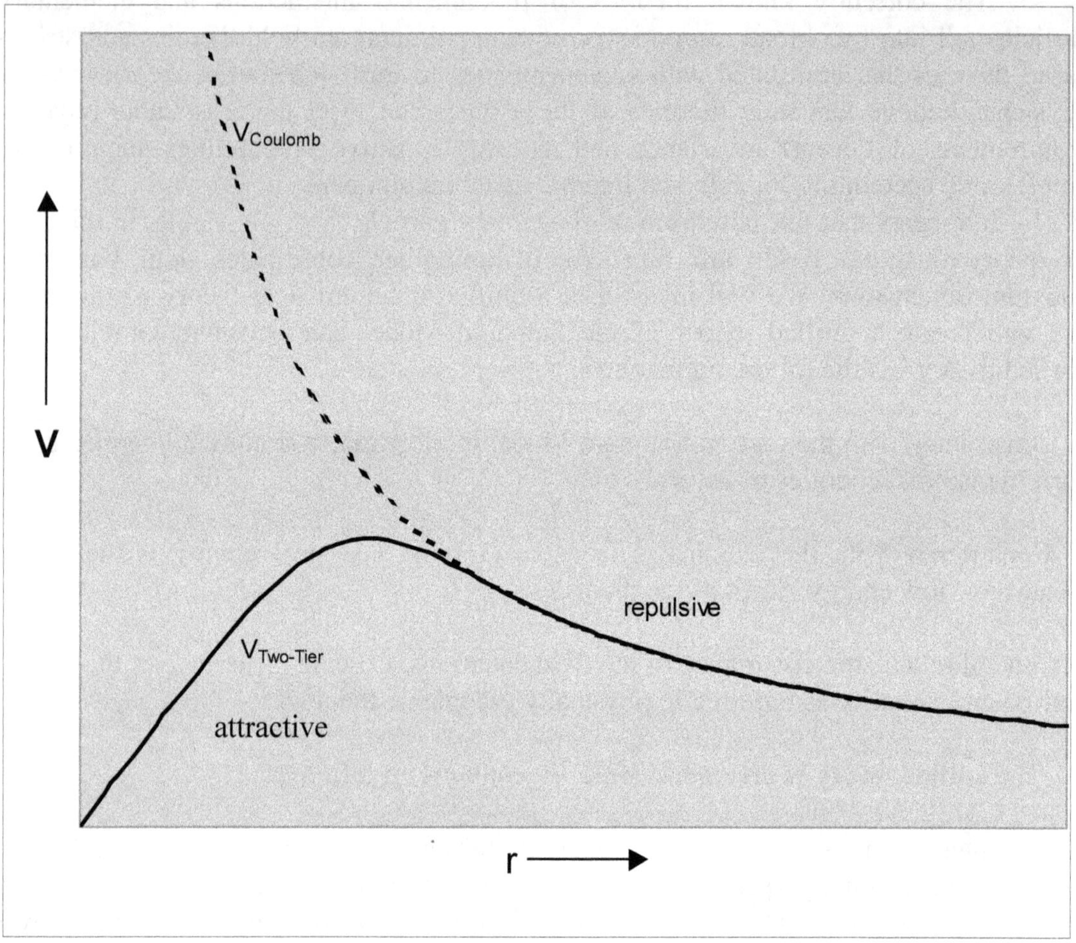

Figure 1.1. Modified electromagnetic Coulomb force between two particles with charges of the same sign (eg. two electrons) in Two-Tier QED and the Two-Tier Standard Model. The potential is repulsive at long distances and attractive at short distances. The potential becomes linear near the origin r = 0 opening up the possibility of unstable doubly charged resonances.

The curve in Fig. 1.1 is the Two-Tier potential between two particles of the same sign. The Two-Tier electromagnetic potential between two singly charged particles of the same sign is:

$$V_{new} = \alpha\Phi(M_c^2\pi r^2)/r \tag{1.18}$$

where α is the fine structure constant, $\Phi(z)$ is the error function, M_c is the mass setting the scale of the new short distance behavior and r is the radial distance. At small distances ($\pi r^2 \ll M_c^{-2}$) we find

$$V_{new} \to \alpha 2\sqrt{\pi}\, M_c^2\, r \tag{1.19}$$

and at large distances ($\pi r^2 \gg M_c^{-2}$) it becomes identical to the Coulomb potential:

$$V_{new} \to V_{Coul} = \alpha/r \tag{1.20}$$

8. The short distance modifications of all four interactions open up the possibility of a more fundamental, smaller set of particles, of which the currently observed particles are bound states. Thus quarks and leptons might be bound states of more fundamental particles.

2. The Standard Model, Gravity and Superstring Theories

In this chapter we consider aspects of the Standard Model and Gravitation, and Superstring theories that are relevant for the discussion of our new unified theory.

Standard Model

The Standard Model unites the ElectroWeak interactions and the Strong interactions in a successful theory that appears to be consistent with the known experimental data at all accessible energies. The theory satisfies the unitarity condition with positive probabilities adding to unity, Lorentz invariance, and renormalizability. There are several variants and proposed extensions of the Standard Model. But the major aspects of the theory are common to all variants.

The most important technical issue that faced the development of the Standard Model was the question of the renormalizability of the unification of the electromagnetic interactions and the weak interactions in the ElectroWeak Theory sector. The issue was resolved by the proof of renormalizability by ‘t Hooft in 1971.

At first glance ElectroWeak Theory, and its electromagnetic sector, Quantum Electrodynamics (QED), appear to have divergences when perturbative calculations are made of transition probabilities, or of physical quantities such as the anomalous magnetic moment of the electron or muon. However these divergences can be isolated into a finite number of infinite quantities (renormalization theory) which in turn can be absorbed into redefinitions of fundamental parameters of the theory (such as the “bare” electric charge or the “bare” mass of the electron.)

For example in QED the “bare” charge e_0 is renormalized by infinite factors to give the “physical” charge e: $e = Z_1^{-1}Z_2\surd Z_3 e_0$ where Z_1, Z_2, and Z_3 are divergent renormalization constants. Numerous authors over the last fifty years or so have commented on the “unnaturalness” of renormalization which, after all, amounts to multiplying infinity by zero (e_0) to obtain a finite observable number, the electric charge e in this case.

The quantum field theory formalism that we shall develop resolves these issues by being divergence-free – no infinities appear because of a short distance modification that appears in all particle propagators. The result is a logically satisfactory theory that avoids divergences independent of the details of the interactions and symmetries, and, in fact, allows a wider range of interactions such as interactions with derivative couplings, which were previously totally unrenormalizable.

Quantization of Gravity

Classical gravity began with Newton’s theory of gravity, which still remains an acceptable theory for most phenomena involving small masses and velocities. In the early twentieth century A. Einstein developed a new theory of gravitation based on geometrical concepts. This theory, the General Theory of Relativity, has been tested and found to be correct as far as we can determine at the classical level.

There have been many attempts to develop a quantum version of the General Theory of Relativity. While part of the overall framework of such a quantum theory is known, all attempts to develop it within the framework of conventional quantum field theory have failed due to the infinite number of divergences that appear when calculations are attempted in perturbation theory.[66] A theory with an infinite number of divergences cannot be handled with the renormalization procedures used to tame the divergences of QED, Yang-Mills theories, or the ElectroWeak Theory; and thus does not have predictive power. The gravitational sector of the unified theory that we will develop does not have these divergences and is in fact divergence-free.

Correspondence Principle

The development of Quantum Mechanics was guided by a principle developed by Bohr called the *Correspondence Principle*. Simply put, this principle states that quantum mechanical systems must approach the behavior of classical systems as the features of the quantum mechanical system become much larger than the quantum of action – Planck's constant h. Classical mechanics is thus the large scale limit of quantum mechanics.

Similarly, the Theory of Special Relativity also has a sort of correspondence principle – the predictions of the theory of Special Relativity for a system must approach the predictions of classical mechanics in the limit that all the velocities of the system become much smaller than the speed of light. Thus classical mechanics is a limiting case of Special Relativity.

Just as these twentieth century theories have a type of correspondence principle in which they must approximate an earlier theory in a certain limit, any new theory of elementary particles must approximate the Standard Model in the currently known range of energies. After all the Standard Model works! Therefore it must have an element of physical truth and a more fundamental theory must embody that truth in the explored range of experimental energies.

Furthermore the extreme accuracy of QED calculations[67] must be matched by any theory purporting to account for elementary particle interactions. At the moment, and apparently for the foreseeable future, Superstring theories have difficulties when attempts are made to find the Standard Model, or something like it, as an approximation at current energies. Capturing the accuracy of QED predictions would seem to be in the distant future, if at all, in Superstring theories.

---

[66] See for example B. S. DeWitt, Phys. Rev. **162**, 1239 (1967), and R. P. Feynman, Acta Physica Polonica **24**, 697 (1963).

[67] T. Kinoshita, "The Fine Structure Constant", Cornell Univ. Preprint CLNS 96/1406 (1996).

String and Superstring Theories

W. Pauli once remarked, "Man should not join together that which God has put asunder." In apparent contradiction (perhaps) to this dictum Superstring theory attempts to unite fermions (half-integer spin particles) and bosons (integer-spin particles) within a larger symmetry so that they can, in a sense, be rotated into one another. While support for this symmetry is currently lacking in elementary particle physics, there may be evidence for Supersymmetry in nuclear physics.

However the price for Supersymmetry in particle physics appears to be quite high: a large number of dimensions and a large number of particles. No evidence exists at the time of this writing for more than four space-time dimensions or for the Supersymmetric partners of the known elementary particles. Thus, at best, Supersymmetry is for the future when experimental energies reach energies where Supersymmetric features are unequivocally seen (if in fact they exist.)

The theory of strings – two-dimensional substructures that constitute elementary particles – is grounded in phenomenology – more particularly, in the Veneziano-Suzuki formula for scattering amplitudes. Y. Nambu and T. Goto developed a vibrating string theory that accounted for the form of the Veneziano-Suzuki formula. So there appears to be some justification for a string model of elementary particles. In this model the elementary particles, which appear as structure-less and point-like at lower energies, can be seen to have a string-like structure at very high energies.

Thus a sub-structure for elementary particles may have some experimental justification.

Logic Does Not Necessarily Bring Progress in Physics

An often-expressed hope in the last two decades of the twentieth century was that some unique Superstring theory would emerge, and that this theory would prove to be the only logically reasonable theory. After thirty years of effort by a large group of extremely talented physicists (perhaps larger than any group of physics theorists devoted to one topic since the Manhattan project) this hope is yet to be realized.

Historically, the movement from level to deeper level to yet deeper level in Physics has not been the result of logical inquiry but rather has been the result of new experiments confronting existing theory wherein the existing theory is found to be wanting.[68] A rational physicist in the eighteenth century would not have conjectured the theory of special relativity as a logical possibility: "Why should mechanics change when the speed of an object becomes large?" A rational physicist in the nineteenth century would not have conjectured the theory of quantum mechanics as a logical possibility: "Why should it be impossible to measure the momentum and position of a particle with arbitrary precision?" These questions did not, and could not have arisen, in a rational, "down to earth" scientist. Furthermore, despite knowing that these theories are correct we do not know WHY they are implemented in Nature. We only know that they are.[69]

---

[68] This historical process of Physics is described in some detail in the book Blaha(2002) by this author.

[69] A possible answer as to why Nature is quantum in nature has recently been proposed by Blaha (2005) based on Gödel's Theorem.

With this historical perspective in mind it appears that we should not expect to arrive at the form of the next deeper level of Physics through reason alone. Nature will most likely surprise us again (and again). Thus experiment is our teacher as we learn more of the nature of matter and space-time.

Goal: Unified Theory Without Renormalization Issues

There are a number of issues confronting elementary particle physics today. Issues such as CP violation, the nature of dark matter and of dark energy, the form of the symmetries of Nature, and the origin of the "numerous" particles and constants in the Standard Model. Practically speaking, perhaps the most important problem at this time is the development of a renormalizable unified theory of all the interactions. We have seen that we can cope with the incomplete unification in the Standard Model between the ElectroWeak interactions and the strong interaction. The Standard Model is renormalizable and thus we can make perturbation theory calculations with confidence and compare the results with experiment. But we cannot make calculations in quantum gravity with confidence. As a result Planck scale physics is totally speculative and we cannot understand the nature of the Big Bang when the universe was contained within a region the size of the Planck length.

With these issues in mind we have developed a unified theory of all the known interactions that is divergence-free. Thus we can perform calculations to any order of perturbation theory in any sector and obtain finite results that can be compared with experiment. The theory has the Standard Model and classical General Relativity as "low energy" limits. At high energies a string-like sub-structure generates a smooth high-energy limit that eliminates the divergences in perturbation theory. Thus there is a correspondence principle for our theory – it has the right "low energy" limits.

The nature of the string-like sub-structure is not dependent on the details of the interactions and symmetries. The interactions in the Standard Model and quantum gravity are can have any polynomial form (in the quantum fields). A variety of other interactions (such as those with derivative couplings) are allowed in this approach which do not affect the finiteness of the theory. Thus the range of possible extensions of the Standard Model is significantly widened.

3. Quantization of Coordinate Systems

Non-commuting Coordinates

Field theories with non-commuting coordinates are currently an active field of study.[70] Investigators are studying gauge theories, and in particular Quantum Electrodynamics, with non-commuting coordinates. Non-commuting coordinates are usually implemented quantum mechanically by positing non-zero commutators for coordinates:

$$[x^i, x^j] = i\theta^{ij} \tag{3.1}$$

New Approach to Non-Commuting Coordinates

In this book we will consider an alternative approach that postulates a q-number coordinate system X^μ with which all particle fields are defined. This coordinate system is realized as a mapping from a more fundamental c-number coordinate system y^ν, which we will call the subspace for want of a better term. We will treat X^μ as a vector of quantum fields, thus realizing a new type of non-commutative coordinates at unequal subspace times.

This approach is radically different from the non-commutative coordinate realizations hitherto discussed in the literature. It has a number of beneficial results to recommend it – the main result is the finiteness of quantum field theories that are defined within its framework. We will explore some of these results in the following chapters.

The X^μ coordinate system, as we define it, has a c-number real part and a q-number imaginary part. Thus particle fields which are normally defined on four-dimensional real space-time will now be defined on a complex four-dimensional space-time where four imaginary dimensions will appear as *Quantum Dimensions*™ embodied in a vector quantum field $Y^\mu(y)$.

$$X_\mu(y) = y_\mu + i\, Y_\mu(y)/M_c^{\,2}$$

The $Y^\mu(y)$ field is a function of the subspace y coordinates. The real part of the space-time dimensions will be taken to be the subspace y coordinates.[71]

---

[70] M. R. Douglas and N. A. Nekrasov, Rev. Mod. Phys. **73**, 977 (2002) and references therein; J. Harvey, hep-th/0102076; M. Hamanaka and K. Toda, hep-th/0211148; N. Seiberg and E. Witten, hep-th/9908142; R. J. Szabo, hep-th/0109162; G. Berrino, S. L. Cacciatori, A. Celi, L. Martucci, and A. Vicini, hep-th/0210171; S. Godfrey and A. Doncheski, DESY eprint 02-195; M. Caravati, A. Devoto, and W. W. Repko, hep-th/0211463; and references within these papers.

[71] In a deeper theory the real part might also be a quantum field that undergoes a condensation to generate c-number coordinates. We will not consider this possibility in this book.

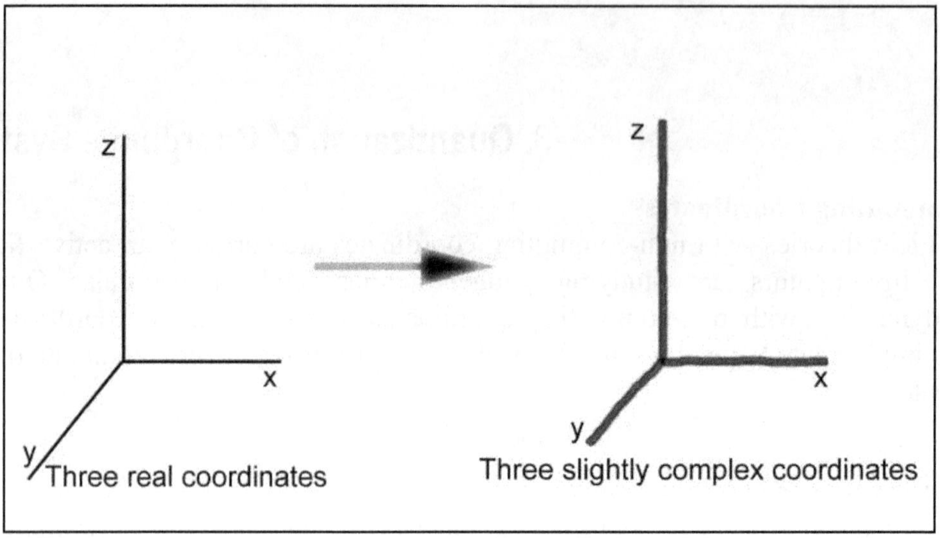

Figure 3.1. The change from purely real space to a slightly complex space with imaginary quantum fluctuations for each spatial axis in the Coulomb gauge of the Y field.

The imaginary part of space-time (which has not been experimentally seen) will simply be the quantum fluctuations of a massless vector quantum field that are suppressed further by a large mass scale – perhaps of the order of the Planck mass – that reduces the imaginary Quantum Dimensions™ to the infinitesimal. The effects of Quantum Dimensions™ only become appreciable in quantum field theory at energies of the order of M_c. At these energies the exponential Gaussian factor in each particle (and ghost) propagator that is generated by the Quantum Dimensions™ serves to make perturbation theory calculations ultra-violet finite – including calculations in Quantum Gravity.

The formalism that we will describe introduces a new form of interaction that does not have the form of the simple polynomial interactions that have hitherto dominated quantum field theories. This form of interaction takes place via the composition of quantum fields and can be called a *Dimensional Interaction*™ or an *Interdimensional Interaction*™ since it affects particle behavior through Quantum Dimensions™.[72]

Quantization Using a C-Number X^μ

We will begin by considering the case of a scalar quantum field theory. We assume a real underlying y subspace. Since X^μ is a set of coordinates, we choose to define a scalar field ϕ as a function of X^μ, which in turn is a function of the y^ν

[72] See the back of title page for the rationale for, and the details of the authorization to use, these trademarked terms.

coordinates. We will provisionally second quantize ϕ treating X^μ as c-number coordinates using a conventional approach.[73]

We assume a Lagrangian specified by eq. 1.10 that leads to the Klein-Gordon equation eq. 1.15. For that Lagrangian formulation, the momentum conjugate to ϕ is:

$$\pi_\phi = \partial L_F / \partial\phi' \equiv \partial L_F / \partial(\partial\phi/\partial X^0) \tag{3.2}$$

Following the canonical quantization procedure, π and ϕ become Hermitian operators with equal time ($X^0 = X^{0\prime}$) commutation rules:

$$[\phi(X), \phi(X')] = [\pi_\phi(X), \pi_\phi(X')] = 0 \tag{3.3}$$

$$[\pi_\phi(X), \phi(X')] = -i\,\delta^3(\mathbf{X} - \mathbf{X}') \tag{3.4}$$

The Hamiltonian is defined by eq. A.112. (Appendix A contains a detailed development of the formalism for the scalar particle case. It was placed there because there are many formal similarities to conventional quantum field and this approach allows us to proceed more quickly to the main points of difference between conventional quantum field theory and Two-Tier quantum field theory in the present chapter. Appendix A also describes a new type of method – the composition of extrema – for the Calculus of Variations. *Equations numbered A.xxx are in Appendix A.*) We assume a metric $\eta_{\mu\nu}$ where $\eta_{00} = +1$, $\eta_{0i} = 0$, and $\eta_{ij} = -1$ for i, j = 1,2,3.

The standard Fourier expansion of the solution to the Klein-Gordon equation (eq. A.34) is:

$$\phi(X) = \int d^3p\, N_m(p)\, [a(p)\, e^{-ip\cdot X} + a^\dagger(p)\, e^{ip\cdot X}] \tag{3.5}$$

where

$$N_m(p) = [(2\pi)^3 2\omega_p]^{-1/2} \tag{3.6}$$

and

$$\omega_p = (\mathbf{p}^2 + m^2)^{1/2} \tag{3.7}$$

The commutation relations of the Fourier coefficient operators are:

$$[a(p), a^\dagger(p')] = \delta^3(\mathbf{p} - \mathbf{p}') \tag{3.8}$$

---

[73] Some texts are: Bogoliubov, N. N., Shirkov, D. V., *Introduction to the Theory of Quantized Fields* (Wiley-Interscience Publishers Inc., New York, 1959); Bjorken, J. D., Drell, S. D., *Relativistic Quantum Fields* (McGraw-Hill, New York, 1965); Huang, K., *Quarks, Leptons & Gauge Fields Second Edition* (World Scientific, River Edge, NJ, 1992); Kaku, M., *Quantum Field Theory* (Oxford University Press, New York, 1993); Weinberg, S., *The Quantum Theory of Fields* (Cambridge University Press, New York, 1995).

$$[a^\dagger(p), a^\dagger(p')] = [a(p), a(p')] = 0 \tag{3.9}$$

The reader will recognize the quantization procedure is formally identical to the standard canonical quantization procedure of a free scalar quantum field.

In the case of spin ½, spin 1 and spin 2 fields the standard quantization procedure *in terms of the X coordinate system* can also be followed in a way similar to the procedure in standard texts. We will see these quantization procedures in the following chapters. In the next section we will quantize the transformation from the y coordinate system to the X coordinate system.

The procedures developed in this section and the following sections may disturb some readers since we are placing operators with Dirac delta functions and using other unusual operator expressions. These concerns should be put at rest when we show that a path integral formulation presented later gives precisely the same results as the present development.

Coordinate Quantization

In this section we will quantize the coordinates X^μ as a vector field defined on a fundamental c-number coordinate system y^ν of the same dimensionality. We will assume the y^ν space is a "normal" flat Minkowski space with three spatial and one time dimensions. Generalizations to spaces with more dimensions are straightforward but will not be considered here.

Thus we will assume X^μ has three spatial dimensions and one time dimension. For reasons primarily of simplicity (primarily to avoid multiple time coordinates) we will assume the X^μ fields are similar to the free electromagnetic vector potential A^μ with the Lagrangian:

$$\mathscr{L}_C = +\tfrac{1}{4}\, M_c^{\,4} F^{\mu\nu} F_{\mu\nu} \tag{3.10}$$

$$F_{\mu\nu} = \partial X_\mu/\partial y^\nu - \partial X_\nu/\partial y^\mu \tag{3.11}$$

where M_c^4 is a mass scale to the fourth power. It is required on dimensional grounds and serves to set the scale for new Physics as we will see later. *Note the sign in eq. 3.10 is not negative – superficially contrary to the conventional electromagnetic Lagrangian. The reason for this difference is that the field part of X^μ is imaginary.* Thus $\mathscr{L}_C$ winds up having the correct sign after taking account of the factors of i in the field strength $F_{\mu\nu}$.

We assume X^μ is complex[74] with the form:

$$X_\mu(y) = y_\mu + i\, Y_\mu(y)/M_c^{\,2} \tag{3.12}$$

[74] Theories of quantum mechanics, and quantum fields, in complex and quaternion spaces have been considered by numerous authors. For example see C. M. Bender, D. C. Brody and H. F. Jones, "Complex Extension of Quantum Mechanics" Phys. Rev. Letters **89**, 270401-1 (2002) and references therein; S. L. Adler and A. C. Millard, "Generalized Quantum Dynamics as Pre-Quantum Mechanics", Princeton Univ. preprint arXiv:hep-th/9508076 (1995) and references therein. These theories are all very different from the theories presented herein.

where $Y_\mu(y)$ is a quantum field, M_c is a mass scale, and the real part is the c-number 4-vector y_μ. If X^μ has this form, then

$$F_{\mu\nu} = i\,(\partial Y_\mu/\partial y^\nu - \partial Y_\nu/\partial y^\mu)/M_c^2 \qquad (3.13)$$

Defining

$$F_{Y\mu\nu} = (\partial Y_\mu/\partial y^\nu - \partial Y_\nu/\partial y^\mu) \qquad (3.14)$$

we see the Lagrangian assumes the form of the conventional electromagnetic Lagrangian:

$$\mathscr{L}_C = -\tfrac{1}{4}\, F_Y^{\mu\nu} F_{Y\mu\nu} \qquad (3.15)$$

This Lagrangian can be used to develop field equations and a canonical quantization that is completely analogous to Quantum Electrodynamics.

Gauge Invariance

The gauge invariance of the Lagrangian allows us to choose a convenient gauge. The gauge invariance of the full Lagrangian

$$\mathscr{L}_s = \mathscr{L}_F(\phi(X), \partial\phi/\partial X^\mu)\, J + \mathscr{L}_C(X^\mu(y), \partial X^\mu(y)/\partial y^\nu) \qquad (A.96)$$

is based on the standard gauge invariance of $\mathscr{L}_C$, and the gauge invariance of $J\mathscr{L}_F$ in the form of translational invariance

$$X^\mu(y) \rightarrow X^\mu(y) + \delta X^\mu(y) \qquad (A.97)$$

for the special case of a translation of X with the form of a gauge transformation:

$$\delta X^\mu(y) = \partial\Lambda(y)/\partial y_\mu$$

In this case eq. A.106 implies

$$\int d^4y\, \Lambda(y)\, \partial\, [\, J\, \partial/\partial X^\mu\, \mathscr{T}_{F\mu\nu}\,]/\partial y_\nu = 0$$

after a partial integration and so gives the differential conservation law:

$$\partial\, [\, J\, \partial/\partial X^\mu\, \mathscr{T}_{F\mu\nu}\,]/\partial y_\nu = 0 \qquad (3.16)$$

since $\Lambda(y)$ is arbitrary. This conservation law is trivially obeyed since, by eq. A.108:

$$\partial/\partial X^\mu \mathscr{T}_{F\mu\nu} = 0 \qquad (A.108)$$

Thus translational invariance in the $\mathscr{L}_F$ sector together with standard gauge invariance in the $\mathscr{L}_C$ sector automatically guarantees Y field gauge invariance of the total Lagrangian. Basically we use the separate invariance of each term of

$$L = \int d^4y\,[\mathscr{L}_F J + \mathscr{L}_C] = \int d^4X\,\mathscr{L}_F + \int d^4y\,\mathscr{L}_C = L_F + L_C$$

under a constant translation $X^\mu \rightarrow X^\mu + \delta X^\mu$ where δX^μ is constant to establish eq. A.108. Then we consider a position dependent translation/gauge transformation to derive eq. 3.16, which taken together with eq. A.108, establishes the invariance under the position dependent translation/gauge transformation eq. A.97.

An alternate approach that leads to the same result is to start with the particle part of the Lagrangian L_F rewritten to be invariant under general coordinate transformations as it must when we generalize to include General Relativity. Since position dependent translations are a form of general coordinate transformation the full theory must be invariant under position dependent translations due to invariance under general coordinate transformations.

Having established invariance under gauge transformations we now choose to use the most convenient gauge – the Coulomb gauge[75]:

$$\partial Y^i/\partial y^i = 0 \qquad (3.17a)$$

which, in the absence of external sources, allows us to set

$$Y^0 = 0 \qquad (3.17b)$$

since Y^0 does not have a canonically conjugate momentum. A conventional treatment leads to the equal time commutation relations:

$$[Y^\mu(\mathbf{y}, y^0), Y^\nu(\mathbf{y}', y^0)] = [\pi^\mu(\mathbf{y}, y^0), \pi^\nu(\mathbf{y}', y^0)] = 0 \qquad (3.18)$$

$$[\pi^j(\mathbf{y}, y^0), Y_k(\mathbf{y}', y^0)] = -i\,\delta^{tr}{}_{jk}(\mathbf{y} - \mathbf{y}') \qquad (3.19)$$

(Note the locations of the j indexes in eq. 3.19 introduce a minus sign.) where

[75] It is also possible to quantize using an indefinite metric that preserves manifest Lorentz covariance as was done by Gupta and Bleuler for the electromagnetic field. We will use the Gupta-Bleuler approach later to establish covariance under special relativity later. Now we opt for manifest positivity and use the Coulomb gauge.

$$\pi^k = \partial \mathscr{L}_C \;/\; \partial Y_k' \qquad (3.20)$$
$$\pi^0 = 0 \qquad (3.21)$$

$$\delta^{tr}{}_{jk}(\mathbf{y} - \mathbf{y}') = \int d^3k \; e^{i\,\mathbf{k}\cdot(\mathbf{y}-\mathbf{y}')}(\delta_{jk} - k_j k_k/\mathbf{k}^2)/(2\pi)^3 \qquad (3.22)$$

$$Y_k' = \partial Y_k/\partial y^0 \qquad (3.23)$$

The Coulomb gauge reveals the two degrees of freedom that are present in the vector potential. The Fourier expansion of the vector potential is:

$$Y^i(y) = \int d^3k \; N_0(k) \sum_{\lambda=1}^{2} \varepsilon^i(k, \lambda)[a(k,\lambda)\, e^{-ik\cdot y} + a^\dagger(k,\lambda)\, e^{ik\cdot y}] \qquad (3.24)$$

where

$$N_0(k) = [(2\pi)^3 2\omega_k]^{-1/2} \qquad (3.25)$$

and (since m = 0)

$$\omega_k = (\mathbf{k}^2)^{1/2} = k^0 \qquad (3.26)$$

with $\vec{\epsilon}(k, \lambda)$ being the polarization unit vectors for $\lambda = 1,2$ and $k^\mu k_\mu = 0$.

The commutation relations of the Fourier coefficient operators are:

$$[a(k,\lambda), a^\dagger(k',\lambda')] = \delta_{\lambda\lambda'}\delta^3(\mathbf{k} - \mathbf{k}') \qquad (3.27)$$
$$[a^\dagger(k,\lambda), a^\dagger(k',\lambda')] = [a(k,\lambda), a(k',\lambda')] = 0 \qquad (3.28)$$

and the polarization vectors satisfy

$$\sum_{\lambda=1}^{2} \varepsilon_i(k, \lambda)\varepsilon_j(k, \lambda) = (\delta_{ij} - k_i k_j/\mathbf{k}^2) \qquad (3.29)$$

It will be convenient to divide the Y field into positive and negative frequency parts:

$$Y^+{}_i(y) = \int d^3k \; N_0(k) \sum_{\lambda=1}^{2} \varepsilon_i(k, \lambda)\; a(k,\lambda)\; e^{-ik\cdot y} \qquad (3.30)$$

and

$$Y^-{}_i(y) = \int d^3k \; N_0(k) \sum_{\lambda=1}^{2} \varepsilon_i(k, \lambda)\; a^\dagger(k,\lambda)\; e^{ik\cdot y} \qquad (3.31)$$

For later use we note the commutator between the positive and negative frequency parts is:

$$[\,Y^-{}_j(y_1), Y^+{}_k(y_2)] = -\int d^3k \; e^{ik\cdot(y_1 - y_2)}\,(\delta_{jk} - k_j k_k/\mathbf{k}^2)/[(2\pi)^3 2\omega_k] \qquad (3.32)$$

Bare ϕ Particle States

We now turn to the ϕ particle states. The creation and annihilation operators can be used to define "bare" free particle states. Bare free particle states are states that are not dressed with coherent states of Y quanta. For example a bare one-particle state of momentum p is

$$|p> = a^\dagger(p)\,|0_\phi> \tag{3.33}$$

with corresponding bare bra state

$$<p| = <0_\phi|\,a(p) \tag{3.34}$$

where the vacuum is defined as usual:

$$a(p)\,|0_\phi> = 0 \tag{3.35}$$

$$<0_\phi|\,a^\dagger(p) = 0 \tag{3.36}$$

Multi-particle bare states can also be defined in the conventional way with products of creation and annihilation operators applied to the vacuum.

Y Fock Space Imaginary Coordinate States

States can also be defines for the quantized Y field. These states will be similar in form to electromagnetic photon states but play a different role in our approach since they are in fact coordinate excitation states for the imaginary part of X^μ. Thus the scalar field (and other particle fields) will exist in a real four-dimensional space with quantum excitations into imaginary Quantum Dimensions™. These excitations become significant at high energies. At the low energies with which we are familiar, space-time appears real; at very high energies space-time becomes slightly complex.

There are two types of imaginary coordinate excitations: 1.) Quantum excitations into Fock states consisting of superpositions of states with a definite finite number of Y "particles" and 2.) Imaginary coordinate excitations into coherent Y states with an "infinite" number of particles. Coherent states can be viewed as representing "classical" fields.

In this section we will consider Y field states with a definite number of excitations ("particles"). The creation and annihilation operators of the Y field can be used to define free particle states. For example a one particle state can be defined by

$$|k, \lambda> = a^\dagger(k, \lambda)\,|0_Y> \tag{3.37}$$

with corresponding bra state

$$<k, \lambda| = <0_Y|\,a(k, \lambda) \tag{3.38}$$

where the "coordinate vacuum" is defined as usual:

$$a(k, \lambda)\,|\,0_Y> = 0 \tag{3.39}$$
$$<0_Y\,|\,a^\dagger(k, \lambda) = 0 \tag{3.40}$$

Multi-particle states can also be defined in the conventional way with products of the creation and annihilation operators applied to the vacuum. The set of all states containing a finite number of "particles" constitutes a Fock space.

A state with a finite number of Y "particles" represents a quantum fluctuation into imaginary Quantum Dimensions™. Such states do not appear in Two-Tier quantum field theory since the Y field is a free field and has no source. Thus they appear only as part of normal particles. A normal particle, such as a ϕ particle, has a coherent state of Y quanta associated with it, which play a role in interactions. The Y coherent state part of a normal particle can be viewed as boring an infinitesimal "hole" into an extra pair of imaginary dimensions in a neighborhood of the particle of a radial extent set by the length M_c^{-1}.

Y Coherent Imaginary Coordinate States

Coherent Y states bring us closer what we might consider to be "classical" imaginary dimensions – dimensions that we can, in principle, experience as we do normal dimensions. Let us define the coherent state[76]

$$|\;y, p> = e^{-\mathbf{p}\cdot\mathbf{Y}^-(y)/M_c^2}\,|\,0_Y> \tag{3.41}$$

This state is an eigenstate of the coordinate operator $Y^+(y')$:

$$Y^+_j(y_1)\;|y_2, p> = -[Y^+_j(y_1), \mathbf{p}\cdot\mathbf{Y}^-(y_2)]/M_c^2\;|y, p> \tag{3.42}$$

$$= -\int d^3k\;[N_0(k)]^2\,e^{ik\cdot(y_2-y_1)}\,(p_j - k_j\mathbf{p}\cdot\mathbf{k}/\mathbf{k}^2)/M_c^2\;|y, p> \tag{3.43}$$

$$= p^i\Delta_{Tij}(y_1 - y_2)/M_c^2\;|y, p> \tag{3.44}$$

where $p^i\Delta_{Tij}(y_1 - y_2)/M_c^2$ is the eigenvalue of $Y^+_j(y_1)$. As we will see in the next chapter, the eigenvalue of Y^+ becomes large as $(y_1 - y_2)^2 \to 0$. Thus the imaginary Quantum Dimensions™ become significant at very short distances, and significantly modify the high-energy behavior of quantum field theories. In particular, Quantum Dimensions™ have a significant effect when

$$(y_1 - y_2)^2 \lessapprox (4\pi^2 M_c^2)^{-1} \tag{3.45}$$

---

[76] Coherent states are well known in the physics literature. See for example T. W. B. Kibble, J. Math. Phys. **9**, 315 (1968) and references therein; V. Chung, Phys. Rev. **140**, B1110 (1965); J. R. Klauder, J. McKenna, and E. J. Woods, J. Math. Phys. **7**, 822 (1966) and references therein.

according to eq. 4.13 in the next chapter. We are assuming the mass scale M_c is very large – perhaps of the order of the Planck mass (1.221×10^{19} GeV/c^2). Thus imaginary Quantum Dimensions™ are far from detectable in today's "low" energy experiments. Their effect is significant in the analysis of the first instants after the Big Bang.[77]

The Dynamical Generation of New Dimensions

Effectively, the imaginary dimensions that we have constructed raise the total number of real and Quantum Dimensions™ to 8 with 6 space dimensions and two time dimensions. As we will see later the requirement of gauge invariance for the quantized Y field reduces the number of time dimensions to one and constrains the six space dimensions to five degrees of freedom giving a 5+1 dimensional space. Since X is a function of y we can also view the four dimensional world that we live in as a four-dimensional surface in a 6-dimensional space-time.

Generation of Quantum Dimensions™ by the ϕ(X) field

The ϕ(X) field generates Quantum Dimensions™ via coherent states from the vacuum. From eq. 3.5 and 3.12 we see

$$\phi(X) = \int d^3p\, N_m(p)\, [a(p)\, e^{-ip\cdot(y + iY/M_c^2)} + a^\dagger(p)\, e^{ip\cdot(y + iY/M_c^2)}] \tag{3.46}$$

with the result

$$\phi(X)\,|0\rangle = \int d^3p\, N_m(p)\, a^\dagger(p)\, e^{ip\cdot(y + iY/M_c^2)}\,|0\rangle \tag{3.47}$$

is a superposition of coherent Y states plus one scalar particle. The vacuum state $|0\rangle$ is the product of the ϕ and Y vacuum states $|0\rangle = |0_Y\rangle|0_\phi\rangle$. We will use $|0\rangle$ in most of the following discussions.

We can also define coherent Y states with total momentum q using the expression:

$$|q\, Y\rangle = \int d^4y\, e^{iq\cdot X(y)}\,|0\rangle = \int d^4y\, e^{iq\cdot(y + iY/M_c^2)}\,|0\rangle \tag{3.48}$$

Expanding the Y part of the exponential in eq. 3.48 gives

$$|q\, Y\rangle = \sum_{n=0}^{\infty}(-1)^n(n!)^{-1}\prod_{j=1}^{n}\left(\int d^3k_j N_0(k_j)\right)\delta^4\!\left(q - \sum_{s=1}^{n} k_s\right)\prod_{r=1}^{n}\sum_{\lambda_r=1}^{2} \mathbf{q}\cdot \boldsymbol{\varepsilon}(k_r, \lambda_r)\, a^\dagger(k_r,\lambda_r)\,|0\rangle \tag{3.49}$$

which indicates that the sum of the Y particle momenta for each term in the expansion is q.

---

[77] Blaha (2004).

Hamiltonian for Particle and Coordinate States

The Hamiltonian for the separable (field Hamiltonian term separate from the Y Hamiltonian term – see Appendix A), coordinate quantized, scalar quantum field theory is:

$$H_s = \int d^3y\ \mathscr{H}_s \tag{A.79}$$

with

$$\mathscr{H}_s = J\mathscr{H}_F + \mathscr{H}_C \tag{A.82}$$

$$\mathscr{H}_F(\phi(X), \pi_\phi, \partial\phi/\partial X^i) = \pi_\phi\,\phi' - \mathscr{L}_F \tag{A.83}$$

$$\mathscr{H}_C(X^\mu(y), \pi_X{}^\mu, \partial X^\mu(y)/\partial y^i, y^\nu) = \pi_X{}^\mu\, X_\mu{}' - \mathscr{L}_C \tag{A.84}$$

$$\mathscr{L}_F = ½\,[\,(\partial\phi/\partial X^\nu)^2 - m^2\phi^2\,] \tag{A.33}$$

$$\mathscr{L}_C = -¼\, M_c{}^4 F_Y{}^{\mu\nu} F_{Y\mu\nu} \tag{3.15}$$

We note

$$\mathscr{H}_F = ½\,[\pi_\phi{}^2 + (\partial\phi/\partial X^i)^2 + m^2\phi^2] \tag{3.50}$$

is the conventional scalar particle Hamiltonian when viewed as a function of the X coordinates. $\mathscr{H}_C$ has the same form as the conventional electromagnetic Hamiltonian when eq. 3.12 is used to specify X in terms of the Y fields.

$$\mathscr{H}_C = ½\,(E_Y{}^2 + B_Y{}^2) \tag{3.51}$$

where

$$E_Y{}^i = -\partial Y^i/\partial y^0 \tag{3.52}$$

$$B_Y{}^i = \varepsilon^{ijk}\,\partial Y_j/\partial y^k \tag{3.53}$$

Using the Fourier expansions of ϕ and X^μ (eqs. 3.5 and 3.24) we obtain the following expression for the normal-ordered Hamiltonian H_S:

$$P_s{}^0 \equiv H_s = \int :\mathscr{H}_s:\, d^3y \tag{3.54}$$

$$H_s = \int d^3p\,(\mathbf{p}^2 + m^2)^{½} a^\dagger(p)a(p) + \int d^3k \sum_{\lambda=1}^{2} (\mathbf{k}^2)^{½} a^\dagger(k,\lambda)a(k,\lambda) \tag{3.55}$$

where : : indicates normal ordering and where we perform a functional integration over X (Note the Jacobian is present within $\mathscr{H}_s$.) for the particle part of the Hamiltonian $\mathscr{H}_F$. The Hamiltonian is manifestly positive definite.

The spatial momentum is specified by

$$P_s^{\,j} = -\int d^3X :\pi_\phi(X)\partial\phi(X)/\partial X_j: + \int d^3y :E_Y^{\,i}\partial Y^i/\partial y_j: \qquad (3.56)$$

$$= \int d^3p\ p^j\ a^\dagger(p)a(p) + \int d^3k \sum_{\lambda=1}^{2} k^j\, a^\dagger(k, \lambda)a(k, \lambda) \qquad (3.57)$$

where the first term in eq. 3.57 follows because of $\int d^3X$ in eq. 3.56. The momentum operator generates displacements in ϕ

$$[P_s^{\,\mu}, \phi(X)] = -i\partial\phi/\partial X_\mu \qquad (3.58)$$

Second Quantized Coordinates

At this point we have developed a formalism for a scalar particle quantum field theory based on our non-commutative coordinates. In the following chapters we will proceed to use this formalism to develop a unified quantum field theory of the known forces of nature.

4. Scalar Two-Tier Quantum Field Theory

It appears then that there must be fundamental changes in our basic formulation of quantum field theory, so that unrenormalized masses and unrenormalized coupling constants can become finite.
T. D. Lee & G. C. Wick[78]

Introduction

In this chapter we will examine a new formulation of quantum field theory that we call *Two-Tier quantum field theory* in more detail for the case of a free scalar particle. This type of quantum field theory incorporates a structure similar to a string-like substructure within a quantum field theoretic framework. In the following chapters we will apply the approach to QED, ElectroWeak Theory, the Standard Model and lastly to a unified model for the known forces of nature.

"Two-Tier" Space

In the preceding chapter we developed quantized coordinates X^μ defined on an underlying c-number coordinates y^ν with the equations:

$$X_\mu(y) = y_\mu + iY_\mu(y)/M_c^{\,2} \qquad (3.12)$$

$$Y^i(y) = \int d^3k\, N_0(k) \sum_{\lambda=1}^{2} \varepsilon^i(k, \lambda)[a(k,\lambda)\, e^{-ik\cdot y} + a^\dagger(k,\lambda)\, e^{ik\cdot y}] \qquad (3.24)$$

We also developed a free scalar quantum field theory with the Fourier expansion:

$$\phi(X) = \int d^3p\, N_m(p)\, [a(p)\, e^{-ip\cdot X} + a^\dagger(p)\, e^{ip\cdot X}] \qquad (3.5)$$

We will now consider the implications of the separable Lagrangian:

$$\mathscr{L}_s = \mathscr{L}_F(\phi(X), \partial\phi/\partial X^\mu)\, J + \mathscr{L}_C(X^\mu(y), \partial X^\mu(y)/\partial y^\nu) \qquad (A.96)$$

where

$$\mathscr{L}_F = \tfrac{1}{2}\, [\, (\partial\phi/\partial X^\nu)^2 - m^2\phi^2\,] \qquad (A.33)$$

and

$$\mathscr{L}_C = -\tfrac{1}{4}\, M_c^{\,4} F_Y^{\ \mu\nu} F_{Y\mu\nu} \qquad (3.10)$$

---

[78] T. D. Lee and G. C. Wick, Phys. Rev. **D2**, 1033 (1970). Lee and Wick's model QED is totally unrelated to our approach.

with

$$F_{Y\mu\nu} = \partial Y_\mu/\partial y^\nu - \partial Y_\nu/\partial y^\mu \tag{3.14}$$

M_c is the mass that sets the scale at which the imaginary part of X^μ becomes significant.

This quantum field theory behaves as a conventional quantum field theory until energies reach the magnitude of M_c. At energies of the order of M_c, and above, the imaginary part of X^μ becomes significant and alters the high-energy behavior of the theory in a major way. This modification leads to the elimination of divergences that normally appear in perturbation theory when interactions are introduced. Yet the low energy behavior of the theory remains the same remains the same as conventional scalar quantum field theory. Thus the precise calculations of QED that have been verified to an amazing degree of accuracy remain valid when a Two-Tier formulation of QED is created (in chapter 5). And the "low energy" results found in other conventional quantum field theories such as ElectroWeak Theory and the Standard Model also are closely approximated by their corresponding Two-Tier versions.

The straightforward use of the above equations[79] (and the canonical quantization described in the preceding chapters) leads to a scalar quantum field with the Fourier expansion:

$$\phi(X) = \int d^3p\, N_m(p)\, [a(p)e^{-ip\cdot(y + iY/M_c^2)} + a^\dagger(p)e^{ip\cdot(y + iY/M_c^2)}] \tag{4.1}$$

using eq. 3.5 above. We note the equal time commutation relations of ϕ and π_ϕ are the same as the conventional equal time commutation relations of a scalar field despite the fact that X^μ and Y^μ are themselves quantum fields since $[Y^\mu(\mathbf{y}, y^0), Y^\nu(\mathbf{y}', y^0)] = 0$ for $\mathbf{y} \neq \mathbf{y}'$. In addition, we note the ϕ and π_ϕ fields are not Hermitian.

The Fourier expansion of ϕ does require one refinement – the exponential terms in X^μ must be *normal ordered* to avoid infinities in the unequal time commutation relations:

$$\phi(X) = \int d^3p\, N_m(p)\, [a(p) :e^{-ip\cdot(y + iY/M_c^2)}: + a^\dagger(p) :e^{ip\cdot(y + iY/M_c^2)}:] \tag{4.2}$$

Since the Hamiltonian as well as other quantities are normal ordered in quantum field theory the additional requirement of normal ordering in the field operator is merely an extension of a standard procedure to a more complex situation and is not disturbing. The unequal time commutation relation of the normal ordered ϕ field is:

$$[\phi(X^\mu(y_1)), \phi(X^\mu(y_2))] = i\Delta(y_1 - y_2) + \mathcal{O}(1/M_c^2) \tag{4.3}$$

where

---

[79] The use of functionals in quantum field theory is, of course, far from new as one can see in texts such as Bogoliubov (1959) (see for example pp. 198-226).

$$\Delta(y_1 - y_2) = -i \int d^3k\, (e^{-ik\cdot(y_1 - y_2)} - e^{ik\cdot(y_1 - y_2)})/[(2\pi)^3 2\omega_k] \qquad (4.4)$$

is a familiar c-number invariant singular function. The additional terms in eq. 4.3 are q-number terms that become significant at very short distances of the order M_c^{-1}. Thus precise measurements of field strengths at larger distances are limited by standard quantum effects as indicated by the commutation relation.

The principle of *microscopic causality* is violated at extremely short distances of the order M_c^{-1} since the commutator (eq. 4.3) is non-zero, in general, for space-like distances of the order of M_c^{-1} due to the q-number terms. This violation is not experimentally measurable now – and for the foreseeable future – and reflects a type of non-locality at extremely short distances.

We will see that the short distance behavior of Two-Tier quantum field theory leads to the elimination of divergences resulting in finite interacting quantum field theories.

Vacuum Fluctuations

While the expectation value of a *conventional* free scalar field $\phi_{conv}(X)$ is zero in a conventional quantum field theory:

$$<0|\phi_{conv}(X)|0> = 0 \qquad (4.5)$$

the vacuum fluctuations of *conventional* scalar quantum field theory are quadratically divergent:

$$<0|\phi_{conv}(X)\phi_{conv}(X)|0> = \int d^3p/[(2\pi)^3 2\omega_p] \qquad (4.6)$$

In "Two-Tier" quantum field theory we find the vacuum expectation value of a free field is zero (like eq. 4.5) *and the expectation value of the square of the field is also zero:*

$$<0|\phi(X)\phi(X)|0> = \int d^3p\, e^{-p^ip^j\Delta_{Tij}(0)/Mc^4}/[(2\pi)^3 2\omega_p] = 0 \qquad (4.7)$$

since the exponential factor in the integral is $-\infty$. The exponent contains

$$\Delta_{Tij}(z) = \int d^3k\, e^{-ik\cdot z}\,(\delta_{ij} - k_ik_j/\mathbf{k}^2)/[(2\pi)^3 2\omega_k] \qquad (4.8)$$

where "T" is for "Two-Tier". Thus *vacuum fluctuations are zero in Two-Tier quantum field theory.* Correspondingly, we will see that renormalization constants are finite in the Two-Tier versions of QED, ElectroWeak Theory, the Standard Model and Quantum Gravity.

The Feynman Propagator

The Feynman propagator for a Two-Tier free scalar quantum field is:

$$i\Delta_F^{TT}(y_1 - y_2) = <0|T(\phi(X(y_1)),\phi(X(y_2)))|0> \tag{4.9}$$

$$\equiv <0|\phi(X(y_1))\phi(X(y_2))|0>\ \theta(y_1^{\ 0} - y_2^{\ 0}) +$$

$$+\ \phi(X(y_2))\phi(X(y_1))|0>\ \theta(y_2^{\ 0} - y_1^{\ 0}) \tag{4.10}$$

Since $X^0 = y^0$ in the Coulomb gauge of the X^μ field there is no ambiguity in the choice of the relevant time variable. A straightforward calculation shows:

$$i\Delta_F^{TT}(y_1 - y_2) = i\int d^4p\ e^{-ip\cdot(y_1 - y_2)}\ R(\mathbf{p},\ y_1 - y_2)/[(2\pi)^4(p^2 - m^2 + i\varepsilon)] \tag{4.11}$$

where

$$R(\mathbf{p},\ y_1 - y_2) = \exp[-\,p^ip^j\Delta_{Tij}(y_1 - y_2)/M_c^{\ 4}] \tag{4.12}$$

$$= \exp\{-p^2[A(v) + B(v)\cos^2\theta]\ /\ [4\pi^2M_c^{\ 4}z^2]\} \tag{4.13}$$

with

$$z^\mu = y_1^{\ \mu} - y_2^{\ \mu} \tag{4.14}$$

$$z = |\mathbf{z}| = |\mathbf{y}_1 - \mathbf{y}_2| \tag{4.15}$$

$$p = |\mathbf{p}| \tag{4.16}$$

$$v = |z^0|/z \tag{4.17}$$

$$A(v) = (1 - v^2)^{-1} + .5v\ \ln[(v - 1)/(v + 1)] \tag{4.18}$$

$$B(v) = v^2(1 - v^2)^{-1} - 1.5v\ \ln[(v - 1)/(v + 1)] \tag{4.19}$$

$$\mathbf{p}\cdot\mathbf{z} = pz\cos\theta \tag{4.20}$$

and $|\mathbf{p}|$ denoting the length of a spatial vector **p** while $|z^0|$ is the absolute value of z^0.

As eq. 4.11 indicates, the Gaussian damping factor R(p, z) for large momentum p is the same for both the positive and negative frequency parts of the Two-Tier Feynman propagator. It is also important to note that R(p, z) does not depend on p^0 (in the Y Coulomb gauge) and thus the integration over p^0 proceeds in the usual way to produce time-ordered positive and negative frequency parts.

Large Distance Behavior of Two-Tier Theories

The large distance behavior of the Two-Tier Feynman propagator approaches the behavior of the conventional Feynman propagator since

$$R(\mathbf{p}, y_1 - y_2) \to 1 \tag{4.21}$$

when $(y_1 - y_2)^2$ becomes much larger than M_c^{-2} as eq. 4.13 shows. Thus the behavior of a conventional quantum field theory naturally emerges at large distance. We will see that the conventional Standard Model is the large distance limit of the Two-Tier Standard Model thus *realizing a form of Correspondence Principle for Quantum Field Theory.* Some features of the conventional Standard Model that depend specifically on the existence of divergences, such as the axial anomaly, will be different in the Two-Tier Standard Model since it is a divergence-free theory.

Short Distance Behavior of Two-Tier Theories

At short distances the Gaussian factor dominates and radically changes the behavior of the Feynman propagator eliminating its short distance singular behavior, and thus paving the way to finite quantum field theories. Near the light cone, $M_c^{-2} \gg -(y_1 - y_2)^2 \to 0$, we can approximate eq. 4.11 with

$$i\Delta_F^{TT}(y_1 - y_2) \approx \int d^3p\, [N(p)]^2\, R(\mathbf{p}, y_1 - y_2) \tag{4.22}$$

since $e^{-ip\cdot(y_1 - y_2)}$ is approximately unity for small $(y_1 - y_2)$. We assume the mass of the ϕ particle is zero or is negligible at high energies so we set $m = 0$ to study the high energy behavior of eq. 4.22. Upon performing the integrations in eq. 4.22 for space-like $(y_1 - y_2)^2$ (and analytically continuing to the time-like regions[80,81]) we find

$$i\Delta_F^{TT}(y_1 - y_2) \approx [z^2M_c^4/(4i\surd A\surd B)]\ \ln[(\surd A + i\surd B)/(\surd A - i\surd B)] \tag{4.23}$$

with A and B defined in eqs. 4.18 and 4.19. As $(y_1 - y_2)^2 \to 0$ from the space-like or time-like side of the light cone we find eq. 4.23 becomes:

$$i\Delta_F^{TT}(y_1 - y_2) \to \pi M_c^4 |(y_1 - y_2)^\mu (y_1 - y_2)_\mu|/8 \tag{4.24}$$

Eq. 4.24 has several noteworthy points:

---

[80] See S. Blaha, "Relativistic Bound State Models with Quasi-Free Constituent Motion", Phys. Rev. **D12**, 3921 (1975) and references therein.

[81] It should be noted that A and B in eq. 4.23 have the same sign for $0 \le v < 1.1243$ thus making for easy analytic continuation across the light cone (which corresponds to $v = 1$ in eqs. 4.18 and 4.19).

1. The propagator is well behaved on the light cone and approaches zero smoothly from both space-like and time-like directions. In contrast, the conventional scalar Feynman propagator diverges as $[(y_1 - y_2)^\mu(y_1 - y_2)_\mu]^{-2}$. This good behavior near the light cone will be seen later for other particle propagators with the net result that the usual infinities found in conventional quantum field theory are absent in Two-Tier quantum field theories.

2. The quadratic form of the propagator in eq. 4.24 is suggestive of attempts to formulate a relativistic harmonic oscillator model of elementary particles[82] and more recent attempts to achieve quark confinement. The fact that the absolute value of the quadratic term appears in eq. 4.24 neatly avoids the common pitfall seen in fully relativistic harmonic oscillator attempts.

3. The quadratic behavior *in coordinate space* of the propagator at short distances is equivalent to a high-energy behavior of

$$p^{-6} \tag{4.25}$$

in momentum space. Thus we get the equivalent *of a higher derivative theory* in Two-Tier quantum field theory at high energies while retaining a positive definite energy spectrum. The problems of negative metric states that have plagued conventional higher derivative quantum field theories are avoided.[83]

String-like Substructure of the Theory

Imaginary Quantum Dimensions™ endow a particle with an extended structure that resembles to some extent the extended structure seen in bosonic string and Superstring theories. For example, Bailin (1994) use the operator[84]

$$V_\Lambda(k) = \int d^2\sigma \sqrt{-h}\, W_\Lambda(\tau, \sigma)\, e^{-ik\cdot X} \tag{4.26}$$

where X^μ is a quantized Fourier expansion of the string fields (see eq. 7.22 of Bailin (1994)).

We note our X^μ coordinate-field has two transverse degrees of freedom due to gauge invariance, which also invites comparison to the bosonic string. A point of difference is that we will create a well-defined quantum field theoretic formulation in conventional space-time that has the Standard Model as its "large distance" behavior

---

[82] H. Yukawa, H., Phys. Rev. **91**, 416 (1953); Y. S. Kim and M. E. Noz, Phys. Rev. **D8**, 3521 (1973) and references therein.

[83] S. Blaha, Phys.Rev. **D10**, 4268 (1974); S. Blaha, Phys.Rev. **D11**, 2921 (1975); S. Blaha, Nuovo Cim. **A49**, :113 (1979); S. Blaha, "Generalization of Weyl's Unified Theory to Encompass a Non-Abelian Internal Symmetry Group" SLAC-PUB-1799, Aug 1976; S. Blaha, "Quantum Gravity and Quark Confinement" Lett. Nuovo Cim. **18**, 60 (1977); Nakanishi, N., Suppl. Prog. Theo. Phys. **51**, 1 (1972); and references therein.

[84] D. Bailin and A. Love, *Supersymmetric Gauge Field Theory and String Theory* (Institute of Physics Publishing, Philadelphia, PA, 1994) page 272.

thus introducing a note of reality that is not (yet?) very apparent in Superstring theories. We see that the interacting quantum field theories based on this approach also have good, finite, short distance behavior just as string theories.

The scalar, and other particles', Feynman propagators can be viewed as describing the propagation of a particle cloaked (accompanied) by a cloud of Y particles (which generates the $R(\mathbf{p}, y_1 - y_2)$ factor in the propagator of eq. 4.11). If we examine the Fourier transform of R(p, z) we see:

$$(2\pi)^4 R(\mathbf{p}, q) = \int d^4z\, e^{iq\cdot z} R(\mathbf{p}, z) = \int d^4z\, e^{iq\cdot z} \exp[-p^i p^j \Delta_{Tij}(z)/M_c^4] \qquad (4.27)$$

and we find

$$R(\mathbf{p},q) = \sum_{n=0}^{\infty} [i(2\pi M_c)^4]^{-n} (n!)^{-1} \prod_{j=1}^{n} [\int d^4k_j\, \theta(k_j^0)(\mathbf{p}^2 - (\mathbf{p}\cdot\mathbf{k}_j)^2/\mathbf{k}_j^2)/(k_j^2 + i\varepsilon)]\, \delta^4(q - \sum_r k_r) \qquad (4.28)$$

which can be interpreted as a "cloud" of Y particles dressing the "bare" particle propagator. (The manifest divergences in eq. 4.28 for R(p, q) are an artifact of the expansion and the subsequent Fourier transformation. They are not present in the $R(\mathbf{p}, y_1 - y_2)$ factor in the propagator of eq. 4.11.) See Fig. 4.1 for the Feynman diagram of the Two-Tier cloaked propagator as compared to the normal scalar particle Feynman propagator. The Two-Tier Feynman propagator is basically a conventional scalar propagator that is modified by coherent Y particle emission.[85]

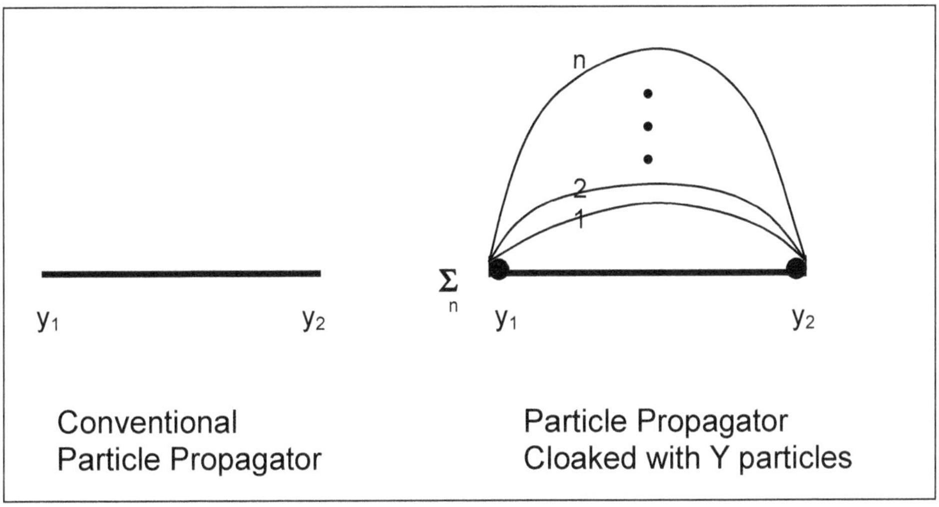

Figure 4.1. Feynman diagram for conventional and cloaked Two-Tier propagators.

[85] T. W. B. Kibble, Phys. Rev. **173**, 1527 (1968) and references therein. In particular see p. 1532 of Kibble's paper.

We note that R(p, q) satisfies the convolution theorem:

$$\int d^4k\, R(\mathbf{p}, k)\, R(\mathbf{p}, q - k) = [R(\mathbf{p}, q)]^2 \qquad (4.29a)$$

or

$$(2\pi)^4 \int d^4z\, e^{iq\cdot z}\, R(\mathbf{p}, z)\, R(\mathbf{p}, z) = [\, \int d^4z\, e^{iq\cdot z}\, R(\mathbf{p}, z)\,]^2 \qquad (4.29b)$$

The proof follows from eq. 4.28 and the Binomial theorem.

Parity

Parity can appear in two guises within the framework of Two-Tier quantum field theory. One can consider a parity operation where the space parts of X^μ are reversed while y^μ is unchanged. Or one can consider a second type of parity where the space parts of y^μ are reversed.

X Parity

Under this form of parity operation y^μ is unchanged while the arguments of ϕ *appear* to change by

$$X^i(y) \rightarrow - X^i(y) \qquad (4.30)$$

$$X^0(y) \rightarrow X^0(y) \qquad (4.31)$$

We will denote the parity operator of this type $\mathscr{P}_X$. Under $\mathscr{P}_X$ the arguments of the scalar quantum field operator ϕ change according to eqs. 4.30-1 so that ϕ transforms as

$$\mathscr{P}_X \phi(\mathbf{X}(y), X^0(y)) \mathscr{P}_X^{-1} = \phi(-\mathbf{X}(y), X^0(y)) \qquad (4.32)$$

From the form of ϕ in eq. 4.2 we see we can implement eq. 4.32 by requiring:

$$\mathscr{P}_X a(\mathbf{p}, p^0) \mathscr{P}_X^{-1} = a(-\mathbf{p}, p^0) \qquad (4.33)$$

$$\mathscr{P}_X a^\dagger(\mathbf{p}, p^0) \mathscr{P}_X^{-1} = a^\dagger(-\mathbf{p}, p^0) \qquad (4.34)$$

$$\mathscr{P}_X X^0(y) \mathscr{P}_X^{-1} = X^0(y) \qquad (4.35)$$

$$\mathscr{P}_X X^i(y) \mathscr{P}_X^{-1} = X^i(y) \qquad (4.36)$$

$$\mathscr{P}_X Y^i(y)\mathscr{P}_X^{-1} = Y^i(y) \tag{4.37}$$

where i = 1, 2, 3.

This parity transformation is analogous to the standard form of parity transformation in conventional quantum field theory. The separable Lagrangian in eq. A.96 (and listed at the beginning of this chapter) is invariant under this parity transformation.

y Parity

This form of parity transformation in which $y^i \to -y^i$ has significant differences from the normal parity transformation. We specify this parity transformation for a scalar quantum field by:

$$\mathscr{P}_y\phi(\mathbf{X}(\mathbf{y}, y^0), X^0(\mathbf{y}, y^0))\mathscr{P}_y^{-1} = \phi(\mathbf{X}(-\mathbf{y}, y^0), X^0(-\mathbf{y}, y^0)) \tag{4.38}$$

This transformation can be implemented through the following set of transformations:

$$\mathscr{P}_y a(\mathbf{p}, p^0)\mathscr{P}_y^{-1} = a(-\mathbf{p}, p^0) \tag{4.39}$$

$$\mathscr{P}_y a^\dagger(\mathbf{p}, p^0)\mathscr{P}_y^{-1} = a^\dagger(-\mathbf{p}, p^0) \tag{4.40}$$

$$\mathscr{P}_y X^0(\mathbf{y}, y^0)\mathscr{P}_y^{-1} = X^0(-\mathbf{y}, y^0) \tag{4.41}$$

$$\mathscr{P}_y Y^i(\mathbf{y}, y^0)\mathscr{P}_y^{-1} = -Y^i(-\mathbf{y}, y^0) \tag{4.42a}$$

$$\mathscr{P}_y a(\mathbf{k}, k^0,1)\mathscr{P}_y^{-1} = a(-\mathbf{k}, k^0,1) \tag{4.42b}$$

$$\mathscr{P}_y a(\mathbf{k}, k^0,2)\mathscr{P}_y^{-1} = -a(-\mathbf{k}, k^0,2) \tag{4.42c}$$

where i = 1,2,3 and assuming: $\boldsymbol{\varepsilon}(\mathbf{k},k^0,1)=-\boldsymbol{\varepsilon}(-\mathbf{k},k^0,1)$ and $\boldsymbol{\varepsilon}(\mathbf{k},k^0,2)=+\boldsymbol{\varepsilon}(-\mathbf{k},k^0,2)$.

Forms of the Parity Transformations

The parity transformations for a scalar particle are

$$\mathscr{P}_X = \exp\{-i\pi\int d^3p\ [a^\dagger(\mathbf{p}, p^0)a(\mathbf{p}, p^0) - a^\dagger(\mathbf{p}, p^0)a(-\mathbf{p}, p^0)]/2\} \tag{4.43a}$$

$$\mathscr{P}_y = \mathscr{P}_X \exp\{-i\pi \int d^3k\ [\sum_{\lambda=1}^{2} a^\dagger(\mathbf{k},k^0,\lambda)a(\mathbf{k},k^0,\lambda) - a^\dagger(\mathbf{k}, k^0, 1)a(-\mathbf{k},k^0,1) +$$

$$+ a^\dagger(\mathbf{k}, k^0, 2)a(-\mathbf{k},k^0,2)]/2\} \tag{4.43b}$$

The separable Lagrangian of eq. A.96 is invariant under these parity transformations.

Charge Conjugation

Charge conjugation is implemented in a way similar to that of conventional quantum field theory. In particular

$$\mathscr{C} X^{\mu}(\mathbf{y}, y^0)\mathscr{C}^{-1} = X^{\mu}(\mathbf{y}, y^0) \tag{4.44}$$

Time Reversal

Since $X^0 = y^0$ in the Y Coulomb gauge in Two-Tier quantum theory the only non-trivial form of time reversal transformation $\mathscr{T}$ is based on $y^0 = -y^0$. This time reversal transformation is similar in part to the conventional time reversal transformation in conventional quantum field theory. Therefore we will define $\mathscr{T}$ as the product of the operation of taking the complex conjugate of all c-numbers times a unitary operator $\mathscr{U}_Y$. Under $\mathscr{T}$ a scalar quantum field operator ϕ transforms as

$$\mathscr{T}\phi(\mathbf{X}(\mathbf{y}, y^0), X^0(\mathbf{y}, y^0))\mathscr{T}^{-1} = \phi(\mathbf{X}(\mathbf{y}, -y^0), X^0(\mathbf{y}, -y^0)) \tag{4.45}$$

From the form of in ϕ eq. 4.2 we see that

$$\mathscr{T} a(\mathbf{p}, p^0)\mathscr{T}^{-1} = a(-\mathbf{p}, p^0) \tag{4.46}$$
$$\mathscr{T} a^{\dagger}(\mathbf{p}, p^0)\mathscr{T}^{-1} = a^{\dagger}(-\mathbf{p}, p^0) \tag{4.47}$$
$$\mathscr{T} X^i(\mathbf{y}, y^0)\mathscr{T}^{-1} = X^i(\mathbf{y}, -y^0) \tag{4.48}$$
$$\mathscr{T} Y^i(\mathbf{y}, y^0)\mathscr{T}^{-1} = -Y^i(\mathbf{y}, -y^0) \tag{4.49a}$$
$$\mathscr{T} a(\mathbf{k}, k^0,1)\mathscr{T}^{-1} = a(-\mathbf{k}, k^0,1) \tag{4.49b}$$
$$\mathscr{T} a(\mathbf{k}, k^0,2)\mathscr{T}^{-1} = -a(-\mathbf{k}, k^0,2) \tag{4.49c}$$

where i = 1,2,3 and assuming: $\boldsymbol{\varepsilon}(\mathbf{k},k^0,1)=-\boldsymbol{\varepsilon}(-\mathbf{k},k^0,1)$ and $\boldsymbol{\varepsilon}(\mathbf{k},k^0,2)=+\boldsymbol{\varepsilon}(-\mathbf{k},k^0,2)$.

The unitary operator $\mathscr{U}_Y$ is given by

$$\mathscr{U}_X = \exp\{-i\pi\int d^3p\ [a^{\dagger}(\mathbf{p}, p^0)a(\mathbf{p}, p^0) - a^{\dagger}(\mathbf{p}, p^0)a(-\mathbf{p}, p^0)]/2\} \tag{4.50a}$$

and

$$\mathscr{U}_y = \mathscr{U}_X \exp\{-i\pi \int d^3k\ [\sum_{\lambda=1}^{2} a^{\dagger}(\mathbf{k},k^0,\lambda)a(\mathbf{k},k^0,\lambda) - a^{\dagger}(\mathbf{k}, k^0, 1)a(-\mathbf{k},k^0,1) + $$
$$+ a^{\dagger}(\mathbf{k}, k^0, 2)a(-\mathbf{k},k^0,2)]/2\} \tag{4.50b}$$

The separable Klein-Gordon Lagrangian (eq. A.96) is invariant under our definition of time reversal.

We note

$$\mathscr{U}_y = \mathscr{P}_y \tag{4.50c}$$

Although the present theory is somewhat more complicated than conventional quantum field theory the overall nature of the $\mathscr{P}$, $\mathscr{C}$, and $\mathscr{T}$ transformations is the same.

5. Interacting Quantum Field Theory – Perturbation Theory

Introduction

The form of quantum field theory that we have developed in chapters 3 and 4 can be used as the basis for new formulations of QED, ElectroWeak Theory, the Standard Model and a divergence-free, unified theory of all the known interactions. The development of these theories requires a number of topics be addressed. This chapter covers perturbation theory. As much as possible, we attempt to retain the features of the standard approach so that the reader will more readily follow the discussion and more readily accept this new formalism. In physics originality is secondary to reality. The perturbation theory that we will develop will be shown to be identical to the perturbation theory that we develop later using a path integral formalism.

Two-Tier theory will be shown to satisfy unitarity in chapter 6 and invariant under special relativity in Appendix B.

An Auxiliary Asymptotic Field

The definition of the asymptotic "free" in and out states is an issue in Two-Tier quantum field theory because the "free particle field" of the theory $\phi(X(y))$ is a "dressed" particle, *ab initio*, since it is cloaked in a cloud of Y particles as discussed in the passage following eq. 4.27.

While one could use $\phi(X(y))$ directly to define in and out asymptotic states it is more convenient initially to introduce a "fictitious" auxiliary asymptotic quantum field $\Phi(y)$ that will represent the equally fictitious "bare ϕ particle" in and out states.

We will consider the case of a scalar field. We define a free, scalar Klein-Gordon particle field with the physical mass m of the physical $\phi(X(y))$ particle.

$$\Phi(y) = \int d^3p\, N_m(p)\, [a(p)\, e^{-ip\cdot y} + a^\dagger(p)\, e^{ip\cdot y}] \tag{5.1}$$

using the creation and annihilation operators of $\phi(X(y))$ (in eq. 4.2). The set of particle states of $\Phi(y)$ has the familiar Fock space form

$$|\, p_1, p_2, \ldots\, p_n> = a^\dagger(p_1,)a^\dagger(p_2)\, \ldots\, a^\dagger(p_n)\,|0> \tag{5.2}$$

with powers of creation operators allowed since Φ particles are bosons. The set of particle states constitutes a complete orthonormal set of states. The corresponding bra states are defined by Hermitian conjugation:

$$<p_1, p_2, \ldots\, p_n| = (|\, p_1, p_2, \ldots\, p_n>)^\dagger \tag{5.3}$$

We note that the energy spectrum of these states is positive definite with the Hamiltonian

$$H_\Phi = P_\Phi{}^0 = \int d^3y\ \tfrac{1}{2}\,[\pi_\Phi{}^2 + (\partial\Phi/\partial X^i)^2 + m^2\Phi^2] \qquad (5.4a)$$

$$= \int d^3p\ (\mathbf{p}^2 + m^2)^{1/2} a^\dagger(p)a(p) \qquad (5.4b)$$

and momentum vector:

$$\mathbf{P}_\Phi = \int d^3p\ \mathbf{p}\ a^\dagger(p)a(p) \qquad (5.5)$$

We will use this set of energy-momentum eigenstates to define asymptotic “in” and “out” states in perturbation theory.

Transformation Between Φ(y) and ϕ(X(y))

For later use in the definition of the perturbation theory expansion, we will determine the transformations between the in and out Φ(y) fields, and the in and out $\phi(X(y))$ fields. Let us define a transformation $W_a(y)$ that transforms in and out Φ(y) fields to in and out $\phi(X(y))$ fields respectively:

$$\phi_a(X(y)) = :W_a(y)\Phi_a(y)W_a^{-1}(y): \qquad (5.6)$$

where the label a = “in” or a = “out”, where : … : signifies normal ordering, and where

$$\Phi_{in}(y) = \int d^3p\ N_m(p)\ [a_{in}(p)\ e^{-ip\cdot y} + a_{in}^\dagger(p)\ e^{ip\cdot y}] \qquad (5.7)$$

$$\Phi_{out}(y) = \int d^3p\ N_m(p)\ [a_{out}(p)\ e^{-ip\cdot y} + a_{out}^\dagger(p)\ e^{ip\cdot y}] \qquad (5.8)$$

$$\phi_{in}(X) = \int d^3p\ N_m(p)\ [a_{in}(p)\ :e^{-ip\cdot(y + iY/M_c^2)}: + a_{in}^\dagger(p)\ :e^{ip\cdot(y + iY/M_c^2)}:] \qquad (5.9)$$

$$\phi_{out}(X) = \int d^3p\ N_m(p)\ [a_{out}(p)\ :e^{-ip\cdot(y + iY/M_c^2)}: + a_{out}^\dagger(p)\ :e^{ip\cdot(y + iY/M_c^2)}:] \qquad (5.10)$$

Note that the transformation eq. 5.6 includes normal ordering. While this transformation may seem strange it is no stranger than the time reversal operator, in which the complex conjugate of all c-number terms is taken in addition to applying a unitary transformation.

In the Coulomb gauge of Y it is easy to show that

$$W_a(y) = \exp(-\mathbf{Y}(y)\cdot\mathbf{P}_{\Phi a}/M_c^2) \qquad (5.11)$$

and

$$W_a^{-1}(y) = \exp(\mathbf{Y}(y)\cdot\mathbf{P}_{\Phi a}/M_c^2) \qquad (5.12)$$

where the label a = "in" or a = "out", where the inner products in the exponentials are the usual spatial vector inner product, and where

$$\mathbf{P}_{\Phi a} = -\int d^3y\, \partial\Phi_a(y)/\partial y^0\, \nabla\Phi_a(y) = \int d^3p\, \mathbf{p}\, a_a^\dagger(p)a_a(p) \qquad (5.12a)$$

is a spatial vector (the Φ spatial momentum operator) that is written solely in terms of $\Phi_a(y)$'s creation and annihilation operators.

In addition to performing the transformation in eq. 5.6 $W_a(y)$ also performs a "translation" in Y^μ:

$$W_a(y)Y^i(y')W_a^{-1}(y) = Y^i(y') + i\Delta^{trij}(y' - y)P_{\Phi a}{}^j/M_c^2 \qquad (5.13a)$$

where

$$i\Delta^{trij}(y' - y) = \int d^3k\, (e^{-ik\cdot(y' - y)} - e^{ik\cdot(y' - y)})(\delta_{jk} - k_jk_k/\mathbf{k}^2)/[(2\pi)^3 2\omega_k] \qquad (5.13b)$$

We note that $W_a(y)$ is not a unitary operator but it is pseudo-unitary:

$$W_a^{-1}(y) = V\, W_a^\dagger(y)\, V^{-1} = V\, W_a(y)\, V^{-1} \qquad (5.14)$$

where

$$V = \exp(-i\pi \sum_{\lambda=1}^{2} \int d^3k\, a^\dagger(k, \lambda)a(k, \lambda)) \qquad (5.15)$$

is a unitary operator with the property

$$V\, Y^j(y)\, V^{-1} = -Y^j(y) \qquad (5.16)$$

for j = 1, 2, 3. We note

$$V^\dagger = V^{-1} = V \qquad (5.17)$$

and thus

$$V^2 = I \qquad (5.18)$$

V will be shown to be a metric operator in the following discussion.[86] We note that the Y "particle" (Hermitian) number operator appears in eq. 5.9 in the expression for V:

[86] P. A. M. Dirac, Proc. R. Soc. London A **180**, 1 (1942); T. D. Lee and G. C. Wick, Nucl. Phys. **B9**, 209 (1969); C. M. Bender, D. C. Brody and H. F. Jones, "Complex Extension of Quantum Mechanics" Phys. Rev. Letters **89**, 270401-1 (2002) and references therein.

$$N_Y = \sum_{\lambda=1}^{2} \int d^3k\, a^\dagger(k, \lambda) a(k, \lambda) \tag{5.19}$$

Thus states with an even number of Y "particles" have a V eigenvalue of one, and states with an odd number of Y "particles" have a V eigenvalue of minus one.

Model Lagrangian with ϕ^4 Interaction

We will develop our perturbation theory using a scalar Lagrangian model with a ϕ^4 interaction term:

$$\mathscr{L}_s = J\mathscr{L}_F + \mathscr{L}_C \tag{5.20}$$

with

$$\mathscr{L}_F = \tfrac{1}{2} [(\partial\phi/\partial X^\nu)^2 - m^2\phi^2] + \mathscr{L}_{Fint} \tag{5.21}$$

and

$$\mathscr{L}_C = -\tfrac{1}{4} F_Y^{\mu\nu} F_{Y\mu\nu} \tag{5.22}$$

with

$$F_{Y\mu\nu} = \partial Y_\mu/\partial y^\nu - \partial Y_\nu/\partial y^\mu \tag{5.23}$$

and

$$\mathscr{L}_{Fint} = \tfrac{1}{4!} \chi_0\, \phi(X(y))^4 + \tfrac{1}{2} (m^2 - m_0^2)\phi^2 \tag{5.24}$$

where J is the Jacobian (as in Appendix A), χ_0 is the bare coupling constant, and m_0 is the bare mass.

The conserved momentum operator is:

$$P_{F\beta} = \int d^3X\, \mathscr{T}_{F0\beta} \tag{5.25}$$

where

$$\mathscr{T}_{F\mu\nu} = -g_{\mu\nu}\mathscr{L}_F + \partial\mathscr{L}_F / \partial(\partial\phi/\partial X_\mu)\, \partial\phi/\partial X^\nu \tag{5.26}$$

is the ϕ field energy-momentum tensor with conservation law (eq. A.110):

$$\partial P_{F\beta}/\partial X^0 = 0 \tag{5.27}$$

due to eq. A.108.

The Hamiltonian density (eq. A.83) is

$$\mathscr{H}_F = \mathscr{T}_{F0\beta} = \mathscr{H}_{F0} + \mathscr{H}_{Fint} \tag{5.28}$$

with

$$\mathscr{H}_{F0} = \tfrac{1}{2} [\pi_\phi^2 + (\partial\phi/\partial X^i)^2 + m^2\phi^2] \tag{5.29}$$

$$\mathscr{H}_{\text{Fint}} = -\,{}^1\!/_{4!}\, \chi_0\, \phi(X(y))^4 + {}^1\!/_2\, (m^2 - m_0^2)\phi(X(y))^2 \tag{5.30}$$

In-states and Out-States

In this section we will develop properties of in-fields and out-fields. We will use a somewhat more complicated procedure to set up the perturbation theory for the S matrix due to the introduction of imaginary coordinates. The procedure can be schematized as:

$$\Phi_{\text{in}}(y) \Rightarrow \phi_{\text{in}}(X(y)) \Rightarrow \phi(X(y)) \Rightarrow \phi_{\text{out}}(X(y)) \Rightarrow \Phi_{\text{out}}(y) \tag{5.31}$$

In-states are constructed using the auxiliary field Φ_{in} which are then effectively transformed into $\phi_{\text{in}}(X(y))$ expressions in order to make contact with our Lagrangian formalism. Then $\phi_{\text{in}}(X(y))$ is related to the interacting field $\phi(X(y))$ as a limit ($y^0 \rightarrow -\infty$). Similarly out-states are constructed using the auxiliary field Φ_{out} which are then expressed in terms of $\phi_{\text{out}}(X(y))$. Then $\phi_{\text{out}}(X(y))$ is related to the interacting field $\phi(X(y))$ using the LSZ limiting process ($y^0 \rightarrow +\infty$).

Since much of the development differs only trivially from the standard treatment in textbooks we will simply "list" relevant equations and let the reader pursue them further in quantum field theory textbooks.

ϕ In-Field

In order to define a perturbation theory for particle scattering we will next specify features of the in-field $\phi_{\text{in}}(X(y))$ and in-field states – the field and states representing physical particles as $X^0 = y^0 \rightarrow -\infty$.

A. The in-field $\phi_{\text{in}}(X(y))$ satisfies the Klein-Gordon equation in the X variable:

$$(\Box_X + m^2)\, \phi_{\text{in}}(X) = 0 \tag{5.32}$$

where

$$\Box_X = (\partial/\partial X^\nu)(\partial/\partial X_\nu)$$

B. Under coordinate displacements and Lorentz transformations $\Phi_{\text{in}}(y)$, $\phi_{\text{in}}(X(y))$, and $\phi(X(y))$ transform in the same way:

$$[P^\mu, \Phi_{\text{in}}(y)] = -\,i\partial\Phi_{\text{in}}/\partial y_\mu \tag{5.33a}$$

$$[P^\mu, \phi_{\text{in}}(X)] = -\,i\partial\phi_{\text{in}}/\partial y_\mu \tag{5.33b}$$

$$[P^\mu, \phi(X)] = -\,i\partial\phi/\partial y_\mu \tag{5.34}$$

with the energy-momentum vector P^μ specified by eq. A.57.

C. We can relate the asymptotic in-field $\phi_{in}(X(y))$ to the interacting field $\phi(X(y))$ using the equation of motion of $\phi(X(y))$

$$(\Box_X + m^2)\,\phi(X) = j(X) \tag{5.35}$$

where j(X) embodies the interaction. Using the physical mass m we find

$$(\Box_X + m^2)\,\phi(X) = j(X) + (m^2 - m_0^2)\phi(X) = j_{tot}(X) \tag{5.36}$$

If the current is taken to be the source of the scattered waves we may write

$$\sqrt{Z}\,\phi_{in}(X(y)) = \phi(X(y)) - \int d^4X(y')\,\Delta_{ret}(y - y')\,j_{tot}(X(y')) \tag{5.37}$$

$$= \phi(X(y)) - \int d^4y'\,J\,\Delta_{ret}(y - y')\,j_{tot}(X(y')) \tag{5.38}$$

where Z is a wave function renormalization constant, J is the Jacobian, and Δ_{ret} is a retarded Green's function.

D. We can define Φ_{in} in-field states with expressions like

$$|\,p_1, p_2, \ldots\, p_n\ in> = a_{in}^\dagger(p_1)a_{in}^\dagger(p_2)\ \ldots\ a_{in}^\dagger(p_n)\,|0> \tag{5.39}$$

with powers of creation operators allowed since Φ_{in} is a boson field. The set of all particle states constitutes a complete orthonormal set of states. The corresponding bra states are defined by Hermitian conjugation:

$$<p_1, p_2, \ldots\, p_n\ in| = (|\,p_1, p_2, \ldots\, p_n\ in>)^\dagger \tag{5.40}$$

ϕ Out-Field

In order to define a perturbation theory for particle scattering we begin by listing aspects of the out-field $\phi_{out}(X(y))$ and out-field states – the field and states representing physical particles as $X^0 = y^0 \rightarrow -\infty$.

A. The out-field $\phi_{out}(X(y))$ satisfies the Klein-Gordon equation in the X variable:

$$(\Box_X + m^2)\,\phi_{out}(X) = 0 \tag{5.41}$$

where

$$\Box_X = (\partial/\partial X^\nu)(\partial/\partial X_\nu)$$

B. Under coordinate displacements and Lorentz transformations $\Phi_{out}(y)$, $\phi_{out}(X(y))$, and $\phi(X(y))$ transform in the same way:

$$[P^\mu, \Phi_{out}(y)] = - i\partial\Phi_{out}/\partial y_\mu \quad (5.42a)$$

$$[P^\mu, \phi_{out}(X)] = - i\partial\phi_{out}/\partial y_\mu \quad (5.42b)$$

$$[P^\mu, \phi(X)] = - i\partial\phi/\partial y_\mu \quad (5.43)$$

with the energy-momentum vector P^μ specified by eq. A.57.

C. We can relate the asymptotic out-field $\phi_{out}(X(y))$ to the interacting field $\phi(X(y))$ using the equation of motion of $\phi(X(y))$ specified by eq. 5.36:

$$\sqrt{Z}\, \phi_{out}(X(y)) = \phi(X(y)) - \int d^4X(y')\, \Delta_{adv}(y - y')\, j_{tot}(X(y')) \quad (5.44)$$

$$= \phi(X(y)) - \int d^4y'\, J\, \Delta_{adv}(y - y')\, j_{tot}(X(y')) \quad (5.45)$$

where Z is a wave function renormalization constant, J is the Jacobian, and Δ_{adv} is an advanced Green's function.

D. We can define Φ_{out} out-field states with expressions like

$$|\, p_1, p_2, \ldots\, p_n\, out> = a_{out}^\dagger(p_1,)a\Phi_{out}^\dagger(p_2)\, \ldots\, a\Phi_{out}^\dagger(p_n)\,|0> \quad (5.46)$$

with powers of creation operators allowed since Φ_{out} is a boson field. The set of all particle states constitutes a complete orthonormal set of states. The corresponding bra states are defined by Hermitian conjugation:

$$<p_1, p_2, \ldots\, p_n\, out| = (|\, p_1, p_2, \ldots\, p_n\, out>)^\dagger \quad (5.47)$$

The Y Field

The Y field in the present model Lagrangian (eq. 5.20) is a free field and thus:

$$Y_{in}(y) = Y_{out}(y) = Y(y) \quad (5.48)$$

The states of the Y field have two general forms: 1) States in a Fock space consisting of particle states that are eigenstates of the Y particle number operator (eq. 5.19); and 2) Coherent states in a non-Fock space of generalized coherent states in an infinite tensor product space.[87]

[87] See Kibble and other references on coherent states.

The coherent ket states that arise in Two-Tier quantum field theory have the general form (eq. 3.41):

$$|y,\ p> = e^{-\mathbf{p}\cdot\mathbf{Y}^{-}(y)/M_c^2}|0> \qquad (3.41)$$

as can be seen from an examination of $\phi_{in}(X(y))$. The corresponding bra state is:

$$<y, p| = (V|\ y, p>)^{\dagger} = <0|e^{+\mathbf{p}\cdot\mathbf{Y}^{+}(y)/M_c^2} \qquad (5.49)$$

with V, the metric operator, reversing the sign of Y in the exponential. The inner product of coherent states is:

$$<y_1\ , p_1|\ y_2, p_2> = \exp[-p_1{}^i p_2{}^j \Delta_{Tij}(y_1 - y_2)/M_c{}^4] \qquad (5.50)$$

showing the set of coherent states is not orthonormal and, in fact, is overcomplete. Comparing eq. 5.50 to eq. 4.12 gives

$$<y_1\ , p|\ y_2, p> = R(p, y_1 - y_2) \qquad (5.50a)$$

The completeness of the set of states for each time y^0 can be verified by examining the projection operator:

$$\mathscr{R}_Y(y^0) = \because \exp[-i \int d^3y\ Y^-{}_i(y)|0><0|\pi^{+i}(y)] \because \qquad (5.51)$$

where

$$\pi^{+i}(y) = -\partial Y^{+i}(y)/\partial y^0 \qquad (5.52)$$

and where $\because$ represents an extended normal ordering operator:

$$\because \ \ldots \ \because$$

which is defined as placing creation operators to the left, projection operators in the center, and annihilation operators to the right. Thus eq 5.51 can be written

$$\mathscr{R}_Y = \sum_n (-i/n!)^n \int d^3y_1 \ldots \int d^3y_n Y^{-j_1}(y_1)Y^{-j_2}(y_2)\ldots Y^{-j_n}(y_n)|0><0|\pi^+{}_{j_1}(y_1)\pi^+{}_{j_2}(y_2)\ldots\pi^+{}_{j_n}(y_n) \qquad (5.53)$$

where we have used the fact that |0><0| is a projection operator, and reduced |0><0| |0><0| ... |0><0| to |0><0| in eq. 5.53. The vacuum state is the product of the Y and ϕ vacuum states:

$$|0> = |0_Y>|0_\phi> \tag{5.53a}$$

We note

$$\mathscr{R}_Y\,(y^0)\,|\mathbf{y}, y^0\;p> = |\mathbf{y}, y^0\;p> \tag{5.54}$$

using eq. 3.22 and $\int d^3y_2\, p^i\, \Delta^{tr}{}_{ij}(y_1 - y_2)Y^{+j}(y_2) = \mathbf{p}\cdot\mathbf{Y}^+(y_1)$. Also

$$\mathscr{R}_Y(y^0)\,|n> = |n> \tag{5.55}$$

where $|n>$ is any Y particle Fock state of finite particle number. In view of eqs. 5.54 and 5.55, we see that $\mathscr{R}_Y$ is the identity operator on the Fock space and the space of generalized coherent Y field states. Thus the set of Y coherent states forms an overcomplete set of states. We will define the S matrix for any combination of Φ Fock space states and coherent Y states. The $\mathscr{R}_Y$ operator can be generalized to include Φ Fock space states:

$$\mathscr{R}_{\Phi Y}(y^0) = \because \exp[-i \int d^3y\; Y^-{}_j(y)\mathscr{R}_\Phi \pi^{+j}(y)] \because \tag{5.56}$$

with

$$\mathscr{R}_\Phi = \sum_n |n><n| \tag{5.57}$$

is a sum over all Φ Fock space states with vacuum state given by eq. 5.53a. Since $\mathscr{R}_\Phi$ is a projection:

$$[\mathscr{R}_\Phi\,]^N = \mathscr{R}_\Phi$$

for any power N, we find:

$$\mathscr{R}_{\Phi Y}(y^0) = \sum_n (-i)^n \int d^3y_1 \ldots \int d^3y_n Y^{-j_1}(y_1)Y^{-j_2}(y_2)\ldots Y^{-j_n}(y_n)\mathscr{R}_\Phi \pi^+{}_{j_1}(y_1)\pi^+{}_{j_2}(y_2)\ldots\pi^+{}_{j_n}(y_n) \tag{5.58}$$

As a result we have

$$\mathscr{R}_{\Phi Y}(y^0)\,|y,\,p;\,n_\Phi> = |y,\,p;\,n_\Phi> \tag{5.59}$$

for any combination of Y coherent states and Φ Fock space states n_Φ. Also

$$\mathscr{R}_{\Phi Y}(y^0)\,|n_\Phi> = |n_\Phi> \tag{5.60}$$

Thus $\mathscr{R}_{\Phi Y}$ is the identity operator on this space – the (over) complete space of in and out states which we will use to define the S matrix of the scalar field theory specified by the Lagrangian eq. 5.20.

S Matrix

Following the standard definition of the S matrix we have:

$$S_{\alpha\beta} = <\alpha\ \text{out}|\beta\ \text{in}> \tag{5.61}$$

$$= <\alpha\ \text{in}|S|\beta\ \text{in}> \tag{5.62}$$

$$|0> = |0\ \text{in}> = |0\ \text{out}> = S|0\ \text{in}> \tag{5.63}$$

$$\Phi_{in}(y) = S\Phi_{out}(y)S^{-1} \tag{5.64}$$

and the other standard properties of the S matrix with the sole exception being the form of the unitarity relation (discussed later).

LSZ Reduction for Scalar Fields

In this section we will determine the reduction formula for the S matrix for scalar ϕ fields. Consider the S matrix element corresponding to an in state of particles β plus one ϕ particle of momentum p, and an out state α:

$$S_{\alpha\beta p} = <\alpha\ \text{out}|\beta p\ \text{in}> \tag{5.65}$$

After standard manipulations we have:

$$S_{\alpha\beta p} = <\alpha - p\ \text{out}|\beta\ \text{in}> - i<\alpha\ \text{out}|\int d^3y\ f_p(y)\ \overleftrightarrow{\partial}_0\ [\Phi_{in}(y) - \Phi_{out}(y)]\ |\beta\ \text{in}> \tag{5.66}$$

where $<\alpha - p\ \text{out}|$ is an out state with a particle of momentum p removed (if present) and where

$$f(y^0)\ \overleftrightarrow{\partial}_0\ g(y^0) = f(y^0)\ \partial g(y^0)/\partial y^0 - \partial f(y^0)/\partial y^0\ g(y^0) \tag{5.67}$$

and

$$f_p(y) = N_m(p)e^{-ip\cdot y} \tag{5.68}$$

with $N_m(p)$ specified by eq. 3.6.

We now express

$$S_{\alpha\beta p} = S_{\alpha - p\beta} - i<\alpha\ \text{out}|\ \int d^3y\ f_p(y)\ \overleftrightarrow{\partial}_0 W^{-1}[\phi_{in}(X(y)) - \phi_{out}(X(y))]W|\beta\ \text{in}> \tag{5.69}$$

using $W(y) = W_{in}(y)$ with

$$\Phi_a(y) = W_a^{-1}(y)\phi_a(X(y))W_a(y) \tag{5.70}$$

where the label a = "in" or a = "out", and where

$$W_a(y) = \exp(-\mathbf{Y}(y)\cdot\mathbf{P}_{\Phi a}/M_c^2) \tag{5.71}$$

and

$$W_a^{-1}(y) = \exp(\mathbf{Y}(y)\cdot\mathbf{P}_{\Phi a}/M_c^2) \tag{5.72}$$

in the Coulomb gauge of Y with $\mathbf{P}_{\Phi a}$ the momentum spatial vector defined by eq. 5.12a.

We note that the interacting $\phi(X(y))$ approaches the in and out fields $\phi_{in}(X(y))$ and $\phi_{out}(X(y))$ in the limit that $y^0 \to -\infty$ and $y^0 \to +\infty$ respectively in the sense of Lehmann, Symanzik and Zimmermann[88] which we *symbolize* as:

$$\phi(X(y)) \to \sqrt{Z}\,\phi_{in}(X(y)) \quad \text{as} \quad y^0 \to -\infty \tag{5.73}$$

$$\phi(X(y)) \to \sqrt{Z}\,\phi_{out}(X(y)) \quad \text{as} \quad y^0 \to +\infty \tag{5.74}$$

with $\sqrt{Z}$ defined in eqs. 5.37 and 5.44. Thus we can rewrite eq. 5.69 as

$$S_{\alpha\beta p} = S_{\alpha-p\beta} + iZ^{-1/2}\,(\lim_{y_0\to+\infty} - \lim_{y_0\to-\infty})\langle\alpha\text{ out}|\int d^3y\, f_p(y)\,\overleftrightarrow{\partial}_0\, W^{-1}\phi(X(y))W\,|\beta\text{ in}\rangle \tag{5.75}$$

which after standard manipulations becomes

$$S_{\alpha\beta p} = S_{\alpha-p\beta} + iZ^{-1/2}\int d^4y\, f_p(y)(\Box_y + m^2)\langle\alpha\text{ out}|\,W(y)^{-1}\phi(X(y))W(y)\,|\beta\text{ in}\rangle \tag{5.76}$$

Eq. 5.76 is similar to the usual LSZ reduction formula except for the appearance of the W(y) operator and its inverse. We note that $W(y) = W_{in}(y)$ still because $\mathbf{P}_{\Phi in}$ is independent of y^0.

Similarly an out ϕ particle can be reduced from an S matrix part. For example,

$$\begin{aligned}\langle\alpha\text{ out}|W^{-1}(y)\phi(X(y))W(y)|\beta\text{ in}\rangle &= \langle\alpha-p'\text{ out}|W^{-1}(y)\phi(X(y))W(y)|\beta-p'\text{ in}\rangle \\ &- i\langle\alpha-p'\text{ out}|\int d^3y'\,[W^{-1}(y')\phi_{in}(X(y'))W(y')W^{-1}(y)\phi(X(y))W(y) - \\ &- W^{-1}(y)\phi(X(y))W(y)W^{-1}(y')\phi_{out}(X(y'))W(y')]\,|\beta\text{ in}\rangle\,\overleftrightarrow{\partial}_0\, f_{p'}^{*}(y')\end{aligned} \tag{5.77}$$

which becomes

[88] H. Lehmann, K. Symanzik and W. Zimmermann, Nuov. Cim., **1**, 1425 (1955); W. Zimmermann, Nuov. Cim., **10**, 567 (1958); O. W. Greenberg, Doctoral Dissertation, Princeton University 1956.

$$<\alpha\ \text{out}|W^{-1}(y)\phi(X(y))W(y)|\beta\ \text{in}> = <\alpha-p'\ \text{out}|\varphi(y)|\beta-p'\ \text{in}> +$$
$$+ iZ^{-1/2}\int d^4y'\ <\alpha-p'\ \text{out}|T(\varphi(y')\varphi(y))|\beta\ \text{in}>\ (\overleftarrow{\Box}_{y'} + m^2)\, f_{p'}^{\ *}(y') \tag{5.78}$$

where the time ordered product is defined with respect to ordering with respect to y^0 and where

$$\varphi(y) = W^{-1}(y)\phi(X(y))W(y) \tag{5.79}$$

These results directly generalize to multi-particle in and out states:

$$<p_1, p_2, \dots p_n\ \text{out}|\ q_1, q_2, \dots q_m\ \text{in}> = \dots <0|T(\varphi(y'_1) \dots \varphi(y'_n)\varphi(y_1) \dots \varphi(y_m))|0> \dots \tag{5.80}$$

thus reducing the development of the perturbation theory of the S matrix to the evaluation of time ordered products such as

$$<0|T(\varphi(y_1) \dots \varphi(y_n))|0> \tag{5.81}$$

6. Perturbation Theory II

The U Matrix

The U matrix for a Two-Tier theory is developed in a way similar to conventional field theory starting from the defining relations:

$$\phi(X(y)) = U^{-1}\phi_{in}(X(y))U \tag{6.1}$$

$$\pi_\phi(X(y)) = U^{-1}\pi_{\phi in}(X(y))U \tag{6.2}$$

From eq. 5.29 we define the free field Hamiltonian

$$H_{F0in}(\phi_{in}, \pi_{\phi in}) = \int d^3X\, \mathscr{H}_{F0}(\phi_{in}, \pi_{\phi in}) \tag{6.3}$$

Noting $X^0 = y^0$ in the Y Coulomb gauge we find

$$\partial\phi_{in}/\partial y^0 = i[H_{F0in}, \phi_{in}(X)] \tag{6.4}$$

$$\partial\pi_{\phi in}/\partial y^0 = i[H_{F0in}, \pi_{\phi in}(X)] \tag{6.5}$$

For the entire Hamiltonian (eq. 5.28) we have

$$\partial\phi/\partial y^0 = i[H_F, \phi(X)] \tag{6.6}$$

$$\partial\pi_\phi/\partial y^0 = i[H_F, \pi_\phi(X)] \tag{6.7}$$

with

$$H_F(\phi, \pi_\phi) = :\int d^3X\, \mathscr{H}_F(\phi, \pi_\phi): \tag{6.8}$$

(Note the *entire* interaction term is normal ordered since d^3X is a q-number. Combining the above equations in the standard way yields a familiar differential equation for the U matrix:

$$i\partial U(y^0)/\partial y^0 = (H_{Fint} + E_0(t))U(y^0) \tag{6.9}$$

where $E_0(t)$ is a c-number function of y^0 that we can set equal to 0 (as it would be cancelled later in any case), and where

$$H_{Fint}(\phi_{in}, \pi_{\phi in}) = :\int d^3X\, \mathscr{H}_{Fint}(\phi_{in}, \pi_{\phi in}): \tag{6.10}$$

with $\mathscr{H}_{Fint}$ given by eq. 5.30. Solving for U gives the familiar time ordered exponential:

$$U(y^0) = T\left(\exp[-i\int_{-\infty}^{t} dy^0\, H_{Fint}]\right) \tag{6.11a}$$

which is a symbolic notation for:

$$U(y^0) = 1 + \sum_{n=1}^{\infty} (-i)^{-n}(n!)^{-1}\int_{-\infty}^{y^0} dy_1{}^0 \ldots \int_{-\infty}^{y^0} dy_n{}^0\, T(H_{Fint}(y_1{}^0) \ldots H_{Fint}(y_n{}^0)) \tag{6.11b}$$

We note for later use that the hermiticity of H_{Fint} is not used in the derivation of eq. 6.11. Thus eq. 6.11 would still hold if H_{Fint} were not Hermitian.

Reduction of Time Ordered φ Products

In the previous chapter we reduced the calculation of the S matrix to the evaluation of time ordered products of the form

$$\tau(y_1, \ldots, y_n) = \langle 0|T(\varphi(y_1) \ldots \varphi(y_n))|0\rangle \tag{6.12}$$

where $\varphi(y)$ is specified by eq. 5.79. Expanding the terms within eq. 6.12 using eq. 5.79 we find

$$\varphi(y_1) \ldots \varphi(y_n) = W^{-1}(y_1)\phi(X(y_1))W(y_1)W^{-1}(y_2)\phi(X(y_2))W(y_2) \ldots W^{-1}(y_n)\phi(X(y_n))W(y_n) \tag{6.13}$$

which can be re-expressed as

$$W^{-1}(y_1)U^{-1}(y_1{}^0)\phi_{in}(X(y_1))U(y_1{}^0)W(y_1)W^{-1}(y_2)U^{-1}(y_2{}^0)\phi_{in}(X(y_2))U(y_2{}^0)W(y_2) \ldots \tag{6.14}$$

using eq. 6.1 and denoting $W_{in}(y)$ as $W(y)$. Defining

$$\mathscr{U}(y_1, y_2) = U(y_1{}^0)W(y_1)W^{-1}(y_2)U^{-1}(y_2{}^0) \tag{6.15}$$

we see eq. 6.14 can be rewritten as

$$W^{-1}(y_1)U^{-1}(y_1{}^0)\phi_{in}(X(y_1))\mathscr{U}(y_1, y_2)\phi_{in}(X(y_2))\,\mathscr{U}(y_2, y_3)\phi_{in}(X(y_3)) \ldots \phi_{in}(X(y_n))U(y_n{}^0)W(y_n) \tag{6.16}$$

From eqs. 5.71 and 5.72

$$\mathscr{U}(y_1, y_2) = U(y_1^0)\exp((\mathbf{Y}(y_2) - \mathbf{Y}(y_1))\cdot\mathbf{P}_{\Phi a}/M_c^2)U^{-1}(y_2^0) \quad (6.17)$$

Defining

$$W(y_1, y_2) = \exp((\mathbf{Y}(y_2) - \mathbf{Y}(y_1))\cdot\mathbf{P}_{\Phi a}/M_c^2) \quad (6.18)$$

and looking ahead to the Wick expansion of the time ordered product of eq. 6.12 we note that the only time ordered products involving $W(y_1, y_2)$ that would appear in the expansion are

$$<0|T(\phi_{in}(X(y))W(y_1, y_2))|0> = 0 \quad (6.19a)$$

$$<0|T(Y(y)W(y_1, y_2))|0> = 0 \quad (6.19b)$$

$$<0|T(\partial Y(y)/\partial y^\mu W(y_1, y_2))|0> = 0 \quad (6.19c)$$

$$<0|T(\partial Y(y)/\partial y^\mu \phi_{in}(X(y)))|0> = 0 \quad (6.19d)$$

$$<0|T(W(y_1, y_2)W(y_3, y_4))|0> = 1 \quad (6.19e)$$

due to the factor of $\mathbf{P}_{\Phi a}$ that appears in $W(y_1, y_2)$. Also

$$<0|T(\phi_{in}(X(y))Y(y_1))|0> = 0 \quad (6.20)$$

due to the $a_{in}(p)$ and $a_{in}^\dagger(p)$ factors appearing in $\phi_{in}(X(y))$.

Thus the $W(y_1, y_2)$ factor in eq. 6.17 may be set to the value one with the result

$$\mathscr{U}(y_1, y_2) \equiv U(y_1^0)U^{-1}(y_2^0) = U(y_1^0, y_2^0) \quad (6.21)$$

where $U(y_1^0, y_2^0)$ is the conventionally defined U matrix satisfying

$$i\partial\, U(y_1^0, y_2^0)/\partial y_1^0 = iH_{Fint}\, U(y_1^0, y_2^0) \quad (6.22)$$

with the boundary condition

$$U(y^0, y^0) = 1 \quad (6.23)$$

This result would still be true if the $W(y_1, y_2)$ exponentials were expanded in their "power series" form.

Then, paralleling the standard approach we find an expression for the U matrix:

$$U(y_1^0, y_2^0) = T\Big(\exp[-i \int_{y_2^0}^{y_1^0} dy'^0{:}d^3X(y')\, \mathscr{H}_{Fint}(\phi_{in}(X(y')), \pi_{\phi in}(X(y'))){:}]\Big)$$

(6.24)

The $U(y_1^0, y_2^0)$ matrix satisfies the conventional multiplication rule:

$$U(y_1^0, y_3^0) = U(y_1^0, y_2^0)U(y_2^0, y_3^0) \qquad (6.25)$$

The inverse of $U(y_1, y_2)$ is

$$U^{-1}(y_1^0, y_2^0) = U(y_2^0, y_1^0) \qquad (6.26)$$

We now return to eq. 6.16, which can now be written in the form:

$$U^{-1}(y^0)U(y^0, y_1^0)\phi_{in}(X(y_1))U(y_1^0, y_2^0)\phi_{in}(X(y_2))U(y_2^0, y_3^0) \ldots \phi_{in}(X(y_n))U(y_n^0, -y^0)U(-y^0)$$

(6.27)

where y^0 is a reference time that is later than all other times, and $-y^0$ is earlier than all the other times, in the time-ordered product. As a result the time-ordered product in eq. 5.80 can be expressed in a symbolic notation as:

$$<0|U^{-1}(y^0)T(\phi_{in}(X(y_1))\phi_{in}(X(y_2)) \ldots \phi_{in}(X(y_n))U(y^0, -y^0))U(-y^0)|0> \qquad (6.28)$$

The analysis of eq. 6.28 as $y^0 \rightarrow \infty$ follows the standard path, which begins by noting

$$U(-y)|0> = \lambda_-|0> \qquad \text{when } y^0 \rightarrow \infty \qquad (6.29a)$$

$$U(y)|0> = \lambda_+|0> \qquad \text{when } y^0 \rightarrow \infty \qquad (6.29b)$$

following a standard textbook proof, which, in turn, leads to:

$$\lambda_-\lambda_+{}^* = <0|T\Big(\exp[+i \int_{-\infty}^{\infty} dy'^0{:}d^3X(y')\mathscr{H}_{Fint}(\phi_{in}(X(y')), \pi_{\phi in}(X(y'))){:}]\Big)|0>$$

(6.30)

$$= \Big[<0|T\Big(\exp[-i \int_{-\infty}^{\infty} dy'^0 d^3X(y')\, \mathscr{H}_{Fint}(\phi_{in}(X(y')), \pi_{\phi in}(X(y')))]\Big)|0>\Big]^{-1}$$

(6.31)

Thus the time ordered product of eq. 6.12, which appears in the evaluation of the S matrix element in eq. 5.80, can be symbolically written as:

$$\tau(y_1, \ldots, y_n) = \frac{<0|T(\phi_{in}(X(y_1)) \ldots \phi_{in}(X(y_n))U(\infty, -\infty))|0>}{<0|T\left(\exp\left[-i\int dy'^0 :d^3X(y')\mathscr{H}_{Fint}(\phi_{in}(X(y')),\pi_{\phi in}(X(y'))):\right]\right)|0>} \tag{6.32}$$

in the limit $y^0 \to \infty$.

The $\int d^3X$ Integration

The integration over the X space coordinates presents the difficulty of a functional integration of a q-number that needs to be properly defined. Since

$$X^\mu(y) = y^\mu + i\,Y^\mu(y)/M_c^{\,2} \tag{3.12}$$

by definition and since, in the Y Coulomb gauge we have $X^0(y) = y^0$ due to $Y^0 = 0$, the classical Jacobian for the transformation from y to X coordinates is the absolute value:

$$J = \left|\varepsilon^{ijk}\left(\delta^{1i} + \frac{i}{M_C^{\,2}}\frac{\partial Y^1}{\partial y^i}\right)\left(\delta^{2j} + \frac{i}{M_C^{\,2}}\frac{\partial Y^2}{\partial y^j}\right)\left(\delta^{3k} + \frac{i}{M_C^{\,2}}\frac{\partial Y^3}{\partial y^k}\right)\right| \tag{6.33}$$

The Jacobian appears in a change of integration variables:

$$\int d^3X = \int d^3y\, J \tag{6.34}$$

and

$$\int d^4X = \int d^4y\, J \tag{6.35}$$

in the Y Coulomb gauge.

A change of variables for c-number coordinate transformations is well known. The situation changes when one set of coordinates are in fact q-numbers. The second quantization of the Y field requires the definition of J to be clarified since the product of fields at the same position is normally undefined. The normal ordering of the interaction Hamiltonian term in eqs. 6.34 and 6.32 resolves the issue. Therefore eq. 6.33 must be considered as inserted within a normal ordered expression.

While normal ordering eliminates the infinities that would otherwise be present, J still presents a problem because it is still effectively part of the interaction term. This situation appears to be unsatisfactory in the present, scalar quantum field theory in which Y is not intended to play a direct dynamical role but rather a passive role as a coordinate. The normal ϕ field is supposed to be the only in, out, and interacting field.

The problem of J is resolved by eqs. 6.19c and 6.19d, which reduces the effect of the derivative terms in eq. 6.33 to zero in the Wick expansion of the time ordered product in eq. 6.32 if no Y quanta appear in or out S matrix states. Thus

$$J \equiv 1 \tag{6.36}$$

As a result the time ordered product (eq. 6.32) becomes:

$$\tau(y_1,\ldots, y_n) = \frac{<0|T(\phi_{in}(X(y_1))\ldots\phi_{in}(X(y_n))\exp[-i\int d^4y'\mathscr{H}_{Fint}(\phi_{in}(X(y')))])\ |0>}{<0|T\left(\exp\ [-i\int d^4y'\mathscr{H}_{Fint}(\phi_{in}(X(y')))]\right)|0>} \tag{6.37}$$

Y In and out states

The Y fields have no interactions and are thus free fields in the model Lagrangian under consideration and in the Two-Tier quantum field theories that we will construct later. Therefore "in" Y quanta are the same as "out" Y quanta.

Since the Lagrangians that we consider do not have interaction terms explicitly containing Y field factors, the S matrix is "block diagonal" in the sense that if an in-state does not contain Y quanta, (or Y coherent states) then out-states will not contain Y quanta (or coherent Y states). The proof is based on the expansion of S matrix elements using Wick's theorem in products of time ordered products of pairs of in field operators. Eqs. 6.19, 6.20 and 6.36, and in particular,

$$<0|\ T(\phi_{in}(X(y_1))Y^j(y_2))\,|0> = 0 \tag{6.39}$$

and

$$<0|T(\phi_{in}(X(y_1))e^{-\mathbf{q\cdot Y^-}(y)/M_c^2})\,|0> = <0|T(\phi_{in}(X(y_1))e^{+\mathbf{q\cdot Y^+}(y)/M_c^2})\,|0> = 0 \tag{6.40}$$

prove S matrix elements with no incoming Y quanta or coherent states will have zero matrix elements to produce outgoing Y quanta or coherent states. In addition any non-zero S matrix element with n incoming Y quanta must have n outgoing Y quanta. For example an incoming state with 5 Y quanta and 2 ϕ particles can only become an outgoing state with 5 Y quanta and two or more ϕ particles. Therefore we have proved the general result:

Theorem 6.I: ***Any non-zero S matrix element has the same number of incoming Y quanta and outgoing Y quanta.***

This theorem is true in any Two-Tier quantum field theory. In order to have a tractable theory we will require all in-states and out-states not *to contain Y quanta or coherent states. All normal in-state and out-state particles will contain factors of* $:e^{\pm p \cdot Y/M_c^2}:$ *in the Fourier expansions of their corresponding fields.*

Unitarity

For many years it has been evident that modified field theories[11, 17, 89] might offer some hope of avoiding the divergences of conventional quantum field theory. Usually these theories suffer from unitarity problems: negative norms and negative probabilities. In the absence of a physically acceptable interpretation of negative probabilities, these theories have been thought to be unsatisfactory.

The Two-Tier type of quantum field theory *superficially* also appears to have a unitarity problem due to the non-Hermitian nature of Two-Tier Hamiltonians. The lack of hermiticity is due entirely to the appearance of iY^μ in the X^μ field coordinates. *In fact Two-Tier quantum field theories satisfy unitarity for physical states. Physical states are defined to consist of any number of normal Two-Tier particles and NO Y quanta.*
Two-Tier interaction Hamiltonians, such as the one in eq. 6.37, are not Hermitian. For example,

$$H_{Fint} = \int d^3y' \mathscr{H}_{Fint}(\phi_{in}(y' + iY(y')/M_c^2)) \tag{6.41}$$

and

$$H_{Fint}^\dagger = \int d^3y' \mathscr{H}_{Fint}(\phi_{in}(y' - iY(y')/M_c^2)) \neq H_{Fint} \tag{6.42}$$

The relation between H_{Fint} and its Hermitian conjugate is

$$H_{Fint} = V\, H_{Fint}^\dagger\, V \tag{6.43}$$

where $V^2 = I$ is the metric operator defined in eqs. 5.15 – 5.18. Thus the S matrix is not unitary; the S matrix is *pseudo-unitary*:

$$S^{-1} = V\, S^\dagger\, V \tag{6.44}$$

and

$$VS^\dagger VS = I \tag{6.45}$$

We will now show that the S matrix is *unitary between physical states.* To prove this point, consider eq. 6.45 between physical states $|i\rangle$ and $\langle f|$ – each consisting of a number of ϕ particles and no Y quanta.

---

[89] S. Blaha, Phys.Rev. **D10**, 4268 (1974); S. Blaha, Phys.Rev. **D11**, 2921 (1975); S. Blaha, Nuovo Cim. **A49**, :113 (1979); S. Blaha, "Generalization of Weyl's Unified Theory to Encompass a Non-Abelian Internal Symmetry Group" SLAC-PUB-1799, Aug 1976; S. Blaha, "Quantum Gravity and Quark Confinement" Lett. Nuovo Cim. **18**, 60 (1977); S. Blaha, "The Local Definition of Asymptotic Particle States" Nuovo Cim. **A49**, 35 (1979) and references therein.

$$\delta_{fi} = \langle f\,|I|\,i\rangle = \langle f\,|VS^{\dagger}VS|\,i\rangle$$

$$= \sum_{n,\,m,\,p} \langle f\,|V|\,p\rangle\langle p\,|S^{\dagger}|\,n\rangle\langle n\,|V|\,m\rangle\langle m\,|S|\,i\rangle$$

$$= \sum_{n,\,m,\,p} \langle f\,|S^{\dagger}|\,m\rangle\langle m\,|S|\,i\rangle \qquad (6.46)$$

since V has the eigenvalue 1 between states consisting of no Y quanta. Due to eqs. 6.19a – 6.19e and 6.20 since there are no incoming Y quanta there are no outgoing Y quanta.

The block diagonality of S (and the diagonality of V) limits the intermediate states |n> and |m> to states containing ϕ particles and no Y quanta – although normalization factors R(**p**, z) will appear (described later) due to the presence of $:e^{\pm p\cdot Y/M_c^2}:$ factors within quantum field Fourier expansions that embody Y coherent state effects. Thus

$$S_{phys}^{\dagger} S_{phys} = I \qquad (6.47)$$

and

$$S_{phys}^{\dagger} = S_{phys}^{-1} \qquad (6.48)$$

proving unitarity between physical states – states consisting of ϕ particles and no Y quanta that are properly normalized. A detailed example is presented starting on page 67.

Finite Renormalization of External Legs

In the previous section we showed the theory satisfies unitarity for states that are properly normalized. However the use of the non-unitary operator W(y) (eq. 5.6) to transform $\Phi_{in}(y)$ fields into $\phi_{in}(X(y))$ fields in the LSZ procedure in eq. 5.69, and related equations, does not preserve the norm of input and output ϕ particle legs. Thus a finite renormalization is needed for each external particle leg in order to have a unitary S-matrix.

We define this renormalization of external legs within the framework of a perturbation theory example in the section beginning on page 67.

Perturbation Expansion

Perturbation theory in Two-Tier quantum field theory is very similar to conventional perturbation theory. The difference is in the form of the propagators, which have a high energy damping factor R(**p**, z) that eliminates infinities that normally appear at high energy in conventional quantum field theories.

In order to develop a feeling for Two-Tier perturbation theory we will calculate a few low order diagrams in the perturbation theory of the model scalar ϕ^4 theory that we have been using as an example in this chapter.

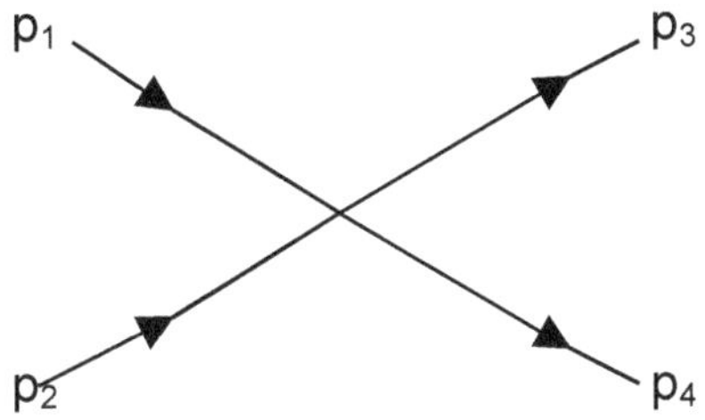

Figure 6.1. Lowest order quartic interaction diagram.

Fig. 6.1 contains the lowest order diagram for the scattering of two ϕ particles into a two ϕ particle out-state. The S matrix element for this diagram is

$$S_1 = i^4(\tfrac{1}{4!}\, i\chi_0) \prod_{j=1}^{4} \int d^4y_j\, d^4y\, f_{Zp_1}(y_1) f_{Zp_2}(y_2) f_{Zp_3}{}^*(y_3) f_{Zp_4}{}^*(y_4)(\Box_{y_1} + m^2)\cdot$$
$$\cdot(\Box_{y_2}+m^2)(\Box_{y_3}+m^2)(\Box_{y_4}+m^2)<0|T(\phi_{in}(X(y_1))\ldots\phi_{in}(X(y_4)):(\phi_{in}(X(y))^4:)|0> \qquad (6.49)$$

with $f_{Zp}(y)$ specified by

$$f_{Zp}(y) = [(2\pi)^3 2p^0 Z_p]^{-1/2}\, e^{-ip\cdot y} \qquad (6.49a)$$

where Z_p is a normalization factor that will be specified later.

Expanding the time ordered product and realizing there are 4! ways of combining the four field factors in the interaction Hamiltonian leads to:

$$S_1 = i^4(i\chi_0) \prod_{j=1}^{4} \int d^4y_j\, d^4y\, f_{Zp_1}(y_1) f_{Zp_2}(y_2) f_{Zp_3}{}^*(y_3) f_{Zp_4}{}^*(y_4)(\Box_{y_1} + m^2)\cdot$$
$$\cdot(\Box_{y_2}+m^2)(\Box_{y_3}+m^2)(\Box_{y_4}+m^2) i\Delta_F{}^{TT}(y_1-y) i\Delta_F{}^{TT}(y_2-y) i\Delta_F{}^{TT}(y_3-y) i\Delta_F{}^{TT}(y_4-y) \qquad (6.50)$$

where

$$i\Delta_F{}^{TT}(y_1 - y_2) = <0|T(\phi(X(y_1)),\phi(X(y_2)))|0> \qquad (6.51)$$

$$= i \int \frac{d^4p\, e^{-ip\cdot(y_1 - y_2)} R(\mathbf{p}, y_1 - y_2)}{(2\pi)^4 (p^2 - m^2 + i\varepsilon)} \tag{6.52}$$

with

$$R(\mathbf{p}, y_1 - y_2) = \exp[-p^i p^j \Delta_{Tij}(y_1 - y_2)/M_c^4] \tag{6.53}$$

(summations are over space indices only in the Y Coulomb gauge) and

$$\Delta_{Tij}(z) = \int d^3k\, e^{-ik\cdot z}(\delta_{ij} - k_i k_j/\mathbf{k}^2)/[(2\pi)^3 2\omega_k] \tag{6.54}$$

From chapter 4 we have:

$$R(\mathbf{p}, y_1 - y_2) = \exp\{-p^2[A(v) + B(v)\cos^2\theta] / [4\pi^2 M_c^4 z^2]\} \tag{6.55}$$

with

$$z^\mu = y_1^\mu - y_2^\mu \tag{6.56}$$

$$z = |\mathbf{z}| = |\mathbf{y_1} - \mathbf{y_2}| \tag{6.57}$$

$$p = |\mathbf{p}| \tag{6.58}$$

$$v = |z^0| / z \tag{6.59}$$

$$A(v) = (1 - v^2)^{-1} + .5v \ln[(v - 1)/(v + 1)] \tag{6.60}$$

$$B(v) = v^2(1 - v^2)^{-1} - 1.5v \ln[(v - 1)/(v + 1)] \tag{6.61}$$

$$\mathbf{p}\cdot\mathbf{z} = pz \cos\theta \tag{6.62}$$

and with $|\mathbf{p}|$ denoting the length of the spatial vector $\mathbf{p}$, while $|z^0|$ is the absolute value of z^0.

We note

$$R(\mathbf{p}, y) = R(\mathbf{p}, -y) \tag{6.62a}$$

for later use.

Letting $y_i = w_i + y$ yields

$$S_1 = i^4(i\chi_0)(2\pi)^4\,\delta^4(p_3 + p_4 - p_1 - p_2)\mathcal{N}^+(p_4)\mathcal{N}^+(p_3)\mathcal{N}(p_2)\mathcal{N}(p_1) \quad (6.63)$$

where

$$\mathcal{N}(p) = iZ_p^{-1/2}\int d^4w\, f_p(w)(\Box + m^2)\Delta_F^{TT}(w) \quad (6.64)$$

$$\mathcal{N}^+(p) = iZ_p^{-1/2}\int d^4w\, f_p{*}(w)(\Box + m^2)\Delta_F^{TT}(w) \quad (6.65)$$

are "normalizations" of the "external legs" – the in and out states due to the Y field cloud around each particle with $Z^{-1/2}$ a renormalization factor to be determined later. In the limit of low momentum ($p \ll M_C$):

$$\mathcal{N}(p) = \mathcal{N}^+(p) \to -iZ_p^{-1/2}[(2\pi)^3\, 2p^0\,]^{-1/2} \quad (6.66)$$

which the reader will note is the standard normalization factor for external scalar field legs in conventional quantum field theory modulo the $Z_p^{-1/2}$ factor. The factor $Z_p^{-1/2}$ performs the finite renormalization of external legs discussed in the preceding unitarity discussion.

Higher Order Diagram With a Loop

We will now consider the simplest one loop scattering diagrams in the scalar ϕ^4 theory.

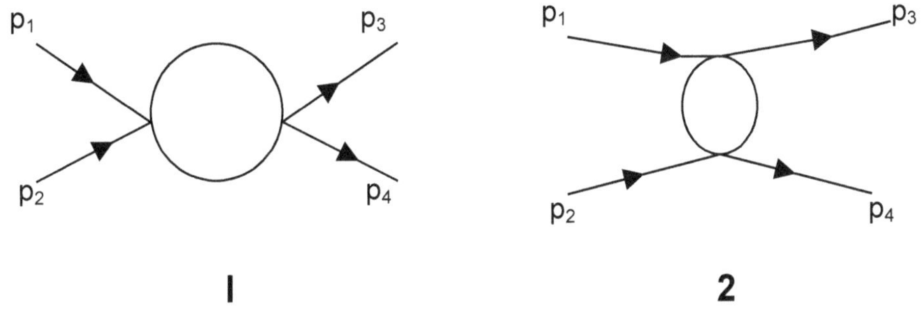

Figure 6.2. Lowest order loop scattering diagrams.

The S matrix element for these diagrams (and some other disconnected diagrams) is contained in

$$S_2 = i^4(1/4!\, i\chi_0)^2 \prod_{j=1}^{4} \int d^4y_j\, d^4y_1'\, d^4y_2'\, f_{Zp_1}(y_1)f_{Zp_2}(y_2)f_{Zp_3}{}^*(y_3)f_{Zp_4}{}^*(y_4)(\Box_{y_1} + m^2)\cdot$$

$$\cdot(\Box_{y_2}+m^2)(\Box_{y_3}+m^2)(\Box_{y_4}+m^2)<0|T(\phi_{in}(X(y_1))\ldots\phi_{in}(X(y_4))$$
$$:(\phi_{in}(X(y_1'))^4::(\phi_{in}(X(y_2'))^4:)|0>/2! \quad (6.67)$$

together with some other disconnected diagrams.

Expanding the time ordered product and keeping only the terms corresponding to Fig. 6.2 gives:

$$S_2 = i^4(i\chi_0)^2/2 \prod_{j=1}^{4} \int d^4y_j\, d^4y_1'\, d^4y_2'\, f_{ZP_1}(y_1)f_{ZP_2}(y_2)f_{ZP_3}{}^*(y_3)f_{ZP_4}{}^*(y_4)\cdot$$
$$\cdot(\Box_{y_1}+m^2)(\Box_{y_2}+m^2)(\Box_{y_3}+m^2)(\Box_{y_4}+m^2)\cdot$$
$$\cdot\{i\Delta_F{}^{TT}(y_1-y_1')i\Delta_F{}^{TT}(y_2-y_1')i\Delta_F{}^{TT}(y_3-y_2')i\Delta_F{}^{TT}(y_4-y_2') +$$
$$+ i\Delta_F{}^{TT}(y_1-y_1')i\Delta_F{}^{TT}(y_2-y_2')i\Delta_F{}^{TT}(y_3-y_1')i\Delta_F{}^{TT}(y_4-y_2')\}\cdot$$
$$\cdot i\Delta_F{}^{TT}(y_1'-y_2')i\Delta_F{}^{TT}(y_1'-y_2') \quad (6.68)$$

Following a similar procedure to the previous calculation yields

$$S_2 = i^4[(i\chi_0)^2/2](2\pi)^4\delta^4(p_3 + p_4 - p_1 - p_2)\mathcal{N}^+(p_4)\mathcal{N}^+(p_3)\mathcal{N}(p_2)\mathcal{N}(p_1)\cdot$$
$$\cdot \int d^4z\, [e^{-i(p_1+p_2)\cdot z} + e^{-i(p_1-p_3)\cdot z}]\, [i\Delta_F{}^{TT}(z)]^2 \quad (6.69)$$

revealing a similar normalization of the external legs to that found in eq. 6.63, and a momentum conserving delta function as in eq. 6.63. The loop integrals have the form:

$$I(q) = \int d^4z\, e^{-iq\cdot z}\, [i\Delta_F{}^{TT}(z)]^2 \quad (6.70)$$

The behavior of the Two-Tier Feynman propagator $\Delta_F{}^{TT}(z)$ was studied at long and short distance in eqs. 4.21-4.24. The large distance behavior of the Two-Tier Feynman propagator $\Delta_F{}^{TT}(z)$ approaches the behavior of the conventional Feynman propagator since

$$R(\mathbf{p}, z) \to 1 \quad (6.71)$$

as $z^2 = z^\mu z_\mu$ becomes much larger than M_c^{-2} ($z^2 \gg M_c^{-2}$) (eq. 6.55). Thus I(q) approaches the standard one loop expression of conventional field theory at large distance (or small momentum). Again we see that Two-Tier *quantum field theory realizes a form of*

Correspondence Principle approaching conventional quantum field theory at large distance.

At short distances the Gaussian factor R(**p**, z) dominates. The Two-Tier Feynman propagator $\Delta_F^{TT}(z)$ is radically different from the conventional Feynman propagator at very short distances (or very high momentum). The singular behavior of the conventional Feynman propagator is replaced with a well-behaved, high-energy (short distance) behavior. Near the light cone $M_c^{-2} \gg z^2 \to 0$ (or $p^2 \gg M_c^2$) we can approximate eq. 6.52 with

$$i\Delta_F^{TT}(z) \approx \int d^3p\, [N(p)]^2\, R(\mathbf{p}, z) \tag{6.72}$$

since $e^{-ip\cdot z}$ is approximately unity for small z. We assume the mass of the ϕ particle is negligible on this scale. Upon performing the integrations (see eq. 4.23 for the exact result) we find eq. 6.72 approaches:

$$i\Delta_F^{TT}(z) \to \pi\, M_c^{\,4}\, |z^2|/8$$

as $z^2 = z^\mu z_\mu \to 0$ from the space-like or time-like side of the light cone where $|\ |$ represents the absolute value.

Therefore I(q) is finite and well-behaved. At high energy ($q^2 \gg M_c^2$)

$$I(q) \sim q^{-8}$$

since the Fourier transform of $\Delta_F^{TT}(z)$ (momentum space) is

$$\Delta_F^{TT}(p) = \int d^4z\, e^{-ip\cdot z}\, \Delta_F^{TT}(z) \ \sim p^{-6}$$

for large p ($p^2 \gg M_c^2$). (Compare the preceding high energy behavior of I(q) with the conventional logarithmically divergent one loop result $I(q) \sim \ln(q^2/\Lambda^2)$ with Λ a cutoff.)

Thus Two-Tier quantum provides the benefits of a higher derivative theory without its drawbacks.

Finite Renormalization of External Particle Legs & Unitarity Example

The renormalization factor $Z_p^{-1/2}$ appearing in eqs. 6.64 and 6.65 that is due to the use of the non-unitary operator W(y) (eq. 5.6) to transform $\Phi_{in}(y)$ fields into $\phi_{in}(X(y))$ fields in the LSZ procedure in eq. 5.69, and related equations, does not preserve the norm of input and output ϕ particle legs. $Z_p^{-1/2}$ performs a finite renormalization for each external particle leg to compensate for the effects of W(y).

The required renormalization is nicely illustrated by considering the unitarity sum in the imaginary part of the preceding example.

The transition matrix T_{fi} is defined in terms of the S matrix by

$$S_{fi} = \delta_{fi} - i \, (2\pi)^4 \, \delta^4(P_f - P_i) \, T^{(+)}{}_{fi}$$

The unitarity condition is

$$T^{(+)}{}_{fi} - T^{(-)}{}_{fi} = -i \sum_n (2\pi)^4 \, \delta^4(P_n - P_i) \, T^{(-)}{}_{fn} \, T^{(+)}{}_{ni} \qquad (6.73)$$

Therefore we see that the first term on the right side of eq. 6.69 gives a transition matrix term:

$$T^{(+)}{}_{2a} = -i[\mathcal{X}_0^2/2]\mathbf{N}^+(p_4)\mathbf{N}^+(p_3)\mathbf{N}(p_2)\mathbf{N}(p_1)\int d^4z \, e^{-iP\cdot z}[i\Delta_F^{TT}(z)]^2 \qquad (6.69a)$$

where $P = p_1 + p_2$. Substituting for $i\Delta_F^{TT}$ (using eq. 4.11) we find that the imaginary part of $T^{(+)}{}_{2a}$ is given by (Note $R(\mathbf{p}, z)$ is real.)

$$T^{(+)}{}_{2a} - T^{(-)}{}_{2a} = -i[\mathcal{X}_0^2/2]\mathbf{N}^+(p_4)\mathbf{N}^+(p_3)\mathbf{N}(p_2)\mathbf{N}(p_1) \int d^4z \, e^{-iP\cdot z} \cdot$$

$$\cdot \left[i \int d^4p \, \theta(p_0) \, \delta(p^2 - m^2) e^{-ip\cdot z} \, R(\mathbf{p}, z)/(2\pi)^3 \right]^2$$

If we express the $R(\mathbf{p}, z)$ factors in terms of their Fourier transforms (see eq. 4.27):

$$R(\mathbf{p}, z) = \int d^4q \, e^{-iq\cdot z} \, R(\mathbf{p}, q)$$

Then we find

$$T^{(+)}{}_{2a} - T^{(-)}{}_{2a} = -i[\mathcal{X}_0^2/2]\mathbf{N}^+(p_4)\mathbf{N}^+(p_3)\mathbf{N}(p_2)\mathbf{N}(p_1) \int d^4z \, e^{-iP\cdot z} \cdot$$

$$\cdot \left[i \int d^4k_1 \, d^4q_1 \theta(k_1^0) \, \delta(k_1^2 - m^2) e^{-ik_1\cdot z} \, e^{-iq_1\cdot z} \, R(\mathbf{k_1}, q_1)/(2\pi)^3 \right] \cdot$$

$$\cdot \left[i \int d^4k_2 \, d^4q_2 \theta(k_2^0) \, \delta(k_2^2 - m^2) e^{-ik_2\cdot z} \, e^{-iq_2\cdot z} \, R(\mathbf{k_2}, q_2)/(2\pi)^3 \right]$$

Performing the integral over z gives

$$T^{(+)}{}_{2a} - T^{(-)}{}_{2a} = +i[\mathcal{X}_0^2/2]\mathbf{N}^+(p_4)\mathbf{N}^+(p_3)\mathbf{N}(p_2)\mathbf{N}(p_1)(2\pi)^4 \cdot$$

$$\cdot \int d^4k_1 d^4q_1 d^4k_2 d^4q_2 \theta(k_1^0) \, \delta(k_1^2 - m^2) \, \theta(k_2^0) \, \delta(k_2^2 - m^2) \cdot$$

$$\cdot R(\mathbf{k_1}, q_1)R(\mathbf{k_2}, q_2) \, \delta^4(P + k_1 + q_1 + k_2 + q_2)/(2\pi)^6$$

Introducing delta functions enables us to re-express this equation as

$$T^{(+)}{}_{2a} - T^{(-)}{}_{2a} = +i[\chi_0{}^2/2]N^+(p_4)N^+(p_3)N(p_2)N(p_1)\int d^4r_1d^4r_2(2\pi)^4\delta^4(P - r_1 - r_2)\cdot$$

$$\cdot \int d^4k_1d^4q_1\,\delta^4(r_1 + k_1 + q_1)\theta(k_1{}^0)\,\delta(k_1{}^2 - m^2)\,R(\mathbf{k_1}, q_1)\cdot$$

$$\cdot\int d^4k_2d^4q_2\theta(k_2{}^0)\,\delta(k_2{}^2 - m^2)\,\delta^4(r_2 + k_2 + q_2)R(\mathbf{k_2}, q_2)/(2\pi)^6$$

which becomes

$$T^{(+)}{}_{2a} - T^{(-)}{}_{2a} = +i[\chi_0{}^2/2]N^+(p_4)N^+(p_3)N(p_2)N(p_1)\int d^4r_1d^4r_2(2\pi)^4\delta^4(P - r_1 - r_2)\cdot$$

$$\cdot \int d^4k_1\theta(k_1{}^0)\,\delta(k_1{}^2 - m^2)\,R(\mathbf{k_1}, -k_1 - r_1)\cdot$$

$$\cdot\int d^4k_2\theta(k_2{}^0)\,\delta(k_2{}^2 - m^2)R(\mathbf{k_2}, -k_2 - r_2)/(2\pi)^6$$

$R(\mathbf{k}_2, -k_2 - r_2)$ can be expressed in terms of its Fourier transform $R(\mathbf{k}_2, z)$ using eq. 4.27.
We can now rewrite the above expression in terms of intermediate states:

$$T^{(+)}{}_{2a} - T^{(-)}{}_{2a} = -i\int d^4r_1d^4r_2\,(2\pi)^4\delta^4(P - r_1 - r_2)\cdot$$

$$\cdot\, i\chi_0N^+(p_4)N^+(p_3)\int d^4k_1\theta(k_1{}^0)\,\delta(k_1{}^2 - m^2)\,[R(\mathbf{k_1}, -k_1 - r_1)/(2\pi)^3]\cdot$$

$$\cdot\int d^4k_2\theta(k_2{}^0)\,\delta(k_2{}^2 - m^2)[R(\mathbf{k_2}, -k_2 - r_2)/(2\pi)^3]i\,\chi_0N(p_2)N(p_1)/2$$

which has the form:

$$T^{(+)}{}_{2a} - T^{(-)}{}_{2a} = -i\int d^4r_1d^4r_2(2\pi)^4\delta^4(P - r_1 - r_2)\left[\int d^4k_1\theta(k_1{}^0)\,\delta(k_1{}^2 - m^2)\,R(\mathbf{k_1}, -k_1 - r_1)/(2\pi)^3\right]\left[\int d^4k_2\theta(k_2{}^0)\,\delta(k_2{}^2 - m^2)R(\mathbf{k_2}, -k_2 - r_2)/(2\pi)^3\right]T^{(-)}{}_{fn}T^{(+)}{}_{ni}/2!$$

where

$$T^{(-)}{}_{fn} = \chi_0N^+(p_4)N^+(p_3)N(r_2)N(r_1)$$

$$T^{(+)}{}_{ni} = \chi_0N^+(r_2)N^+(r_1)N(p_2)N(p_1)$$

if

$$N^{+}(p) = N(p) = 1 \tag{6.74}$$

Eq. 6.74 implies the (finite) external leg renormalization must be

$$Z_p = -\left[\int d^4w\, f_p(w)(\Box + m^2)\Delta_F{}^{TT}(w)\right]^2 \tag{6.74a}$$

by 6.64 and 6.65. Thus all external legs must be "lopped off."

The result is a theory that satisfies the unitarity condition (eq. 6.73) as shown in the above detailed discussion.

If we define

$$\mathfrak{N}(r) = \int d^4k\, \theta(k^0)\delta(k^2 - m^2)R(\mathbf{k}, -k - r) \tag{6.75a}$$

$$= (2\pi)^{-4}\int d^4k\, d^4z\, \theta(k^0)\delta(k^2 - m^2)\, e^{-i(k + r)\cdot z}\, R(\mathbf{k}, z) \tag{6.75b}$$

then the Two-Tier completeness expression becomes:

$$S_{fi} = \sum_n (2\pi)^{-3n}(n!)^{-1}\int\left(\prod_{j=1}^{n} d^4r_j \mathfrak{N}(r_j)\right)S_{fn}S_{ni}{}^{\dagger}\,\delta^4(P_n - \sum_{k=1}^{n} r_k) \tag{6.75c}$$

This expression reflects the fact that ϕ particles are surrounded by a "cloud" of Y quanta. Thus we have achieved unitarity! For small momenta $r_j \ll M_c$, we find $\mathfrak{N}(r_j) \simeq \theta(r_j^0)\delta(r_j^2 - m^2)$ (eq. 6.75b with $R(k, q) \simeq 1$.) $\theta(r_j^0)\delta(r_j^2 - m^2)$ is the form seen in conventional quantum field theory. For large momenta $\mathfrak{N}(r_j)$ is very different.

General Form of Propagators

In this chapter we have considered a scalar Two-Tier quantum field theory. We have seen that the Two-Tier Feynman propagator is well behaved near the light cone resulting in a finite ϕ^4 theory. This finite ϕ^4 theory approximates the results of conventional ϕ^4 theory at low energy thus implementing a correspondence principle: *At low energy results in Two-Tier quantum field theory approach the corresponding results of the corresponding conventional quantum field theory.*

The observations on Two-Tier field theory made in this chapter generally apply to Two-Tier versions of Quantum Electrodynamics, ElectroWeak Theory and the Standard Model as well as Two-Tier Quantum Gravity:

1. At low energy ($p^2 \ll M_c^2$ or large distances $z^2 \gg M_c^{-2}$) the Two-Tier quantum field theory is the same as the corresponding conventional quantum field theory to good approximation. (Correspondence Principle)

2. At high energy ($p^2 \gg M_c^2$ or short distances: $z^2 \ll M_c^{-2}$) Two-Tier quantum field theories (of physical interest) are well-behaved and finite.

3. Two-Tier quantum field theories (of physical interest) satisfy unitarity and Lorentz invariance (and in the case of quantum gravity their dynamical equations satisfy the requirements of general relativity).

The generality of these results is based on:

1. The expansion of the S matrix in time ordered products of field operators.
2. Wick's Theorem
3. The general form of all particle propagators in Two-Tier quantum field theories. All particle Feynman propagators have the form:

$$iG_F^{TT}{}_{\ldots}(y_1 - y_2) = <0|T(\chi_{\ldots}(X(y_1)),\chi_{\ldots}(X(y_2)))|0> \tag{6.76}$$

$$= \int d^4p\, iG_{F\ldots}(p)e^{-ip\cdot(y_1 - y_2)} R(\mathbf{p}, y_1 - y_2) \tag{6.77}$$

where $iG_{F\ldots}(p)$ is the conventional momentum space $\chi_{\ldots}$ particle propagator, and where … represents the relevant tensor and matrix indices. $R(\mathbf{p}, y_1 - y_2)$ introduces a damping factor in each particle propagator that eliminates divergences.

Scalar Particle Propagator

The Two-Tier propagator for the case of a free scalar particle is:

$$i\Delta_F^{TT}(y_1 - y_2) = <0|T(\phi(X(y_1)),\phi(X(y_2)))|0> \tag{6.51}$$

$$= i \int \frac{d^4p\, e^{-ip\cdot(y_1 - y_2)} R(\mathbf{p}, y_1 - y_2)}{(2\pi)^4 (p^2 - m^2 + i\varepsilon)} \tag{6.52}$$

Since the mass m is not relevant at high energy we set $m = 0$. This enables us to obtain a more tractable expression for the propagator. After some manipulation the massless scalar propagator can be represented as:

$$i\Delta_F^{TT}(z) = -\beta[16\pi^3(AB)^{1/2}]^{-1} \int_{-\infty}^{\infty} dy_1 \int_{-\infty}^{\infty} dy_2 \cdot$$

$$\cdot\{\theta(z_0)\exp[-\beta((y_1 - z_0)^2B + (y_2 + z)^2A)/(4AB)] +$$

$$+ \theta(-z_0)\exp[-\beta((y_1 + z_0)^2B + (y_2 - z)^2A)/(4AB)]\}/(y_1^2 - y_2^2) \tag{6.78}$$

with $\beta = 4\pi^2M_c^4\mathbf{z}^2$. Using

$$(y_1^2 - y_2^2)^{-1} = -0.5 \int_0^\infty dq_1 \int_{-\infty}^\infty dq_2\, \theta(q_1^2 - q_2^2)\exp[iq_1y_1 - iq_2y_2] \tag{6.79}$$

we obtain the representation

$$i\Delta_F^{TT}(z^\mu) = (8\pi^2)^{-1} \int_0^\infty dq_1 \int_{-\infty}^\infty dq_2\, \theta(q_1^2 - q_2^2)\,\cdot$$

$$\cdot \exp\{iq_1|z_0| + iq_2 z - [A'q_1^2 + B'q_2^2]/[\beta'(\mathbf{z}^2 - z_0^2)]\} \tag{6.80}$$

where $|z_0|$ is the absolute value of z_0, $z^2 - z_0^2 = -z^\mu z_\mu$ and

$$A = A'/(1 - v^2) \tag{6.81}$$

$$B = B'/(1 - v^2) \tag{6.82}$$

$$\beta = 4\pi^2 M_c^4 \mathbf{z}^2 = \beta' z^2 \tag{6.83}$$

with $\mathbf{z} = |\vec{\mathbf{z}}|$ – the magnitude of the spatial vector $\vec{\mathbf{z}}$, and A and B given by eqs. 6.60 – 6.61.

The representation of $i\Delta_F^{TT}$ in eq. 6.80 is particularly useful in determining its low energy ($\ll M_c$), and its high energy ($\gg M_c$) behavior. The low energy behavior is governed by the linear terms in the exponential in eq. 6.80 since $\beta'(\mathbf{z}^2 - z_0^2)$ is very large in this limit:

$$i\Delta_F^{TT}(z^\mu)_{low} \simeq (8\pi^2)^{-1} \int_0^\infty dq_1 \int_{-\infty}^\infty dq_2\, \theta(q_1^2 - q_2^2)\exp\{iq_1|z_0| + iq_2 z\} \tag{6.84}$$

$$= [4\pi^2(\mathbf{z}^2 - z_0^2)]^{-1} \tag{6.85}$$

$$= i\Delta_F(z^\mu)$$

equaling the exact massless, free, spin 0 Feynman propagator of conventional quantum field theory.

In the high energy limit when $\beta'(\mathbf{z}^2 - z_0^2)$ is small since $\mathbf{z}^2 \approx z_0^2$ (i.e. near the light cone), the quadratic terms in the exponential in eq. 6.80 dominate and $A' \simeq B'$. We then find

$$i\Delta_F^{TT}(z^\mu)_{high} \simeq (8\pi^2)^{-1} \int_0^\infty dq_1 \int_{-\infty}^\infty dq_2 \theta(q_1^{\,2}-q_2^{\,2}) \exp\{A'(q_1^{\,2} + q_2^{\,2})/[\beta'(\mathbf{z}^2-z_0^{\,2})]\} \tag{6.86}$$

$$= \pi M_c^{\,4} \,|\, (\mathbf{z}^2 - z_0^{\,2}) \,|\, /8 \tag{6.87}$$

as in eq. 4.24. As pointed out earlier, eq. 6.87 corresponds to k^{-6} behavior in momentum space:

$$i\Delta_F^{TT}(k)_{high} \backsim k^{-6} \tag{6.87a}$$

Spin ½ Particle Propagator

For the case of a free, spin ½ particle the propagator is:

$$iS_F^{TT}(y_1 - y_2) = <0\,|\,T(\bar{\psi}(X(y_1))\psi(X(y_2)))\,|\,0> \tag{6.88}$$

$$= i \int \frac{d^4p\, e^{-ip\cdot(y_1 - y_2)} (\not{p} + m)\, R(\mathbf{p}, y_1 - y_2)}{(2\pi)^4\, (p^2 - m^2 + i\varepsilon)}$$

Again setting m = 0 we find a convenient representation in the form:

$$S_F^{TT}(z^\mu) = i(8\pi^2)^{-1} \int_0^\infty dq_1 \int_{-\infty}^\infty dq_2\, \theta(q_1^{\,2} - q_2^{\,2})(\epsilon(z_0)q_1\gamma_0 - q_2\vec{\mathbf{z}}\cdot\vec{\gamma}/z)\cdot$$

$$\cdot \exp\{iq_1|z_0| + iq_2 z - [A'q_1^{\,2} + B'q_2^{\,2}]/[\beta'(\mathbf{z}^2 - z_0^{\,2})]\} \tag{6.89}$$

using the same symbols and notation as eq. 6.80, and with $\epsilon(z_0) = +1$ if $z_0 \geq 0$ and -1 otherwise.

The representation of S_F^{TT} in eq. 6.89 is useful in determining its low energy ($\ll M_c$), and high energy ($\gg M_c$) behavior. The low energy behavior is governed by the linear terms in the exponential in eq. 6.89 since $\beta'(\mathbf{z}^2 - z_0^{\,2})$ is large in this limit:

$$S_F^{TT}(z^\mu)_{low} \simeq (8\pi^2)^{-1} \int_0^\infty dq_1 \int_{-\infty}^\infty dq_2\, \theta(q_1^{\,2} - q_2^{\,2})(\epsilon(z_0)q_1\gamma_0 - q_2\vec{\mathbf{z}}\cdot\vec{\gamma}/z)\cdot$$

$$\cdot \exp\{iq_1|z_0| + iq_2 z\} \tag{6.90}$$

$$= \not{z}[2\pi^2(\mathbf{z}^2 - z_0^{\,2})^2]^{-1} \tag{6.91}$$

$$= S_F(z^\mu)$$

equaling the exact massless, spin ½ Feynman propagator of conventional quantum field theory. If we had not set m = 0 initially, we would have obtained the usual massive, spin ½ Feynman propagator.

In the high energy limit when $\beta'(\mathbf{z}^2 - z_0^2)$ is small since $\mathbf{z}^2 \approx z_0^2$ (i.e. near the light cone), the quadratic terms in the exponential in eq. 6.89 dominate and $A' \simeq B'$. We then find

$$S_F^{TT}(z^\mu)_{high} \simeq (8\pi^2)^{-1} \int_0^\infty dq_1 \int_{-\infty}^\infty dq_2 \theta(q_1^2 - q_2^2)(\in(z_0)q_1\gamma_0 - q_2\vec{\mathbf{z}}\cdot\vec{\gamma}/z)\cdot$$

$$\cdot \exp\{A'(q_1^2 + q_2^2)/[\beta'(\mathbf{z}^2 - z_0^2)]\} \tag{6.92}$$

$$= i(8\pi^2)^{-1}\{\mathbf{z}^{-1}(\mathbf{z}^2 - z_0^2)^{3/2}2^{3/2}\pi^{7/2}M_c^6 z_0\gamma_0 - $$
$$- 4i(\mathbf{z}^2 - z_0^2)^2\pi^5 M_c^8 \vec{\mathbf{z}}\cdot\vec{\gamma})\} \tag{6.93}$$

The leading momentum dependence of the Fourier transform of $S_F^{TT}(z^\mu)_{high}$ is

$$S_F^{TT}(p)_{high} \backsim M_c^6 p^{-7}\gamma_0 \tag{6.94}$$

Massless Spin 1 Particle Propagator

The Two-Tier Feynman propagator for the case of a free, massless, spin 1, gauge field particle (coupled to a conserved current) such as a photon is:

$$iD_F^{TT}(z)_{\mu\nu} = -i \int \frac{d^4p\, e^{-ip\cdot z}\, g_{\mu\nu} R(\mathbf{p}, y_1 - y_2)}{(2\pi)^4 (p^2 + i\varepsilon)} \tag{6.95}$$

The form of eq. 6.95 is the same as the scalar particle propagator multiplied by $-g_{\mu\nu}$. As a result we have the representation:

$$iD_F^{TT}(z)_{\mu\nu} = -(8\pi^2)^{-1} \int_0^\infty dq_1 \int_{-\infty}^\infty dq_2\, \theta(q_1^2 - q_2^2)\, g_{\mu\nu}\cdot$$

$$\cdot \exp\{iq_1|z_0| + iq_2 z - [A'q_1^2 + B'q_2^2]/[\beta'(\mathbf{z}^2 - z_0^2)]\} \tag{6.96}$$

As before in the scalar particle case, the low energy behavior is governed by the linear terms in the exponential in eq. 6.96 since $\beta'(\mathbf{z}^2 - z_0^2)$ is very large in this limit:

$$iD_F^{TT}(z)_{\mu\nu low} \simeq -g_{\mu\nu}(8\pi^2)^{-1} \int_0^\infty dq_1 \int_{-\infty}^\infty dq_2\, \theta(q_1^2 - q_2^2)\exp\{iq_1|z_0| + iq_2 z\} \tag{6.97}$$

$$= -g_{\mu\nu}[4\pi^2(\mathbf{z}^2 - z_0^2)]^{-1} \tag{6.98}$$

$$= -ig_{\mu\nu}\Delta_F(z)$$

equaling the exact free, massless, spin 1 Feynman gauge field propagator of conventional quantum field theory.

In the high energy limit when $\beta'(\mathbf{z}^2 - z_0^2)$ is small since $\mathbf{z}^2 \approx z_0^2$ (i.e. near the light cone), the quadratic terms in the exponential in eq. 6.96 dominate, and $A' \simeq B'$. We then find

$$iD_F^{TT}(z)_{\mu\nu high} \simeq -(8\pi^2)^{-1} \int_0^\infty dq_1 \int_{-\infty}^\infty dq_2 \theta(q_1^2 - q_2^2) g_{\mu\nu} \exp\{ A'(q_1^2 + q_2^2)/[\beta'(\mathbf{z}^2 - z_0^2)]\} \tag{6.99}$$

$$= -g_{\mu\nu}\pi\, M_c^4\ |(\mathbf{z}^2 - z_0^2)|/8 \tag{6.100}$$

Eq. 6.100 corresponds to k^{-6} behavior in momentum space:

$$iD_F^{TT}(k)_{\mu\nu high} \backsim g_{\mu\nu} M_c^4 k^{-6} \tag{6.101}$$

Spin 2 Particle Propagator

The Two-Tier propagator for the case of a free, massless, spin 2 particle such as a graviton is:

$$i\Delta_{F2}^{TT}(z)_{\mu\nu\rho\sigma} = i \int \frac{d^4p\, e^{-ip\cdot z}\, b_{\mu\nu\rho\sigma}(p) R(\mathbf{p}, y_1 - y_2)}{(2\pi)^4\, (p^2 + i\varepsilon)} \tag{6.102}$$

in an appropriate gauge where $b_{\mu\nu\rho\sigma}(p)$ is a tensor that is independent of the coordinates. We can express eq. 6.102 in the form:

$$i\Delta_{F2}^{TT}(z)_{\mu\nu\rho\sigma} = (8\pi^2)^{-1} \int_0^\infty dq_1 \int_{-\infty}^\infty dq_2\, \theta(q_1^2 - q_2^2)\, \tilde{b}(z_0, z, q_1, q_2)_{\mu\nu\rho\sigma} \cdot$$

$$\cdot\exp\{ iq_1|z_0| + iq_2 z - [A'q_1^2 + B'q_2^2]/[\beta'(\mathbf{z}^2 - z_0^2)]\} \tag{6.103}$$

where $\tilde{b}(z_0, z, q_1, q_2)_{\mu\nu\rho\sigma}$ is a tensor generated from the $b_{\mu\nu\rho\sigma}(p)$ tensor.

As before in the scalar particle case, the low energy behavior is governed by the linear terms in the exponential in eq. 6.103 since $\beta'(\mathbf{z}^2 - z_0^2)$ is very large in this limit and we find that the covariant piece[90] behaves like:

$$i\Delta_{F2}{}^{TT}(z)_{\mu\nu\rho\sigma\text{lowCov}} \simeq \tilde{\tilde{b}}_{\mu\nu\rho\sigma}(8\pi^2)^{-1}\int_0^\infty dq_1 \int_{-\infty}^\infty dq_2\ \theta(q_1{}^2 - q_2{}^2)\exp\{\, iq_1|z_0| + iq_2 z\} \tag{6.104}$$

$$= \tilde{\tilde{b}}_{\mu\nu\rho\sigma}[4\pi^2(\mathbf{z}^2 - z_0^2)]^{-1} \tag{6.105}$$

$$= i\Delta_F(z^\mu)\ \tilde{\tilde{b}}_{\mu\nu\rho\sigma}$$

where

$$\tilde{\tilde{b}}_{\mu\nu\rho\sigma} = \tfrac{1}{2}\,[\eta_{\mu\rho}\eta_{\nu\sigma} + \eta_{\mu\sigma}\eta_{\nu\rho} - \eta_{\mu\nu}\eta_{\rho\sigma}] \tag{6.106}$$

so that the expression in eq. 6.105 equals the corresponding covariant piece of the exact free, massless, spin 2 Feynman propagator of conventional quantum field theory.

In the high energy limit when $\beta'(\mathbf{z}^2 - z_0^2)$ is small since $\mathbf{z}^2 \approx z_0^2$ (i.e. near the light cone), the quadratic terms in the exponential in eq. 6.103 dominate, and $A' \simeq B'$. We then find

$$i\Delta_{F2}{}^{TT}(z)_{\mu\nu\rho\sigma\text{high}} \simeq (8\pi^2)^{-1}\int_0^\infty dq_1 \int_{-\infty}^\infty dq_2\,\theta(q_1{}^2 - q_2{}^2)\ \tilde{b}(z_0, z, q_1, q_2)_{\mu\nu\rho\sigma}\,\cdot$$

$$\cdot\ \exp\{A'(q_1{}^2 + q_2{}^2)/[\beta'(\mathbf{z}^2 - z_0^2)]\} \tag{6.107}$$

and the covariant piece behaves like

$$i\Delta_{F2}{}^{TT}(z)_{\mu\nu\rho\sigma\text{highCov}} \simeq \tilde{\tilde{b}}_{\mu\nu\rho\sigma}\pi M_c{}^4|(\mathbf{z}^2 - z_0^2)|/8 \tag{6.108}$$

The coordinate space behavior of eq. 6.108 corresponds to k^{-6} behavior in momentum space:

$$i\Delta_{F2}{}^{TT}(k)_{\mu\nu\rho\sigma\text{highCov}} \backsim \tilde{\tilde{b}}_{\mu\nu\rho\sigma}\,k^{-6} \tag{6.109}$$

The high-energy behavior of the spin 2 propagator in momentum space results in a Two-Tier theory of quantum gravity that has no high-energy divergences and is thus finite. See chapter 9 for a detailed discussion.

---

[90] S. Weinberg, Phys. Rev. **135**, B1049 (1964); Phys. Rev. **138**, B988 (1965).

7. Two-Tier Quantum Electrodynamics

Formulation

There have been numerous attempts to develop a finite theory of Quantum Electrodynamics (QED). Among the noteworthy attempts are the Lee-Wick[91] formulation of QED and the Johnson-Baker-Willey model.[92] The unification of the electromagnetic interaction with the weak interaction in the ElectroWeak Theory, and the proof that it is renormalizable, has switched the focus of interest away from QED. However the extremely precise experimental tests of QED, which are among the most accurate measurements made by science, and the impressive agreement[93] with the theoretical predictions of QED, make QED of interest in its own right.

In this chapter we will describe the formulation of Two-Tier QED. We will see that it is finite, and yet it is in complete agreement with the highly accurate calculations of QED if M_c is sufficiently large – such as of the order of the Planck mass. We will also see a modification of the Coulomb potential in the Two-Tier model that makes possible the existence of (unstable) bound states of particles with the same charge such as a two electron bound state. The Two-Tier QED Coulomb potential is linear at ultra-short distances and zero at r =0.

A New Quantum Electrodynamics with Non-Commuting Coordinates

Two-Tier QED is formulated in a way that is similar to conventional QED and captures the excellent results of QED while making the theory finite. We will consider the case of QED for electrons. The results apply directly to any charged spin ½ field and with a few changes to charged particles of other spin. The Two-Tier QED Lagrangian that we will investigate is:

$$\mathscr{L} = J\mathscr{L}_F + \mathscr{L}_C(X^\mu(y), \partial X^\mu(y)/\partial y^\nu, y) \tag{7.1}$$

with J the Jacobian and

$$\mathscr{L}_F = \bar{\psi}(X(y))((i\nabla\!\!\!/_X - e_0 A\!\!\!/(X(y)) - m_0)\psi(X(y)) - {}^1\!/_4\, F^{\mu\nu}(X(y))F_{\mu\nu}(X(y)) \tag{7.2}$$

with

[91] T. D. Lee and G. C. Wick, Phys. Rev. **D2**, 1033 (1970); T. D. Lee and G. C. Wick, Nucl. Phys. **B9**, 209 (1969) and references therein.

[92] S. Blaha, "An Approximate Calculation of the Eigenvalue Function in Massless Quantum Electrodynamics", Phys.Rev. **D9**, 2246 (1974) and references therein.

[93] T. Kinoshita, "The Fine Structure Constant", Cornell University preprint CLNS 96/1406 (1996); V. W. Hughes and T. Kinoshita, Rev. Mod. Phys. **71**, S133 (1999).

$$F_{\mu\nu}(X(y)) = \partial A_\mu(X(y))/\partial X^\nu - \partial A_\nu(X(y))/\partial X^\mu \tag{7.3}$$

and

$$\mathscr{L}_C(X^\mu(y), \partial X^\mu(y)/\partial y^\nu, y) = -\tfrac{1}{4} F_Y^{\mu\nu} F_{Y\mu\nu} \tag{7.4}$$

$$F_{Y\mu\nu} = \partial Y_\mu/\partial y^\nu - \partial Y_\nu/\partial y^\mu \tag{7.5}$$

We note the Lagrangian $\mathscr{L}_F$ has the form of the conventional electromagnetic Lagrangian[94] except for the functional dependence on X(y).

Since the Lagrangian in eq. 7.2 is separable we will follow the same procedure as we did for the scalar field theory in the development of eqs. A.74 – A.112. Thus we obtain the Hamiltonian (as in eq. A.112 for the scalar case):

$$H_F = :\int d^3X\, (\mathscr{H}_{F0} + \mathscr{H}_{Fint}): \tag{7.6}$$

where

$$\mathscr{H}_{F0} = \bar{\psi}(X(y))(i\nabla\!\!\!/_X - m_0)\psi(X(y)) + \tfrac{1}{2}(\mathbf{E}^2 + \mathbf{B}^2) \tag{7.7}$$

$$\mathscr{H}_{Fint} = e_0\bar{\psi}(X(y))A\!\!\!/(X(y))\psi(X(y)) \tag{7.8}$$

with **E** being the electric field and **B** being the magnetic field:

$$E^i = -\partial A^i/\partial y^0$$

$$B^i = \varepsilon^{ijk}\, \partial A_j/\partial y^k$$

The field equations are:

$$(i\nabla\!\!\!/_X - e_0 A\!\!\!/(X(y)) - m_0)\psi(X(y)) = 0 \tag{7.9}$$

and

$$\partial(F^{\mu\nu}(X(y)))/\partial X^\nu = e_0\bar{\psi}(X(y))\gamma^\mu\psi(X(y)) \tag{7.10}$$

The Y Field

The X^μ coordinate field is related to the Y^μ field via:

$$X_\mu(y) = y_\mu + i\, Y_\mu(y)/M_c^{\,2} \tag{3.12}$$

The Y field has the free Lagrangian eq. 7.4 (based on eqs. 3.10 – 3.14), and the Hamiltonian

---

[94] We follow the conventions of Kaku (1993), and Bjorken (1995). It will be evident that the proof that Two-Tier QED is finite will not require detailed knowledge of specific conventions.

$$\mathscr{H}_C = \tfrac{1}{2}\,(E_Y^{\,2} + B_Y^{\,2}) \tag{7.11}$$

where

$$E_Y^{\,i} = -\partial Y^i/\partial y^0 \tag{7.12}$$

$$B_Y^{\,i} = \varepsilon^{ijk}\,\partial Y_j/\partial y^k \tag{7.13}$$

The Y field equations are

$$\partial F_Y^{\,\mu\nu}(y)/\partial y^\nu = 0 \tag{7.14}$$

The quantization of the Y field in the Coulomb gauge is described in chapter 3. We will use the Y Coulomb gauge throughout our discussions of various Two-Tier quantum field theories.

Quantization of the Free Dirac Field

The quantization procedure is formally identical to that of conventional QED. The standard equal time anti-commutation relations for the spin ½ field are:

$$\{\psi_\alpha(X), \psi_\beta(X')\} = \{\pi_{\psi\alpha}(X), \pi_{\psi\beta}(X')\} = 0 \tag{7.15}$$

$$\{\pi_{\psi\alpha}(X), \psi_\beta(X')\} = i\,\delta_{\alpha\beta}\,\delta^3(\mathbf{X} - \mathbf{X}') \tag{7.16}$$

where α and β are the spinor indices and where

$$\pi_{\psi\alpha}(X) = i\,\psi_\alpha^{\dagger}(X) \tag{7.17}$$

The spin ½ field can be expanded in a Fourier series:

$$\psi(X(y)) = \sum_{\pm s}\int d^3p\; N^d_m(p)\,[b(p,s)u(p,s)\;{:}e^{-ip\cdot(y + iY/M_c^2)}{:} + + d^\dagger(p,s)v(p,s)\;{:}e^{ip\cdot(y + iY/M_c^2)}{:}] \tag{7.18}$$

$$\psi^\dagger(X(y)) = \sum_{\pm s}\int d^3p\; N^d_m(p)\,[b^\dagger(p,s)\bar{u}(p,s)\gamma^0\;{:}e^{+ip\cdot(y + iY/M_c^2)}{:} + + d(p,s)\bar{v}(p,s)\gamma^0\;{:}e^{-ip\cdot(y + iY/M_c^2)}{:}] \tag{7.19}$$

where

$$N^d_m(p) = [m/((2\pi)^3E_p)]^{1/2} \tag{7.20}$$

and

$$E_p = (\mathbf{p}^2 + m^2)^{1/2} \tag{7.21}$$

The commutation relations of the Fourier coefficient operators are:

$$\{b(p,s), b^\dagger(p',s')\} = \delta_{ss'}\delta^3(\mathbf{p} - \mathbf{p}') \tag{7.22}$$

$$\{d(p,s), d^\dagger(p',s')\} = \delta_{ss'}\delta^3(\mathbf{p} - \mathbf{p}') \tag{7.23}$$

$$\{b(p,s), b(p',s')\} = \{d(p,s), d(p',s')\} = 0 \tag{7.24}$$

$$\{b^\dagger(p,s), b^\dagger(p',s')\} = \{d^\dagger(p,s), d^\dagger(p',s')\} = 0 \tag{7.25}$$

$$\{b(p,s), d^\dagger(p',s')\} = \{d(p,s), b^\dagger(p',s')\} = 0 \tag{7.26}$$

$$\{b^\dagger(p,s), d^\dagger(p',s')\} = \{d(p,s), b(p',s')\} = 0 \tag{7.27}$$

The spinors u(p,s) and v(p,s) are defined in the conventional way (as in Kaku (1993), and in Bjorken (1965)).

Quantization of the Electromagnetic Field

The gauge invariance of Two-Tier QED can be seen by examining the field equation

$$(i\nabla\!\!\!\!/_X - e_0 A\!\!\!/(X(y)) - m_0)\psi(X(y)) = 0 \tag{7.9}$$

and considering the effect of a gauge gauge transformation:

$$A^\mu(X(y)) \rightarrow A^\mu(X(y)) - \partial\Lambda(X(y))/\partial X_\mu \tag{7.28}$$

The field equation eq. 7.10 remains unchanged and the change in eq. 7.9 can be accommodated by a change of phase of the Dirac field:

$$\psi(X(y)) \rightarrow \exp(ie_0\Lambda(X(y))\ \psi(X(y)) \tag{7.29}$$

The only novelty is that $\Lambda(X(y))$ in general becomes a complex q-number quantity at extremely short distances since X is a complex q-number.

The gauge invariance of the Lagrangian eq. 7.2 allows us to choose a convenient gauge. It appears that the most convenient gauge is the Coulomb gauge[95]:

$$\partial A^i/\partial X^i = 0 \tag{7.30}$$

---

[95] It is also possible to quantize in Two-Tier QED using an indefinite metric that preserves manifest Lorentz covariance as was done by Gupta and Bleuler for conventional QED. See Heitler (1954) or Bogoliubov (1959).

where the sum is over spatial components labeled with i. We also set

$$A^0 = 0 \tag{7.31}$$

in the absence of an external source.
A conventional treatment leads to the equal time commutation relations:

$$[A^\mu(X(y)), A^\nu(X(y'))] = [\pi_A{}^\mu(X(y)), \pi_A{}^\nu(X(y'))] = 0 \tag{7.32}$$

$$[\pi_A{}^j(X(y)), A_k(X(y'))] = -i\,\delta^{tr}{}_{jk}(\mathbf{X}(y) - \mathbf{X}(y')) \tag{7.33}$$

(Note the locations of the j component label in eq. 7.33 introduce a minus sign.) where

$$\pi_A{}^k = \partial\mathscr{L}_F/\partial\dot{A}_k \tag{7.34}$$

$$\pi_A{}^0 = 0 \tag{7.35}$$

$$\delta^{tr}{}_{jk}(X(y)- X(y')) = \int d^3k\, e^{i\,\mathbf{k}\cdot(\mathbf{X}(y) - \mathbf{X}(y))}(\delta_{jk} - k_j k_k/\mathbf{k}^2)/(2\pi)^3 \tag{7.36}$$

$$\dot{A}_k = \partial A_k/\partial X^0 \tag{7.37}$$

The Coulomb gauge reveals the two transverse degrees of freedom that are present in the vector potential. The Fourier expansion of the vector potential is:

$$A^i(X(y)) = \int d^3k\, N_0(k) \sum_{\lambda=1}^{2}\varepsilon^i(k, \lambda)[a(k,\lambda)\, e^{-ik\cdot X(y)} + a^\dagger(k,\lambda)\, e^{ik\cdot X(y)}] \tag{7.38}$$

where

$$N_0(k) = [(2\pi)^3 2\omega_k]^{-1/2} \tag{7.39}$$

(m = 0) and

$$\omega_k = (\mathbf{k}^2)^{1/2} = k^0 \tag{7.40}$$

with $\epsilon(k, \lambda)$ being the polarization unit vectors for $\lambda = 1, 2$.

The commutation relations of the Fourier coefficient operators are:

$$[a(k,\lambda), a^\dagger(k',\lambda')] = \delta_{\lambda\lambda'}\delta^3(\mathbf{k} - \mathbf{k}') \tag{7.41}$$

$$[a^\dagger(k,\lambda), a^\dagger(k',\lambda')] = [a(k,\lambda), a(k',\lambda')] = 0 \tag{7.42}$$

and the polarization vectors satisfy

$$\sum_{\lambda=1}^{2}\varepsilon_i(\mathrm{k}, \lambda)\varepsilon_j(\mathrm{k}, \lambda) = (\delta_{ij} - k_i k_j/\mathbf{k}^2) \tag{7.43}$$

Particle propagators

The electron and photon Feynman propagators differ from the conventional QED propagators by having the Gaussian factor R(**p**, z) in their Fourier expansions:

$$iS_F^{TT}(y_1 - y_2) = <0|T(\bar{\psi}(X(y_1))\psi(X(y_2)))|0> \tag{7.44}$$

where the time ordering is with respect to y_1^0 and y_2^0. Expanding the free fields leads to the Fourier representation:

$$iS_F^{TT}(y_1 - y_2) = i\int \frac{d^4p\, e^{-ip\cdot(y_1 - y_2)}\, (\not{p} + m)\, R(\mathbf{p}, y_1 - y_2)}{(2\pi)^4\, (p^2 - m^2 + i\varepsilon)} \tag{7.45}$$

with the Gaussian factor R(**p**, z) specified in eq. 6.53. The photon propagator is

$$iD_F^{trTT}(y_1 - y_2)_{\mu\nu} = <0|T(A_\mu(X(y_1))A_\nu(X(y_2)))|0> \tag{7.46}$$

$$= -ig_{\mu\nu}\int \frac{d^4k\, e^{-ik\cdot(y_1 - y_2)}\, R(\mathbf{k}, y_1 - y_2)}{(2\pi)^4\, (k^2 + i\varepsilon)} \tag{7.47}$$

plus gauge terms and minus the Coulomb term. The presence of the Gaussian factor R(**p**, z) results in a theory of QED that has no divergences and thus is finite. See eqns. 6.94 and 6.101 for their large momentum behavior.

Coulomb Interaction

The Coulomb interaction in Two-Tier QED is different at short distances from the conventional Coulomb interaction. The Coulomb potential between singly charged (same sign) particles in Two-Tier QED is:

$$V_{TT}(y_1 - y_2) = e^2\int \frac{d^4p\, e^{-ip\cdot(y_1 - y_2)}\, R(\mathbf{p}, y_1 - y_2)}{(2\pi)^4\, \mathbf{k}^2} \tag{7.48}$$

$$= \alpha\, \Phi(M_c^{\,2}\pi r^2)\, \delta(y_1^{\,0} - y_2^{\,0})/r \qquad (7.49)$$

where α is the fine structure constant, $\Phi(z)$ is the error function, M_c is the mass setting the scale of the short distance behavior, and $r = \surd(\mathbf{y}_1 - \mathbf{y}_2)^2$ is the radial distance. At small distances ($\pi r^2 \ll M_c^{-2}$) the Two-Tier potential becomes linear in r:

$$V_{TT} \rightarrow 2\alpha r\surd\pi\, M_c^{\,2}\delta(y_1^{\,0} - y_2^{\,0}) \qquad (7.50)$$

and at large distances ($\pi r^2 \gg M_c^{-2}$) the Two-Tier potential approaches the conventional Coulomb potential:

$$V_{TT} \rightarrow V_{Coul} = \alpha\, \delta(y_1^{\,0} - y_2^{\,0})/r \qquad (7.51)$$

using the error function normalization $\Phi(\infty) = 1$. The modified Coulomb potential V_{TT} of eq. 7.49 (modulo the delta function in time) is plotted in Fig. 7.5 using $M_c = 200$ GeV/c^2 and Fig 7.6 using M_c = Planck mass. At large distances the Coulomb potential (which has been verified experimentally with great precision) can be approximated arbitrarily closely by Two-Tier QED by simply letting M_c become larger. Conceivably M_c can be as large as the Planck mass (1.221×10^{19} GeV/c^2) or even larger. Thus conventional QED is the "large" distance limit of Two-Tier QED.

The short distance behavior of the Two-Tier Coulomb potential opens the possibility of quasi-bound states of particles of the same sign such as a two electron bound state. The normally repulsive potential has a linear behavior near $r = 0$ and a potential barrier before becoming like the conventional Coulomb potential at larger distances. A pair of electrons, if localized within the linear region of the potential, would be bound but the "bound state" would quickly decay via electron tunneling through the barrier. States of this type conceivably might have existed in the first instants after the Big Bang and influenced the earliest evolution of the universe. Creating these dilepton states does not appear to be feasible if M_c is extremely large.

Asymptotic States

The development of perturbation theory in chapter 6 applies to the Two-Tier theory of QED with only superficial changes.

First we note that the form of the photon propagator has exactly the form of eq. 7.47 since terms proportional to k_μ or k_ν that would appear in the evaluation of eq. 7.46 do not contribute due to current conservation. In addition the instantaneous Coulomb interaction cancels the remaining Coulomb-like term appearing in the evaluation of eq. 7.46.

Thus we are left with the electron propagator (eq. 7.45) and the effective photon propagator (eq. 7.47). The formalism and role of the Y field is the same as in the scalar Two-Tier quantum field theory considered earlier.

In-states and Out-states

The dependence of the Lagrangian, and the particle fields in particular, on X^μ rather than directly on the coordinates y^μ leads to a "fuzziness" of the definition of asymptotic particle states that we have chosen to resolve with the construction of an asymptotic free field for each particle species of the "normal" sort. In the scalar case we defined an auxiliary field $\Phi_{in}(y)$ using the creation and annihilation operators of the free scalar field $\phi_{in}(X(y))$. In actuality $\Phi_{in}(y) \equiv \phi_{in}(y)$. The change from the argument y^μ to $X^\mu(y)$ is a form of translation that can be implemented using the (non-unitary) exponentiated momentum operator as we did in eq. 5.70:

$$\phi_a(y) \equiv \Phi_a(y) = W_a^{-1}(y)\phi_a(X(y))W_a(y)$$

for a = "in" or "out." *The benefit from this approach is a clean simple definition of asymptotic particle states of definite momentum (and spin etc.).* We will follow the same strategy in Two-Tier QED.

Fermion In-states and Out-States

In this section we will develop properties of fermion in-fields and out-fields. The LSZ procedure can be schematized as:

$$\psi_{in}(y) \Rightarrow \psi_{in}(X(y)) \Rightarrow \psi(X(y)) \Rightarrow \psi_{out}(X(y)) \Rightarrow \psi_{out}(y) \qquad (7.52)$$

In-states are constructed using $\psi_{in}(y)$ which is then transformed into $\psi_{in}(X(y))$ in order to make contact with our Lagrangian formalism. The interacting field $\psi(X(y))$ is related to $\psi_{in}(X(y))$ using the standard LSZ limiting ($y^0 \to -\infty$) process. Similarly out-states are constructed using $\psi_{out}(y)$ which is transformed into $\psi_{out}(X(y))$. Again the interacting field $\psi(X(y))$ is related to $\psi_{out}(X(y))$ as part of the familiar LSZ limiting ($y^0 \to +\infty$) process.

Since much of the development differs only trivially from the standard treatment in textbooks we will simply list relevant equations and let the reader pursue them further in quantum field theory introductory textbooks.

ψ In-Field

In order to define a perturbation theory for particle scattering we will use a free Dirac field $\psi_{in}(y)$ that satisfies

$$(i\nabla\!\!\!/_y - m)\psi_{in}(y) = 0 \qquad (7.53)$$

where

$$\nabla\!\!\!/_y = \gamma^\nu\partial/\partial y^\nu$$

Defining the Fourier expansion for the "bare" and "cloaked" fermion fields:

$$\psi_{in}(y) = \sum_{\pm s}\int d^3p\ N^d{}_m(p)\ [b_{in}(p,s)u(p,s)\ e^{-ip\cdot y} + d_{in}{}^\dagger(p,s)v(p,s)\ e^{ip\cdot y}] \quad (7.54a)$$

$$\psi_{in}(X(y)) = \sum_{\pm s}\int d^3p\ N^d{}_m(p)\ [b_{in}(p,s)u(p,s):e^{-ip\cdot X(y)}: + d_{in}{}^\dagger(p,s)v(p,s):e^{ip\cdot X(y)}:] \quad (7.54b)$$

we can define ψ_{in} in-field states with expressions like

$$|(p_ns_n), \ldots, (p_1s_1); (\bar{p}_m\bar{s}_m), \ldots (\bar{p}_1\bar{s}_1)\ in\rangle = b_{in}{}^\dagger(p_ns_n) \ldots b_{in}{}^\dagger(p_1s_1)\ d_{in}{}^\dagger(\bar{p}_m\bar{s}_m) \ldots d_{in}{}^\dagger(\bar{p}_1\bar{s}_1)|0\rangle \quad (7.55)$$

for n electrons and m positrons. The development parallels the conventional development of fermion in-states.

ψ Out-Field

Similarly we can define fermion out-states for the free field $\psi_{out}(y)$. Again we will use a free Dirac field $\psi_{out}(y)$ that satisfies

$$(i\nabla\!\!\!/_y - m)\psi_{out}(y) = 0 \quad (7.56)$$

Defining the Fourier expansions for the "bare" and "cloaked" fermion fields:

$$\psi_{out}(y) = \sum_{\pm s}\int d^3p\ N^d{}_m(p)\ [b_{out}(p,s)u(p,s)\ e^{-ip\cdot y} + d_{out}{}^\dagger(p,s)v(p,s)\ e^{ip\cdot y}] \quad (7.57a)$$

$$\psi_{out}(X(y)) = \sum_{\pm s}\int d^3p\ N^d{}_m(p)\ [b_{out}(p,s)u(p,s):e^{-ip\cdot X(y)}: + d_{out}{}^\dagger(p,s)v(p,s):e^{ip\cdot X(y)}:] \quad (7.57b)$$

we can define bare ψ_{out} out-field states with expressions like

$$|(p_ns_n), \ldots, (p_1s_1); (\bar{p}_m\bar{s}_m), \ldots (\bar{p}_1\bar{s}_1)\ out\rangle = b_{out}{}^\dagger(p_ns_n) \ldots b_{out}{}^\dagger(p_1s_1)\cdot \cdot d_{out}{}^\dagger(\bar{p}_m\bar{s}_m) \ldots d_{out}{}^\dagger(\bar{p}_1\bar{s}_1)|0\rangle \quad (7.58)$$

for n electrons and m positrons. The development again parallels the conventional development of fermion out-states.

Photon In and Out States

In this section we will develop properties of photon in-fields and out-fields. The LSZ procedure can be schematized as:

$$A_{in}{}^\mu(y) \Rightarrow A_{in}{}^\mu(X(y)) \Rightarrow A^\mu(X(y)) \Rightarrow A_{out}{}^\mu(X(y)) \Rightarrow A_{out}{}^\mu(y) \quad (7.59)$$

In-states are constructed using a “bare” field $A_{in}^{\mu}(y)$ which are then transformed into $A_{in}^{\mu}(X(y))$ in order to make contact with our Lagrangian formalism. The interacting field $A^{\mu}(X(y))$ is related to $A_{in}^{\mu}(X(y))$ using the standard LSZ limiting ($y^0 \to -\infty$) process. Similarly out-states are constructed using $A_{out}^{\mu}(y)$ which are transformed into $A_{out}^{\mu}(X(y))$. Again the interacting field $A^{\mu}(X(y))$ is related to $A_{out}^{\mu}(X(y))$ as part of the familiar LSZ limiting ($y^0 \to +\infty$) process.

The Fourier expansions of the free “bare” and “cloaked” photon in and out fields are

$$A_a^{\,i}(y) = \int d^3k\, N_0(k) \sum_{\lambda=1}^{2} \varepsilon^i(k, \lambda)[a_a(k,\lambda)\, e^{-ik\cdot y} + a_a^{\,\dagger}(k,\lambda)\, e^{ik\cdot y}] \tag{7.60a}$$

and

$$A_a^{\,i}(X(y)) = \int d^3k\, N_0(k) \sum_{\lambda=1}^{2} \varepsilon^i(k, \lambda)[a_a(k,\lambda){:}e^{-ik\cdot X(y)}{:} + a_a^{\,\dagger}(k,\lambda){:}e^{ik\cdot X(y)}{:}] \tag{7.60b}$$

where a = “in” or ‘out.”
We can define bare photon in-field states with expressions like

$$|(k_n\lambda_n), \ldots, (k_1\lambda_1)\ \text{in}\rangle = a_{in}^{\,\dagger}(k_n\lambda_n) \ldots a_{in}^{\,\dagger}(k_1\lambda_1)|0\rangle \tag{7.61}$$

for n photons. The development parallels the conventional development of photon in-states as does the definition of photon out-states.

S Matrix

The S matrix is defined in a familiar way by

$$\psi_{in}(y) = S\psi_{out}(y)S^{-1} \tag{7.62}$$

$$A^{\mu}{}_{in}(y) = SA^{\mu}{}_{out}(y)S^{-1} \tag{7.63}$$

and the other standard properties of the S matrix with the sole exception being the form of the unitarity relation (which was discussed in the previous chapter).

LSZ Reduction

In this section we will determine the reduction formula for fermions and photons for the S matrix in Two-Tier QED.

Dirac Fields

Consider a charged Dirac particle such as an electron. The S matrix element corresponding to an in-state: β plus one Dirac particle of momentum p and spin s, and an out state α can be represented by

$$S_{\alpha\beta_{ps}} = \langle\alpha\ \text{out}|\beta\ (ps)\ \text{in}\rangle \tag{7.64}$$

which becomes

$$S_{\alpha\beta_{ps}} = <\alpha - (ps)\ out|\beta\ in> + <\alpha\ out|\int d^3y\ U_{ps}(y)[\psi_{in}^{\dagger}(y) - \psi_{out}^{\dagger}(y)]|\beta\ in> \tag{7.65}$$

through standard manipulations where $<\alpha - (ps)\ out|$ is an out state with a particle of momentum p and spin s removed (if present) and where

$$U_{ps}(y) = \{m/[(2\pi)^3E_p]\}^{1/2}u(p,s)e^{-ip\cdot y} \tag{7.66a}$$

and for later use

$$V_{ps}(y) = \{m/[(2\pi)^3E_p]\}^{1/2}v(p,s)e^{ip\cdot y} \tag{7.66b}$$

Eq. 7.65 can be reexpressed as

$$S_{\alpha\beta ps}=S_{\alpha-ps\beta}+<\alpha\ out|\int d^3yU_{ps}(y)W_{QED}^{-1}[\psi_{in}^{\dagger}(X(y))-\psi_{out}^{\dagger}(X(y))]W_{QED}|\beta\ in> \tag{7.67}$$

using $W_{QED}(y) = W_{QEDin}(y)$ with

$$\psi_a(y) = W_{QEDa}^{-1}(y)\psi_a(X(y))W_{QEDa}(y) \tag{7.68a}$$

$$\psi_a^{\dagger}(y) = VW_{QEDa}^{-1}(y)\psi_a^{\dagger}(X(y))W_{QEDa}(y)V \tag{7.68b}$$

where the label a = "in" or a = "out", and where

$$W_{QEDa}(y) = \exp(-\mathbf{Y}(y)\cdot\mathbf{P}_{QEDa}/M_c^2) \tag{7.69}$$

and

$$W_{QEDa}^{-1}(y) = \exp(\mathbf{Y}(y)\cdot\mathbf{P}_{QEDa}/M_c^2) \tag{7.70}$$

in the Coulomb gauge of Y with $\mathbf{P}_{QEDa}$ being the spatial momentum vector for the free fermion and photon fields defined by

$$\mathbf{P}_{QEDa} = \sum_{\pm s} \int d^3p\ \mathbf{p}\ [b_a^{\dagger}(p,s)b_a(p,s) + d_a^{\dagger}(p,s)d_a(p,s)] + \sum_{\lambda=1}^{2} \int d^3k\ \mathbf{k}\ a_a^{\dagger}(k,\lambda)a_a(k,\lambda) \tag{7.71}$$

using the free fermion and photon creation and annihilation operators.

We note that $W_{QEDa}(y)$ is not a unitary operator – a similar situation to that of the scalar particle quantum field theory – but is pseudo-unitary:

$$W_{QEDa}^{-1}(y) = V\, W_{QEDa}^{\dagger}(y)\, V^{-1} \tag{7.72}$$

where (letting $a_Y^{\dagger}(k, \lambda)$ and $a_Y(k, \lambda)$ represent the creation and annihilation operators of the Y field) V is given by

$$V = \exp(-i\pi \sum_{\lambda=1}^{2} \int d^3k\, a_Y^{\dagger}(k, \lambda) a_Y(k, \lambda)) \tag{7.73}$$

V is a unitary operator with the property

$$V\, Y^j(y)\, V^{-1} = -Y^j(y) \tag{7.74}$$

for j = 1, 2, 3. We note (as in the scalar field discussion)

$$V^{\dagger} = V^{-1} = V \tag{7.75}$$

and thus

$$V^2 = I \tag{7.76}$$

V is a metric operator in the sense of Dirac as discussed earlier. We also note:

$$X^{\mu}(y) = V[X^{\mu}(y)]^{\dagger} V^{-1} \tag{7.77}$$

$$\psi_a^{\dagger}(X(y)) = V[\psi_a(X(y))]^{\dagger} V^{-1} \tag{7.78}$$

$$A^{\mu}{}_a(X(y)) = V[A^{\mu}{}_a(X(y))]^{\dagger} V^{-1} \tag{7.79}$$

for a = "in" or "out." These properties are required for eqs. 7.68a and 7.68b to hold.

The interacting $\psi(X(y))$ field approaches the in and out fields $\psi_{in}(X(y))$ and $\psi_{out}(X(y))$ in the limit that $y^0 \to -\infty$ and $y^0 \to +\infty$ respectively in the sense of Lehmann, Symanzik and Zimmermann which we *symbolize* as:

$$\psi(X(y)) \to \surd Z_2\, \psi_{in}(X(y)) \quad \text{as} \quad y^0 \to -\infty \tag{7.72}$$

$$\psi(X(y)) \to \surd Z_2\, \psi_{out}(X(y)) \quad \text{as} \quad y^0 \to +\infty \tag{7.73}$$

with $\surd Z_2$ the wave function renormalization constant. Using eqs. 7.72 and 7.73 and following the standard LSZ reduction procedure leads to:

$$S_{\alpha\beta ps} = S_{\alpha-ps\beta} - iZ_2^{-1/2} \int d^4y \langle \alpha \text{ out} | W_{QED}^{-1} \bar{\psi}(X(y)) W_{QED} | \beta \text{ in} \rangle (-i\overleftarrow{\nabla\!\!\!\!/}_y - m) U_{ps}(y) \tag{7.74}$$

Eq. 7.74 is similar to the usual LSZ reduction formula for a fermion extracted from an in-state except for the appearance of the W(y) operator and its inverse. We note that $W(y) = W_{in}(y)$ still because $\mathbf{P}_{QEDin}$ is independent of y^0.

The expressions for the other possible reductions of a fermion and its anti-particle are:

1. Reduction of an anti-particle from an in-state

$$iZ_2^{-1/2}\int d^4y\bar{V}_{\bar{p}\bar{s}}(y)(i\not\nabla_y - m)<\alpha\ \text{out}|W_{QED}^{-1}\psi(X(y))W_{QED}|\beta\ \text{in}> \quad (7.75)$$

2. Reduction of a particle from an out-state

$$-iZ_2^{-1/2}\int d^4y\bar{U}_{p's'}(y)(i\not\nabla_y - m)<\alpha\ \text{out}|W_{QED}^{-1}\psi(X(y))W_{QED}|\beta\ \text{in}> \quad (7.76)$$

3. Reduction of an anti-particle from an out-state

$$iZ_2^{-1/2}\int d^4y <\alpha\ \text{out}|W_{QED}^{-1}\bar{\psi}(X(y))W_{QED}|\beta\ \text{in}>\overleftarrow{(-i\not\nabla_y - m)}V_{\bar{p}'\bar{s}'}(y) \quad (7.77)$$

where

$$V_{ps}(y) = \{m/[(2\pi)^3E_p]\}^{1/2}v(p,s)e^{ip\cdot y} \quad (7.78)$$

Electromagnetic Field

The LSZ reduction of a photon from an S matrix element:

$$S_{\alpha\beta k\lambda} = <\alpha\ \text{out}|\beta\ \gamma(k\lambda)\ \text{in}> \quad (7.79)$$

begins with

$$S_{\alpha\beta k\lambda} = <\alpha - \gamma(k\lambda)\ \text{out}|\beta\ \text{in}> - iZ_3^{-1/2}\int d^3y\ A_{k\lambda}^{\mu *}(y)<\alpha\ \text{out}|\,[A_{in}^{\mu}(y) - A_{out}^{\mu}(y)]\,|\beta\ \text{in}> \quad (7.80)$$

where Z_3 is a normalization constant and

$$A_{k\lambda}^{\mu}(y) = [(2\pi)^3\omega_k]^{-1/2}e^{-ik\cdot y}\ \varepsilon^{\mu}(k, \lambda) \quad (7.81)$$

Using the LSZ symbolic notation we see

$$A^{\mu}(X(y)) \to \sqrt{Z_3}\ A_{in}^{\mu}(X(y)) \quad \text{as} \quad y^0 \to -\infty \quad (7.82)$$

$$A^{\mu}(X(y)) \to \sqrt{Z_3}\ A_{out}^{\mu}(X(y)) \quad \text{as} \quad y^0 \to \infty \quad (7.83)$$

and

$$A_a^{\mu}(y) = W_{QEDa}^{-1}(y)A_a^{\mu}(X(y))W_{QEDa}(y) \tag{7.84}$$

where a = "in" or "out". Next we arrive at the reduction expression following steps that parallel the scalar field reduction:

$$S_{\alpha\beta k\lambda} = <\alpha - \gamma(k\lambda)\ \text{out}\,|\,\beta\ \text{in}> - iZ_3^{-1/2}\int d^4y\ A_{k\lambda}^{\mu *}(y)\Box_y<\alpha\ \text{out}\,|\,W_{QED}^{-1}A^{\mu}(X(y))W_{QED}\,|\,\beta\ \text{in}> \tag{7.85}$$

Time Ordered Products and Perturbation Theory

By repeated application of the LSZ procedure outlined above, an S matrix element is reduced to the vacuum expectation value of time ordered products of fermion and photon fields.

$$<\alpha\ \text{out}\,|\,\beta\ \text{in}> = \ldots <0\,|\,T(\ldots W^{-1}(y_m)U^{-1}(y_m^{\ 0})\psi_{in}(X(y_m))U(y_m^{\ 0})W(y_m)\ldots W^{-1}(y_n)U^{-1}(y_n^{\ 0})\bar{\psi}_{in}(X(y_n))U(y_n^{\ 0})W(y_n)\ldots W^{-1}(y_p)U^{-1}(y_p^{\ 0})A_{in}^{\ \mu}(X(y_p))U(y_p^{\ 0})W(y_p)\ldots)\,|\,0>\ldots \tag{7.86}$$

Following the same development of the U matrix as described in chapter 6 with minor changes in details leads to the time ordered product for S matrix elements:

$$\tau(y_1,\ldots, y_n) = \frac{<0\,|\,T(\ldots\psi_{in}(X(y_m))\ldots\bar{\psi}_{in}(X(y_n))\ldots A_{in}^{\ \mu}(X(y_p))\ldots\exp[-i\int d^4y'\mathscr{H}_{FintQED}])\,|\,0>}{<0\,|\,T(\exp[-i\int d^4y'\mathscr{H}_{FintQED}])\,|\,0>} \tag{7.87}$$

where the QED interaction Hamiltonian is

$$\mathscr{H}_{FintQED} = e_0{:}\bar{\psi}_{in}(X(y))\not{A}_{in}(X(y))\,\psi_{in}(X(y)){:} \tag{7.88}$$

plus mass counter terms.

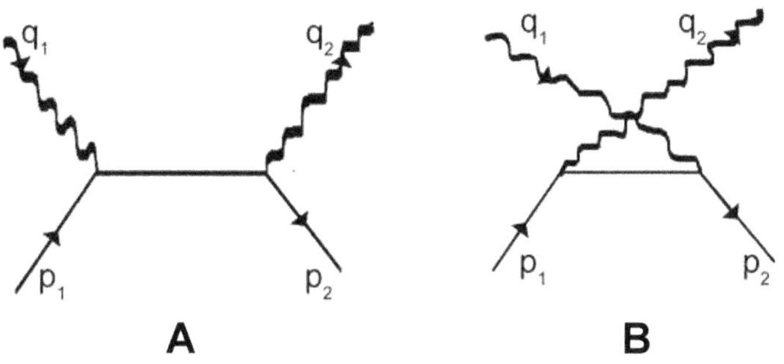

Figure 7.1. Lowest order elastic photon-electron scattering diagrams.

Example - Photon-Electron Elastic Scattering

In order to illustrate perturbative calculations in Two-Tier QED we will calculate the lowest order photon-electron elastic scattering S matrix element (Fig. 7.1).

We will see that at large distances (where the momenta are $\ll M_c$) the result is the same as the conventional QED calculation. At short distances (where the momenta are $\gg M_c$) the result differs markedly due to the effects of the Y field. The S matrix element containing the contribution of these diagrams is

$$S_{\gamma e} = (Z_2Z_3)^{-1}e_0^{\,2}\prod_{j=1}^{6}\int d^4y_j\, \bar{U}_{p_2s_2}(y_4)A_{k_1\lambda_1}^{\ \ \mu_1}(y_2)A_{k_2\lambda_2}^{\ \ \mu_2 *}(y_3)\Box_{y_2}\Box_{y_3}\cdot$$

$$\cdot(i\not\nabla_{y_4} - m)<0\,|\,T(\bar{\psi}_{in}(X(y_1))\,\psi_{in}(X(y_4))A_{in}^{\ \mu_1}(X(y_2))A_{in}^{\ \mu_2}(X(y_3))\cdot$$

$$\cdot:\bar{\psi}_{in}(X(y_5))\not{A}_{in}(X(y_5))\,\psi_{in}(X(y_5))::\bar{\psi}_{in}(X(y_6))\not{A}_{in}(X(y_6))\,\psi_{in}(X(y_6)):$$

$$)\,|\,0>(-i\overleftarrow{\not\nabla}_{y_1} - m)U_{p_1s_1}(y_1) \tag{7.89}$$

Diagram A in Fig. 7.1 corresponds to the Wick expansion:

$$S_{\gamma eA} = (Z_2Z_3)^{-1}e_0^{\,2}\prod_{j=1}^{6}\int d^4y_j\, A_{k_1\lambda_1}^{\ \ \mu_1}(y_2)A_{k_2\lambda_2}^{\ \ \mu_2 *}(y_3)\bar{U}_{p_2s_2}(y_4)\{$$

$$(i\not\nabla_{y_4}-m)iS_F^{\ TT}(y_4 - y_6)\gamma^{\nu}iS_F^{\ TT}(y_6 - y_5)\gamma^{\mu}iS_F^{\ TT}(y_5 - y_1)(-i\overleftarrow{\not\nabla}_{y_1}-m)\}$$

$$U_{p_1s_1}(y_1)\,\Box_{y_2}\Box_{y_3}\, iD_F^{\ trTT}(y_5 - y_2)_{\mu_1\mu}iD_F^{\ trTT}(y_3 - y_6)_{\mu_2\nu} \tag{7.90}$$

After some manipulation eq. 7.90 can be placed in the form:

$$S_{\gamma eA} = (Z_2Z_3)^{-1}e_0^{\ 2}\,(2\pi)^4\delta^4(p_2 + k_2 - p_1 - k_1)\bar{u}(p_2,s_2)\mathscr{S}_L^{\ TT}(p_2)\,\not{\epsilon}(k_2,\lambda_2)\cdot$$

$$\cdot\mathscr{S}^{TT}(p_1 + k_1)\,\not{\epsilon}(k_1,\lambda_1)\mathscr{S}_R^{\ TT}(p_1)u(p_1,s_1)\mathscr{D}^{\ TT}(k_1)\mathscr{D}^{TT}(k_2) \qquad (7.91)$$

where

$$\mathscr{S}_L^{\ TT}(p) = iN_{fp}\int d^4y\; e^{ip\cdot y}\,(i\nabla\!\!\!/_y - m)\; S_F^{\ TT}(y) \qquad (7.92)$$

$$\mathscr{S}^{TT}(p) = i\int d^4y\; e^{ip\cdot y}\, S_F^{\ TT}(y) \qquad (7.93)$$

$$\mathscr{S}_R^{\ TT}(p) = iN_{fp}\int d^4y\; e^{ip\cdot y}\, S_F^{\ TT}(y)\; (i\overleftarrow{\nabla\!\!\!/_y} - m) \qquad (7.94)$$

$$\mathscr{D}^{\ TT}(k) = iN_{\gamma k}\int d^4y\; e^{ik\cdot y}\,\Box_y\; D_F^{\ trTT}(y)_{00} \qquad (7.95)$$

with

$$N_{fp} = \{m/[(2\pi)^3E_p]\}^{1/2} \qquad (7.96)$$

and

$$N_{\gamma k} = [(2\pi)^3\omega_k]^{-1/2} \qquad (7.97)$$

The factors $\mathscr{S}_L^{\ TT}(p_2)$, $\mathscr{S}_R^{\ TT}(p_1)$, and $\mathscr{D}^{\ TT}(k_1)$ and $\mathscr{D}^{\ TT}(k_2)$ serve to normalize the in and out particle legs. They are a consequence of the dressing of the legs by Y particle "clouds." At long distance (low momentum) they approach the corresponding values of conventional QED:

$$\mathscr{S}_L^{\ TT}(p) \rightarrow iN_{fp} \qquad (7.98)$$

$$\mathscr{S}_R^{\ TT}(p) \rightarrow iN_{fp} \qquad (7.99)$$

$$\mathscr{D}^{TT}(k) \rightarrow iN_{\gamma k} \qquad (7.100)$$

if $p, k \ll M_c$.

Here again we must "lop off" external legs as in eqs. 6.74-6.75 to achieve a theory satisfying unitarity. We leave this as an exercise for the reader.

Deep e-p Inelastic Scattering Partons?

The form of eq. 7.93 shows the fermion propagator factor $\mathscr{S}^{TT}(p)$ is not the simple form of a free fermion propagator. Rather it consists of a fermion traveling within a "stream" of free Y quanta. This picture is reminiscent of Feynman's picture of deep inelastic e-p scattering in which the proton is viewed as a stream of partons. This question will be addressed in a future publication.

External Leg "Normalizations"

The external leg factors $\mathscr{S}_L^{TT}(p)$, $\mathscr{S}_R^{TT}(p)$ and $\mathscr{D}^{TT}(k)$ change the normalization of the external legs due to the Y particle "cloud" surrounding each particle. W*e must "lop off" external legs as in eqs. 6.74-6.75 to achieve a theory satisfying unitarity. We leave this as an exercise for the reader. In this section we examine external leg factors to see the effect of the Y quanta cloud around particles.*

In order find their form for large momenta we will first evaluate the large momentum limit of the fermion propagator as $z^2 \to 0$ (the light cone). Starting from eq. 7.45 one can show that (space-like limit)

$$iS_F^{TT}(z) \to \gamma^0 \in(z^0) M_c^3 [-M_c^2 \pi z^2/2]^{3/2} + \mathscr{O}(z^5) \tag{7.101}$$

where $z^2 = z_0^2 - \mathbf{z}^2$. Therefore on dimensional grounds we see that

$$\mathscr{S}^{TT}(p) \backsim \gamma^0 M_c^6 p^{-7} + \mathscr{O}(p^{-9}) \tag{7.102}$$

as p gets very large ($p \gg M_c$). At low momenta ($p \ll M_c$) the standard momentum space form of the fermion propagator is found:

$$\mathscr{S}^{TT}(p) \backsim (\not{p} - m + i\epsilon)^{-1} \tag{7.103}$$

Similarly we find

$$\mathscr{S}_L^{TT}(p) \backsim M_c^6 p^{-6} + \mathscr{O}(p^{-8}) \tag{7.104}$$

$$\mathscr{S}_R^{TT}(p) \backsim M_c^6 p^{-6} + \mathscr{O}(p^{-8}) \tag{7.105}$$

as p gets very large ($p \gg M_c$), and

$$\mathscr{D}^{TT}(k) \backsim M_c^4 k^{-4} \tag{7.106}$$

from eqs. 7.47 and 4.24 as k gets very large ($k \gg M_c$).

For small momenta compared to M_c we find the usual normalization:

$$\mathscr{S}_L^{TT}(p) = \mathscr{S}_R^{TT}(p) = iN_{fp} \tag{7.107}$$

$$\mathscr{D}^{TT}(k) = iN_{\gamma k} \tag{7.108}$$

Thus for low momenta (p, k $\ll M_c$) $S_{\gamma eA}$ yields the standard result of QED while at large momenta (p, k $\gg M_c$) we see a high power of inverse momentum showing the well behaved nature of the theory at short distances.

Renormalization of Two-Tier QED

Two-Tier QED is a finite quantum field theory satisfying the unitarity condition. The degree of divergence of a Feynman diagram term in *conventional QED* is

$$D = 4k - 2b - f \tag{7.109}$$

where

k = the number of internal momentum integrations
b = the number of internal photon lines
f = the number of internal electron lines

Many diagrams are thus divergent in conventional QED and a renormalization program must be followed to achieve a theory with all divergences formally absorbed into renormalizations of the fundamental parameters of the theory. Despite the success of this approach and the excellent agreement of QED with experiment the presence of divergences in QED is logically unsatisfactory and suggests QED, and its successors ElectroWeak Theory and the Standard Model, are at best interim theories. Numerous attempts have been made to modify QED in order to eliminate its divergences. Some noteworthy attempts include the Lee-Wick formulation of QED and the Johnson-Baker-Willey model of QED. None of these attempts have succeeded for one reason or another.[96]

The degree of divergence formula is different in Two-Tier QED. It demonstrates there are no divergences in Two-Tier QED. Thus it would be more aptly named the "degree of convergence." The formula is:

$$D^{TT} = 4k - 6b - 7f \tag{7.110}$$

with k, b and f as above. The coefficient of b is 6 in Two-Tier QED because the Two-Tier photon propagator behaves as k^{-6} at high momentum (See eq. 7.47 for the photon propagator. Eq. 4.25 shows its high momentum behavior). The coefficient of f is 7 in

---

[96] Lee, T. D. and Wick, G. C., Phys. Rev. **D2**, 1033 (1970); T. D. Lee and G. C. Wick, Nucl. Phys. **B9**, 209 (1969) and references therein; M. Baker and K. Johnson, Phys. Rev. **D3**, 2516 (1971); M. Baker and K. Johnson, Phys. Rev. **D8**, 1110 (1973); K. Johnson, M. Baker, and R. Willey, Phys. Rev. **136**, B1111 (1964); K. Johnson, R. Willey, and M. Baker, Phys. Rev. **163**, 1699 (1967); S. Blaha,, Phys. Rev. **D9**, 2246 (1974) and references therein; S. Adler, Phys. Rev. **D5**, 3021 (1972).

Two-Tier QED because the Two-Tier fermion propagator behaves as k^{-7} at high momentum (see eq. 7.102).

We note the degree of Divergence in Two-Tier QED is always negative – indicating a finite theory!

$$D^{TT} < 0 \tag{7.111}$$

For example the degree of divergence in Two-Tier QED of the lowest order in α (see Fig. 7.2): i) vacuum polarization diagram is $D^{TT} = -10$, ii) fermion self-energy is $D^{TT} = -9$ and iii) electromagnetic vertex correction is $D^{TT} = -16$.

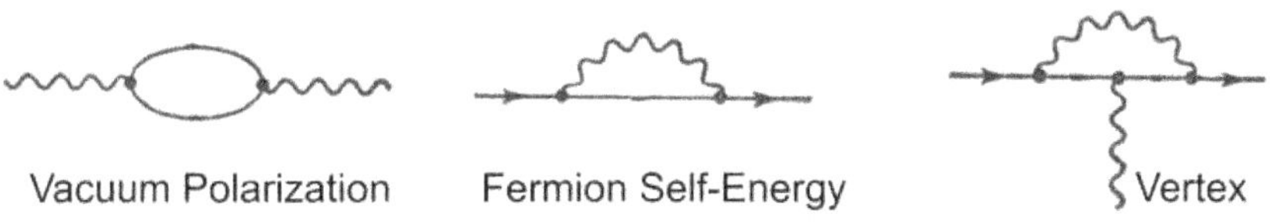

Figure 7.2 Some low order (normally divergent) diagrams for the vacuum polarization, fermion self-energy and electromagnetic vertex correction.

It is easy to see that all Two-Tier QED Feynman diagrams are ultra-violet finite.

Unitarity of Two-Tier QED

The remaining major issue is unitarity. Two-Tier QED satisfies the unitarity condition between physical states. The argument demonstrating unitarity parallels the discussion of unitarity for scalar ϕ^4 quantum field theory in chapter 6.

Two-Tier QED *superficially* appears to have a unitarity problem due to the non-Hermitian nature of its Hamiltonian. The lack of hermiticity is entirely due to the appearance of iY^μ in the X^μ field coordinates. Thus the interaction Hamiltonian in eq. 7.88 is not Hermitian:

$$H_{FintQED} = \int d^3y' \mathscr{H}_{FintQED}(\bar{\psi}_{in}(y' + iY(y')/M_c^2), A_{in}^{\mu}(y' + iY(y')/M_c^2), \psi_{in}(y' + iY(y')/M_c^2)) \tag{7.112}$$

and

$$H_{FintQED} \neq H_{FintQED}^{\dagger} = \int d^3y' \mathscr{H}_{FintQED}(\bar{\psi}_{in}(y' - iY(y')/M_c^2), A_{in}^{\mu}(y' - iY(y')/M_c^2), \psi_{in}(y' - iY(y')/M_c^2)) \tag{7.113}$$

The metric operator (eq. 7.73) establishes the relation between $H_{FintQED}$ and its Hermitian conjugate is

$$H_{FintQED} = V\, H_{FintQED}^{\dagger}\, V \tag{7.114}$$

Thus the Two-Tier QED S matrix is not unitary – S is pseudo-unitary in general:

$$S^{-1} = V\, S^{\dagger}\, V \tag{7.115}$$

which implies

$$VS^{\dagger}VS = I \tag{7.116}$$

The Two-Tier *QED S matrix satisfies unitarity between physical asymptotic states* which in Two-Tier QED are states consisting of charged fermions, and photons. The proof is identical to eqs. 6.46 – 6.48.

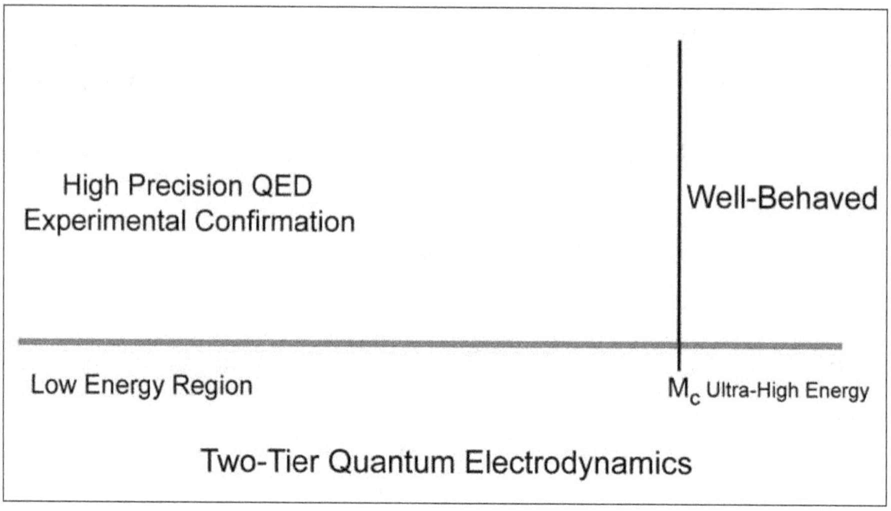

Figure 7.3. A Schematic View of Two-Tier QED.

Correspondence Principle for Two-Tier QED

The Two-Tier QED theory that we have developed has the behavior of conventional QED at "low energies." Since M_c sets the scale at which deviations from the normal QED results become significant we can set M_c to an extremely high value such as the Planck mass and obtain as close an agreement with conventional QED as we wish. Thus Two-Tier QED implements a type of Correspondence Principle with conventional QED as its "low energy" limit.

Consistency with Precision QED Experiments

Since we can obtain results as close to conventional QED with a sufficiently high choice of M_c, Two-Tier QED has the same excellent agreement[97] with experiment as conventional QED.

[97] See, for example, T. Kinoshita, "The Fine Structure Constant", Cornell University preprint CLNS 96/1406 (1996); V. W. Hughes and T. Kinoshita, Rev. Mod. Phys. **71**, S133 (1999) and references therein.

Calculation Rules for Feynman Diagrams

This section describes the procedure for the calculation of S matrix terms corresponding Feynman diagrams in Two-Tier QED and other quantum field theories. By design the approach parallels the perturbative approaches of conventional QED and other quantum field theories. The examples considered earlier in this chapter, and the preceding chapter, illustrate Two-Tier perturbation theory.

Procedure: Follow the conventional procedure to form the expression for each Feynman diagram in the perturbative calculation. Then replace each propagator in each expression with the corresponding Two-Tier propagator as specified below. Then evaluate the expression.

Feynman Propagators and their Two-Tier Propagator Equivalents

Two-Tier propagators are denoted with the superscript "TT".

Spin 0 Propagator Case

$$\Delta_F(p) = (p^2 - m^2 + i\varepsilon)^{-1} \qquad \rightarrow \qquad \int d^4z\, e^{+ip\cdot z}\, \Delta_F^{TT}(z) \qquad (7.120)$$

where $\Delta_F^{TT}(z)$ is given by eq. 4.9.

Spin 1 Photon Propagator Case

$$D_F(p)_{\mu\nu} = -g_{\mu\nu}(p^2 + i\varepsilon)^{-1} \qquad \rightarrow \qquad \int d^4z\, e^{+ip\cdot z}\, D_F^{TT}(z)_{\mu\nu} \qquad (7.121)$$

where $D_F^{TT}(z)_{\mu\nu}$ is given by eq. 7.47.

Spin ½ Fermion Propagator Case

$$S_F(p) = (\not{p} - m + i\varepsilon)^{-1} \qquad \rightarrow \qquad \int d^4z\, e^{+ip\cdot z}\, S_F^{TT}(z) \qquad (7.122)$$

where $S_F^{TT}(z)$ is given by eq. 6.88.

Spin 2 Massless Boson (graviton) Propagator Case

$$\Delta_{F2}(p)_{\mu\nu\rho\sigma} = b_{\mu\nu\rho\sigma}(p^2 - m^2 + i\varepsilon)^{-1} \quad \rightarrow \quad \int d^4z\, e^{+ip\cdot z}\, \Delta_{F2}^{TT}(z)_{\mu\nu\rho\sigma} \qquad (7.123)$$

where $\Delta_{F2}^{TT}(z)_{\mu\nu\rho\sigma}$ the graviton propagator.

Two-Tier Coulomb Potential vs. Conventional Coulomb Potential

The familiar Coulomb potential is (for two particles of opposite unit electric charge):

$$V_{Coulomb} = -\,a/\,|\mathbf{r}| \qquad (7.124)$$

The Two-Tier QED Coulomb potential (eq. 7.49) is:

$$V_{\text{Two-TierCoul}} = -\,a\Phi(M_c^{\,2}\pi\,|\mathbf{r}|^2)/\,|\mathbf{r}| \qquad (7.125)$$

where $\Phi(x)$ is the error function.[98] At small distances ($\pi r^2 \ll M_c^{-2}$)

$$V_{\text{Two-TierCoul}} \rightarrow -\,2a\surd\pi\; M_c^{\,2}\,|\mathbf{r}| \qquad (7.126)$$

a linear potential, and at large distances ($\pi r^2 \gg M_c^{-2}$)

$$V_{\text{Two-TierCoul}} \rightarrow V_{\text{Coulomb}} \qquad (7.127)$$

The Two-Tier Coulomb potential has a minimum at

$$M_c^{\,2}\pi\,|\mathbf{r}|^2 = 1 \qquad (7.128)$$

[98] W. Magnus and F. Oberhettinger, *Formulas and Theorems for the Special Functions of Mathematical Physics* (Chelsea Publishing Co., New York, 1949) page 96.

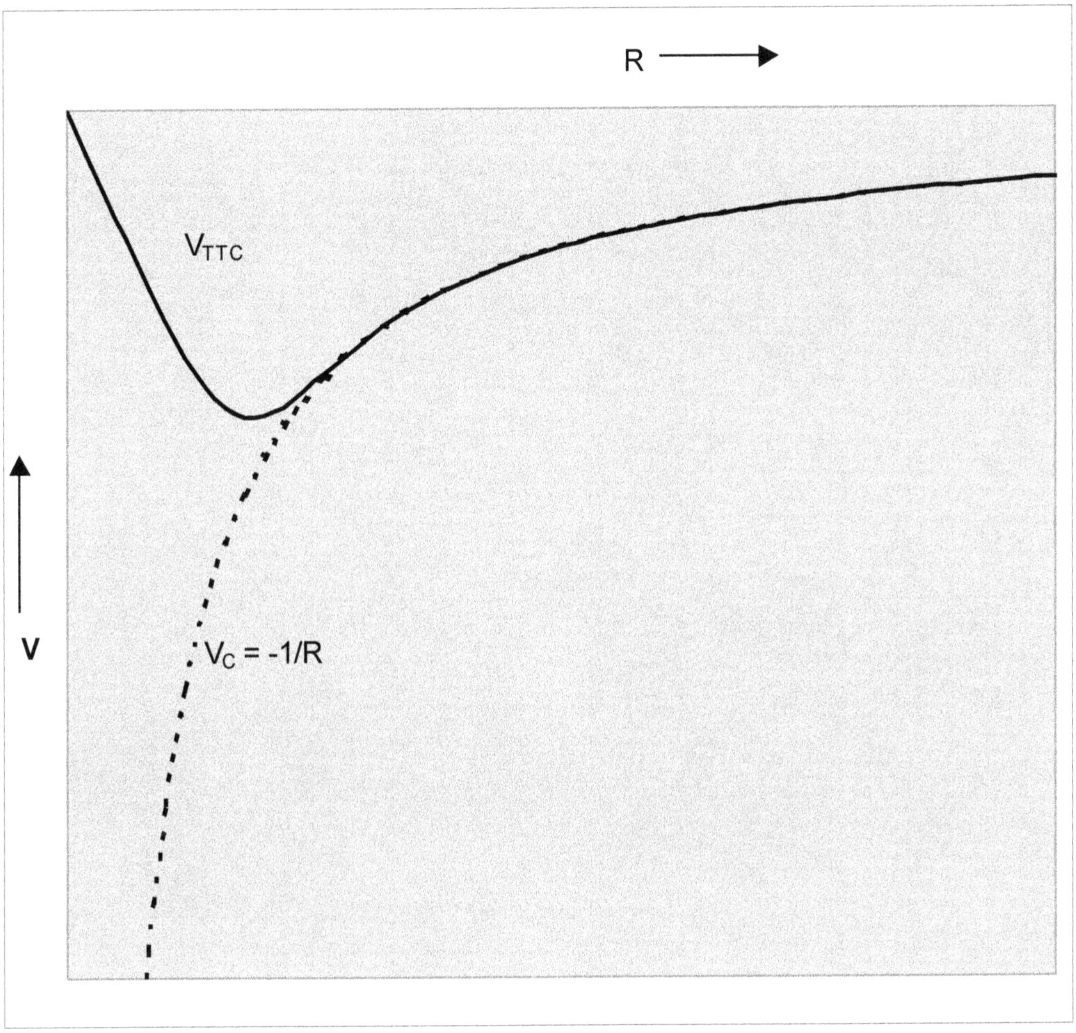

Figure 7.4. Plot of the form of the Two-Tier Coulomb "attractive" potential between particles of opposite unit charge divided by αM_c: $V_{TTC} = V_{Two\text{-}TierCoul}/(\alpha M_c)$. V_{TTC} is dimensionless. The dotted line is the conventional Coulomb attractive potential divided by αM_c: $V_c = V_{Coulomb}/(\alpha M_c) = 1/R$. Note the Two-Tier Coulomb force between particles of opposite charge is repulsive at short distance.

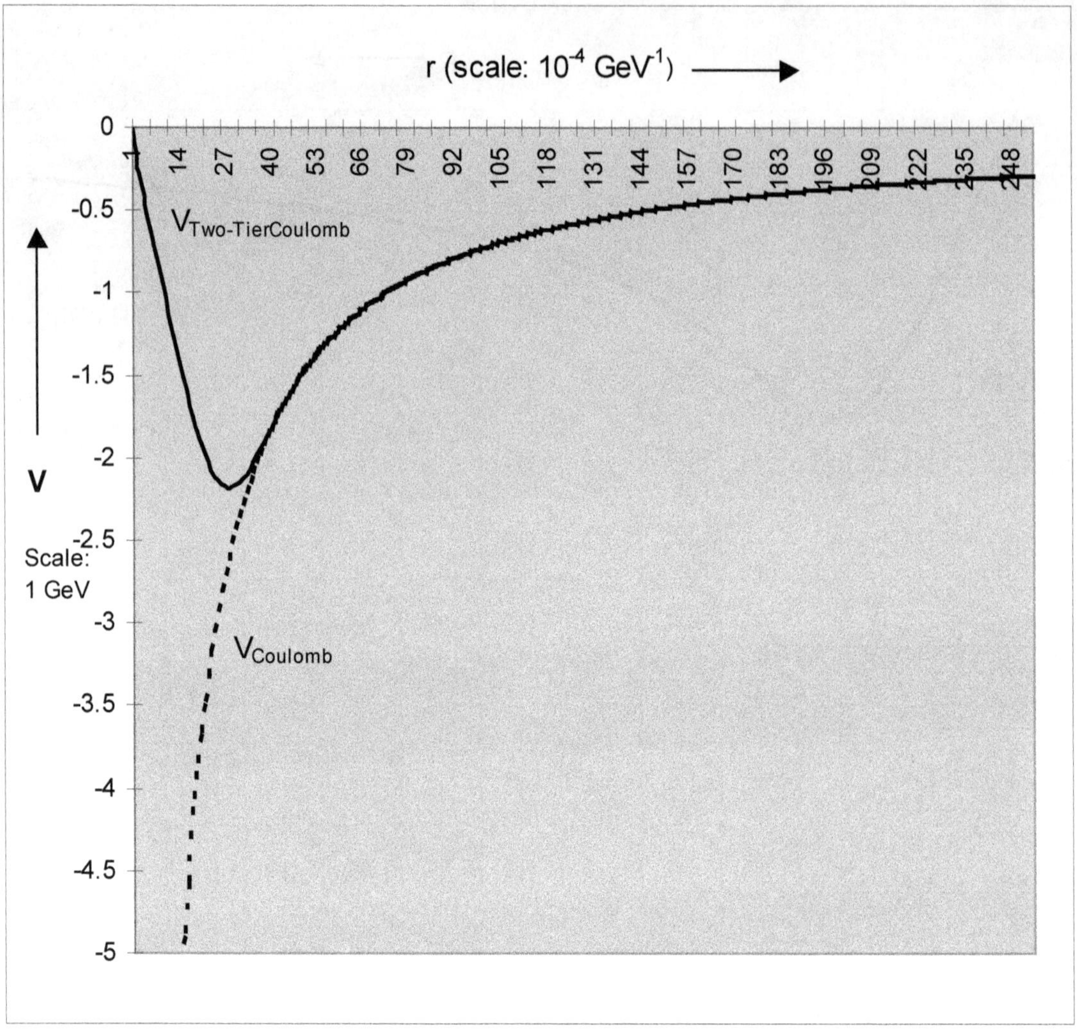

Figure 7.5. Two-Tier Coulomb Potential compared to conventional Coulomb potential for M_c = 200 GeV/c^2. Radial distance is measured in units of 10^{-4} GeV^{-1}. The potential energy for two opposite unit charges is measured in GeV units.

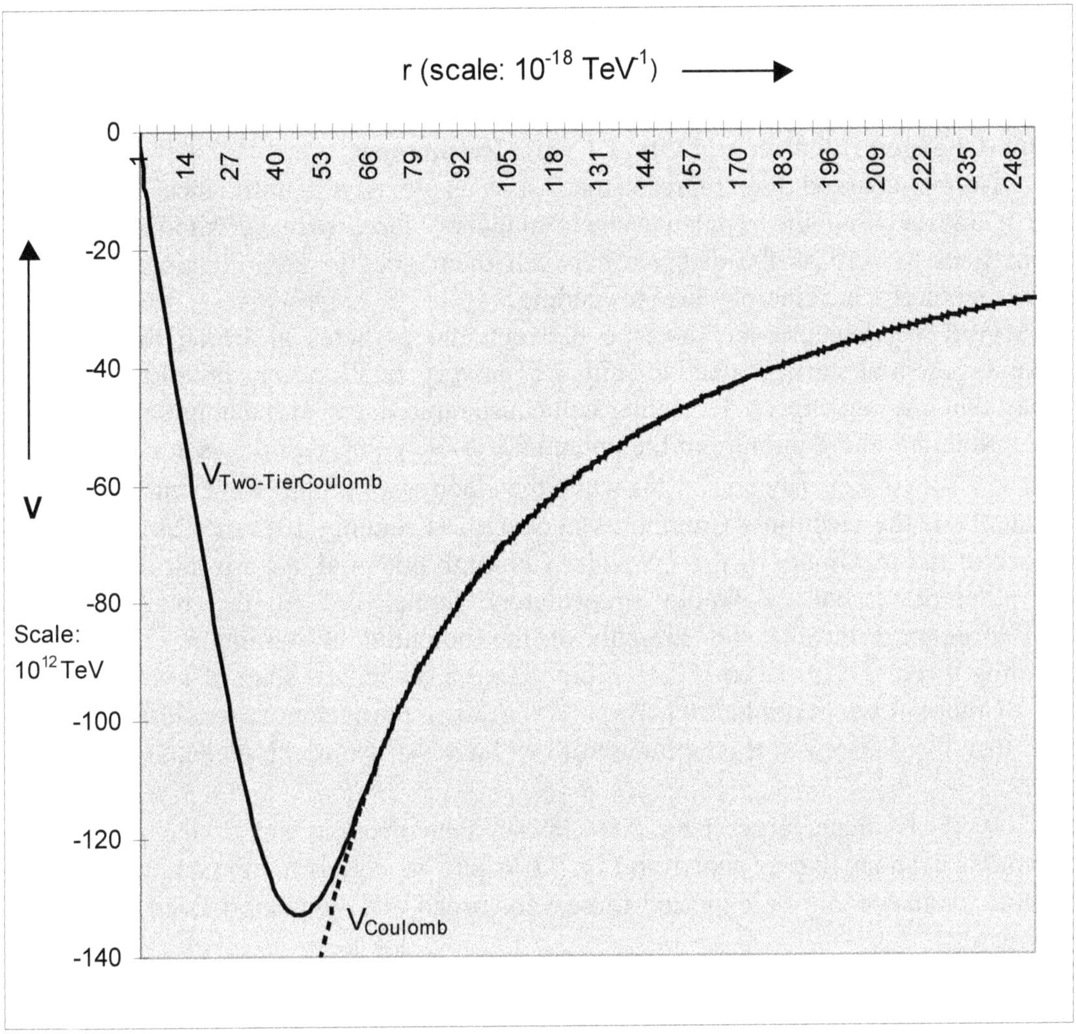

Figure 7.6. Two-Tier Coulomb Potential compared to conventional Coulomb potential for $M_c = M_{planck} = 1.22 \times 10^{19}$ GeV/c^2. Radial distance is in units of 10^{-18} TeV^{-1}. The potential energy for two opposite unit charges is measured in units of 10^{12} TeV.

At the minimum $V_{\text{Two-TierCoul}}$ has the value:

$$V_{\text{Two-TierCoulMIN}} = -\,.8427\alpha\sqrt{\pi}\ M_c \qquad (7.129)$$

If we define distance in terms of the mass scale

$$|\mathbf{r}| = R/M_c \qquad (7.130)$$

then

$$V_{\text{Two-TierCoul}} = -\,\alpha M_c \Phi(\pi R^2)/R \qquad (7.131)$$

Fig. 7.4 displays a plot of $V_{TTC} = V_{Two\text{-}TierCoul}/(\alpha M_c)$ showing the general form of the Two-Tier Coulomb potential for particles of opposite unit charge. We also plot $V_C = V_{Coulomb}/(\alpha M_c) = 1/R$.

Doubly-Charged Dilepton and Other Exotic Resonances

Doubly charged dilepton resonances such as e^-e^- or $\mu^-\mu^-$ resonances are possible since it appears that the repulsive electromagnetic force between similarly charged leptons goes to zero as the distance between them goes to zero creating a potential barrier that might, in principle, lead to binding.

However because the distance between the particles at which the repulsive Coulomb potential starts to decline (Fig. 7.5) is very small, a very broad resonance is the best that one can expect. Tunneling will cause rapid decay of the bound state.

Note that the minimum of the potential, for $M_c = 200$ GeV/c^2, is at $r = 2.8 \times 10^{-5}$ GeV^{-1} $\equiv 5.5 \times 10^{-19}$ cm (by eq. 7.128) while the electron Compton wavelength is 3.861×10^{-13} m. Thus the electron's Compton wavelength is roughly 10^6 times larger than the distance of the maximum of the Two-Tier Coulomb potential. As a result the electrons in a dilepton resonance would immediately tunnel out of the binding region (*zitterbewegung*!) despite the strength of the potential at minimum: 21.7 TeV/c^2 according to eq. 7.129. Even if one could create a resonance state of two τ^- leptons with τ Compton wavelengths of 1.135×10^{-14} cm the τ wavelength would be 10^5 times larger than the distance of the minimum of the Two-Tier Coulomb potential if $M_c = 200$ GeV/c^2.

If M_c is much larger than 200 GeV/c^2 then the location of the peak of the potential is even smaller as shown in Fig. 7.6 where M_c equals the Planck mass. Thus a dilepton resonance can be expected to be very broad and very short-lived if indeed it exists at all.

Other exotic resonances might be possible involving the strong force which also interchanges repulsion and attraction at very short distances.

8. Two-Tier Standard Model Theory

Introduction

The Standard Model is a unified model of the electromagnetic, weak and strong interactions. It is formed from ElectroWeak Theory and an SU(3) color gauge theory of the strong interaction called Quantum Chromodynamics (QCD). Each part is separately renormalizable, as is the combined theory. This ad hoc theory of all known interactions except gravitation has been remarkably successful in accounting for the experimental data at energies currently accessible at accelerators. There are a number of phenomena that are not understood within the framework of the Standard Model such as dark matter and dark energy, some CP violation phenomena, and the spin dependence observed in certain experiments. There is also an appreciation of the ad hoc nature of the unification of the strong and ElectroWeak sectors. They are "just glued together". An overall symmetry encompassing both sectors (broken as it would be) would be more reassuring and more elegant. However the theory works, and appears to work well, for most particle physics phenomena.

Superstring theory has become a seriously considered alternative to the Standard Model. It appears to offer a finite theoretical framework without divergences. It offers the hope of a rational grand unification. It is mathematically interesting and elegant. However, it has almost no experimental support. It predicts many particles that remain to be found. It requires additional dimensions that are not evidenced. It does not have a demonstrable, satisfactory "low energy limit" that resembles the Standard Model although ad hoc Superstring-like Standard Model variants have been studied. It has the problem of too many possibilities. After thirty years of effort one cannot point to a Superstring theory and say it is THE Superstring theory, and here is the method of symmetry-breaking that results in the Standard Model. Considering that far more theoretical effort has been expended on Superstring theory by more theorists then on any other theory of physics, its physical justification, its nature, and its contact with experiment is still not well understood.

In the hope of developing a sound theory that clearly makes contact with known experimental data, that has a direct connection to the Standard Model, that is finite (no divergences), and that can be directly extended to encompass gravitation, we have developed a Two-Tier formulation of the Standard Model and its variants. It is admittedly far more modest in its scope than Superstring theory. Yet it allows us to have a finite theory – a major goal of Superstring theory – and to address cosmological issues (Blaha(2004)) by its easy unification with Two-Tier quantum gravity (next chapter) – also a finite theory without divergences.

Thus we have a theory that accounts for all known interactions and yet incrementally extends known successful theories – in a significant way. With (perhaps) the exception of General Relativity, the growth of particle physics theory in the twentieth-century has been largely incremental.

The nature of the mechanism that Two-Tier quantum field theory uses to avoid divergences has the important feature that it does not depend on "magic cancellations", does not depend on details of the symmetries, and does not depend on details of the form of the theories. Thus all the variations and extensions of the Standard Model have an equivalent Two-Tier version that is finite. This feature gives the theorist the flexibility to modify and extend the Two-Tier Standard Model without worrying about the introduction of new divergences.

The other major advantage is that the theory is very much like the world we see at current energies – four dimensions and no large number of new particles. It leaves open the door for new interactions and new particles IF they are found at higher energies. Thus it is not a showstopper. Nor is it a prediction of a vast desert extending to the highest energies. It gives a finite, unified theory with an "Open Door" policy for the future.

Formulation of Two-Tier Standard Models

There are many excellent texts[99] on the Standard Model. Therefore this author sees no reason to recapitulate the features of the Standard Model. We will focus on the form and features of a generic Two-Tier version of the Standard Model.

We will assume that the form of the Standard Model Lagrangian is:

$$\mathscr{L}_{\rm SM} = {\rm J}\mathscr{L}_{\rm F}^{\rm SM}({\rm X}^\mu) + \mathscr{L}_{\rm C}({\rm X}^\mu({\rm y}), \partial {\rm X}^\mu({\rm y})/\partial {\rm y}^\nu) \tag{8.1}$$

where $\mathscr{L}_{\rm F}^{\rm SM}$ is the complete "normal" quantum field theory Lagrangian for the Standard Model variant under consideration and J is a Jacobian (eq. A.21). *All particle fields in* $\mathscr{L}_{\rm F}^{SM}$ *are assumed to be functions of the* X^μ *coordinate only. The dependence of the particle fields on the "underlying" coordinates* y^μ *is assumed to be solely through* X^μ. The Lagrangian $\mathscr{L}_{\rm SM}$ is a separable Lagrangian of the type of eq. A.26 embodying the composition of extrema described in Appendix A. Scalar particle examples of separable Lagrangians appear in eqs. A.96, 3.10 and 3.15. A scalar particle ϕ^4 theory with a separable Lagrangian is specified by eqs. 5.20 – 5.24 and decribed in detail in chapters 5 and 6. The separable Lagrangian Two-Tier version of QED is specified by eqs. 7.1 – 7.5 and described in some detail in chapter 7.

In all of these cases the coordinate part of the Lagrangian $\mathscr{L}_{\rm C}$ as

[99] Some of the many excellent texts: P. Ramond, *Journeys Beyond the Standard Model* (Westview Press, New York, 1999); W. N. Cottingham and D. A. Greenwood, *An Introduction to the Standard Model of Particle Physics* (Cambridge University Press, New York, 1999); K. Huang, *Quarks, Leptons and Gauge Fields* (World Scientific, River Edge, NJ 1992); T. P, Cheng and L. F. Li, *Gauge Theories of the Elementary Particles* (Oxford University Press, New York, 1988); J. Donoghue, E. Golowich and B. Holstein, *Dynamics of the Standard Model* (Cambridge University Press, New York, 1994).

$$\mathscr{L}_C = -\tfrac{1}{4}\, F_Y^{\mu\nu} F_{Y\mu\nu} \tag{3.15}$$

with

$$F_{Y\mu\nu} = \partial Y_\mu / \partial y^\nu - \partial Y_\nu / \partial y^\mu \tag{3.14}$$

The field equations of the theory are found using the conventional approach. The Hamiltonian, energy-momentum tensor and conserved quantities are also found using a conventional approach as illustrated in chapters 2, 3 and 4. The canonical quantization procedure is also followed. Thus the application of these procedures to the Standard Model will yield the standard results with the sole difference being that all fields are functions of X^μ and all derivatives of fields are derivatives with respect to X^μ. A perturbation theory along the lines of chapters 5 and 6 can then be directly developed.

The net result is that the free field Feynman propagators in a Feynman diagram approach to perturbative calculations are each replaced with the corresponding Two-Tier propagator. All coordinate space integrations are over the y coordinates.

Rule 1: *If the conventional coordinate space Feynman propagator for a free particle has the form:*

$$G\ldots(z) = \int d^4p\, e^{-ip\cdot z}\, G\ldots(p)/(2\pi)^4 \tag{8.2}$$

where ... represents any space-time indices, spin indices and internal symmetry indices, then the equivalent coordinate space (y – coordinates) Two-Tier propagator is

$$G\ldots^{TT}(z) = \int d^4p\, e^{-ip\cdot z}\, G\ldots(p) R(p, z)/(2\pi)^4 \tag{8.3}$$

where R(p, z) is specified in eqs. 4.12 – 4.20.

The vertex expressions remain identically the same as in the conventional Standard Model. All integrations are done in the y-coordinate space as shown in the perturbation discussion leading to eq. 6.37 and the perturbation theory calculations at the end of chapter 6 and in chapter 7.

Procedure to Construct the Two-Tier Equivalent of a Conventional Feynman Diagram Expression

After we form the expression for a Feynman diagram in a conventional Standard Model variant, we can construct the Two-Tier equivalent by taking *every* particle or ghost propagator factor of the form:

$$G\ldots(p)$$

in momentum space, and replacing it with

$$G\ldots^{TT}(p) = \int d^4z\, e^{ip\cdot z}\, G\ldots^{TT}(z)$$
$$= \int d^4z\, e^{ip\cdot z} \int d^4p'\, e^{-ip'\cdot z}\, G\ldots(p')R(p', z)/(2\pi)^4$$

The resulting expression is the Two-Tier expression for the Feynman diagram.

"Low Energy" Behavior of the Two-Tier Standard Model

Since $R(p, z) \cong 1$ for $p \ll M_c$ the Two-Tier Standard Model is virtually identical to the conventional, corresponding Standard Model variant for energies much less than M_c. Since M_c can be very large – perhaps equal to the Planck mass or larger, the Two-Tier Standard Model predictions at current energies can be made arbitrarily close to the "low energy" results of the corresponding Standard Model. The Standard Model is a limiting case of the Two-Tier Standard Model – thus implementing a *Correspondence Principle*.

"High Energy" Behavior of the Two-Tier Standard Model

At high energies (short distances) where p is of the order of M_c or larger, R(p, q) provides a Gaussian damping factor that makes all perturbation theory calculations ultra-violet finite. This has been shown in exhaustive detail for fermions and bosons in earlier chapters.

Massive Vector Bosons – No Divergence Problems

The only new case is the case of massive vector bosons. The propagator has the general form:

$$D_{VM}(k)_{\mu\nu} = -(g_{\mu\nu} - k_\mu k_\nu/m^2)(k^2 - m^2 + i\varepsilon)^{-1} \tag{8.4}$$

in momentum space in conventional quantum field theory modulo internal symmetry indices.

The Two-Tier quantum field theory equivalent is:

$$D_{VM}^{TT}(z)_{\mu\nu} = \int d^4k\, e^{-ip\cdot z}\, D_{VM}(k)_{\mu\nu} R(k, z)/(2\pi)^4 \tag{8.5}$$

in y-coordinate space. The leading momentum space behavior at high-energy of D_{VM}^{TT} is generated by the $g_{\mu\nu}$ term (unlike the situation in conventional quantum field theory):

$$D_{VM}^{TT}(k)_{\mu\nu} \sim g_{\mu\nu}k^{-6} \tag{8.6}$$

as in eq. 6.101 for the massless vector boson case. Thus a Higgs mechanism is not needed to give vector bosons mass while maintaining renormalizability in the ElectroWeak sector.

Parenthetically we note that the Two-Tier version of the ghost propagators appearing in the gauge theory sectors is also handled by Rule 1. These propagators have the same high energy leading momentum behavior as massless scalar bosons, k^{-6}.

Thus the leading high-energy behavior of all Two-Tier propagators of Standard Model particles and ghosts is a high negative power of momentum. As a result all perturbative calculations are ultra-violet finite.

Path Integral Formulation

Until now we have been viewing quantum field theory calculations in terms of conventional, Feynman diagram-based, perturbation theory. The appearance of a number of internal symmetries in the Standard Model (usually SU(2)⊗SU(1)⊗SU(3)) that are implemented with Yang-Mills gauge fields often makes it convenient to use a path integral formalism. Since there are many excellent introductions to the path integral formalism in relation to Yang-Mills fields and the Standard Model we will confine our discussion to aspects specifically related to the Two-Tier formalism for quantum field theory that we have been developing.

We will consider the case of a Two-Tier scalar particle quantum field theory initially for the sake of simplicity, and then consider Two-Tier Yang-Mills gauge field theories. The primary quantity in the path integral formulation is

$$Z(J) = \langle 0^+ | 0^- \rangle \tag{8.8}$$

$$= N \int D^4Y D\phi \exp\{ i\int d^4y\, [\mathscr{L} + j_\mu(y)Y^\mu(y) + J(y)\phi(X)]\} \tag{8.9}$$

$$= \langle 0 | T\big(\exp\{i\int d^4y\, [\mathscr{L} + j_\mu(y)Y^\mu(y) + J(y)\phi(X)]\}\big) | 0 \rangle \tag{8.10}$$

with X defined by eq. 3.12. Eq. 8.9 expresses the path integral as a product of integrals over classical field points, and eq. 8.10 provides an operator formulation of the path integral. N is a normalization factor. (As earlier we use units where $\hbar = 1$.) We note that $\langle 0^+ | 0^- \rangle$ is the probability amplitude that the vacuum state at $y^0 = -\infty$ transitions to the vacuum state at $y^0 = +\infty$. The external sources $j_\mu(y)$ and $J(y)$ are arbitrary c-number functions of the respective variables with the restriction: as $y^0 \equiv Y^0 \rightarrow \pm\infty$ (in the Y Coulomb gauge that we are using) both external sources approach zero:

$$\lim_{y^0 \rightarrow \pm\infty} j_\mu(y) = \lim_{y^0 \rightarrow \pm\infty} J(y) = 0 \tag{8.11}$$

The functional derivatives satisfy

$$\delta J(y_1)/\delta J(y_2) = \delta^4(y_1 - y_2) \tag{8.12}$$

and

$$\delta j_\mu(y_1)/\delta j^\nu(y_2) = g_{\mu\nu}\delta^4(y_1 - y_2) \tag{8.13}$$

General Procedure

In the general case we will assume that the Lagrangian (with external sources added) has the form:

$$\mathscr{L} = \mathcal{J}(Y, y)[\mathscr{L}_{\text{Fint}} + \sum_i \mathscr{L}_{\text{F0i}}] + \mathscr{L}_C \tag{8.14}$$

where $\mathscr{L}_{\text{Fint}}$ is the "interaction Lagrangian" that contains all non-quadratic field operator product terms, $\mathscr{L}_{\text{F0i}}$ is the "free field" Lagrangian for particle species i, $\mathscr{L}_C$ is the (free) Lagrangian for the Y^μ Abelian gauge field described in detail in earlier chapters, and $\mathcal{J}(Y, y)$ is the Jacobian for the transformation of y^μ coordinates to X^μ coordinates. (We now use script $\mathcal{J}$ for the Jacobian because the sources use the symbol J.) $\mathscr{L}_{\text{Fint}}$ and $\mathscr{L}_{\text{F0i}}$ are functions of the fields corresponding to the physical particles of the theory. Each field is solely a function of X^μ and all derivatives are with respect to X^μ as developed in the Two-Tier models of preceding chapters:

$$X_\mu(y) = y_\mu + i\, Y_\mu(y)/M_c^{\,2} \tag{3.12}$$

In the present discussion we will deal directly with the Y^μ field and use eq. 3.12 as a notational convenience. We begin with eqs. 8.9 and 8.14 for the case of n scalar Klein-Gordon particles, which we write as:

$$Z(J) = N\, \{\exp\{ i\int d^4y\, \mathcal{J}(-i\delta/\delta j^\nu(y)\,, y)\mathscr{L}_{\text{Fint}}(-i\,\delta/\delta J_1(y), \ldots, -i\,\delta/\delta J_n(y))\}\cdot$$

$$\cdot\prod_k [\, \int D\phi_k \exp\{ i\int d^4y\, [\mathcal{J}(-i\delta/\delta j^\nu(y), y)\mathscr{L}_{\text{F0k}} + J_k(y)\phi_k(X)]\} \,|_{j_\mu = 0}\,]\cdot$$

$$\cdot\int D^4Y \exp\{ i\int d^4y[\mathscr{L}_C + j_\mu(y)Y^\mu(y)]\}\} \,\Big|_{j_\mu = 0} \tag{8.15}$$

using functional derivatives with respect to the sources. We note that we set $j^\mu = 0$ *after* evaluating each of the n free field factors in eq. 8.15. The physical justification for this procedure is that we wish the free field propagators to be truly independent free field propagators and not depend on external sources, and not be convoluted together via the Y^μ field. The result will be a simpler, physically reasonable, perturbation theory in which each ϕ particle is truly a free field (independent of the other ϕ fields) that emits Y quanta and then absorbs all its emitted Y quanta. (An *alternative, different*, theory would allow a ϕ particle to emit a Y quantum that would then be absorbed by a different ϕ particle – thus ϕ particles could interact via the exchange of Y quanta. The exchange of Y quanta would make these seemingly free fields actually interacting fields – contrary to our starting assumption that the $\mathscr{L}_{\text{F0i}}$ terms each define a free field. This type of theory would also be a calculational nightmare since a ϕ particle emits an infinite superposition of Y quanta. Thus the lowest order diagram in ϕ^4 perturbation would have an infinite number of terms. We therefore will not consider this possibility in the hope that Nature opted for our Two-Tier theory.)

We will now evaluate eq. 8.15 in stages since a number of novel features appear in its evaluation although the path integral evaluations themselves are conventional. We start by performing the Y integration. The Y field is a free Abelian gauge field. We choose the same Coulomb gauge as in previous chapters $\nabla\cdot \mathbf{Y} = 0$ and $Y^0 = 0$. Consider

$$Z_Y(j_\mu) = \int D^4Y \exp\{ i\int d^4y[\mathscr{L}_C + j_\mu(y)Y^\mu(y)]\} \tag{8.16}$$

with $\mathscr{L}_C$ given by eq. 3.15. After some manipulations we find

$$Z_Y(j_\mu) = \exp\{ -i\int d^4y_1d^4y_2\, j^i(y_1)D_{Fij}(y_1 - y_2)j^j(y_2)/2\} \tag{8.17}$$

with

$$D_{Fij}(y_1 - y_2) = \int d^4k\, e^{-ik\cdot(y_1 - y_2)}\, (\delta_{ij} - k_ik_j/\mathbf{k}^2)/[(2\pi)^4(k^2 + i\varepsilon)] \tag{8.18}$$

the spatial components of the massless vector boson Feynman propagator in the Coulomb gauge.

We now evaluate one of the factors in the exponentiated n free boson Lagrangian:

$$Z_{\phi_k}(J_k) = [\int D\phi_k \exp\{ i\int d^4y\, [\mathcal{J}(-i\delta/\delta j^\nu(y), y)\mathscr{L}_{F0k} + J_k(y)\phi_k(X)]\}\, Z_Y(j_\mu)]\,\Big|_{j_\mu = 0} \tag{8.19}$$

The free Klein-Gordon Lagrangian is:

$$\mathscr{L}_{F0k} = \tfrac{1}{2}\, [\, (\partial\phi_k/\partial X^\nu)^2 - m_k^{\,2}\phi_k^{\,2}] \tag{8.20}$$

As a result the classical action can be written as

$$S_{0k}[\phi_k] = \int d^4y\, \mathcal{J}(-i\delta/\delta j^\nu(y), y)\mathscr{L}_{F0k} \tag{8.21}$$

$$= -\tfrac{1}{2}\int d^4X_{op}\, \phi_k(X_{op})(\Box_X + m_k^{\,2})\phi_k(X_{op}) \tag{8.22}$$

under the usual assumptions of good behavior at infinity and utilizing the Jacobian factor with

$$X_{op\mu}(y) = y_\mu + M_c^{-2}\, \delta/\delta j^\mu(y) \tag{8.23}$$

and with

$$\Box_X = (\partial/\partial X_{op}{}^{\nu})(\partial/\partial X_{op\nu}) \tag{8.24}$$

Inserting eq. 8.21 in eq. 8.19 we obtain:

$$Z_{\phi_k}(J_k) = [\int D\phi_k \exp\{i\int d^4X_{op}[-\tfrac{1}{2}\phi_k(X_{op})(\Box_X + m_k{}^2)\phi_k(X_{op}) +$$

$$+ J_k(y)\phi_k(y)/\mathcal{J}(-i\delta/\delta j^{\nu}(y), y)]\}\, Z_Y(j_\mu)]\,|_{j_\mu = 0} \tag{8.25}$$

The (Gaussian in ϕ_k) path integral in eq. 8.25 can be performed (see below) using standard techniques to yield:

$$Z_{\phi_k}(J_k) = [\exp\{-i\int d^4X_{op}(y_1)d^4X_{op}(y_2)[J_k(y_1)/\mathcal{J}(-i\delta/\delta j^{\nu}(y_1), y_1)]\cdot$$

$$\cdot\Delta_{Fk}{}^{TT}(y_1 - y_2)[J_k(y_2)/\mathcal{J}(-i\delta/\delta j^{\nu}(y_2), y_2)]/2\}\, Z_Y(j_\mu)]\,|_{j_\mu = 0}$$

$$= \exp\{-i\int d^4y_1 d^4y_2 J_k(y_1)\Delta_{Fk}{}^{TT}(y_1 - y_2)J_k(y_2)/2\} \tag{8.26}$$

after changing the integration variables in the exponential using the Jacobian with $\Delta_{Fk}{}^{TT}(y_1 - y_2)$ given by eqs. 6.51-52. The added index k serves to identify the propagator $\Delta_{Fk}{}^{TT}(y_1 - y_2)$ as having the mass m_k.

The derivation of eq. 8.25 uses functional techniques in a straightforward way. We can best derive the form of $\Delta_{Fk}{}^{TT}(y_1 - y_2)$ in momentum space. The operator $(\Box_X + m_k{}^2)^{-1}$ can be represented via

$$(\Box_X + m_k{}^2)^{-1}g(X_{op}(y_1))Z_Y(j_\mu) = (\Box_X + m_k{}^2)^{-1}\int d^4p\{e^{-ip\cdot X_{op}(y_1)}g(p)/(2\pi)^4\}Z_Y(j_\mu)$$

$$= -\int d^4p\{e^{-ip\cdot X_{op}(y_1)}g(p)/[(2\pi)^4(p^2 - m_k{}^2 + i\varepsilon)]\}Z_Y(j_\mu)$$

$$= -\int d^4X_{op}(y_2)\Delta_{Fkop}{}^{TT}(X_{op}(y_1) - X_{op}(y_2))g(X_{op}(y_2))Z_Y(j_\mu)$$

$$= -\int d^4X_{op}(y_2)\Delta_{Fkop}{}^{TT}(X_{op}(y_1) - X_{op}(y_2))Z_Y(j_\mu)g(X(y_2))$$

We now calculate

$$\Delta_{Fkop}{}^{TT}(X_{op}(y_1) - X_{op}(y_2))Z_Y(j_\mu) = \int d^4p\ \{e^{-ip\cdot(y_1 - y_2)}/[(2\pi)^4(p^2 - m_k{}^2 + i\varepsilon)]\}\cdot$$

$$\cdot \exp(-ip^{\nu}[\delta/\delta j^{\nu}(y_1) - \delta/\delta j^{\nu}(y_2)]/M_c^{2}) Z_Y(j_\mu)$$

$$= Z_Y(j_\mu) \int d^4p \, \{e^{-ip\cdot(y_1-y_2)}/[(2\pi)^4 (p^2 - m_k^{\,2} + i\varepsilon)]\}\cdot$$

$$\cdot \exp(-ip^i p^j[D_{Fij}(y_1 - y_2) + D_{Fij}(y_2 - y_1)]/(2M_c^{\,4})) \tag{8.27}$$

where $D_{ij}(y_1 - y_2)$ is given by eq. 8.18. The last factor in eq. 8.27 is the Gaussian damping factor that appears in Two-Tier propagators as repeatedly seen earlier – for example in eq. 6.53 and 6.54:

$$R(\mathbf{p}, y_1 - y_2) = \exp(-ip^i p^j[D_{Fij}(y_1 - y_2) + D_{Fij}(y_2 - y_1)]/(2M_c^{\,4})) \tag{8.28}$$

Eq. 8.26 then follows with the Two-Tier scalar Klein-Gordon Feynman propagator.

Therefore the n scalar Klein-Gordon particle path integral of eq. 8.15 simplifies to

$$Z(J) = N\,\{\exp\{\, i\textstyle\int d^4y\, \mathcal{J}(-i\delta/\delta j^{\nu}(y), y)\, \mathscr{L}_{Fint}(-i\delta/\delta J_1(y), \ldots, -i\delta/\delta J_n(y))\}\cdot$$

$$\cdot\exp\{-\sum_k [\, i\textstyle\int d^4y_1 d^4y_2 J_k(y_1)\Delta_{Fk}^{\ TT}(y_1 - y_2) J_k(y_2)/2]\}\, Z_Y(j_\mu)\,\}\Big|_{j_\mu = 0} \tag{8.29}$$

The only dependence on functional derivatives with respect to j^μ in eq. 8.29 is in the Jacobian of the interaction Lagrangian term. It is readily seen that the Jacobian effectively reduces to one if there is no external physical source for Y particles (as we have assumed.) Consider

$$\begin{aligned}\mathcal{J}(-i\delta/\delta j^{\nu}(y), y) Z_Y(j_\mu) = \{&\varepsilon^{ijk}(\delta_{1i} + M_C^{-2}(\partial/\partial y^i)\delta/\delta j^1(y))\cdot \\ &\cdot(\delta_{2j} + M_C^{-2}(\partial/\partial y^j)\, \delta/\delta j^2(y))\cdot \\ &\cdot(\delta_{3k} + M_C^{-2}(\partial/\partial y^k)\, \delta/\delta j^3(y)) Z_Y(j_\mu)\}\end{aligned} \tag{8.30a}$$

Since the interaction Hamiltonian – including the q-number Jacobian – is normal-ordered (eq. 6.8) the functional derivatives in eq. 8.30a can only apply to $Z_Y(j_\mu)$. Therefore

$$\begin{aligned}\mathcal{J}(-i\delta/\delta j^{\nu}(y), y) Z_Y(j_\mu) &= Z_Y(j_\mu)\varepsilon^{ijk}\big(\delta_{1i} + M_C^{-2}(\partial/\partial y^i)(-i/2)[\textstyle\int d^4y_2 D_{F1a}(y - y_2) j^a(y_2) + \\ &+ \textstyle\int d^4y_1\, j^a(y_1) D_{Fa1}(y_1 - y)]\big)\cdot\big(\delta_{2j} + M_C^{-2}(\partial/\partial y^j)\,(-i/2)[\textstyle\int d^4y_2\, D_{F2a}(y - y_2) j^a(y_2) + \\ &+ \textstyle\int d^4y_1\, j^a(y_1) D_{Fa2}(y_1 - y)]\big)\cdot\big(\delta_{3k} + M_C^{-2}(\partial/\partial y^k)\,(-i/2)[\textstyle\int d^4y_2\, D_{F3a}(y - y_2) j^a(y_2) +\end{aligned}$$

$$+ \int d^4y_1\, j^a(y_1) D_{Fa3}(y_1 - y)]) \tag{8.30a}$$

After setting $j_\mu = 0$ *for the case of no incoming or outgoing Y quanta* we see

$$\mathcal{J}(-i\delta/\delta j^\nu(y), y) \equiv 1 \tag{8.30b}$$

This result is consistent with our physical notion that the integration, in itself, does not generate dynamical effects. *Otherwise* a c-number interaction Lagrangian term would generate a non-trivial interacting quantum field theory effects – contrary to our physical expectations. Eq. 8.30b eliminates this physically unreasonable possibility.

Thus eq. 8.29 becomes

$$Z(J) = N \exp\{ i\int d^4y\, \mathscr{L}_{Fint}(-i\delta/\delta J_1(y), \ldots, -i\delta/\delta J_n(y))\}\cdot$$

$$\cdot\exp\{-i\int d^4y_1 d^4y_2 \sum_k J_k(y_1)\Delta_{Fk}{}^{TT}(y_1 - y_2)J_k(y_2)/2]\} \tag{8.31}$$

which gives the same perturbation theory found earlier using canonical quantum field theory that is built on the U matrix expansion.

Derivative Coupling Case

Eq. 8.31 is based on the assumption that derivative couplings do not appear in the interaction Lagrangian. (There are no derivative couplings in the Standard Model but there are derivative couplings in quantum gravity so we must deal with this issue if we wish to create a unified theory.) If the interaction Lagrangian does contain derivatives of scalar fields with respect to the X variable then the interaction Lagrangian has a superficial dependence on the functional derivative with respect to j^μ, which we can symbolize in the modified path integral expression:

$$Z(J) = N\, \{\exp\{ i\int d^4y\, \mathscr{L}_{Fint}(\partial/\partial(y^\nu + M_c^{-2}\delta/\delta j_\nu(y)), -i\delta/\delta J_1(y), \ldots, -i\delta/\delta J_n(y))\}\cdot$$

$$\cdot\exp\{-\sum_k[i\int d^4y_1 d^4y_2 J_k(y_1)\Delta_{Fk}{}^{TT}(y_1 - y_2)J_k(y_2)/2]\}\, Z_Y(j_\mu)\, \}\,\Big|_{j_\mu = 0} \tag{8.32}$$

obtained from eq. 8.31.

The j^μ dependence now appears only in the interaction Lagrangian. For good reason we will now show the j^μ dependence of the interaction Lagrangian in eq. 8.32 can be eliminated.

Our approach will be to separate the coordinate dependence in the propagator into two parts: the coordinate dependence in the Gaussian factor, and the coordinate dependence in the $e^{ip\cdot x}$ factor. We can then express the derivatives of fields in the interaction Lagrangian as derivatives with respect to the coordinates in the $e^{ip\cdot x}$ factor appearing in integral representations of the Two-Tier propagator when a perturbative expansion of the path integral solution is made.

We begin by noting that

$$i\Delta_{Fk}^{TT}(y_1 - y_2) = <0|T(\phi(X(y_1)),\phi(X(y_2)))|0> \tag{6.51}$$

$$= i\int \frac{d^4p\, e^{-ip\cdot(y_1 - y_2)}\, R(\mathbf{p}, y_1 - y_2)}{(2\pi)^4 (p^2 - m_k^2 + i\varepsilon)} \tag{6.52}$$

with

$$R(\mathbf{p}, y_1 - y_2) = \exp[-p^i p^j \Delta_{Tij}(y_1 - y_2)/M_c^4] \tag{6.53}$$

(summations are over space indices only in the Y Coulomb gauge) and

$$\Delta_{Tij}(z) = \int d^3k\, e^{-ik\cdot z}\, (\delta_{ij} - k_i k_j/\mathbf{k}^2)/[(2\pi)^3 2\omega_k] \tag{6.54}$$

We now define *a more general* Two-Tier propagator by introducing a distinction between the spatial dependence of the Gaussian and exponential terms:

$$i\Delta_{Fk}^{TT}(y_1 - y_2, z) = i\int \frac{d^4p\, e^{-ip\cdot(y_1 - y_2)}\, R(\mathbf{p}, z)}{(2\pi)^4 (p^2 - m_k^2 + i\varepsilon)} \tag{8.33}$$

We then note that

$$\partial i\Delta_{Fk}^{TT}(y_1 - y_2)/\partial X^\mu = \partial i\Delta_{Fk}^{TT}(y_1 - y_2, z)/\partial y_1^\mu \big|_{z=y_1-y_2} \tag{8.34}$$

comparing eqs. 6.51 and 6.52 with eq. 8.33. As a result we can write eq. 8.32 symbolically as:

$$Z(J) = N\,\{\exp\{ i\int d^4y\, \mathscr{L}_{Fint}(\partial/\partial y^\nu, -i\delta/\delta J_1(y), \ldots, -i\delta/\delta J_n(y))\}\cdot$$

$$\cdot\exp\{-\sum_k [\, i\int d^4y_1 d^4y_2 J_k(y_1)\Delta_{Fk}^{TT}(y_1 - y_2, z) J_k(y_2)/2]\}\} \big|_{z=y_1-y_2} \tag{8.35}$$

Eq. 8.35 is interpreted as the following:

1. For a given process take appropriate functional derivatives of Z(J) with respect to $J_1(y), \ldots, J_n(y)$.

2. Then expand the exponential factors in a perturbation series applying any derivatives with respect to y in $\mathscr{L}_{\text{Fint}}$. Do not perform any of the $\int d^4y_1 d^4y_2$ integrals.

3. Then set $z = y_1 - y_2$ in each $\Delta_{Fk}{}^{TT}(y_1 - y_2, z)$ propagator.

4. Lastly perform all $\int d^4y_1 d^4y_2$ integrals.

Thus we thus achieve a path integral formulation that is very similar to the corresponding expression in conventional field theory – the only difference is the form of the free field propagators, which now contain a Gaussian factor.

ϕ^4 Theory Path Integral Formulation Example

We will now consider the case of Two-Tier scalar field theory using the specific example of the ϕ^4 Lagrangian of eqs. 5.20 – 5.24:

$$\mathscr{L}_{\text{Fint}}(X^\mu) = \tfrac{1}{4!}\,\chi_0\,\phi(X)^4 + \tfrac{1}{2}\,(m^2 - m_0{}^2)\phi(X)^2 \tag{8.36}$$

with

$$\mathscr{L}_{F0} = \tfrac{1}{2}\,[\,(\partial\phi/\partial X^\nu)^2 - m^2\phi^2] \tag{8.37}$$

In this theory eq. 8.31 can be written as:

$$Z(J) = N\,\exp\{i\int d^4y\,\mathscr{L}_{\text{Fint}}(-i\delta/\delta J(y))\}\,\exp\{-i\int d^4y_1 d^4y_2 J(y_1)\Delta_{Fk}{}^{TT}(y_1 - y_2)J(y_2)/2\} \tag{8.38}$$

The perturbation theory generated from eq. 8.38 is the same as conventional perturbation theory except for the differing free field propagators.

Two-Tier Yang-Mills Gauge Fields

Two-Tier Yang-Mills gauge field theories have many similarities to conventional Yang-Mills theories.[100] We assume the reader in familiar with internal symmetries and the conventional Yang-Mills formulation.

General Rule: All gauge fields and derivatives of gauge fields, as well as group properties and other features such as the Faddeev-Popov method, are expressed solely in terms of the X coordinate system (which in turn is a function of the y coordinates).

We note all matter fields are assumed to be functions of X coordinates only as done in previous chapters. The general rule is implemented by defining:

1. The covariant derivative of any matter field $\Psi(X)$ by

---

[100] C. N. Yang and R. L. Mills, Phys. Rev. **96**, 191 (1954).

$$D^{\mu}\Psi(X) = [\partial/\partial X_{\mu} + igA^{\mu}(X)]\Psi(X) \tag{8.39}$$

with $A^{\mu}(X)$ being an element of a Lie algebra:

$$A^{\mu}(X) = A_a^{\ \mu}(X)L_a \tag{8.40}$$

where L_a is a generator of a Lie algebra (a and b are internal symmetry indexes) with commutation relations:

$$[L_a, L_b] = ic_{ab}^{\ \ c}L_c \tag{8.41}$$

with $c_{ab}^{\ \ c}$ being real numbers called the *structure constants* of the Lie algebra.

2. The field strengths are defined as the commutator of covariant derivatives:

$$F^{\mu\nu} = [D^{\mu},D^{\nu}] = \partial A^{\nu}(X)/\partial X_{\mu} - \partial A^{\mu}(X)/\partial X_{\nu} + ig[A^{\mu}(X), A^{\nu}(X)] \tag{8.42}$$

3. The Lagrangian for a Yang-Mills gauge field interacting with a matter field $\psi(X)$ has the form:

$$\mathscr{L}_{YM} = [\mathscr{L}_F^{YM}(X^{\mu}) + \mathscr{L}_F^{Matter}(X^{\mu})]\, J + \mathscr{L}_C(X^{\mu}(y), \partial X^{\mu}(y)/\partial y^{\nu}) \tag{8.43}$$

where J is the Jacobian and

$$\mathscr{L}_F^{YM}(X^{\mu}) = -\tfrac{1}{4}\, F_a^{\ \mu\nu}\, F_{a\mu\nu} \tag{8.44}$$

with $F^{\mu\nu} = F_a^{\ \mu\nu}L_a$ and

$$\mathscr{L}_F^{Matter}(X^{\mu}) = \mathscr{L}_F^{Matter}(\psi(X), D^{\mu}\psi(X)) \tag{8.45}$$

and $\mathscr{L}_C$ is defined as previously and specifies the Y field evolution.

The generalization to multiple gauge fields interacting with multiple matter fields is direct. The overall form of the Standard Model Lagrangian is:

$$L = \int d^4y\, \{J[\mathscr{L}_F^{Matter}(X^{\mu}) + \mathscr{L}_F^{GaugeFields}(X^{\mu}) + \mathscr{L}_F^{Higgs}(X^{\mu})] + \\ + \mathscr{L}_C(X^{\mu}(y), \partial X^{\mu}(y)/\partial y^{\nu})\} \tag{8.46}$$

where J is the Jacobian for the transformation from y to X coordinate integration. Eq. 8.46 can be rewritten as

$$L = \int d^4X[\mathscr{L}_F^{Matter}(X^\mu) + \mathscr{L}_F^{GaugeFields}(X^\mu) + \mathscr{L}_F^{Higgs}(X^\mu)] +$$
$$+ \int d^4y \mathscr{L}_C(X^\mu(y), \partial X^\mu(y)/\partial y^\nu) \quad (8.47)$$

It is clear from eq. 8.47 that the conventional Standard Model equations of motion and canonical quantization procedure emerge in the Two-Tier formulation if all expressions are written as functions of the X coordinates as shown previously in our discussions of separable Lagrangians. The second quantization of X as a function of the y coordinates leads to the Gaussian factor in all free particle propagators (except the Y propagator).

Thus the gauge field and matter field parts of the Lagrangian, the gauge field transformation laws and other related operations solely depend on the X coordinate system as stated in the general rule above. The quantization and field equations are the same as the conventional case – except that they are specified solely in terms of the X coordinate system.

Path Integral Formulation and Faddeev-Popov Method

We now turn to the Two-Tier path integral formulation of Yang-Mills gauge theories and in particular to the Two-Tier version of the Faddeev-Popov method. The Two-Tier path integral for a gauge field can be written symbolically as:

$$Z(J^\mu) = N \int DADY \Delta_{FP}(A)\delta(F(A)) \exp\{i\int d^4y\ [\mathscr{L} +$$
$$+ j_\mu(y)Y^\mu(y) + J^\mu(y)A_\mu(X)]\}\ |_{j_\mu = 0} \quad (8.48)$$

where $\delta(F(A))$ specifies the gauge and $\Delta_{FP}(A)$ is the Faddeev-Popov determinant. The Lagrangian is

$$\mathscr{L} = J\mathscr{L}_F^{YM}(X^\mu) + \mathscr{L}_C(X^\mu(y), \partial X^\mu(y)/\partial y^\nu) \quad (8.49)$$

with $\mathscr{L}_F^{YM}$ specified by eq. 8.44 and J the Jacobian for the transformation from X coordinates to y coordinates. The Faddeev-Popov determinant may be calculated in the standard way. First we note that the delta function fixing the gauge can be written as a delta function in the gauge times a determinant:

$$\delta(F(A^\omega)) = \delta(\omega - \omega_0)\ |\det \delta F(A_\mu{}^\omega(X))/\delta\omega(X)|^{-1}\ \Big|_{F(A)=0} \quad (8.50)$$

where ω_0 is a reference gauge, where

$$A_\mu{}^\omega(X) = A_\mu(X) + \partial\omega(X)/\partial X^\mu \quad (8.51)$$

and where

$$\Delta_{FP}(A) = \left| \det \delta F(A_\mu{}^{\omega}(X))/\delta\omega(X) \right| \Big|_{F(A)=0} \tag{8.52}$$

We will choose the Lorentz gauge to evaluate the Faddeev-Popov determinant:

$$F_a(A) = \partial A_a{}^{\mu}(X)/\partial X^{\mu} = 0 \tag{8.53}$$

Under an infinitesimal gauge transformation:

$$A_{a\mu}{}^{\omega}(X) = A_{a\mu}(X) + g^{-1}\partial\omega_a/\partial X^{\mu} + c_{ab}{}^{c}\,\omega_b(X)A_c{}^{\mu}(X) \tag{8.54}$$

we find

$$F_a(A_\mu{}^{\omega}(X)) = \partial(A_{a\mu}(X) + g^{-1}\partial\,\omega_a(X)/\partial X^{\mu} + c_{ab}{}^{c}\,\omega_b(X)A_c{}^{\mu}(X))/\partial X^{\mu}$$

$$= g^{-1}\,\Box_X\,\omega_a(X) + c_{ab}{}^{c}\,\partial\,\omega_b(X)/\partial X^{\mu}\;A_c{}^{\mu}(X) \tag{8.55}$$

Thus

$$\delta F_a(A_\mu{}^{\omega}(X))/\delta\omega_b(X) = g^{-1}\,\delta_{ab}\Box_X + c_{ab}{}^{c}A_c{}^{\mu}(X)\partial/\partial X^{\mu} \tag{8.56}$$

and

$$\Delta_{FP}(A) = \left| \det\,(g^{-1}\,\delta_{ab}\Box_X + c_{ab}{}^{c}A_c{}^{\mu}(X)\partial/\partial X^{\mu}) \right| \Big|_{F(A)=0} \tag{8.57}$$

where | … | represent absolute value.

We note the Two-Tier Faddeev-Popov determinant is solely a function of the X coordinates. Thus we can follow the standard procedure and rewrite the determinant as a path integral over anti-commuting c-number fields with a ghost Lagrangian that is solely a function of X – just like the gauge particles and other particles in Two-Tier theories:

$$\Delta_{FP}(A) = \int D\chi^{*}D\chi\,\exp[\,i\int d^4X\,\mathscr{L}^{ghost}(X^{\mu})] \tag{8.58}$$

where

$$\mathscr{L}^{ghost}(X^{\mu}) = \chi_a{}^{*}(X)[\delta_{ab}\Box_X + g\,c_{ab}{}^{c}A_c{}^{\mu}(X)\partial/\partial X^{\mu}]\chi_b(X) \tag{8.59}$$

Thus the complete path integral is

$$Z(J^{\mu}) = N\int DAD\chi^{*}D\chi DY\delta(F(A))\,\exp\{i\int d^4y\,[\mathscr{L} +$$

$$+ \, j_\mu(y)Y^\mu(y) + J^\mu(y)A_\mu(X)]\} \,|_{j_\mu = 0} \qquad (8.60)$$

where $\delta(F(A))$ specifies the gauge and

$$\mathscr{L} = J\mathscr{L}_F^{YM}(X^\mu) + J\mathscr{L}^{ghost}(X^\mu) + \mathscr{L}_C(X^\mu(y), \partial X^\mu(y)/\partial y^\nu) \qquad (8.61)$$

with $\mathscr{L}_F^{YM}$ specified by eq. 8.44.

At this point it is obvious that we can follow almost identical steps as we did in the scalar particle case starting from eq. 8.9 and obtain an expression similar to eq. 8.38. The result is a perturbation theory for the Yang-Mills gauge field that is identical to the usual theory except that the free propagators for the gauge fields, *and the ghost fields*, acquire the Gaussian factor R(p,z) as stated earlier in the discussions of eqs. 6.52, 6.88, 6.95, 6.102, and 7.120-123.

Two-Tier Massive Vector Fields

Massive vector fields have been a problem for perturbation theory due to the $k_\mu k_\nu/m^2$ term appearing in the free field propagator. This term makes conventional interacting quantum field theories of massive vector bosons non-renormalizable. The Higgs mechanism is used in ElectroWeak theory to evade the non-renormalizability of massive gauge fields. It gives mass to the vector bosons mediating the weak force while maintaining the renormalizability of the theory. It also implements symmetry breaking.

In Two-Tier quantum field theory a massive vector boson does not create renormalization issues. For example the Two-Tier version of the Weinberg-Salam model of the electromagnetic and weak forces (with massive vector bosons) is finite! Thus the need for spontaneous symmetry breaking to give mass to vector bosons in ElectroWeak theory is not present. Higgs particles are not needed (although they may exist. Their existence is an experimental question – not a theoretical necessity for a Two-Tier ElectroWeak theory with massive vector bosons.)

Massive Vector Boson Propagator

The massive free vector particle propagator in <u>conventional</u> quantum field theory has the representation:

$$i\Delta_{FV}(y_1 - y_2)_{\mu\nu} = -i \int \frac{d^4k \, e^{-ik\cdot(y_1 - y_2)} \, (g_{\mu\nu} - k_\mu k_\nu/m^2)}{(2\pi)^4 \, (k^2 - m^2 + i\varepsilon)} \qquad (8.62)$$

A Two-Tier massive vector boson theory can be constructed in a straightforward way from the Lagrangian:

$$\mathscr{L} = J[-\tfrac{1}{4} F_V^{\mu\nu}(X(y))F_{V\mu\nu}(X(y)) - \tfrac{1}{2} m^2 V^\mu V_\mu] + \mathscr{L}_C(X^\mu(y), \partial X^\mu(y)/\partial y^\nu, y) \tag{8.63}$$

with J the Jacobian and with

$$F_{V\mu\nu}(X(y)) = \partial V_\mu(X(y))/\partial X^\nu - \partial V_\nu(X(y)) /\partial X^\mu \tag{8.64}$$

and the usual Two-Tier Y Lagrangian terms

$$\mathscr{L}_C(X^\mu(y), \partial X^\mu(y)/\partial y^\nu, y) = -\tfrac{1}{4} F_Y^{\mu\nu}F_{Y\mu\nu} \tag{7.4}$$

$$F_{Y\mu\nu} = (\partial Y_\mu/\partial y^\nu - \partial Y_\nu /\partial y^\mu) \tag{7.5}$$

Following steps similar to the previously considered scalar particle, fermion and massless vector particle (photon) cases we find the Two-Tier massive vector boson Feynman propagator to be:

$$i\Delta_{FV}{}^{TT}(y_1 - y_2)_{\mu\nu} = -i \int \frac{d^4k\, e^{-ik\cdot(y_1 - y_2)} (g_{\mu\nu} - k_\mu k_\nu/m^2)\, R(\mathbf{k}, y_1 - y_2)}{(2\pi)^4 (k^2 - m^2 + i\varepsilon)} \tag{8.65}$$

No Need for Higgs Mechanism in ElectroWeak Theory

The leading coordinate space dependence at high energy (short distance) of the Fourier transform of $\Delta_{FV}{}^{TT}$ is the same as $\Delta_F{}^{TT}$ since it comes from the $g_{\mu\nu}$ term in $\Delta_F{}^{TT}$

$$\Delta_{FV}{}^{TT}(y_1 - y_2)_{\mu\nu} \sim g_{\mu\nu} (y_1 - y_2)^2 \tag{8.66}$$

which in momentum space is equivalent to

$$\Delta_{FV}{}^{TT}(p)_{\mu\nu} \sim g_{\mu\nu}\, p^{-6} \tag{8.67}$$

The $k_\mu k_\nu/m^2$ term appearing in the free vector field Two-Tier propagator actually is of higher order in the large momentum limit (short distance):

$$\int \frac{d^4k\, e^{-ik\cdot(y_1 - y_2)} k_\mu k_\nu/m^2)\, R(\mathbf{k}, y_1 - y_2)}{(2\pi)^4 (k^2 - m^2 + i\varepsilon)} \sim (y_1 - y_2)^4 \tag{8.68}$$

which corresponds to momentum space behavior of

$$p^{-8} \tag{8.69}$$

Two-Tier propagators such as the propagator in eq. 8.65 have the feature that the Gaussian factor "inverts" the high-energy behavior of the terms in the numerator of the integrand: terms with higher powers of momentum are less significant at short distances. The term with the lowest power of momentum generates the leading behavior at high energy (short distances).

Thus Two-Tier massive vector boson theories are ultra-violet convergent and do not constitute a problem as they do in conventional quantum field theory. *Therefore there is no need for the Higgs mechanism in the ElectroWeak sector of the Standard Model in order to obtain a renormalizable theory. Ordinary massive vector bosons can be used and the resulting Two-Tier theory is finite!*

General Short Distance (High Momentum) Behavior of Two-Tier Propagator

The higher the power of the momentum in the numerator of the integrand of a Two-Tier propagator, the more convergent the large momentum behavior of the Fourier transform of the Two-Tier propagator.

The short distance behavior of a term with n factors of momentum in a Two-Tier propagator has the leading short distance coordinate space behavior:

$$\int \frac{d^4k\, e^{-ik\cdot(y_1 - y_2)}\, k_{\mu_1} k_{\mu_2} \cdots k_{\mu_n}\, R(\mathbf{k}, y_1 - y_2)}{(2\pi)^4\, (k^2 - m^2 + i\varepsilon)} \sim (y_1 - y_2)^{2+n} \tag{8.70}$$

which corresponds to the high energy behavior:

$$p^{-6-n} \tag{8.71}$$

In contrast to conventional quantum field theory the more powers of momentum in the numerator of a Two-Tier propagator, the better the short distance behavior!

Higgs Particles

A previous section shows that the Higgs Mechanism is not needed in the Two-Tier ElectroWeak sector of the Standard Model. Massive vector bosons such as **W**'s would not make the ElectroWeak theory non-renormalizable. *As we have seen a Two-Tier ElectroWeak theory with massive vector bosons is finite.*

Nevertheless we would like to point out a Two-Tier version of the Higgs particle sector of the Standard Model can be defined that largely parallels the conventional treatment of the Higgs sector. Higgs particles continue to be of interest since they may play a role in the origin of particle masses and symmetry breaking.

The Two-Tier scalar Higgs field Lagrangian terms (plus $\mathscr{L}_C$) can be written:

$$\mathscr{L}_{\text{Higgs}} = J[D_\mu\phi^\dagger D^\mu\phi - V(\phi(X))] + \mathscr{L}_C \tag{8.72}$$

$$V(\phi(X)) = m_0^2\phi^\dagger(X(y))\phi(X(y)) + \lambda[\phi^\dagger(X(y))\phi(X(y))]^2 + \\ + G_c[\bar{L}(X)\phi(X)R(X) + \bar{R}(X)\phi^\dagger(X)R(X)] \tag{8.73}$$

with the covariant derivative defined with the usual B_μ and W_μ^a gauge fields of the Standard Model, and L representing a left-handed fermion isodoublet, and R representing a right-handed fermion isosinglet. We note all items in eqs. 8.72 and 8.73 are written solely as functions of the X coordinates with the exception of $\mathscr{L}_C$, which is the Lagrangian term for the Y field.

The conventional effective potential method can be followed to implement the Higgs mechanism. In particular we may write

$$\phi(X(y)) = <\phi> + \eta(X) \tag{8.74}$$

with $<\phi>$ the vacuum expectation value, which is a constant by translational invariance. Then vector bosons can acquire mass via the Higgs mechanism. The quantum part of the Higgs field has a Two-Tier propagator with p^{-6} behavior for large momentum. Thus all sectors of the Standard Model wind up with Two-Tier propagators and all perturbative calculations are finite.

General Form of the Two-Tier Standard Model Path Integral

The general form of the path integral for the Two-Tier version of the Standard Model is:

$$Z(J) = N\, \{\exp\{\, i\int d^4y\, \mathscr{L}_{\text{Fint}}(\partial/\partial y^\nu, -i\delta/\delta J_1(y), \ldots, -i\delta/\delta J_n(y))\}\cdot$$

$$\cdot \exp\{-\sum_k [\, i\int d^4y_1 d^4y_2 J_k(y_1)\Delta_{Fk}^{TT}(y_1 - y_2, z)J_k(y_2)/2]\}\}\,|_{z=y1-y2} \tag{8.75}$$

where the sum over k is a sum over all matter fields, gauge fields, Higgs fields, and ghost fields. The index k, and indices on the functional derivatives, represent all space-time indices and internal symmetry indices that are relevant for each particle. We assume the total number of particle and ghost fields is n. Notice all dependence on the Y field has been "integrated away."

Also any derivatives with respect to y appearing in the interaction Lagrangian are applied to Two-Tier propagators using the method represented by eqs. 8.33 and 8.34. This procedure results in the same momentum polynomials for Feynman diagrams in a Two-Tier version as appear in the corresponding conventional theory. Thus we have the same algebraic structure (both in the momentum polynomials and internal symmetries) in the Two-Tier version of a conventional theory.

We note that the large distance behavior of the Two-Tier theory is the same as the Standard Model in all respects including gauge symmetries. Deviations from the conventional Standard Model results only appear at extremely high energies of the order of M_c.

Renormalization - Finite

Since all particle (and ghost) propagators are Two-Tier propagators in the Two-Tier Standard Model, the Two-Tier Standard Model yields finite results to all orders in perturbation theory. We note the large momentum behavior of the various types of particle propagators in the Two-Tier Standard Model is:

Fermion Propagators

$$p^{-7}$$

Vector Boson (Gauge Field) Propagators

$$p^{-6}$$

Ghost Propagators

$$p^{-6}$$

Higgs Particle Propagators

$$p^{-6}$$

Thus all Feynman diagrams are highly ultra-violet convergent and the Two-Tier Standard Model is finite. This result is independent of the details of the internal symmetries, particle spectrum, and particle masses.

Unitarity

The Two-Tier Standard Model *superficially* appears to have a unitarity problem due to the non-Hermitian nature of its Hamiltonian. The lack of hermiticity is due entirely to the appearance of iY^μ in the X^μ field coordinates.

Thus the Two-Tier Standard Model interaction Hamiltonian is not Hermitian:

$$H_{Fint} = \int d^3y' \mathscr{H}_{Fint}(y' + iY(y')/M_c^2) \tag{8.76}$$

and

$$H_{Fint}^{\dagger} = \int d^3y' \mathscr{H}_{Fint}(y' - iY(y')/M_c^2) \neq H_{Fint} \tag{8.77}$$

The relation between H_{Fint} and its Hermitian conjugate is

$$H_{Fint} = V\, H_{Fint}^{\dagger}\, V \tag{8.78}$$

where $V^2 = I$ is the metric operator defined in eqs. 5.16 – 5.18. Eq. 8.78 implies that the Two-Tier Standard Model S matrix is not unitary. The Two-Tier Standard Model S matrix is pseudo-unitary:

$$S^{-1} = V\, S^{\dagger}\, V \tag{8.79}$$

Therefore

$$S^{\dagger} VS = V \tag{8.80}$$

Two-Tier Standard Model S matrix satisfies unitarity between physical asymptotic states – states consisting of only physical particles: leptons, quarks, photons, W and Z particles, gluons, and Higgs particles (if they exist). Put another way: physical states can consist of any set of particles in the Two-Tier Standard Model except ghosts and Y particles. The proof is identical to eqs. 6.46 – 6.48.

Anomalies

The axial anomaly (Adler-Bell-Jackiw anomaly) follows from the linear divergence of a fermion triangle graph (Fig. 8.1) in the conventional Standard Model. All higher order terms are divergence-free. These terms do not contribute to the axial anomaly. Thus the axial anomaly can properly be regarded as an artifact of the regularization of the divergence of the fermion triangle diagram.

In Two-Tier theory the axial anomaly does not appear to be present. Fermion triangle diagrams in Two-Tier quantum field theories are finite. Thus the source of the anomaly in conventional theories is absent in Two-Tier theories.

A massless Dirac field theory is formally invariant under a chiral transformation implying a conserved axial-vector current. The Two-Tier axial-vector current is

$$j_5^{\mu}(X(y)) = \bar{\psi}(X(y))\gamma^{\mu}\gamma_5\psi(X(y)) \tag{8.84}$$

with formal conservation law:

$$\partial j_5^{\mu}(X(y))/\partial X^{\mu} = 2m\ j_5(X(y)) = 2m\ \bar{\psi}(X(y))\gamma_5\psi(X(y)) \tag{8.85}$$

Eq. 8.85 implies

$$\partial j_5^{\mu}(X(y))/\partial X^{\mu} = 0 \tag{8.86}$$

in the limit $m \to 0$. The question we now address is whether eq. 8.86 holds in Two-Tier perturbation theory – perhaps in the same form as the conventional axial anomaly:

$$\partial j_5^{\mu}(X(y))/\partial X^{\mu} = 2m\ j_5(X(y)) + \alpha_0(4\pi)^{-1}\varepsilon^{\mu\nu\alpha\beta}F_{\alpha\beta}F_{\mu\nu} \quad ? \tag{8.87}$$

where α_0 is the unrenormalized fine structure constant.

The simplest manifestation of the axial anomaly in conventional field theory is the fermion triangle diagram, which we will now examine in Two-Tier quantum field theory. As stated earlier, the Two-Tier triangle diagram is finite and zero unlike the conventional quantum field theory result. *Thus the axial anomaly does not appear to exist in Two-Tier quantum field theory. The axial anomaly is a result of the divergence of the triangle diagram in conventional quantum field theory.*

The absence of the anomaly reflects the absence of divergences in Two-Tier quantum field theory, which preserves chiral invariance. Unlike Pauli-Villars regularization, for example, the finiteness of Two-Tier theory follows from the Gaussian factors. Unlike the dimensional regularization approach (where there is no equivalent to γ_5), Two-Tier theory can use the normal γ_5 matrix.

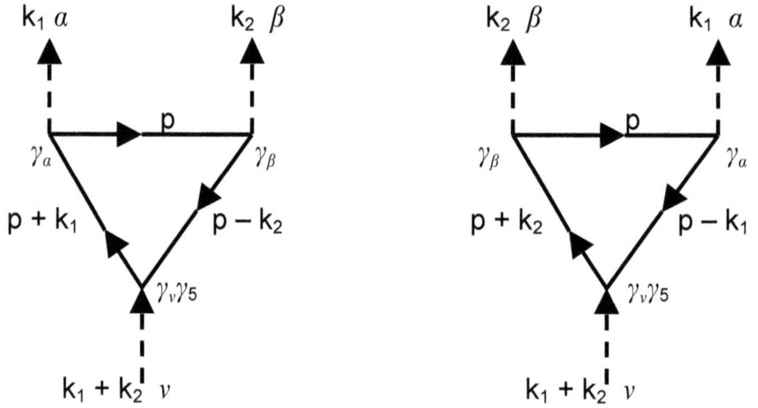

Figure 8.1. The V-V-A triangle diagrams.

The expression for the Two-Tier triangle diagrams is:

$$T_{\alpha\beta\nu}(k_1, k_2) = S_{\alpha\beta\nu}(k_1, k_2) + S_{\beta\alpha\nu}(k_2, k_1) \tag{8.88}$$

where

$$S_{\alpha\beta\nu}(k_1, k_2)\delta^4(k_1 + k_2 - q) = -iN\int d^4y_1 d^4y_2 d^4y_3\, e^{ik_1\cdot y_1 + ik_2\cdot y_2 - iq\cdot y_3} \cdot$$
$$\cdot \mathrm{Tr}\{S_F^{TT}(y_1 - y_3)\gamma_\alpha S_F^{TT}(y_2 - y_1)\gamma_\beta S_F^{TT}(y_3 - y_2)\gamma_\nu\gamma_5\} / (2\pi)^4 \tag{8.89}$$

where N is a constant, and where $S_F^{TT}(z)$ is specified by eq. 7.45. We now define the Fourier transform:

$$S_F^{TT}(z) = -i\int d^4p\, e^{-ip\cdot z}\, \mathscr{S}^{TT}(p) \big/ (2\pi)^4 \tag{8.90}$$

where $\mathscr{S}^{TT}(p)$ defined by eq. 7.93. We then substitute the Fourier transform in eq. 8.89 and perform the coordinate integrations to obtain:

$$S_{\alpha\beta\nu}(k_1, k_2) = N\int d^4p\, Tr\{\mathscr{S}^{TT}(p + k_1)\gamma_\alpha \mathscr{S}^{TT}(p)\gamma_\beta \mathscr{S}^{TT}(p - k_2)\gamma_\nu\gamma_5\} \big/ (2\pi)^4$$

We note that

$$k_1^{\alpha} T_{\alpha\beta\nu}(k_1, k_2) \neq 0 \tag{8.91}$$
$$k_2^{\beta} T_{\alpha\beta\nu}(k_1, k_2) \neq 0 \tag{8.92}$$
$$(k_1 + k_2)^{\nu} T_{\alpha\beta\nu}(k_1, k_2) \neq 0 \tag{8.93}$$

in Two-Tier quantum field theory because the conservation laws are expressed with respect to the X coordinates – not the y coordinates. Thus since k_1^{α} corresponds $\partial/\partial y_\alpha$, and not $\partial/\partial X_\alpha$ there is no reason for eqs. 8.91-93 to be zero. However at "large distances" relative to M_c^{-1} we see

$$k_1^{\alpha} T_{\alpha\beta\nu}(k_1, k_2) \cong 0 \tag{8.94}$$
$$k_2^{\beta} T_{\alpha\beta\nu}(k_1, k_2) \cong 0 \tag{8.95}$$
$$(k_1 + k_2)^{\nu} T_{\alpha\beta\nu}(k_1, k_2) \cong 0 \tag{8.96}$$

to very good approximation since the Gaussian damping factor in the fermion propagators is approximately unity and thus the Two-Tier expression becomes essentially the same as the conventional field theory expression.

On the other hand at very short distances the anomaly appears to be absent since Two-Tier theory is very well behaved at high energy with

$$\mathscr{S}^{TT}(p) \backsim \gamma^0 M_c^{\,6}\, p^{-7} + \mathscr{O}(p^{-9}) \tag{7.102}$$

As a result we see

$$k_1^{\alpha} T_{\alpha\beta\nu}(k_1, k_2) \sim p^{4-21} \sim p^{-17} \tag{8.97}$$

as $p \to \infty$ is highly convergent. Thus there is no high energy divergence unlike conventional field theory where the integral is linearly divergent. And so no anomaly is generated.

Asymptotic Freedom and Quark Confinement

Two-Tier quantum field theory is totally consistent with the Standard Model at currently accessible energies. Thus if the Standard Model Quantum Chromodynamics (QCD) sector is asymptotically free, and also gives quark confinement at large distance (compared to M_c^{-1}), then similar statements would also be true in the Two-Tier version of QCD.

We assume M_c is extremely large – much beyond current energies – and possibly of the order of the Planck mass. Fig. 8.2 depicts the various regions in Two-Tier QCD assuming a very large M_c.

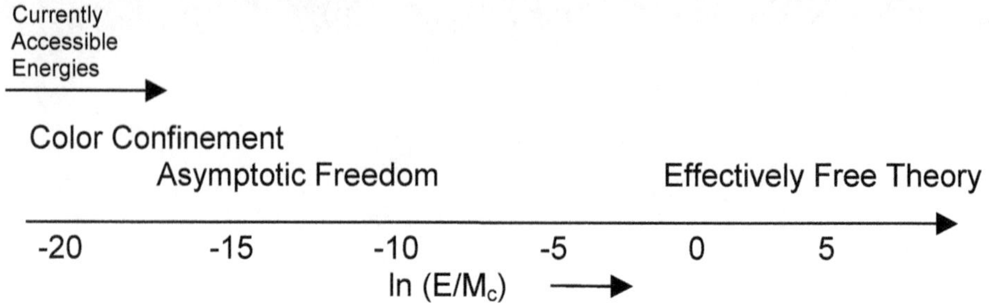

Figure 8.2. A depiction of the Two-Tier QCD regions as a function of the logarithm of the energy assuming M_c is of the order of the Planck mass.

Two-Tier Coordinates

We have developed a physical picture of Two-Tier coordinate systems in which we define two sets of related coordinates. This picture views ordinary real 4-dimensional space as a low energy approximation to a complex 4-dimensional space that only becomes apparent at ultra-high energies with the scale set by M_c. The imaginary part of the coordinates is based on the excitations of a quantum field Y^μ. The real part of the coordinates is the familiar 4-dimensional c-number coordinate system.

Another way to view the Two-Tier X and y coordinate systems is to view them as defining a four-dimensional hyperplane in a six (or eight) dimensional real space-time. See Blaha (2004) for a detailed discussion.

It would also be interesting to consider an extension of the theory to the case where both the real and imaginary parts of the coordinate system are quantum fields. In this case the c-number coordinates of daily experience would emerge as a "condensation", or spontaneous symmetry breaking, phenomena.

One could also generalize the Y quantum field to a non-Abelian gauge field, which, properly handled, could be the origin of internal symmetries such as QCD SU(3) using a Kaluza-Klein approach within a general relativistic *vierbein* framework. This approach is an alternative to the compactification of dimensions that is a cornerstone of SuperString theories.

9. Two-Tier Quantum Gravity: Finite!

Introduction

There are numerous excellent books and monographs on classical gravity and a large literature on quantum gravity.[101] Therefore our discussion will assume the reader is familiar with classical General Relativity and aware of attempts to create quantum theories of gravity.

We will begin by establishing the general form of Two-Tier classical General Relativity and then proceed to define a quantization procedure. We will work in Minkowski space with three space and one time dimension. The flat-space metric $\eta_{\alpha\beta}$ is defined as diagonal with $\eta_{00} = 1$ and $\eta_{ij} = -\delta_{ij}$ for i, j = 1, 2, 3.

Two-Tier General Relativity

In developing Quantum Gravity we will make the same *ansatz* that we have made throughout our development of Two-Tier quantum field theories: all field expressions are functions of the X coordinate field system, which in turn are functions of the "ordinary" y space-time coordinate system. Two-Tier Theory of Quantum Gravity is invariant under special relativistic transformations. The dynamical field equations, which are strictly functions of the X coordinates, are covariant under general relativistic transformations. The rationale for these assumptions is described in detail in chapter 10.

We define the proper time differential $d\tau$ as

$$d\tau^2 = g_{\mu\nu}(X(y))dX^\mu dX^\nu \tag{9.1}$$

where, as usual,

$$X^\mu(y) = y^\mu + i\, Y^\mu(y)/M_c^{\,2} \tag{3.12}$$

Thus eq. 9.1 could be written:

---

[101] H. Weyl, *Space, Time, Matter* (Dover, New York, 1950); L. D. Landau and E. M. Lifshitz, *The Classical Theory of Fields*, (Addison-Wesley, New York, 1962); S. Weinberg, *Gravitation and Cosmology*, (John Wiley & Sons, New York, 1972); C. W. Misner, K. S. Thorne and J. A. Wheeler, *Gravitation*, (W. H. Freeman, San Francisco, 1973); B. S. DeWitt, Phys. Rev. **162**, 1239 (1967), **162**, B1195 (1967); R. P. Feynman, Acta Physica Polonica **24**, 697 (1963); S. Deser and P. van Nieuwenhuizen, Phys. Rev. Letters **32**, 245 (1974); S. Deser, H.-S. Tsao and P. van Nieuwenhuizen, "One Loop Divergences of the Einstein-Yang-Mills System", Brandeis Univ. preprint (1974); S. Weinberg, Phys. Rev. **138**, B988 (1965); L. Smolin, *Three Roads to Quantum Gravity*, (Basic Books, New York, 2001); L. Smolin, "How Far are We From the Quantum Theory of Gravity", (Univ. Waterloo preprint (2003) and references therein; T. Thiemann, "Lectures on Loop Quantum Gravity", Preprint AEI-2002-087, Albert Einstein Insitute, Golm, Germany (2002) and references therein; A. pais and G. E. Uhlenbeck, Phys. Rev. **79**, 145 (1950); G. E. Uhlenbeck, "Lecture Notes on General Relativity", The Rockefeller University (1967), unpublished; S. Blaha, "Generalization of Weyl's Unified Theory to Encompass a Non-Abelian Internal Symmetry Group" SLAC-PUB-1799, Aug 1976; S. Blaha, "Quantum Gravity and Quark Confinement" Lett. Nuovo Cim. **18**, 60 (1977); R. Utiyama, Phys. Rev. **101**, 1597 (1956); T. W. B. Kibble, J. Math. Phys. **2**, 212 (1961); R. Arnowitt, S. Deser, and C. W. Misner, Phys. Rev. **117**, 1595 (1960); and references therein.

$$d\tau^2 = g_{\mu\nu}(X(y))(\eta^\mu{}_\alpha + iM_c^{-2}\partial Y^\mu/\partial y^\alpha)(\eta^\nu{}_\beta + iM_c^{-2}\partial Y^\nu/\partial y^\beta)dy^\alpha dy^\beta \quad (9.2)$$

The inverse of $g_{\mu\nu}$, denoted $g^{\nu\lambda}$, satisfies

$$g_{\mu\nu}(X(y))g^{\nu\lambda}(X(y)) = \delta_\mu{}^\lambda \quad (9.3)$$

Since the algebraic manipulation of the tensor indices is the same as that of the conventional theory of gravitation the Two-Tier affine connection is:

$${}_X\Gamma^\sigma{}_{\lambda\mu} = \tfrac{1}{2}\, g^{\nu\sigma}\{\partial g_{\mu\nu}/\partial X^\lambda + \partial g_{\lambda\nu}/\partial X^\mu - \partial g_{\lambda\mu}/\partial X^\nu\} \quad (9.4)$$

The Two-Tier Riemann-Christoffel curvature tensor is:

$${}_X R^\lambda{}_{\mu\nu\kappa} \equiv \partial_X\Gamma^\lambda{}_{\mu\nu}/\partial X^\kappa - \partial_X\Gamma^\lambda{}_{\mu\kappa}/\partial X^\nu + {}_X\Gamma^\alpha{}_{\mu\nu}\,{}_X\Gamma^\lambda{}_{\kappa\alpha} - {}_X\Gamma^\alpha{}_{\mu\kappa}\,{}_X\Gamma^\lambda{}_{\nu\alpha} \quad (9.5)$$

and the Two-Tier Ricci tensor is

$${}_X R_{\mu\nu} = {}_X R^\alpha{}_{\mu\alpha\nu} \quad (9.6)$$

The Two-Tier curvature scalar is

$${}_X R = g^{\mu\nu}\,{}_X R_{\mu\nu} \quad (9.7)$$

We also define

$${}_X R_{\lambda\mu\nu\kappa} = g_{\lambda\alpha}\,{}_X R^\alpha{}_{\mu\nu\kappa} \quad (9.8)$$

with the result

$${}_X R_{\lambda\mu\nu\kappa} = \tfrac{1}{2}[\partial^2 g_{\lambda\nu}/\partial X^\kappa\partial X^\mu - \partial^2 g_{\mu\nu}/\partial X^\kappa\partial X^\lambda - \partial^2 g_{\lambda\kappa}/\partial X^\nu\partial X^\mu + \partial^2 g_{\mu\kappa}/\partial X^\nu\partial X^\lambda] +$$

$$+\, g_{\alpha\beta}[{}_X\Gamma^\alpha{}_{\nu\lambda}\,{}_X\Gamma^\beta{}_{\mu\kappa} - {}_X\Gamma^\alpha{}_{\kappa\lambda}\,{}_X\Gamma^\beta{}_{\mu\nu}] \quad (9.9)$$

We denote the fact that all quantities in eqs. 9.4 – 9.8 are only functions of X by placing a left subscript X on each quantity.

The algebraic properties, and the Bianchi identities, satisfied by ${}_X R_{\lambda\mu\nu\kappa}$ in the Two-Tier theory of gravitation are identical to those of the conventional theory with all derivatives being with respect to X.

The Two-Tier version of Einstein's field equations is:

$${}_X R_{\mu\nu} - \tfrac{1}{2}\, g_{\mu\nu}\,{}_X R = -8\pi G\, T_{\mu\nu} \quad (9.10)$$

where G is Newton's gravitational constant (6.674×10^{-11} $m^3kg^{-1}s^{-2}$) and $T_{\mu\nu}$ is the energy-momentum tensor – also strictly a function of X. It is convenient to define the coupling constant

$$\kappa = \sqrt{4\pi G} \tag{9.11}$$

Lagrangian Formulation

We will now formulate a Two-Tier Quantum Gravity theory following the same ansatz that we have used throughout this book.

Unified Standard Model and Quantum Gravity Lagrangian

We define the Lagrangian, and action, for the unified quantum field theory of Gravitation and the Standard Model as

$$L_{\text{Unified}} = \int d^4y\, \mathscr{L}_{\text{Unified}} \tag{9.12}$$

$$\mathscr{L}_{\text{Unified}} = J\sqrt{g(X)}\, \left(\mathscr{L}_F^{\text{Grav}}(X^\mu) + \mathscr{L}_F^{\text{SM}}(X^\mu)\right) + \mathscr{L}_C \tag{9.13}$$

with

$$\mathscr{L}_F^{\text{Grav}}(X^\mu) = (2\kappa^2)^{-1}{}_X R \tag{9.14}$$

where $\mathscr{L}_F^{\text{SM}}$ is the complete "normal" Quantum Field theory Lagrangian for the Standard Model version under consideration written in a general covariant form, g(X) is the absolute value of the determinant of $g_{\mu\nu}$, and J is the Jacobian of eq. A.21. *All particle fields in $\mathscr{L}_F^{SM}$ are assumed to be functions of the X^μ coordinate only. The dependence of the particle fields on the "underlying" coordinates y^μ is assumed to be solely through X^μ.* The Lagrangian $\mathscr{L}_{\text{Unified}}$ is a separable Lagrangian of the type of eq. A.26 embodying the composition of extrema described in Appendix A.

As in all of cases that we have considered, we have specified the coordinate part of the Lagrangian $\mathscr{L}_C$ as

$$\mathscr{L}_C = -\tfrac{1}{4}\, F_Y^{\mu\nu} F_{Y\mu\nu} \tag{3.15}$$

with

$$F_{Y\mu\nu} = \partial Y_\mu/\partial y^\nu - \partial Y_\nu/\partial y^\mu \tag{3.14}$$

and

$$F_Y^{\mu\nu} = \eta^{\mu\alpha}\eta^{\nu\beta} F_{Y\alpha\beta} \tag{9.15}$$

Why Are the Y Field Dynamics Independent of the Gravitational Field?

It is evident from eqs. 9.12-3 and 9.15 that the Y field is truly free and, in particular, does not depend on the gravitational field as represented by $\sqrt{g}$ and $g_{\mu\nu}$. Our rationale for this formulation is described in detail in chapter 10. For the moment it suffices to make the following remarks. The Y field is a quantum field at each point in space-time including regions with ultra-strong gravitational fields such as the neighborhoods of Black Holes. If Y were to depend on the gravitational field then the Y field could be appreciable in such regions and might even be a "classical" field. In this case we would have new dimensions, albeit imaginary, for which no evidence exists.

Furthermore, the Y part of the Lagrangian establishes a functional relation between the imaginary Y coordinates and the real space-time y coordinates. The Principle of Equivalence applies only to real coordinates and has not been shown to apply to imaginary Quantum Dimensions™. Thus there is no reason to require the Y part of the action to be invariant under general coordinate transformations.

Lastly, the non-invariance of the Y part of the action under general coordinate transformations effectively creates an "absolute" coordinate system – actually a class of "absolute coordinate systems" – namely the class of inertial reference frames that are related to each other by special relativistic transformations. This feature does not conflict with our knowledge of the universe. The universe appears to be almost flat. The large-scale distribution of masses is responsible for this flatness. The flatness, or flattened space if it is slightly curved, together with Mach's Principle (inertial forces are absent in the reference frame determined by the distribution of masses in the universe) selects a preferred class of local reference frames – local inertial reference frames. Since space is almost flat, or flat, these local reference frames occupy a large volume (if we exclude regions with intense gravitational fields.) We can define the Y field within this class of local inertial reference frames in each locale and establish a satisfactory quantum field theory. Thus we have a dynamics defined in the variable X, which we require to be covariant under general coordinate transformations, and a local "ground state" that "breaks" general coordinate invariance down to special relativistic invariance. See chapter 10 for a more complete discussion.

No "Space-time Foam"

The fact that our unified theory of the known forces of Nature *self-consistently* has a weak gravitational field at high energies (the graviton sector is finite to all orders in perturbation theory) supports the formulation of eq. 9.12-3. Gravity becomes weaker at ultra-short distances. Therefore space-time is not quantum foam at ultra-short distances but rather smooth and flat *a là* special relativity – consistent with our formulation.

Quantum Gravity – Scalar Particle Model Lagrangian

While the application of the Two-Tier approach to the unified theory is a straightforward extension of the concepts and approaches described in the preceding chapters, it is useful to consider a simplified model that minimizes the tensorial verbiage so that the concepts and features might better stand out. The procedure differs only in detail from the case of gauge fields.

The introduction of spinor fields requires the use of a Two-Tier *vierbein* formalism, which is straightforward to develop. A Two-Tier *vierbein* field $e^{\mu}{}_{a}$ is a function of X, $e^{\mu}{}_{a}(X)$, with $g_{\mu\nu} = e_{\mu a}(X)e_{\nu}{}^{a}(X)$ where the index a is an index of a flat tangent space defined at each space-time point. The Two-Tier formulation of a *vierbein* theory is similar to the other Two-Tier formulations that we have considered and will not be developed here.

Thus we will consider the Lagrangian model for a scalar particle field interacting with the $g_{\mu\nu}$ gravitational field:

$$L_{GS} = \int d^4y \sqrt{g(X)}\, \mathscr{L}_{GS} + \mathscr{L}_C \tag{9.16}$$

$$\mathscr{L}_{GS} = J\, \mathscr{L}_F^{Grav}(X^\mu) + J\, \mathscr{L}_{F\phi}(X^\mu) \tag{9.17}$$

with covariant versions of eqs. 5.21 and 5.24:

$$\mathscr{L}_{F\phi} = \tfrac{1}{2}\, [\, g^{\mu\nu}\partial\phi/\partial X^\mu\, \partial\phi/\partial X^\nu - m^2\phi^2] + \mathscr{L}_{F\phi int} \tag{9.18}$$

$$\mathscr{L}_{F\phi int} = \tfrac{1}{4!}\, \chi_0\, \phi(X(y))^4 + \tfrac{1}{2}\, (m^2 - m_0^2)\phi^2 \tag{9.19}$$

A Justifiable Weak Field Approximation for Quantum Gravity

Many discussions of quantizing conventional gravity make a weak field approximation for the gravity sector which, in view of divergences in the resulting quantum field theory, are impossible to justify:

$$g_{\mu\nu} = \eta_{\mu\nu} + \kappa h_{\mu\nu} \tag{9.20}$$

where $\eta_{\mu\nu}$ is the flat space-time metric and $h_{\mu\nu}$ is a "small" deviation ($\langle h_{\mu\nu}\rangle \ll 1$) from the flat space-time metric.

The Two-Tier formulation of quantum gravity is finite and the effective field becomes increasingly weaker at short distances. Thus the weak field approximation becomes *more accurate* at short distances:

$$g_{\mu\nu}(X(y)) \simeq \eta_{\mu\nu} + \kappa h_{\mu\nu}(X(y)) \tag{9.21a}$$

At short distances space-time can be considered approximately flat (except possibly in the neighborhood of singularities) with quantum fluctuations embodied in $h_{\mu\nu}$. Thus eq. 9.21a is reasonable within the context of Two-Tier Quantum Gravity.

To first order in $h_{\mu\nu}$ the square root of the absolute value of the determinant of the metric tensor is:

$$\sqrt{g(X)} \simeq 1 + \tfrac{1}{2}\, \kappa h^\sigma{}_\sigma(X(y)) \tag{9.21b}$$

Quantization of Quantum Gravity – Scalar Particle Model

We now proceed to quantize gravity based on the linearization of the gravitational field equations in the weak field approximation. Assuming eq. 9.21a and keeping terms to first order in $h_{\mu\nu}$ gives the affine connection:

$$ {}_X\Gamma^{\sigma}{}_{\mu\nu} = \tfrac{1}{2}\,\kappa\eta^{\sigma\alpha}[\partial h_{\alpha\nu}/\partial X^{\mu} + \partial h_{\alpha\mu}/\partial X^{\nu} - \partial h_{\mu\nu}/\partial X^{\alpha}] + \mathcal{O}(h^2) \qquad (9.22) $$

and the Ricci tensor:

$$ {}_X R_{\mu\nu} = \partial_X\Gamma^{\lambda}{}_{\lambda\mu}/\partial X^{\nu} - \partial_X\Gamma^{\lambda}{}_{\mu\nu}/\partial X^{\lambda} + \mathcal{O}(h^2) \qquad (9.23) $$

Thus the linearized gravitation lagarangian terms are

$$ L^{Grav} = \int d^4y\, \sqrt{g(X)}\, J\mathscr{L}_F^{Grav}(X^{\mu}) \rightarrow L^{Grav}{}_{linear} = \int d^4y\, J\mathscr{L}^{Grav}{}_{linear}(X^{\mu}) \qquad (9.24) $$

The scalar particle Lagrangian terms become

$$ L^{\phi} = \int d^4y\sqrt{g(X)}J\mathscr{L}_{F\phi} \rightarrow \int d^4yJ\{[\tfrac{1}{2}(\eta^{\mu\nu}\partial_{\mu}\phi\partial_{\nu}\phi - m^2\phi^2) + \mathscr{L}_{F\phi int}] + $$

$$ + \tfrac{1}{2}\kappa h^{\mu\nu}\partial_{\mu}\phi\partial_{\nu}\phi + \tfrac{1}{4}\,\kappa h(\eta^{\mu\nu}\partial_{\mu}\phi\partial_{\nu}\phi - m^2\phi^2) + $$

$$ + \tfrac{1}{2}\,\kappa h\mathscr{L}_{F\phi int}\} \qquad (9.25) $$

with the notation $h = h^{\sigma}{}_{\sigma}$ and using

$$ \partial_{\mu} \equiv \partial/\partial X^{\mu} \qquad (9.26) $$

$\eta^{\mu\nu}$ and $\eta_{\mu\nu}$ are used to raise and lower indices in keeping with the linearized, weak field approximation.

The Y terms in the Lagrangian are (as previously):

$$ L^Y = \int d^4y\mathscr{L}_C = -\tfrac{1}{4}\int d^4y\eta^{\mu\nu}\eta^{\alpha\beta}F_{Y\mu\alpha}F_{Y\nu\beta} \qquad (9.27) $$

We will lump the higher order terms (in h) in the gravity part of the Lagrangian, and the scalar particle part of the Lagrangian, into

$$ L_{Higher} = \int d^4y\, J\mathscr{L}_{Higher}(h, \phi) \qquad (9.28) $$

Thus the complete lagragian for a scalar particle interacting with gravitons is

$$L_{GS} = L^{Grav}{}_{linear} + L^{\phi}{}_{linear} + L^{Y} + L_{Higher} \tag{9.29}$$

The linearized gravitational Lagrangian term $L^{Grav}{}_{linear}$ generates the field equations:

$$\Box_X h_{\mu\nu} + \partial_\nu\partial_\mu h - \partial_\alpha\partial_\nu h^\alpha{}_\mu - \partial_\alpha\partial_\mu h^\alpha{}_\nu = \kappa S_{\mu\nu} \tag{9.30}$$

where

$$\partial_\mu S^\mu{}_\nu = ½\, \partial_\nu S^\sigma{}_\sigma \tag{9.31}$$

to 0th order in h and where

$$\Box_X = (\partial/\partial X^\nu)(\partial/\partial X_\nu) \tag{9.32}$$

The most general coordinate transformation that maintains the weakness of the gravitational field has the form:

$$y^a \rightarrow\ y'^a = y^a + \chi^a(X(y)) \tag{9.33}$$

This transformation induces a gauge transformation in $h_{\mu\nu}$ to:

$$h'_{\mu\nu} = h_{\mu\nu} - \partial_\mu\chi_\nu - \partial_\nu\chi_\mu \tag{9.34}$$

It is easy to verify that eq. 9.30 is satisfied by $h'_{\mu\nu}$ if it is satisfied by $h_{\mu\nu}$.

Let us assume that we perform a gauge transformation making $h_{\mu\nu}$ traceless:

$$h^\sigma{}_\sigma = 0 \tag{9.35}$$

and choose the gauge

$$\partial^\mu h_{\mu\nu} = 0 \tag{9.36}$$

then eq. 9.30 becomes the wave equation:

$$\Box_X h_{\mu\nu} = \kappa S_{\mu\nu} \tag{9.37}$$

Another gauge transformation of the free field $h_{\mu\nu}$ (if $S_{\mu\nu}= 0$) makes

$$h_{\mu 0} = h_{0\mu} = 0 \tag{9.38}$$

while retaining

$$h_{\mu\nu} = h_{\nu\mu} \tag{9.39}$$

The general solution[102] for the free field $h_{\mu\nu}$ (with $S_{\mu\nu}= 0$ in eq. 9.37) can be expressed as a Fourier expansion:

$$h_{\mu\nu}(X(y)) = \int d^3k\, N_0(k) \sum_{\lambda=1}^{2} \varepsilon_{\mu\nu}(k, \lambda)[a(k,\lambda)\, e^{-ik\cdot X} + a^\dagger(k,\lambda)\, e^{ik\cdot X}] \tag{9.40}$$

where $\lambda = 1,2$ labels the ± 2 helicity states, and where $N_0(k)$ is specified by eq. 3.25. The equal time ($y'^0 = y^0$) commutation relations are:

$$[h_{\mu\nu}(X(y)), h_{\alpha\beta}(X(y'))] = [\pi_{\mu\nu}(X(y)), \pi_{\alpha\beta}(X(y'))] = 0 \tag{9.41}$$

$$[h_{\alpha\beta}(X(y')), \pi_{\mu\nu}(X(y))] = i\, \mathscr{D}_{\alpha\beta\mu\nu}(\mathbf{X}(y) - \mathbf{X}(y')) \tag{9.42}$$

for μ, $\nu = 1, 2, 3$ and where

$$\pi_{\mu\nu}(X(y)) = \partial h_{\mu\nu}(X(y))/\partial y^0 \tag{9.43}$$

in the Y Coulomb gauge where $X^0 = y^0$. $\mathscr{D}_{\alpha\beta\mu\nu}$ is specified by:

$$\mathscr{D}_{\alpha\beta\mu\nu}(X(y)- X(y')) = \int d^3k\, e^{i\,\mathbf{k}\cdot(\mathbf{X}(y) - \mathbf{X}(y'))}\, \Pi_{\alpha\beta\mu\nu}(\mathbf{k})/(2\pi)^3 \tag{9.44}$$

$$\Pi_{\alpha\beta\mu\nu}(\mathbf{k}) = \tfrac{1}{2}\, [(\delta_{\alpha\mu} - k_\alpha k_\mu/\mathbf{k}^2)(\delta_{\beta\nu} - k_\beta k_\nu/\mathbf{k}^2) + (\delta_{\alpha\nu} - k_\alpha k_\nu/\mathbf{k}^2)(\delta_{\beta\mu} - k_\beta k_\mu/\mathbf{k}^2) -$$

$$- (\delta_{\alpha\beta} - k_\alpha k_\beta/\mathbf{k}^2)(\delta_{\mu\nu} - k_\mu k_\nu/\mathbf{k}^2)] \tag{9.45}$$

where α, β, μ, $\nu = 1, 2, 3$.

The "transverse" graviton propagator can be represented as a time-ordered product of field operators:

$$i\Delta_{F2}{}^{TT}(y_1 - y_2)_{\lambda\tau\rho\sigma} = <0|T(h_{\lambda\tau}(X(y_1)), h_{\rho\sigma}(X(y_2)))|0> \tag{9.46}$$

$$= -i \int \frac{d^4k\, e^{-ik\cdot(y_1 - y_2)}\, b_{\lambda\tau\rho\sigma}(k) R(\mathbf{k}, y_1 - y_2)}{(2\pi)^4\, (k^2 + i\varepsilon)}$$

[102] S. Weinberg, Phys. Rev. **135**, B1049 (1964); Phys. Rev. **138**, B988 (1965)

where $R(\mathbf{k}, y_1 - y_2)$ is the Gaussian factor appearing in propagators throughout Two-Tier theories, $\mathbf{k}$ is a spatial 3-vector, and where $b_{\mu\nu\rho\sigma}(k)$ is a function of k only:

$$b_{\alpha\beta\mu\nu}(k) = \tfrac{1}{2}[(\eta_{\alpha\mu} - k_\alpha k_\mu/\mathbf{k}^2)(\eta_{\beta\nu} - k_\beta k_\nu/\mathbf{k}^2) + (\eta_{\alpha\nu} - k_\alpha k_\nu/\mathbf{k}^2)(\eta_{\beta\mu} - k_\beta k_\mu/\mathbf{k}^2) -$$

$$- (\eta_{\alpha\beta} - k_\alpha k_\beta/\mathbf{k}^2)(\eta_{\mu\nu} - k_\mu k_\nu/\mathbf{k}^2)] \qquad (9.47)$$

where $\alpha, \beta, \mu, \nu = 0, 1, 2, 3$.

The quantum gravitational interaction also has an "instantaneous" part (similar to the instantaneous Coulomb interaction of QED) in addition to the transverse interaction embodied in eq. 9.46. This "instantaneous" interaction contains the Newtonian potential (described later) as its large distance limit. The sum of the instantaneous interaction and the transverse interaction gives the total gravitational interaction.

The above graviton propagator has the form given in eq. 6.102. The calculation of the leading behavior is the same as that of the Two-Tier scalar boson propagator except for the presence of factors such as $\eta_{\rho\sigma}$. The leading momentum dependence of the graviton propagator in momentum space is

$$i\Delta_{F2}{}^{TT}(p)_{\lambda\tau\rho\sigma} \backsim p^{-6} \qquad (9.48)$$

The graviton vertices in Two-Tier Quantum Gravity will be described within the framework of the path integral formulation.

Quantum Gravity–Scalar Particle Model Path Integral

A path integral formalism can be developed for Two-Tier Quantum Gravity interacting with matter fields. In this section we will consider the case of a matter field consisting of massive scalar bosons with a quartic interaction. The path integral formalism that we develop is similar to that of Yang-Mills theories in the previous chapter.

The Two-Tier path integral for a Quantum Gravity–Scalar Particle Theory can be written as:

$$Z(J, J^{\mu\nu}) = N \int D\phi Dh DY \Delta_{FPG}(h)\delta(F(h)) \exp\{i\int d^4y[\mathscr{I}(\mathscr{L}^{Grav}{}_{linear}(X^\mu) +$$

$$+ \mathscr{L}^{\phi}{}_{linear}(X^\mu) + \mathscr{L}_{Higher}(h, \phi)) + \mathscr{L}_C(X, y) +$$

$$+ j_\mu(y)Y^\mu(y) + J(y)\phi(X) + J^{\mu\nu}(y)h_{\mu\nu}(X)]\} \,\Big|_{j_\mu = 0} \qquad (9.49)$$

where $\delta(F(h))$ specifies the gauge as a functional delta function, and $\Delta_{FPG}(h)$ is the corresponding Faddeev-Popov determinant. $\mathcal{J}$ is the Jacobian for the transformation from y coordinates to X coordinates. The Faddeev-Popov determinant $\Delta_{FPG}(h)$ can be calculated in the standard way. First we note

$$\delta(F(h^{\chi})) = \delta(\chi - \chi_0) \left| \det \delta F(h_{\mu\nu}{}^{\chi}(X))/\delta\chi(X) \right|^{-1} \Big|_{F(h)=0} \tag{9.50}$$

where

$$h_{\mu\nu}{}^{\chi} = h_{\mu\nu} - \partial_\mu \chi_\nu - \partial_\nu \chi_\mu \tag{9.34}$$

Then

$$\Delta_{FPG}(h) = \left| \det \delta F(h^{\chi}(X))/\delta\chi(X) \right| \Big|_{F(h)=0} \tag{9.51}$$

We will choose the gauge of eq. 9.36 to evaluate the Faddeev-Popov determinant. Under an infinitesimal gauge transformation of the form:

$$h_{\mu\nu}{}^{\chi}(X) = h_{\mu\nu}(X) - \partial_\mu \chi_\nu - \partial_\nu \chi_\mu \tag{9.52}$$

which preserves the weak field nature of $h_{\mu\nu}$, we find

$$F_\nu(h^{\chi}) = \partial^\mu (h_{\mu\nu}(X) - \partial_\mu \chi_\nu - \partial_\nu \chi_\mu)$$

$$= -\Box_X \chi_\nu(X) - \partial_\nu \partial^\mu \chi_\mu \tag{9.53}$$

Thus

$$\delta F_\mu(h^{\chi}(X))/\delta\chi^\nu(X) = -\eta_{\mu\nu}\Box_X - \partial_\mu\partial_\nu \tag{9.54}$$

and

$$\Delta_{FP}(A) = \left| \det(-\eta_{\mu\nu}\Box_X - \partial_\mu\partial_\nu) \right| \Big|_{F(h)=0} \tag{9.55}$$

We note the Two-Tier Faddeev-Popov determinant is solely a function of the X coordinates. The determinant only introduces an overall multiplicative constant that can be absorbed into the normalization constant N. This fact becomes evident if we follow the standard procedure and rewrite the determinant as a path integral over anti-commuting c-number fields with a ghost Lagrangian. Then we see that the ghost does not interact with the other fields and thus only generates an overall multiplicative constant that can be absorbed in N:

$$\Delta_{FPG}(h) = \int Dc^* Dc \exp[\, i\int d^4X\, \mathscr{L}^{ghost}(X^\mu)] \tag{9.56}$$

where

$$\mathscr{L}^{ghost}(X^\mu) = c^{\mu *}(X)[\eta_{\mu\nu}\Box_X + \partial_\mu\partial_\nu]c^\nu(X) \tag{9.57}$$

We now go through the same analysis as we did in the ϕ^4 theory path integral example and the Yang-Mills path integral example (with some superficial differences). First we integrate the linear part of the Y field Lagrangian as we did previously. Then we integrate the linear part of the ϕ field Lagrangian as done previously. Lastly we integrate the linear part of the gravitation Lagrangian to obtain the path integral for the perturbative expansion with the result:

$$Z(J, J^{\mu\nu}) = N \,\{\exp[i\int d^4y \mathscr{L}_{Higher}(\partial/\partial y^\nu, -i\delta/\delta J^{\mu\nu}(y), -i\delta/\delta J(y))]\cdot$$

$$\cdot \exp[-\tfrac{1}{2}\, i\int d^4y_1 d^4y_2 J^{\mu\nu}(y_1)\Delta_{F2}{}^{TT}(y_1 - y_2, z)_{\mu\nu\rho\sigma} J^{\rho\sigma}(y_2)]\cdot$$

$$\cdot \exp[-\tfrac{1}{2}\, i\int d^4y_1 d^4y_2 J(y_1)\Delta_F{}^{TT}(y_1 - y_2, z)J(y_2)]\} \Big|_{z=y_1-y_2} \tag{9.58}$$

There are two issues that arise in the development of eq. 9.58:

1.) The integral over y in $\int d^4y\mathscr{L}_{Higher}$ which began as the integral $\int d^4y \mathscr{J} \mathscr{L}_{Higher} = \int d^4X\mathscr{L}_{Higher}$ in eq. 9.49; and

2.) The handling of derivatives with respect to X in $\mathscr{L}_{Higher}$.

These are resolved by the following respective observations:

1.) See the discussions following eqs. 6.34 and 8.29 that apply here as well without change.

2.) See the discussion following eq. 8.32, which applies here with only superficial changes. In particular we note that the derivative with respect to X of the graviton propagator (eq. 9.46-7) is specified by the following:

$$\partial i\Delta_{F2}{}^{TT}(y_1 - y_2)_{\lambda\tau\rho\sigma}/\partial X^\mu(y_1) = \partial[i\Delta_{F2}{}^{TT}(y_1 - y_2, z)_{\lambda\tau\rho\sigma}]/\partial y_1{}^\mu \Big|_{z = y_1 - y_2} \tag{9.59}$$

where

$$i\Delta_{F2}{}^{TT}(y_1 - y_2, z)_{\lambda\tau\rho\sigma} = -i\int\frac{d^4k\, e^{-ik\cdot(y_1 - y_2)}\, b_{\lambda\tau\rho\sigma}(k)R(\mathbf{k}, z)}{(2\pi)^4\,(k^2 + i\varepsilon)} \tag{9.60}$$

Thus

$$\frac{\partial\, i\Delta_{F2}{}^{TT}(y_1 - y_2)_{\lambda\tau\rho\sigma}}{\partial X^\mu(y_1)} = -i\int\frac{d^4k\, e^{-ik\cdot(y_1 - y_2)}\,(-ik_\mu) b_{\lambda\tau\rho\sigma}(k)R(\mathbf{k}, y_1 - y_2)}{(2\pi)^4\,(k^2 + i\varepsilon)} \tag{9.61}$$

Therefore derivatives with respect to X in the interaction Lagrangian terms can be replaced by derivatives with respect to y if the graviton propagator is generalized to eq. 9.60. After taking all derivatives with respect to y, we set z equal to the respective $y_1 - y_2$ (actually the difference of the appropriate variables) in each propagator with results similar to eq. 9.61.

$$Z(J, J^{\mu\nu}) = N\,\{\exp[i\int d^4y\,\mathscr{L}_{Higher}(\partial/\partial y^\nu, -i\delta/\delta J^{\mu\nu}(y), -i\delta/\delta J(y))]\cdot$$
$$\cdot \exp[-\tfrac{1}{2}\, i\int d^4y_1 d^4y_2 J^{\mu\nu}(y_1)\Delta_{F2}{}^{TT}(y_1 - y_2, z)_{\mu\nu\rho\sigma} J^{\rho\sigma}(y_2)]\cdot$$
$$\cdot \exp[-\tfrac{1}{2}\, i\int d^4y_1 d^4y_2 J(y_1)\Delta_F{}^{TT}(y_1 - y_2, z) J(y_2)]\}\,\Big|_{z=y_1 - y_2} \tag{9.58a}$$

To be precise eq. 9.58a is interpreted as executing the following steps:

1. For a given process take appropriate functional derivatives of Z(J) with respect to J and $J^{\mu\nu}$.

2. Then expand the exponential factors in a perturbation series applying any derivatives with respect to y in $\mathscr{L}_{Higher}$. Do not perform any of the $\int d^4y_1 d^4y_2$ integrals.

3. Then set $z = y_1 - y_2$ in each $\Delta_{Fk}{}^{TT}(y_1 - y_2, z)$ and $\Delta_{F2}{}^{TT}(y_1 - y_2, z)_{\mu\nu\rho\sigma}$ propagator.

4. Lastly perform all $\int d^4y_1 d^4y_2$ integrals.

Thus we achieve a path integral formulation that is very similar to the corresponding expression in conventional field theory – the only difference is in the form of the free field propagators, which each now contain a Gaussian factor. The net consequence is that graviton vertices result in exactly the same polynomials in momenta as the conventional theory.

Thus Two-Tier gravity generates a perturbative expansion identical to conventional quantum gravity except that each graviton propagator has a Gaussian damping factor $R(\mathbf{k}, y_1 - y_2)$. At low energies the tree diagrams of conventional gravity theory emerge to good approximation in Two-Tier gravity. All diagrams with loops converge. Thus Two-Tier gravity is finite.

Finiteness of Quantum Gravity–Scalar Particle Model

Two-Tier Quantum Gravity perturbation theory is finite. Calculations are highly convergent at large momentum ($\gtrsim M_c$). At low momentum the Two-Tier theory is similar to conventional gravity – particularly for tree diagrams and other convergent diagrams in conventional quantum gravity.

For pure *conventional* Quantum Gravity DeWitt[103] finds the superficial degree of divergence of a diagram to be:

$$D = -2L_i + 2\sum_n V_n + 4K \qquad (9.62)$$

where L_i is the number of internal lines, V_n is the number of n-pronged vertices, and K is the number of independent momentum integrations. DeWitt further points out

$$K = L_i - \sum_n V_n + 1 \qquad (9.63)$$

Thus the superficial degree of divergence of a conventional Quantum Gravity diagram is:

$$D = 2(K + 1) \qquad (9.64)$$

for $K \geq 1$, displaying an ever increasing degree of divergence as the order of the diagram increases.

In the case of *Two-Tier Quantum Gravity* the superficial degree of divergence of a diagram is:

$$D_{TT} = -6L_i + 2\sum_n V_n + 4K \qquad (9.65)$$

(from eq. 9.48) with the result (taking account of eq. 9.63):

$$D_{TT} = -2L_i - 2\sum_n V_n + 2 \qquad (9.66)$$

[103] B. S. DeWitt, Phys. Rev. **162**, 1239 (1967).

Since any diagram with a loop has $L_i \geq 1$ and $\sum_n V_n \geq 1$ we see that $D \leq -2$. Thus *all* diagrams are convergent and *the Two-Tier formulation of Quantum Gravity theory is finite to any degree in perturbation theory. The addition of arbitrary species of other Two-Tier fields – matter and gauge fields – does not introduce divergences in the combined Two-Tier theory.*

Unitarity of Quantum Gravity–Scalar Particle Model

The Two-Tier Quantum Gravity – Scalar Particle Model *superficially* appears to have a unitarity problem due to the non-Hermitian nature of its Hamiltonian. The lack of hermiticity is due entirely to the appearance of iY^μ in the X^μ field coordinates.

Thus interaction Lagrangian is not Hermitian:

$$L_{Higher} = \int d^3y' \mathscr{L}_{Higher}(y' + iY(y')/M_c^2) \tag{9.67}$$

and

$$L_{Higher}^\dagger = \int d^3y' \mathscr{L}_{Higher}(y' - iY(y')/M_c^2) \neq L_{Higher} \tag{9.68}$$

The relation between L_{Higher} and its Hermitian conjugate is

$$L_{Higher} = V\, L_{Higher}^\dagger\, V \tag{9.69}$$

where $V^2 = I$ is the metric operator defined in eqs. 5.16 – 5.18. By eq. 6.37 we see as a result that the Two-Tier S matrix is not unitary – it is pseudo-unitary:

$$S^{-1} = V\, S^\dagger\, V \tag{9.70}$$

Therefore

$$S^\dagger V S = V \tag{9.71}$$

The S matrix satisfies the unitarity condition between physical asymptotic states – states consisting of only scalar ϕ particles and gravitons. The proof is identical in form to eqs. 6.46 – 6.48. The S matrix of the unified theory of the Standard Model and Quantum Gravity can be similarly shown to satisfy the unitarity condition.

[Section 1.8 at the beginning of this volume shows that unitarity can be established by suitably normalizing the S-matrix. Comment added to this volume]

The Mass Scale M_c

The mass scale of Two-Tier theories is set by M_c. This mass scale cannot be ascertained with any degree of certainty at current, experimentally accessible, accelerator energies. Cosmic ray data also does not seem to give any clues as to the value of M_c. It appears that M_c is probably above 10^3 GeV/c^2 and may be of the order of (or equal to) the Planck mass:

$$M_{planck} = \sqrt{\hbar c/G} = 1.22 \times 10^{19}\ \mathrm{GeV}/c^2 \qquad (9.75)$$

If M_c is of the 1,000 GeV/c^2 or larger the differences between its predictions at current accelerator energies and the predictions of conventional renormalized perturbation theory will be negligible. Actually a much lower value of M_c would still be consistent with the current stringent QED theoretical predictions as well as other predictions of conventional renormalized perturbation theory.

Planck Scale Physics

A finite theory of Quantum Gravity can provide information on the issues that have been of concern for many years – including the short distance behavior of the gravitational metric and ultra-small Black Holes.

Quantum Foam

Some theorists have conjectured that the classical view of smooth, almost flat space-time does not hold in the quantum regime at energies of the order of the Planck mass. Suggestions that space-time dissolves into quantum foam have appeared.

The finite Two-Tier formulation of Quantum Gravity is well-behaved at short distances and suggests that the quantum behavior of gravity and space-time in the short distance limit does not have limitless quantum fluctuations that result in a foam-like space-time picture.

Measurement of the Quantum Gravity Field

A number of conceptual problems have been raised about the effects of quantized General Relativity. Two-Tier Quantum Gravity seems to resolve these issues.

Measurement of Time Intervals

Wigner[104] has studied the measurement of time intervals in General Relativity and sees a problem in the measurement of extremely short intervals. According to Wigner: the measurement of a time interval in a region of space requires the measurement of the length of time required for an event to happen. The measurement requires an accurate clock. But the accuracy of the clock is limited by the energy-time uncertainty relation:

$$\Delta E \Delta t \geq \hbar \qquad (9.76)$$

Thus the uncertainty in the clock's time measurement is related to the uncertainty in the clock's energy which is, in turn, related to the uncertainty in the clock's mass:

$$\Delta E = (\Delta m)c^2 \qquad (9.77)$$

[104] E. P. Wigner, Rev. Mod. Phys. **29**, 255 (1957); J. Math. Phys. **2**, 207 (1961).

To obtain "infinite" accuracy the uncertainty (fluctuations) in the clock's mass must be infinite and thus the clock's mass must be infinite. Infinite fluctuations in the clock's mass will produce corresponding infinite fluctuations in the gravitational field.

$$\Delta h \propto \Delta E \qquad \text{(in conventional General Relativity)} \tag{9.78}$$

As a result the notions of space-time and time intervals (which depend on the geometry through General Relativity) become uncertain. Thus, according to Wigner and others, the concept of time intervals and space-time points becomes questionable.

The Two-Tier version of Quantum Gravity offers a potential way out of this dilemma. The gravitational force becomes stronger as one goes to shorter distances (higher energies) down to a distance (up to an energy) whose scale is set by M_c. At shorter distances (higher energies) the gravitational force becomes weaker and declines to zero at zero distance. Thus at very high energy the gravitational field fluctuations (Δh) are at worst inversely proportional to the energy (and probably decline by a higher power of inverse energy.) (The same considerations would apply if one chooses to consider fluctuations in the Riemann-Christoffel symbols.)

$$\Delta h < c_1/E < c_1/(\Delta E) \qquad \text{(in Two-Tier Quantum Gravity)} \tag{9.79}$$

Thus Wigner's conclusion does not hold in the Two-Tier version of Quantum Gravity as gravitational fluctuations actually become smaller at energies above a critical energy whose scale is set by M_c.

In fact, combining eqs. 9.79 and 9.76 we see

$$c_1 \Delta t/\Delta h \geq \hbar \tag{9.80}$$

at sufficiently high energy. Therefore the time uncertainty Δt, and the gravitational field fluctuations Δh, can both decrease while maintaining the energy-time uncertainty relation. *Thus the notion of a space-time point "is saved" in Two-Tier quantum gravity.*

Vacuum Fluctuations in the Gravitation Fields

While the expectation value of the free graviton field $h_{\mu\nu\text{conv}}(X)$ is zero in a conventional quantum field theoretic approach:

$$<0|\,h_{\mu\nu\text{conv}}(X)\,|0> = 0 \tag{9.81}$$

the vacuum fluctuations of the *conventional* quantum graviton field is quadratically divergent since

$$<0|\,h_{\mu\nu\text{conv}}(X) h_{\alpha\beta\text{conv}}(X)\,|0> = \int d^3p\; b'_{\mu\nu\alpha\beta}(p)/[(2\pi)^3\, 2\omega_p] = \infty \tag{9.82}$$

where $b'_{\mu\nu\alpha\beta}(p)$ is a rational function of the momentum p.

In Two-Tier quantum field theory we find

$$<0|h_{\mu\nu}(X)h_{\alpha\beta}(X)|0> = \int d^3p\, b'_{\mu\nu\alpha\beta}(p)\, e^{-p^ip^j\Delta_{Tij}(0)}/[(2\pi)^3 2\omega_p] = 0 \tag{9.83}$$

since the exponential factor in the integrand is $-\infty$. The exponent contains

$$\Delta_{Tij}(z) = \int d^3k\, e^{-ik\cdot z}(\delta_{ij} - k_ik_j/\mathbf{k}^2)/[(2\pi)^3 2\omega_k] \tag{4.8}$$

Thus the vacuum fluctuations of $h_{\mu\nu}$ are zero in "Two-Tier" quantum field theory.

The Two-Tier Gravitational Potential vs. Newton's Gravitational Potential

The familiar gravitational potential of Newton is:

$$V_{Newton} = -G/|\mathbf{r}| \tag{9.84}$$

The Two-Tier gravitational potential is:

$$V_{Two\text{-}Tier} = -G\Phi(M_c^2\pi|\mathbf{r}|^2)/|\mathbf{r}| \tag{9.85}$$

where $\Phi(y)$ is the error function.[105] It can be calculated in Two-Tier Quantum Gravity from Two-Tier Quantum Gravity propagator terms similar to corresponding terms in the Two-Tier photon propagator that led to the Two-Tier Coulomb potential (eqs. 7.48 – 7.51). At small distances ($\pi r^2 \ll M_c^{-2}$)

$$V_{Two\text{-}Tier} \rightarrow -G2\sqrt{\pi}\, M_c^2|\mathbf{r}| \tag{9.86}$$

a linear potential, and at large distances ($\pi r^2 \gg M_c^{-2}$)

$$V_{Two\text{-}Tier} \rightarrow V_{Newton} = -G/|\mathbf{r}| \tag{9.87}$$

the Newtonian potential.

The Two-Tier gravitational potential has a minimum at

$$M_c^2\pi r_{MIN}^2 = 1 \tag{9.88}$$

At the minimum $V_{Two\text{-}Tier}$ has the value:

---

105 W. Magnus and F. Oberhettinger, *Formulas and Theorems for the Special Functions of Mathematical Physics* (Chelsea Publishing Co., New York, 1949) page 96.

$$V_{\text{Two-TierMIN}} = -\ .8427G\sqrt{\pi}\ M_c \qquad (9.89)$$

Figs. 9.1 – 9.2 display plots of $V_{\text{Two-Tier}}$ for M_c = 1 TeV/c^2, and M_c = 1.22 10^{19} GeV/c^2 = $G^{-½}$ – the Planck mass.

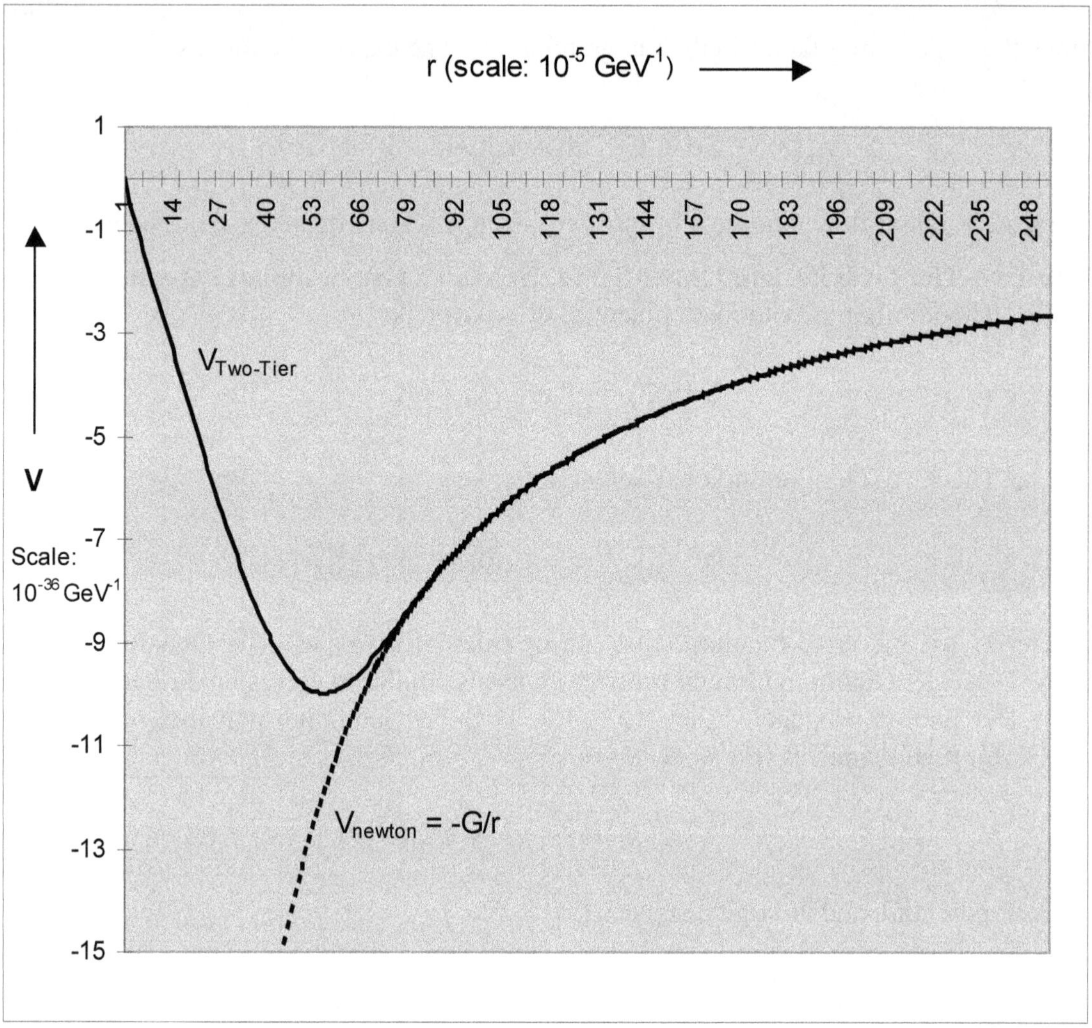

Figure 9.1. Plot of Two-Tier gravitational potential for M_c = 1 TeV/c^2 and Newton's gravitational potential. The potentials are measured in units of 10^{-36} GeV^{-1}. The radial distance is measured in units of 10^{-5} GeV^{-1}. The plot of the Two-Tier potential shows the force of gravity is repulsive for small r < 5.7 × 10^{-4} GeV^{-1}.

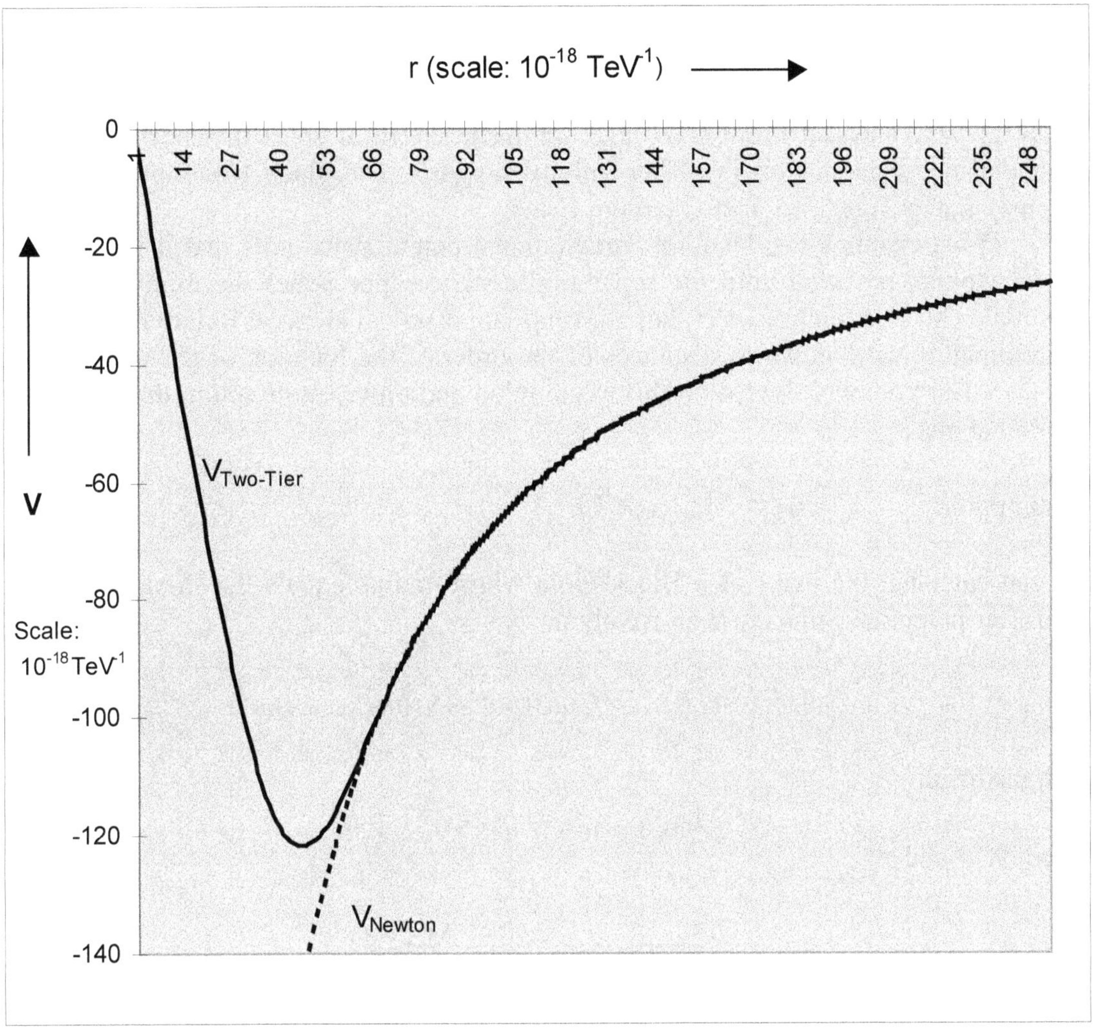

Figure 9.2. Plot of Two-Tier gravitational potential for $M_c = 1.22 \times 10^{19}$ GeV/c^2 (the Planck mass) and Newton's gravitational potential. Potentials are measured in units of 10^{-18} TeV^{-1}. The radial distance is measured in units of 10^{-18} TeV^{-1}.

Black Holes

The existence of microscopic Black Holes has been the subject of much speculation. It appears that arbitrarily small Black Holes can exist in classical General Relativity. The divergences associated with the short distance behavior of its conventional quantization raise the possibility of additional singular behavior at short distances as well.

On the other hand, in Two-Tier Quantum Gravity, at short distances, when the distance scale becomes less than M_c^{-1} (and thus the energy scale becomes greater than M_c), the Two-Tier gravitational force grows smaller and become zero in the limit of zero distance or infinite energy. The preceding figures (Figs. 9.1 – 9.2) show the Two-Tier gravitational potential linearly approaches zero at short distances unlike the Newtonian gravitational potential which approaches $-\infty$ as r approaches zero. (The transverse gravitational propagator also approaches zero at short distances.) Thus the short distance behavior of Two-Tier gravity suggests that Black Holes of ultra-small size may not exist in Two-Tier Quantum Gravity.

If we examine the Two-Tier gravitational potential we note that it is similar to the Newtonian potential until the separation distance approaches the minimum of the potential. Thus we might expect that conventional classical General Relativity would be approximately valid down to distances of the order of the location of the minimum of the Two-Tier potential. Based on this assumption and on the assumption that M_c equals the Planck mass:

Assumption: $$M_c = M_{Planck} = G^{-1/2} \tag{9.90}$$

we can calculate the mass of a Black Hole whose radius equals the minimum of the Two-Tier potential. From eq. 9.88 we obtain

$$r_{MIN} = (G/\pi)^{1/2} = r_{BlackHole} = 2GM_{BlackHoleMIN} \tag{9.91}$$

with the result

$$M_{BlackHoleMIN} = (4\pi G)^{-1/2} = \kappa^{-1} \tag{9.92}$$

by eq. 9.11 and

$$M_{BlackHoleMIN} \cong .282\ M_{Planck} \tag{9.93}$$

or 6.15×10^{-6} grams.This lower limit on Black Hole mass is substantially greater than the collision energy than can be achieved in any current particle accelerator. Thus the production of ultra-small Black Holes in particle accelerators is unlikely.

Since corrections to conventional quantum gravity are at most of the order of M_c^{-2} it appears that the value of $M_{BlackHoleMIN}$ is consistent with the approximate validity of classical expression for a Black Hole radius. We note

$$(M_{BlackHoleMIN}/M_c)^2 \cong .0795 \tag{9.94}$$

and so corrections to eq. 9.93 would be very small.

10. Curved Space-time Generalization of Two-Tier Quantum Gravity

Inertial Reference Frames & Absolute Space-time

The concept of a flat, absolute space-time can be defined in two ways:

1. There exists a specific reference frame, an *absolute reference frame*, with space-time coordinates that we will denote as y^μ for $\mu = 0, 1, 2, 3$. Any reference frame whose space-time coordinates y'^μ are related to the y^μ coordinates by equations of the form:

$$y'^j = R^j_{\ i}y^i + v^j y^0 + c^j \tag{10.1}$$

$$y'^0 = y^0 + c^0 \tag{10.2}$$

where $R^j_{\ i}$ is a constant, real, orthogonal matrix, and where v^j and c^μ are constants for $j = 1, 2, 3$ and $\mu = 0, 1, 2, 3$ is an equivalent inertial reference frame. The set of these reference frames is called the set of *inertial reference frames*. The form of the equations of motion for a set of point-like particles is the same in any inertial reference frame. Thus these reference frames are physically equivalent.

2. There is a class of reference frames called inertial reference frames whose coordinates are related by equations of the form of eqs. 10.1 and 10.2. The form of the equations of motion for a set of point-like particles is the same in any inertial reference frame and for slowly moving particles have the form of Newton's equations without inertial forces. No inertial reference frame has any special significance.

Current cosmological data suggest that space is almost flat, or flat. Thus we can establish either local inertial reference frames (curved space case) or global inertial reference frames (flat space case) as experiment eventually will indicate.

Since the form of the equations of motion is the same in all local inertial frames it would appear that there is no way to physically distinguish between definitions 1 and 2. However, the observed characteristics of cosmic background radiation (CBR), of the redshift – distance relationship, and of the cosmological X-ray background effectively define a preferred local reference frame in each spatial locale. Effectively the CBR plays the role of an *aether selecting a preferred local inertial reference frame in each spatial locale*. The set of all such preferred inertial reference frames for the universe effectively defines an Absolute Reference Frame. IF space is truly flat, then the Absolute Reference Frame consists of one inertial reference frame.

The views of physicists on the question of an absolute reference frame have oscillated over time. Newton chose the first definition and talked of "absolute space" in the famous quote, "Absolute space in its own nature, and with regard to anything external, always remains similar and immovable." Mach challenged Newton's view and enunciated what became known as Mach's Principle postulating that there existed a class of inertial frames (second type of definition) that were defined by the mass distribution, and its movement, of the universe. Einstein established an "intermediate" position between Newton and Mach in his General Theory of Relativity. General Relativity implicitly defines the equivalent of an absolute space and shows that the presence of masses is not required since inertial frames exist in the empty space solutions of the equations of general relativity.

However the dynamical equations of General Relativity are covariant under any general relativistic transformation. *One can view General Relativity as a theory embodying invariance under general relativistic transformations with the invariance "broken" by a class of equivalent local "ground states" called local inertial reference frames that are determined by Mach's Principle, and eqs. 10.1 and 10.2 – an analogue of spontaneous broken symmetry.*

In this chapter we shall examine the Newtonian, Machian, and Einsteinian views in more detail and discuss them in relation to Two-Tier quantum field theory.

Newtonian Mechanics Embodies an Absolute Space-time

Newton developed his formulation of mechanics with equations of motion for groups of point-like particles. These equations had the same form in all inertial reference frames. He then asserted that the class of inertial reference frames was selected because they were either at rest or in a state of constant velocity with respect to a particular reference frame that corresponded to *absolute space*. He postulated the existence of a *physical* absolute space partly for theological reasons.

Mach's Principle Embodies an Absolute Space-time

Ernst Mach disagreed with Newton's concept that the basis for the special properties of inertial reference frames was their relation to the reference frame of an absolute space. Following Leibniz and others he proposed another view that is now called *Mach's Principle*.

An example[106] that illustrates Mach's thinking is:

> Consider a universe consisting of two identical spheres, not necessarily in close proximity, upon each of which an ant is stationary on the sphere's equator. Assume the spheres are rotating with respect to each other with parallel axes of rotation. The ant on each sphere sees the ant

[106] Mach (1991).

on the other sphere rotating with the sphere. Questions: which ant experiences an "upward" centrifugal force? Which ant experiences a Coriolis force (which is proportional to the angular velocity of rotation)?

These questions do not have answers in the universe of two spheres subject only to Newton's laws according to Ernst Mach. Either sphere could be considered to be the non-rotating sphere and a corresponding valid coordinate system defined in which the other sphere would be rotating.

Mach resolved this issue by noting the universe is very large and populated with large masses in all directions at great distance. In his view the distribution and motion of all the masses in the universe defined a preferred reference frame, although it is not usually portrayed in that manner. Mach would then have resolved the two spheres issue by saying the rotation of each sphere must be determined relative to the distribution and motion of the rest of the matter in the *real* universe.

As R. H. Dicke has remarked,[107] "If one were to remove this matter [at great distances], then according to Mach, the inertial force would disappear. If one were to reduce the matter to negligible proportions, there would be striking changes in local inertial effects. To summarize, according to Mach's point of view, we should interpret inertial effects as a consequence of interactions of matter at great distances in the universe with accelerated bodies in the laboratory."

Although many physicists, including Einstein, were strongly influenced by Mach's arguments many physicists were also uneasy about Mach's Principle. As Eddington[108] remarked, on the use of matter at infinity to define inertial frames, "the main feeling seems to be that it is unsatisfactory to have certain conditions prevailing in the world, which can be traced away to infinity and so have, as it were, their source at infinity; and there is a desire to find some explanation of the inertial frame as built up through conditions at a finite distance."

Thus Mach's Principle was viewed with mixed feelings both before, and long after, Einstein's development of his general theory of relativity.

General Relativity Embodies an Absolute Space-time

Einstein, himself, was strongly influenced by Mach's arguments. Mach's Principle was certainly on his mind when he was formulating his equivalence principle and the general theory of relativity. To some extent he felt his theory embodied Mach's Principle. However in the view of almost all observers General Relativity only partly implements Mach's Principle. General Relativity *also* embodies aspects of an absolute space-time.

R. H. Dicke,[109] points out:

"Einstein's theory is not relativistic in the Machian sense. In his theory, space has physical properties and constitutes a physical structure even in the absence of all matter.

[107] R. H. Dicke, lecture entitled *The Many Faces of Mach* (1963).
[108] Eddington (1995) p. 157.
[109] R. H. Dicke, lecture entitled *The Many Faces of Mach* (1963).

… general relativity does not appear to describe Mach's principle properly. This can be seen by noting that, in the absence of all matter, the metric tensor describes a flat space and this flat space possesses inertial properties. Even Schwarzschild's famous solution is unsatisfactory, from the point of view of Mach. As one moves to infinity, and the mass source (the source of inertial forces according to Mach) disappears in the distance, the space becomes flat and continues to possess inertial properties in contradiction with the expectations of Mach. …

We have … the return to the idea that we are dealing with an absolute space-time. From the viewpoint of Synge, general relativity describes the geometry of an absolute space. According to him, certain things are measurable about this space in an absolute way. There exist curvature invariants that characterize this space, and one can, in principle, measure these invariants. Bergmann has pointed out that the mapping of these invariants throughout space is, in a sense, a labeling of the points of this space with invariant labels (independent of coordinate system). These are concepts of an absolute space, and we have here a return to the old notions of an absolute space."

The Case for Absolute Space-time

With the very strong case for absolute space-time made by Synge, Dicke and Bergmann, three of the great general relativists, we now ask whether a return to absolute space-time is in order in the sense of definition 2 above.

The arguments for an absolute space-time are:

1. It is embodied in the two successful theories of mechanics: Newtonian mechanics and general relativity.

2. It supports a local definition of physics consistent with the spirit of Riemannian geometry, general relativity and quantum field theory.

3. **Experimental data is consistent with the existence of absolute rotation. As Eddington[110] notes, "The great stumbling block for a philosophy which denies absolute space is the experimental detection of absolute rotation." Most interestingly, current cosmological experimental data suggests space is very close to flat if not flat. Thus we are living in an absolute reference frame if the universe is considered in the large (with local masses averaged over cosmological distances.)**

4. It appears to be impossible to construct a theory of classical mechanics that is consistent with experiment that does not explicitly, or implicitly, embody absolute space-time in the form of definitions 1 or 2.

---

[110] Eddington (1995) p. 152.

Experimental Determination of an Absolute Reference Frame

Experimentally we have found that space is close to flat although it appears to have enough curvature to form a closed space. It may be flat.

Experimentally we have also found that the observed characteristics of cosmic background radiation (CBR), of the redshift – distance relationship, and of the X-ray background effectively define a preferred local reference frame. The near flatness of space on large distance scales suggests this preferred local reference frame is almost an Absolute Minkowski reference frame. Thus current experiment has found an absolute reference frame – the only question is whether it is local or global. *As Peebles (1993) points out*[111]

"Blackbody radiation can appear isotropic only in one frame of motion. An observer moving relative to this frame finds that the Doppler shift makes the radiation hotter than average in the direction of motion, cooler in the backward direction. That means CBR acts as an aether, giving a local definition for preferred motion. ... In the standard interpretation, the same preferred comoving rest frame is defined by the CBR, the redshift-distance relation for galaxies, and the X-ray background. ... The evidence[112] *is that the frames are consistent to perhaps 300 km* s^{-1}*."*

Thus we have an experimental definition of an almost flat, or flat, absolute reference frame. We conclude with Synge:[113] "The Principle of Equivalence performed the essential office of midwife at the birth of general relativity, but, as Einstein remarked, the infant would never have gotten beyond its long clothes had it not been for Minkowski's concept. I suggest that the midwife be now buried with appropriate honours and the facts of an absolute space-time faced."

In the preceding chapters we have defined a unified quantum field theory that embodies the notion of an absolute inertial reference frame (or a set of local preferred inertial reference frames that apply to large locales) in a more direct way than classical general relativity. We defined X coordinates and a Y field in the preferred inertial reference frame of a locale, and then defined Two-Tier theories that are invariant under special relativistic transformation to other inertial reference frames.

Curved Space-time Generalization of Two-Tier Quantum Gravity

Thus the preceding chapters developed a divergence-free theory of scalar particles and quantum gravity in a flat space-time. In this section we show that a curved space-time version of Two-Tier quantum field theories including quantum gravity can be developed along the lines pioneered by DeWitt and collaborators. Two-Tier curved space-time quantum field theory is based on a mapping from a flat space-time parameterized by y coordinates to a curved space-time parameterized by X coordinates.

---

[111] Eddington (1995) p. 151-2.

[112] M. Aaronson et al, Astrophysical Journal **302**, 536 (1986); R. A. Shafer and A. C. Fabian, in *Early Evolution of the Universe and its Present Structure*, ed. G. O. Abell and G. Chincarini, p. 333 (1983); Rubin (19878).

[113] Synge (1960) pp. ix-x.

The physical picture of the mapping can be visualized using the simple example of a sphere of radius one in three-dimensional space with a coordinate system on the sphere and two planes – one above the sphere and one below it – each with its own flat space coordinate system. Both planes are assumed to be parallel to the disk defined by the crossection of the sphere bounded by the equator of the sphere. A minimum of two coordinate patches are needed to cover a sphere in three dimensions since it necessarily has coordinate singularities.

Let us place a rectangular coordinate system on the top plane. Points on this plane can be mapped onto its northern hemisphere of the sphere in a simple one-to-one fashion. Similarly a rectangular coordinate system can be placed on the bottom plane which can be mapped in a one to one fashion onto the southern hemisphere of the sphere. The top and bottom planes each have a two-dimensional coordinate system that we can choose to be a Cartesian coordinate system in both cases. We will label the coordinates on the top plane x_t^1 and x_t^2, and the points on the bottom plane as x_b^1 and x_b^2. Each plane has a flat space metric $g_{tij} = g_{bij} = \delta_{ij}$ for i, j = 1,2 with δ_{ij} the Kronecker delta. **In addition, just for concreteness, we will place the origin of the top plane coordinate system vertically above the north pole of the sphere, and the origin of the bottom plane coordinate system vertically below the south pole of the sphere.**

If we place the sphere at the center of a three dimensional, coordinate system then the points on the sphere (x, y, z) all satisfy:

$$x^2 + y^2 + z^2 = 1 \tag{10.3}$$

We can define coordinates u^1 and u^2 for each hemisphere on the surface of the sphere with equations of the form:

$$x_n = f_{1n}(u_n^1, u_n^2) \tag{10.4}$$
$$y_n = f_{2n}(u_n^1, u_n^2) \tag{10.5}$$
$$z_n = f_{3n}(u_n^1, u_n^2) \tag{10.6}$$

for the northern hemisphere, and

$$x_s = f_{1s}(u_s^1, u_s^2) \tag{10.7}$$
$$y_s = f_{2s}(u_s^1, u_s^2) \tag{10.8}$$
$$z_s = f_{3s}(u_s^1, u_s^2) \tag{10.9}$$

for the southern hemisphere.

In addition, we choose $u_n^1 = u_n^2 = 0$ at the North Pole and $u_s^1 = u_s^2 = 0$ at the South Pole. The surface of the sphere is curved and each (u^1, u^2) coordinate system has a metric, g_{nij} and g_{sij} for i, j = 1,2 respectively, and a non-zero curvature tensor R_{nijkl} and R_{sijkl}.

Now we are allowed to define a simple map of points on the northern hemisphere of the sphere to points on the top plane such as:

$$x_t^{\ 1} = u_n^{\ 1} \tag{10.10}$$
$$x_t^{\ 2} = u_n^{\ 2} \tag{10.11}$$

and of points on the southern hemisphere of the sphere to points on the bottom plane:

$$x_b^{\ 1} = u_s^{\ 1} \tag{10.12}$$
$$x_b^{\ 2} = u_s^{\ 2} \tag{10.13}$$

Thus we can specify the location of events on the sphere on our planes. Note that eqs. 10.4 – 10.9 are *not* a coordinate transformation of the (u^1, u^2) coordinate systems on the sphere and thus the plane can have a different (flat) metric from the sphere.

The preceding example can be simplified by using a cylinder enclosing the sphere instead of two planes. The cylinder, which is a flat surface technically, is aligned so that its axis is parallel to, and centered on, the north-south axis of the sphere. Then a map can be made from points on the sphere to points on the cylinder that is similar to a Mercator projection, or from points on the sphere to the cylinder that maps the poles to the ends of the cylinder at + and – infinity.

The preceding discussion shows a clear analogy to our map from the y Minkowski space-time to the curved X space-time using

$$X^\mu = y^\mu + i\, Y^\mu(y)/M_c^{\ 2} \tag{10.14}$$

modulo the imaginary term. The y Minkowski space-time has a flat space-time in which we are allowed to choose the Minkowski metric $\eta_{\mu\nu}$. The curved X space-time has an appropriate metric $g_{\mu\nu}(X)$ that can only be transformed to locally inertial coordinates with perhaps a Minkowski metric in the neighborhood of a point. The additional imaginary term does not alter this picture except that the curved X space-time is now a slightly complex manifold in complex space-time.

Therefore we conclude that our Two-Tier quantum field theoretic formalism that is erected on eq. 10.14, where the real part of the X space-time was flat, can be extended to curved space-time while maintaining eq. 10.14 if the y space-time consists of coordinate patches analogous to the two planes (or the cylinder) in the example of the sphere. The difference is that we now use a curved space-time background metric $g_{\mu\nu}(X)$ instead of $\eta_{\mu\nu}$ throughout the Lagrangian with the exception of L^Y (eq. 9.27).

In L^Y we continue to use $\eta_{\mu\nu}$ as the metric. As a result L^Y breaks the invariance of the complete Lagrangian under general coordinate transformations. Thus an implicit absolute space-time is implied – as it is implicitly in classical General Relativity and in cosmological experiments. This consequence is not disturbing and is physically acceptable for the following reasons:

1. As Bergmann and Synge point out classical general relativity implicitly embodies an absolute space-time.

2. Experiment shows that space in the large (of the order of the Hubble length) is nearly flat although space does appear to be closed. CBR, and other, experimental data suggests that an absolute reference frame exists.

Thus our universe does appear to be in a state of broken general coordinate transformation invariance. Two-Tier quantum field theory in curved space-time is not in contradiction with our previous classical general relativistic theories or with our experimental knowledge of the universe. *The full Lagrangian theory L is invariant under special relativity.* $L - L^Y$ *is formally invariant under general coordinate transformations in the X coordinates.*

Why Are the Y Field Dynamics Independent of the Gravitational Field?

It is evident that the Y^a field is a truly free field in our formulation. In particular, it does not depend on, or interact directly with, the gravitational field as represented by $\sqrt{g}$ and $g_{\mu\nu}$ factors. On the other hand, these quantities depend on the Y^a field through their dependence on the variable X^μ.

Thus the role of Y^a is strictly that of coordinates, and of a field that is parameterized by a set of inertial frame coordinates y^μ. The arguments of Mach supplemented by the arguments of Bergmann and Synge show that a de facto absolute reference frame exists (actually it is the set of inertial reference frames). Therefore we can chose to formulate our theory in an inertial reference frame and require that the theory only be invariant under Lorentz transformations to other inertial reference frames.

In this context it is allowed to have one or more fields like Y^a whose dynamics are not invariant under general coordinate transformations. It is reasonable to require the particle and gravitational dynamical equations be covariant under general coordinate transformations in X. *Thus a part of the dynamics is invariant under Lorentz transformations – the* Y^a *sector – but this part of the dynamics is not directly observable; and a part of the dynamics – the observable part – is invariant under general coordinate transformations.*

Some reasons for having a free Y^a field are:

1. It is required to avoid divergences that would appear in perturbation theory if the Y^a were allowed to interact with gravitons. For example an hhYY interaction term causes a divergence to appear by generating a Y particle loop in graviton-graviton scattering.
2. If the Y^a particle interacted with gravity then measurable, classical Y^a fields could be generated in regions with ultra-strong gravitational fields such as the neighborhoods of Black Holes. In this case we would have new dimensions, albeit imaginary, for which no experimental evidence currently exists.

3. The Principle of Equivalence has only been shown to apply on the classical level for real coordinates. Any quantization that uses Minkowskian coordinates, or quasi-Minkowskian coordinates, causes general coordinate transformation invariance to be abandoned *ab initio* in the quantum regime.

11. A Unified Quantum Field Theory of the Known Forces of Nature

Formulation of Unified Theory

The unification of QED and weak interactions in ElectroWeak Theory interrelated the theories within the framework of an overall SU(2)⊗U(1) symmetry and thus was significantly more than merely "gluing" the theories together.

The unification of ElectroWeak theory with Quantum Chromodynamics (QCD) into the Standard Model was a direct combination of these theories in which the symmetries of each respective theory were directly combined: SU(2)⊗U(1) symmetry from ElectroWeak Theory and SU(3) from QCD to produce the Standard Model with SU(2)⊗U(1)⊗SU(3) symmetry. ElectroWeak Theory was "glued together" with QCD to produce the Standard Model without an underlying rationale. Nevertheless, the Standard Model is a renormalizable theory that accounts for the vast majority of experimental data.

The present work has two goals: 1.) To make QED, ElectroWeak Theory, QCD and Quantum Gravity finite and thus remove a major long term defect in Quantum Field Theory, and 2.) To create a unified theory of the known forces of Nature from these pieces. Item 1 has been achieved in the preceding chapters using Two-Tier Quantum Field Theory. This chapter discusses item 2.

In this chapter we propose a finite unified theory of the Standard Model (and any of its variants) and Quantum Gravity within the framework of Two-Tier Quantum Field Theory. As we saw in the case of the Standard Model our unified theory amounts to gluing together the Two-Tier version of the Standard Model with the Two-Tier version of Quantum Gravity. Therefore we regard the theory as provisional in the sense that a deeper unification remains to be formulated.

Nevertheless the Two-Tier Unified Theory may be of some importance beyond the satisfaction of having a finite theory of Nature. It might be a starting point for a deeper, more unified theory. It can be used to address cosmological questions such as the state of the universe immediately after the Big Bang when the size of the universe was of the order of the Planck length or smaller – see Blaha (2004). The interactions in the unified theory become weaker at short distances and thus low order perturbation theory becomes a better approximation to the exact results.

Unified Two-Tier Standard Model and Quantum Gravity Lagrangian

We define the Lagrangian, and action, for the Two-Tier unified quantum field theory of gravitation and the Standard Model as

$$L_{\text{Unified}} = \int d^4y\, \mathscr{L}_{\text{Unified}} \tag{11.1}$$

$$\mathcal{L}_{\text{Unified}} = J\sqrt{g(X)}\,(\mathcal{L}_F^{\text{Grav}}(X) + \mathcal{L}_F^{\text{SM}}(X)) + \mathcal{L}_C \quad (11.2)$$

with

$$\mathcal{L}_F^{\text{Grav}}(X) = (2\kappa^2)^{-1}{}_X R(X) \quad (11.3)$$

where $\mathcal{L}_F^{\text{SM}}$ is the complete "normal" Quantum Field theory Lagrangian for the Standard Model variant under consideration written in a general coordinate covariant form, g(X) is the absolute value of the determinant of $g_{\mu\nu}$, and J is the Jacobian of eq. A.21. *All particle fields in* $\mathcal{L}_F^{SM}$ *are assumed to be functions of the X coordinate only. The dependence of the particle fields on the "underlying" coordinates* y^μ *is assumed to be solely through* X^μ. The Lagrangian $\mathcal{L}_{\text{Unified}}$ is a separable Lagrangian of the type of eq. A.26 embodying the composition of extrema described in Appendix A.

As in all cases considered we define the coordinate part of the Lagrangian $\mathcal{L}_C$ as

$$\mathcal{L}_C = -\tfrac{1}{4}\, F_Y^{\mu\nu} F_{Y\mu\nu} \quad (11.4)$$

with

$$F_{Y\mu\nu} = \partial Y_\mu/\partial y^\nu - \partial Y_\nu/\partial y^\mu \quad (11.5)$$

and

$$F_Y^{\mu\nu} = \eta^{\mu\alpha}\eta^{\nu\beta} F_{Y\alpha\beta} \quad (11.6)$$

The development of the physics embodied in the Lagrangian proceeds along the lines described in the previous chapters, except that the gravitational sector must be in the form of a *vierbein* theory since the full theory contains spin 1/2 particles.

"Low Energy" Behavior

The low energy sector of the unified theory is defined as the sector with momenta whose values are much less than M_c. It is clear from the discussions of the previous chapters that the low energy sector of the unified theory is effectively identical to that of the corresponding conventional quantum field theory if M_c is sufficiently large.

In addition the low energy behavior of the Two-Tier unified theory in the QED sector closely approximates the results of QED calculations which have been found to agree well with experiment to an extremely high degree of accuracy.

Thus the Two-Tier unified theory satisfies a Correspondence Principle in the Standard Model sector: The low energy behavior of the Two-Tier Standard Model sector is the same as the behavior of the conventional Standard Model to a high degree of approximation. In addition the Two-Tier *vierbein* Quantum Gravity sector tree diagrams are the same as the conventional *vierbein* Quantum Gravity tree diagrams to a high degree of accuracy.

Negative Degree of Divergence – A Finite Unified Theory

The previous discussions of the perturbation theory of matter fields, gauge fields and gravitons show that the theory is finite.

Unitarity

The unitarity discussions of the various sectors of the unified theory in previous chapters show the unified theory satisfies unitarity. As long as Y excitations are not allowed in in-states they will not appear in out-states. The S matrix is block diagonal and unitary within the physical asymptotic states sector.

Appendix A. Composition of Extrema in the Calculus of Variations

A New Paradigm in the Calculus of Variations

The Calculus of Variations has a long and venerable history in Physics and Mathematics. Many problems in Physics and Mathematics have been treated with approaches based on techniques in the Calculus of Variations (see the references at the end of the book). In this book we have developed a unified quantum field theory of the known forces of nature based on a new type of problem, or paradigm, in the Calculus of Variations. One way of viewing the spectrum of problems in the calculus of variations is the following progression.

A Classification of Variational Problems

1. Variational problems in a Euclidean, or Minkowski, flat space such as the minimal distance between two points or the extrema of a field theory Lagrangian.

2. Variational problems seeking extrema on a curved surface such as the shortest distance between points on the surface of a sphere.

The development in this book suggests a third and fourth, possibility, that to the author's knowledge, has not been addressed in the literature:

3. Variational problems where the extrema are determined on a surface that is itself defined as an extremum. The discussions in this book exemplify this paradigm.

4. Variational problems where the extrema are determined on a surface that is itself defined as an extremum that depends on the extrema on the surface. More simply put the extrema, and the surface upon which they are defined, are jointly determined and are interrelated. Fortunately, our unified theory does not use this paradigm. A future theory might.

In the unified theory that we will develop all particle fields including the graviton field are defined as a mapping of a Minkowski space-time y to a "particle" space-time X with the mapping determined as an extremum of a variation of a fundamental field (a type 3 variational problem in the above classification). Our theory could be generalized to include a back-reaction of the particle fields on the fundamental field (a type 4 variational problem in the above classification). We will not discuss this possibility in this book.

Simple Physical Example – Strings On Springs

In this section we will describe a simple physical example that illustrates a variational problem of type 3 in the Calculus of Variations. We view it as a composition of extrema. (This problem can be addressed using other calculus of Variations techniques.) The approach used in the solution of this problem is similar to the approach used in Two-Tier quantum field theory.

A Strings on Springs Mechanics Problem

Consider a long string or bar that can oscillate (undulate) in a direction perpendicular to its length. Further assume that one end of this bar or string is attached to a spring that cause the entire bar or string to oscillate back and forth in a direction parallel to its long side. This configuration is illustrated in Fig. A.1.

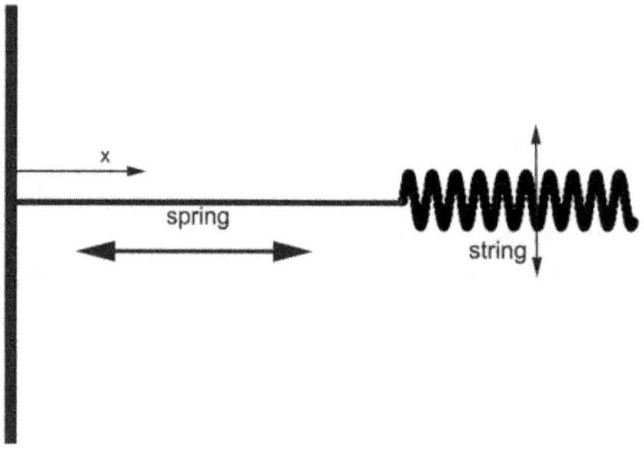

Figure A.1. An oscillating string attached to a spring.

Let x denote the distance to a point on the string when the spring is at equilibrium. If 2π times the frequency of the spring is ω_1, then the location of this point when the spring is oscillating is

$$X(t) = x + A \sin(\omega_1 t + \phi_1) \tag{A.1}$$

where ϕ_1 is a phase, and A is the amplitude of the spring oscillation. Then the vertical displacement of a traveling wave on the *string* can take the form

$$\psi(t) = B \sin(\omega_2 t - k_2(x + A \sin(\omega_1 t + \phi_1)) + \phi_2) \tag{A.2}$$

where B is the amplitude of the string wave, and k_2, ω_2 and ϕ_2 are the parameters of the string wave. These simple mechanical formulae are well known. But they lead to an interesting new application of the ideas of the Calculus of Variations.

Suppose we treat X as an independent variable with X given by eq. (A.1), and with eq. (A.2) written as:

$$\psi(t) = B \sin(\omega_2 t - k_2 X + \phi_2) \tag{A.3}$$

Defining

$$\psi = \psi(X(t), t) \tag{A.4}$$

we can specify the dynamics of the above motion by finding the extrema of

$$I = \int \mathscr{L}_\psi \, dX(t) + \int \mathscr{L}_X \, dt \tag{A.5}$$

where the Lagrangian terms are

$$\mathscr{L}_\psi = \tfrac{1}{2} \{ \mu \, (\partial\psi/\partial t)^2 - Y \, (\partial\psi/\partial X)^2 \} \tag{A.6}$$

with μ and Y being constants, and

$$\mathscr{L}_X = \tfrac{1}{2} \{ m(\partial X/\partial t)^2 - k(X - x)^2 \} \tag{A.7}$$

where m and k are constants, and where x is a parameter. Applying Hamilton's Principle, and performing independent variations of X and ψ yields the Lagrangian equations:

$$\frac{\partial \mathscr{L}_\psi}{\partial \psi} - \frac{\partial}{\partial X}\frac{\partial \mathscr{L}_\psi}{\partial(\partial\psi/\partial X)} - \frac{\partial}{\partial t}\frac{\partial \mathscr{L}_\psi}{\partial(\partial\psi/\partial t)} = 0 \tag{A.8}$$

and

$$\frac{\partial \mathscr{L}_X}{\partial X} - \frac{\partial}{\partial t}\frac{\partial \mathscr{L}_X}{\partial(\partial X/\partial t)} = 0 \tag{A.9}$$

The resulting equations of motion are:

$$\mu \, \partial^2\psi/\partial t^2 - Y \, \partial^2\psi/\partial X^2 = 0 \tag{A.10}$$

and

$$m \, \partial^2 X/\partial t^2 + k(X - x) = 0 \tag{A.11}$$

with the solutions given in eqs. A.1 and A.2.

The procedure that we use to obtain these results may look a bit strange but they illustrate a type 3 problem in the Calculus of Variations involving the composition of extrema—the composition of an extremum that specifies a manifold in a space (possibly including all of space in a $R^n \rightarrow R^n$ mapping) with an extremum determining a function on that manifold. The procedure is described in detail in the next section.

The Composition of Extrema – Lagrangian Formulation

In this section we will explore the general case of the composition of extrema for fields. We will discuss the case of a scalar field ϕ that is a function of a vector field X^μ in a D-dimensional space with coordinate variables that we will denote as y^μ. (The discussion for other types of fields is a straightforward extension of this discussion.) Thus

$$\phi = \phi(X) \tag{A.12}$$

and

$$X^\mu = X^\mu(y) \tag{A.13}$$

We assume that the dynamics can be described by a Lagrangian formulation using an extension of Hamilton's principle:

$$I = \int \mathscr{L} d^4y \tag{A.14}$$

with

$$\mathscr{L} = \mathscr{L}(\phi(X), \partial\phi/\partial X^\nu, X^\mu(y), \partial X^\mu(y)/\partial y^\nu, y) \tag{A.15}$$

If we perform a standard variation[114] in ϕ for fixed y (and thus fixed X) we find

$$\delta I = \int [\delta\phi\, \partial\mathscr{L}/\partial\phi + \delta(\partial\phi/\partial X^\nu)\, \partial\mathscr{L}/\partial(\partial\phi/\partial X^\nu)]\, d^4y \tag{A.16}$$

We can rewrite the variation in the derivative of ϕ as

$$\delta(\partial\phi/\partial X^\nu) = \partial(\delta\phi)/\partial X^\nu \tag{A.17}$$

$$= \partial y^\mu/\partial X^\nu\, \partial(\delta\phi)/\partial y^\mu \tag{A.18}$$

with an implied summation over repeated indices. After substituting eq. A.18 in eq. A.16, and performing an integration by parts (and discarding the surface term which is assumed to yield zero in the standard fashion) we obtain:

$$\delta I = \int \delta\phi\, \{\partial\mathscr{L}/\partial\phi - \partial/\partial y^\mu\, [\partial\mathscr{L}/\partial(\partial\phi/\partial X^\nu)\, \partial y^\mu/\partial X^\nu)\,]\}\, d^4y$$

114 Bogoliubov, N. N., & Shirkov, D. V., Volkoff, G. M. (tr), *Introduction to the Theory of Quantized Fields* (Wiley-Interscience, New York, 1959); Goldstein H., *Classical Mechanics* (Addison-Wesley, Reading, MA 1965).

Since the variation of $\delta\phi$ is arbitrary we conclude

$$\partial\mathscr{L}/\partial\phi - \partial/\partial y^\mu\, [\partial\mathscr{L}/\partial(\partial\phi/\partial X^\nu)\, \partial y^\mu/\partial X^\nu)] = 0 \qquad \text{(A.19a)}$$

The second term in eq. A.19a shows the effect of the dependence of ϕ on the field X, $\phi = \phi(X)$, rather than directly on the coordinate system y.

Similarly we can perform a variation in X^μ and obtain

$$\partial\mathscr{L}/\partial X^\mu - \partial/\partial y^\nu\, [\partial\mathscr{L}/\partial(\partial X^\mu/\partial y^\nu)] = 0 \qquad \text{(A.19b)}$$

The X field defines a "manifold" or, more properly, specifies a transformation from $R^n \rightarrow R^n$. If we make standard assumptions about the mapping: that it is continuous and piece-wise invertible, then we can establish the following lemmas:

Lemma 1: *If the transformation $X^\mu = X^\mu(y)$ is a transformation from $R^n \rightarrow R^n$ that is of class C′ and piece-wise invertible, then*

$$\frac{\partial}{\partial y^\nu}\frac{\partial y^\nu}{\partial X^\mu} = \frac{\partial \ln J}{\partial X^\mu} \qquad \text{(A.20)}$$

where

$$J = |\partial(X)/\partial(y)| \qquad \text{(A.21)}$$

is the absolute value of the Jacobian of the transformation.

Proof:
Consider two equivalent forms of an integral:

$$I = \int \mathscr{L} J\, d^4y = \int \mathscr{L}\, d^4X$$

where $\mathscr{L}$ is specified as in eq. A.15. Then the first expression for I leads to eq. A.19a which can be written in the form

$$\partial\mathscr{L}/\partial\phi - \partial/\partial X^\mu\, [\partial\mathscr{L}/\partial(\partial\phi/\partial X^\mu)] - \partial\mathscr{L}/\partial(\partial\phi/\partial X^\mu)\{\partial[J\partial y^\nu/\partial X^\mu]/\partial y^\nu\} = 0$$

Using the second expression for I above we obtain the following equation by variation in ϕ:

$$\partial\mathscr{L}/\partial\phi - \partial/\partial X^\mu\, [\partial\mathscr{L}/\partial(\partial\phi/\partial X^\mu)] = 0$$

Comparing these two expressions and realizing that $\partial[J\partial y^\nu/\partial X^\mu]/\partial y^\nu$ is totally independent of ϕ and its derivatives leads us to conclude

$$\partial\,[J\partial y^\nu/\partial X^\mu]/\partial y^\nu \;= 0 \qquad \text{(A.22)}$$

It is a general relationship for a transformation between X and y based on continuity and piece-wise invertibility. After a few elementary manipulations eq. A.22 can be rewritten in the form of eq. A.20. ■

Lemma 2: *If the transformation* $X^\mu = X^\mu(y)$ *is a transformation from* $R^n \rightarrow R^n$ *that is of class C′ and piece-wise invertible and* $\mathscr{L} = \mathscr{L}(\phi(X), \partial\phi/\partial X^\nu, X^\mu(y), \partial X^\mu(y)/\partial y^\nu, y)$, *then*

$$\partial\mathscr{L}/\partial(\partial\phi/\partial X^\nu)\; \partial y^\mu/\partial X^\nu = \partial\mathscr{L}/\partial(\partial\phi/\partial y^\mu) \qquad \text{(A.23)}$$

Proof:

Let us express $\mathscr{L}$ as a power series in derivatives of ϕ:

$$\mathscr{L} = \sum_{n=0} a_{n\mu_1\mu_2\ldots\mu_n}(\phi(X), X^\mu(y), \partial X^\mu(y)/\partial y^\nu, y) \prod_{j=1}^{n} \partial\phi/\partial X^{\mu_j}$$

which can rewritten using piece-wise invertibility as

$$\mathscr{L} = \sum_{n=0} a_{n\mu_1\mu_2\ldots\mu_n}(\phi(X), X^\mu(y), \partial X^\mu(y)/\partial y^\nu, y) \prod_{j=1}^{n} \partial\phi/\partial y^{\nu_j}\; \partial y^{\nu_j}/\partial X^{\mu_j}$$

Taking the derivative of this equation with respect to $\partial\phi/\partial y^\mu$ immediately yields the result. ■

Eq. A.23 enables us to rewrite eq. A.19a as:

$$\partial\mathscr{L}/\partial\phi \;- \partial/\partial y^\mu\; [\partial\mathscr{L}/\partial(\partial\phi/\partial y^\mu)] = 0 \qquad \text{(A.24)}$$

which is as one would expect.

In order to get a feeling for the effect of eq. A.19a we will look at a simple example where we specify the relation of the X and y variables directly. Then we will look at the composition of extrema where the transformation between X and y is itself determined as an extremum solution.

Example: a hyperplane

We assume eq. A.19b yields the transformation:

$$X^i = ay^i \qquad \text{for } i = 1,2,3$$
$$X^0 = 0$$

Then eq. A.19a becomes

$$\partial\mathscr{L}/\partial\phi - \partial/\partial y^i\,[\partial\mathscr{L}/\partial(\partial\phi/\partial y^i)] = 0 \qquad \text{(A.25)}$$

with the time derivative disappearing. Effectively the variation of ϕ on the hyperplane $X^0 = 0$ is determined by the differential equation generated by A.25. On this hyperplane the transformation between the X and y variables is invertible.

Coordinate Transformation Determined as an Extremum Solution

We now develop a formalism that determines a mapping from space onto itself as the solution of an extremum problem and also determines the dynamics of one or more fields as a function of this mapping. To this author's knowledge this area in the Calculus of Variations – the determination of an extremum on a manifold where the manifold itself is determined by an extremum – has not been previously explored. We will also develop a Hamiltonian formulation. Then we will proceed to quantize the theory.

Separable Lagrangian Case

Although there are many forms that the composition of extrema could take, one fairly general form that is directly useful in quantum field theory applications is based on a Lagrangian that can be split into two parts which we will call a *separable Lagrangian*:

$$\mathscr{L} = \mathscr{L}_F\, J + \mathscr{L}_C(X^\mu(y), \partial X^\mu(y)/\partial y^\nu, y) \qquad \text{(A.26)}$$

where J is defined in eq. A.21, where $\mathscr{L}_F$ contains all the dynamics of the fields and their interactions, and where $\mathscr{L}_C$ defines the coordinate mapping as an extremum solution. The procedure to determine the differential equations that specify the mapping, and the field equations that specify field interactions and evolution, is to vary in the coordinates X^μ and in the fields independently, using Hamilton's Principle. The extrema are to be determined for

$$I = \int \mathscr{L}\, d^4y \qquad \text{(A.27)}$$

We will begin by considering the case of one scalar field:

$$\mathscr{L}_F = \mathscr{L}_F(\phi(X), \partial\phi/\partial X^\nu) \qquad \text{(A.28)}$$

and

$$\mathscr{L}_C = \mathscr{L}_C(X^\mu(y), \partial X^\mu(y)/\partial y^\nu, y) \tag{A.29}$$

Eq. A.27 can be written in the form:

$$I = \int \mathscr{L}_F(\phi(X), \partial\phi/\partial X^\nu)\, dX + \int \mathscr{L}_C(X^\mu(y), \partial X^\mu(y)/\partial y^\nu, y)\, d^4y \tag{A.30}$$

using the Jacobian to transform to an integral over dX in the first term. A standard variation of ϕ and the application of Hamilton's Principle yields

$$\partial\mathscr{L}_F/\partial\phi - \partial/\partial X^\mu\, [\partial\mathscr{L}_F/\partial(\partial\phi/\partial X^\mu)] = 0 \tag{A.31}$$

reflecting the fact that ϕ is a function of X^μ only, with X^μ a function of the y coordinates.

Next we perform a variation of X^μ determining the mapping from y → X as an extremum of the integral in eq. A.27. We note the piece-wise invertibility of the coordinate mapping $X^\mu(y)$ allows us to write the Jacobian J as a function of y^μ only. A standard variation of X^μ and the application of Hamilton's Principle yields

$$\partial\mathscr{L}_C/\partial X^\mu - \partial/\partial y^\nu\, [\partial\mathscr{L}_C/\partial(\partial X^\mu/\partial y^\nu)] = 0 \tag{A.32}$$

Klein-Gordon Example

The Klein-Gordon scalar field theory furnishes us with a simple example of the application of the preceding development. The Lagrangian is

$$\mathscr{L}_F = \tfrac{1}{2}\, [\, (\partial\phi/\partial X^\nu)^2 - m^2\phi^2\,] \tag{A.33}$$

From eq. A.31 we obtain the field equation:

$$(\Box + m^2)\, \phi(X) = 0 \tag{A.34}$$

where

$$\Box = \partial/\partial X^\nu\, \partial/\partial X_\nu \tag{A.34a}$$

A Fourier representation of the solution of eq. A.34 is:

$$\phi(X) = \int dp\, \delta(p^2 - m^2)\theta(p^0)\, [A(p)\, e^{-ip\cdot X} + A(p)^*\, e^{ip\cdot X}] \tag{A.35}$$

where A(k) is a function of k and * indicates complex conjugation.

The determination of $X^\mu(y)$ depends on the Lagrangian $\mathscr{L}_C$ and the solutions of eq. A.3A. If we chose

$$\mathscr{L}_C = -\tfrac{1}{2}\,(\partial X^\mu/\partial y^\nu)^2 \qquad (A.36)$$

Then we obtain the equation

$$\Box\, X^\mu = 0 \qquad (A.37)$$

with the solution

$$X^\mu = \int dk\, \delta(k^2)\theta(k^0)\,[a^\mu(k)\, e^{-ik\cdot y} + a^\mu(k)^*\, e^{ik\cdot y}] \qquad (A.38)$$

where $a^\mu(k)$ are complex vector functions of k in general. (We ignore positivity issues for the moment.) Substitution of eq. A.38 in eq. A.35 yields an expression with a form reminiscent of bosonic string expressions.[115] We will take up this point later in subsequent chapters.

The Composition of Extrema – Hamiltonian Formulation

The previous section established a Lagrangian formulation of dynamics based on the composition of extrema. In this section we will develop an equivalent Hamiltonian formulation. We will assume a Minkowskian space-time with X^0 and y^0 playing the role of the time coordinates in the respective coordinate systems.

Initially, we will assume a scalar field ϕ with a Lagrangian of the form in eq. A.15 and define canonical momenta with

$$\Pi_\phi = \partial\mathscr{L}/\partial\dot{\phi} \equiv \partial\mathscr{L}/\partial(\partial\phi/\partial X^\mu)\, \partial y^0/\partial X^\mu \qquad (A.39)$$

$$\Pi_X{}^\mu = \partial\mathscr{L}/\partial\dot{X}_\mu \qquad (A.40)$$

where

$$\dot{\phi} = \partial\phi/\partial y^0 \equiv \partial\phi/\partial X^\mu\, \partial X^\mu/\partial y^0 \qquad (A.41)$$

$$\dot{X}^\mu = \partial X^\mu/\partial y^0 \qquad (A.42)$$

Then we define the Hamiltonian density as

$$\mathscr{H} = \Pi_\phi\, \dot{\phi} + \Pi_X{}^\mu\, \dot{X}_\mu - \mathscr{L}(\phi(X), \partial\phi/\partial X^\nu, X^\mu(y), \partial X^\mu(y)/\partial y^\nu, y) \qquad (A.43)$$

and the Hamiltonian

$$H = \int \mathscr{H}\, d^3y \qquad (A.44)$$

---

[115] See for example Polchinski (1998) and Bailin (1994).

The Hamiltonian density has the general form

$$\mathscr{H} = \mathscr{H}(\phi(X), \partial\phi/\partial X^i, \Pi_\phi, X^\mu(y), \partial X^\mu(y)/\partial y^j, \Pi_X{}^\mu, y^\nu) \qquad \text{(A.45)}$$

for the case of one scalar field where the indices i and j represent space coordinates. Time coordinates are assigned index value 0.

If we calculate the differential change in H using eq. A.45 we obtain

$$dH = \int \{ \partial\mathscr{H}/\partial\phi \, d\phi + \partial\mathscr{H}/\partial\Pi_\phi \, d\Pi_\phi - \partial/\partial y^\nu[\partial\mathscr{H}/\partial(\partial\phi/\partial X^i)\partial y^\nu/\partial X^i]d\phi +$$

$$+ \partial\mathscr{H}/\partial X^\mu \, dX^\mu + \partial\mathscr{H}/\partial\Pi_X{}^\mu \, d\Pi_X{}^\mu - \partial/\partial y^j \, [\partial\mathscr{H}/\partial(\partial X^\mu/\partial y^j)] \, dX^\mu \} d^3y$$

$$\text{(A.46)}$$

after some partial integrations. (Repeated indices indicate summations. Indices labeled i and j indicate space coordinates. Greek indices include all space-time components of a variable.)

Expressing the differential in H using eq. A.43 we obtain

$$dH = \int dy \, \{ \Pi_\phi \, d\dot{\phi} + \dot{\phi} \, d\Pi_\phi - \partial\mathscr{L}/\partial\phi \, d\phi - \partial\mathscr{L}/\partial(\partial\phi/\partial X^\mu)d(\partial\phi/\partial X^\mu) +$$

$$+ \Pi_X{}^\mu \, d\dot{X}^\mu + \dot{X}^\mu \, d\Pi_X{}^\mu - \partial\mathscr{L}/\partial X^\mu \, dX^\mu - \partial\mathscr{L}/\partial(\partial X^\mu/\partial y^j)d(\partial X^\mu/\partial y^j) \}$$

$$\text{(A.47a)}$$

After some manipulations we find

$$dH = \int \{ \dot{\phi} \, d\Pi_\phi + \dot{X}_\mu d\Pi_X{}^\mu - \partial/\partial y^0 \, \Pi_\phi \, d\phi - \partial/\partial y^0 \, \Pi_X{}^\mu \, dX_\mu \} \, dy$$

$$\text{(A.47b)}$$

using the equations of motion eqs. A.19a and A.19b.

Comparing eqs A.46 and A.47 we obtain Hamilton's equations in the case of the composition of extrema:

$$\dot{\phi} = \partial\mathscr{H}/\partial\Pi_\phi \qquad \text{(A.48a)}$$

$$\dot{\Pi}_\phi = -\partial\mathscr{H}/\partial\phi + \partial/\partial y^\nu \, [\partial\mathscr{H}/\partial(\partial\phi/\partial X^i) \, \partial y^\nu/\partial X^i] \qquad \text{(A.48b)}$$

$$\dot{X}_\mu = \partial\mathscr{H}/\partial\Pi_X{}^\mu \qquad \text{(A.48c)}$$

$$\dot{\Pi}_X{}^\mu = -\partial\mathscr{H}/\partial X^\mu + \partial/\partial y^j \, [\partial\mathscr{H}/\partial(\partial X^\mu/\partial y^j)] \qquad \text{(A.48d)}$$

where

$$\dot{\Pi}_\phi = \partial\, \Pi_\phi / \partial y^0 \tag{A.49a}$$

$$\dot{\Pi}_X{}^\mu = \partial \Pi_X{}^\mu / \partial y^0 \tag{A.49b}$$

Translational Invariance

If the Lagrangian of a field theory has no explicit dependence on the coordinates then one expects translational invariance accompanied by a conservation law for an energy-momentum stress tensor. We will show this is the case for Lagrangians implementing the composition of extrema. We assume a Lagrangian without an explicit dependence on the coordinates y^ν:

$$\mathscr{L} = \mathscr{L}(\phi(X), \partial\phi/\partial X^\nu, X^\mu(y), \partial X^\mu(y)/\partial y^\nu) \tag{A.50}$$

Under an infinitesimal displacement,

$$y'^\nu = y^\nu + \epsilon^\nu \tag{A.51a}$$

$$\delta\phi = \phi(X(y + \epsilon)) - \phi(X(y))$$

$$= \epsilon^a\, \partial\phi/\partial y^a \tag{A.51b}$$

$$\delta X^\mu = \epsilon^a\, \partial X^\mu/\partial y^a \tag{A.51c}$$

$$\delta(\partial\phi/\partial X^\mu) = \epsilon^a\, \partial(\partial\phi/\partial y^a)/\partial X^\mu \tag{A.51d}$$

$$\delta(\partial X^\mu/\partial y^\nu) = \epsilon^a\, \partial(\partial X^\mu/\partial y^a)/\partial y^\nu \tag{A.51e}$$

and the Lagrangian changes by

$$\delta\mathscr{L} = \epsilon^a\, \partial\mathscr{L}/\partial y^a \tag{A.52}$$

The change can also be expressed in terms of the changes in the fields, their derivatives and the mapping X^μ:

$$\delta\mathscr{L} = \partial\mathscr{L}/\partial\phi\, \delta\phi + \partial\mathscr{L}/\partial(\partial\phi/\partial X^\mu)\, \delta(\partial\phi/\partial X^\mu) + \partial\mathscr{L}/\partial X^\mu\, \delta X^\mu + \\ + \partial\mathscr{L}/\partial(\partial X^\mu/\partial y^\nu)\, \delta(\partial X^\mu/\partial y^\nu) \tag{A.53}$$

Combining eqs. A.51, A.52 and A.53 we obtain (after some manipulations):

$$\in^{\nu} \partial/\partial y_{\mu} \mathscr{T}_{\mu\nu} = 0 \tag{A.54}$$

where

$$\mathscr{T}_{\mu\nu} = -g_{\mu\nu} \mathscr{L} + \partial\mathscr{L}/\partial(\partial\phi/\partial X^{\delta})\, \partial y_{\mu}/\partial X^{\delta}\, \partial\phi/\partial y^{\nu} + \partial\mathscr{L}/\partial(\partial X^{\delta}/\partial y_{\mu})\partial X^{\delta}/\partial y^{\nu} \tag{A.55a}$$

or, alternately using Lemma 2,

$$\mathscr{T}_{\mu\nu} = - g_{\mu\nu} \mathscr{L} + \partial\mathscr{L}/\partial(\partial\phi/\partial y_{\mu})\, \partial\phi/\partial y^{\nu} + \partial\mathscr{L}/\partial(\partial X^{\delta}/\partial y_{\mu})\, \partial X^{\delta}/\partial y^{\nu} \tag{A.55b}$$

Since $\in^{a}$ is an arbitrary displacement we obtain the conservation law:

$$\partial/\partial y_{\mu} \mathscr{T}_{\mu\nu} = 0 \tag{A.56}$$

Eq. A.56 implies the energy-momentum vector

$$P_{\beta} = \int d^3y\, \mathscr{T}_{0\beta} \tag{A.57}$$

is conserved. We note

$$\partial/\partial y^0 P_{\beta} = 0 \tag{A.58}$$

since eq. A.56 and A.57 can be used to obtain the integral of a divergence, which results in zero.

The Hamiltonian (eqs. A.43-44) is

$$H = P_0 \tag{A.59}$$

We note for later use that the total energy, H, which is conserved, contains a term that represents the energy in the X^{μ} mapping. Thus energy can be exchanged in principle between the ϕ field sector and the X^{μ} sector.

Lorentz Invariance and Angular Momentum Conservation

We can also verify Lorentz invariance and obtain the form of the conserved angular momentum by considering the effect of an infinitesimal Lorentz transformation. We will consider the case of a scalar field ϕ.

Under an infinitesimal Lorentz transformation ($\in_{\mu\nu} = -\in_{\nu\mu}$):

$$y'_{\mu} = y_{\mu} + \delta y_{\mu} = y_{\mu} + \in_{\mu\nu} y^{\nu} \tag{A.60a}$$

$$\delta\phi = \phi(X(y')) - \phi(X(y))$$

$$= \epsilon^{\mu\nu}\, y_\nu\, \partial\phi/\partial X^a\, \partial X^a/\partial y^\mu \tag{A.60b}$$

$$\delta X^\mu = S^\mu{}_a X^a(y') - X^\mu(y) \tag{A.60c}$$

$$= \epsilon^\mu{}_a X^a(y) + \partial X^\mu/\partial y^\beta\, \delta y^\beta \tag{A.60d}$$

where $S^\mu{}_a$ is the matrix for the Lorentz transformation of a vector. (If X^μ were a gauge field then an additional operator gauge term would have to be added to eq. A.60d.)

The Lagrangian changes by

$$\delta\mathscr{L} = \epsilon^{\mu\nu}\, y_\nu \partial\mathscr{L}/\partial y^\mu \tag{A.61}$$

under the infinitesimal Lorentz transformation. The change in the Lagrangian can also be expressed as:

$$\delta\mathscr{L} = \partial\mathscr{L}/\partial\phi\, \delta\phi + \partial\mathscr{L}/\partial(\partial\phi/\partial X^\mu)\, \delta(\partial\phi/\partial X^\mu) + \partial\mathscr{L}/\partial X^\mu\, \delta X^\mu + \\ + \partial\mathscr{L}/\partial(\partial X^\mu/\partial y^\nu)\, \delta(\partial X^\mu/\partial y^\nu) \tag{A.62}$$

Combining eqs. A.61 and A.62, and substituting and simplifying terms leads to:

$$\epsilon_{\mu\nu}\, \partial/\partial y^\sigma \mathscr{M}^{\sigma\mu\nu} = 0 \tag{A.63}$$

where

$$\mathscr{M}^{\sigma\mu\nu} = (g^{\mu\sigma}y^\nu - g^{\nu\sigma}y^\mu)\mathscr{L} + \partial\mathscr{L}/\partial(\partial\phi/\partial X^a)\, \partial y^\sigma/\partial X^a\, (y^\mu\partial\phi/\partial y_\nu - y^\nu\partial\phi/\partial y_\mu) + \\ + \partial\mathscr{L}/\partial(\partial X^\delta/\partial y^\sigma)\, (g^{\delta\nu}X^\mu - g^{\delta\mu}X^\nu + y^\mu\, \partial X^\delta/\partial y_\nu - y^\nu\, \partial X^\delta/\partial y_\mu) \tag{A.64}$$

The conserved angular momentum is:

$$M^{\mu\nu} = \int d^3y\, \mathscr{M}^{0\mu\nu} \tag{A.65}$$

with

$$\partial M^{\mu\nu}/\partial y^0 = 0 \tag{A.66}$$

The angular momentum density can be written in the familiar form:

$$\mathscr{M}^{\sigma\mu\nu} = y^\mu\, \mathscr{T}^{\sigma\nu} - y^\nu\, \mathscr{T}^{\sigma\mu} + \partial\mathscr{L}/\partial(\partial X^\delta/\partial y^\sigma)\, (g^{\delta\nu}X^\mu - g^{\delta\mu}X^\nu) \tag{A.67}$$

taking account of the vector nature of X^μ. The spatial part of $M^{\mu\nu}$ is the angular momentum.

Internal Symmetries

We will now consider the case of a set of scalar fields ϕ_r in a Lagrangian with an internal symmetry. Under a local transformation

$$\phi_r(X) \rightarrow \phi_r(X) - i\epsilon\lambda_{rs}\,\phi_s(X) \qquad \text{(A.68)}$$

If the Lagrangian is invariant under this transformation, then

$$\delta\mathcal{L} = 0 = \partial\mathcal{L}/\partial\phi_r\delta\phi_r + \partial\mathcal{L}/\partial(\partial\phi_r/\partial X^\alpha)\,\delta(\partial\phi_r/\partial X^\alpha) \qquad \text{(A.69)}$$

Using the equation of motion eq. A.19a satisfied by all the components ϕ_r we obtain a conserved current:

$$\mathcal{J}^\nu = -i\,\partial\mathcal{L}/\partial(\partial\phi_r/\partial X^\delta)\,\partial y^\nu/\partial X^\delta\,\lambda_{rs}\,\phi_s \qquad \text{(A.70)}$$

which satisfies

$$\partial\mathcal{J}^\nu/\partial y^\nu = 0 \qquad \text{(A.71)}$$

The conserved charge is

$$Q = \int d^3y\,\mathcal{J}^0 \qquad \text{(A.72)}$$

$$\partial Q/\partial y^0 = 0 \qquad \text{(A.73)}$$

Separable Lagrangians

We now consider the case of a separable Lagrangian such as in eq. A.26. Adopting the definitions:

$$\phi' = \partial\phi/\partial X^0 \qquad \text{(A.74)}$$

$$X_\mu' = \partial X_\mu/\partial y^0 \qquad \text{(A.75)}$$

We define canonical momenta as

$$\pi_\phi = \partial\mathcal{L}/\partial\phi' \equiv \partial\mathcal{L}/\partial(\partial\phi/\partial X^0) \qquad \text{(A.76)}$$

$$\pi_X{}^\mu = \partial\mathcal{L}/\partial X_\mu' \equiv \partial\mathcal{L}/\partial(\partial X_\mu/\partial y^0) \qquad \text{(A.77)}$$

We now define the separable Hamiltonian density as

$$\mathscr{H}_s = J\pi_\phi \phi' + \pi_X{}^\mu X_\mu{}' - \mathscr{L}_s \tag{A.78}$$

where J is the Jacobian (eq. A.21) and

$$H_s = \int \mathscr{H}_s \, d^3y \tag{A.79}$$

The separable Lagrangian (from eq. A.26) is:

$$\mathscr{L}_s = \mathscr{L}_F(\phi(X), \partial\phi/\partial X^\mu)\, J + \mathscr{L}_C(X^\mu(y), \partial X^\mu(y)/\partial y^\nu, y) \tag{A.80}$$

In the case of one scalar field the separable Hamiltonian density has the general form

$$\mathscr{H}_s = \mathscr{H}_S(\phi(X), \pi_\phi, \partial\phi/\partial X^i, X^\mu(y), \pi_X{}^\mu, \partial X^\mu(y)/\partial y^j, y^\nu) \tag{A.81}$$

where the indices i and j indicate spatial components. In particular, the terms in the separable Hamiltonian are:

$$\mathscr{H}_s = \mathscr{H}_F J + \mathscr{H}_C \tag{A.82}$$

with

$$\mathscr{H}_F(\phi(X), \pi_\phi, \partial\phi/\partial X^i) = \pi_\phi \phi' - \mathscr{L}_F \tag{A.83}$$

$$\mathscr{H}_C(X^\mu(y), \pi_X{}^\mu, \partial X^\mu(y)/\partial y^j, y^\nu) = \pi_X{}^\mu X_\mu{}' - \mathscr{L}_C \tag{A.84}$$

where J is the absolute value of the Jacobian defined in A.21.

We now define the time integral of H as we did in eq. A.14 when considering the Lagrangian formulation:

$$G = \int dy^0 \, H_s \tag{A.85}$$

Thus G is an integral over all space-time coordinates. Using G we can develop a Hamiltonian formulation. First we calculate the differential change in G. Using eqs. A.81-2 and A.85 we obtain

$$\begin{aligned} dG = \int \{ & J\, \partial\mathscr{H}_F/\partial\phi \, d\phi + J\, \partial\mathscr{H}_F/\partial\pi_\phi \, d\pi_\phi + \\ & + J\, \partial\mathscr{H}_F/\partial(\partial\phi/\partial X^j)\, d(\partial\phi/\partial X^j) + \partial\mathscr{H}_C/\partial X^\mu \, dX^\mu + \\ & + \partial\mathscr{H}_C/\partial\pi_X{}^\mu \, d\pi_X{}^\mu + \partial\mathscr{H}_C/\partial(\partial X^\mu/\partial y^j)\, d(\partial X^\mu/\partial y^j) \} \, d^4y \end{aligned} \tag{A.86}$$

with summations implied by repeated indices. (Index labels i and j label spatial coordinates only; Greek indices label space-time coordinates.) Rewriting dG as two integrals and performing partial integrations yields:

$$dG = \int d^4X \{\partial\mathscr{H}_F/\partial\phi\, d\phi + \partial\mathscr{H}_F/\partial\pi_\phi d\pi_\phi - \partial/\partial X^j[\partial\mathscr{H}_F/\partial(\partial\phi/\partial X^j)]\, d\phi\} +$$
$$+ \int d^4y \{\partial\mathscr{H}_C/\partial X^\mu\, dX^\mu + \partial\mathscr{H}_C/\partial\pi_X{}^\mu\, d\pi_X{}^\mu - \partial/\partial y^j[\partial\mathscr{H}_C/\partial(\partial X^\mu/\partial y^j)]\, dX^\mu\} \qquad (A.87)$$

Alternately, expressing the differential in G using eqs. A.82-4 we obtain

$$dG = \int d^4X \{\pi_\phi\, d\phi' + \phi' d\pi_\phi - \partial\mathscr{L}_F/\partial\phi\, d\phi - \partial\mathscr{L}_F/\partial(\partial\phi/\partial X^\mu) d(\partial\phi/\partial X^\mu)\} +$$
$$+ \int d^4y \{\pi_{X\mu}\, dX^{\mu\prime} + X^{\mu\prime}\, d\pi_{X\mu} - \partial\mathscr{L}_C/\partial X^\mu\, dX^\mu - \partial\mathscr{L}_C/\partial(\partial X^\mu/\partial y^j) d(\partial X^\mu/\partial y^j)\} \qquad (A.88)$$

which becomes

$$dG = \int d^4X \{-\pi_\phi'\, d\phi + \phi'\, d\pi_\phi\} + \int d^4y \{-\pi_{X\mu}'\, dX^\mu + X^{\mu\prime}\, d\pi_{X\mu}\} \qquad (A.89)$$

using the equations of motion eqs. A.31-2.

Comparing eqs A.87 and A.89 we obtain Hamilton's equations for the case of the composition of extrema for a separable Lagrangian:

$$\phi' = \partial\mathscr{H}_F/\partial\pi_\phi \qquad (A.90)$$

$$\pi_\phi' = -\partial\mathscr{H}_F/\partial\phi + \partial/\partial X^j\, [\partial\mathscr{H}_F/\partial(\partial\phi/\partial X^j)] \qquad (A.91)$$

$$X_\mu' = \partial\mathscr{H}_C/\partial\pi_X{}^\mu \qquad (A.92)$$

$$\pi_{X\mu}' = -\partial\mathscr{H}_C/\partial X^\mu + \partial/\partial y^j\, [\partial\mathscr{H}_C/\partial(\partial X^\mu/\partial y^j)] \qquad (A.93)$$

where

$$\pi_\phi' = \partial\, \pi_\phi/\partial X^0 \qquad (A.94)$$

$$\pi_{X\mu}' = \partial\, \pi_{X\mu}/\partial X^0 \qquad (A.95)$$

Notice $\mathscr{L}_F$, $\mathscr{H}_F$ and π_ϕ have precisely the same form, as a function of X^μ, as one sees in a conventional field theory formalism. Yet X^μ is a mapping/function of the coordinates y. In reality, it can be viewed as a field as we shall see.

Separable Lagrangians and Translational Invariance

The general rule for conventional Lagrangians is: if a Lagrangian has no explicit dependence on the coordinates then translational invariance follows accompanied by a conservation law for an energy-momentum tensor. We will show that this rule needs modification for separable Lagrangians that implement the composition of extrema.

Consider the Lagrangian:

$$\mathscr{L}_s = J\,\mathscr{L}_F(\phi(X), \partial\phi/\partial X^\mu) + \mathscr{L}_C(X^\mu(y), \partial X^\mu(y)/\partial y^\nu) \qquad (A.96)$$

in which the X^μ play a dual role as both fields and coordinates. Let us consider a variation in X^μ:

$$X^\mu(y) \rightarrow X^\mu(y) + \delta X^\mu(y) \qquad (A.97)$$

where $\delta X^\mu(y)$ is an arbitrary function of y that vanishes at the endpoints of the integration region of the integral. The action is:

$$I = \int \mathscr{L}_s d^4y \qquad (A.98)$$

We will show that a variation in $X^\mu(y)$ leads to a conserved energy-momentum tensor. But we will use integrals of the Lagrangian density since it provides a simpler derivation of the result. Under the variation of eq. A.97 we find

$$\delta\phi = \phi(X(y)+ \delta X^\mu(y)) - \phi(X(y))$$

$$= \delta X^\mu\, \partial\phi/\partial X^\mu \qquad (A.99a)$$

$$\delta(\partial\phi/\partial X^\nu) = \delta X^\mu\, \partial(\partial\phi/\partial X^\mu)/\partial X^\nu \qquad (A.99b)$$

$$\delta(\partial X^\mu/\partial y^\nu) = \partial(\delta X^\mu)/\partial y^\nu \qquad (A.99c)$$

The integral in eq. A.98 changes by

$$\delta I = \int d^4y\, \delta\mathscr{L}_s = \int d^4y\, [\delta(J\mathscr{L}_F) + \delta\mathscr{L}_C] \qquad (A.100a)$$

which becomes:

$$\delta I = \int d^4y\, [\delta X^\mu\, \partial(J\mathscr{L}_F)/\partial X^\mu + \partial(\delta X^\mu \partial\mathscr{L}_C\, /\partial(\partial X^\mu/\partial y^\nu))/\partial y^\nu] \quad (A.100b)$$

due to the equations of motion of X^μ (eq. A.19b) in X^μ's role. Since the second term is a total divergence its contribution to δI is zero. Thus we can express eq. A.100b as:

$$\delta I = \int d^4y\, [\, J\, \delta\mathscr{L}_F + \mathscr{L}_F\, \delta J\,] \quad (A.101)$$

realizing that the Jacobian J depends on y and thus X:

$$\delta J = \delta X^\mu\, \partial J/\partial X^\mu \quad (A.102)$$

A partial integration gives

$$\mathscr{L}_F\, \delta J = \delta X^\mu\, \partial(\, J\mathscr{L}_F)/\partial X^\mu - \delta X^\mu\, J\, \partial\mathscr{L}_F/\partial X^\mu \quad (A.103)$$

Evaluating $\delta\mathscr{L}_F$ we find:

$$\delta\mathscr{L}_F = \partial\mathscr{L}_F/\partial\phi\, \delta\phi + \partial\mathscr{L}_F\, /\partial(\partial\phi/\partial X^\mu)\, \delta(\partial\phi/\partial X^\mu) \quad (A.104)$$

which gives

$$\delta\mathscr{L}_F = \delta X^\nu\, \partial/\partial X^\mu\, [\partial\mathscr{L}_F/\partial(\partial\phi/\partial X^\mu)\, \partial\phi/\partial X^\nu\,] \quad (A.105)$$

using the equations of motion eq. A.31, and using eq. A.99b. Combining eqs. A.100, A.101, A.103 and A.105 we obtain:

$$\int d^4y\, J\, \delta X^\nu\, \partial/\partial X_\mu\, \mathscr{T}_{F\mu\nu} = \int d^4X\, \delta X^\nu\, \partial/\partial X_\mu\, \mathscr{T}_{F\mu\nu} = 0 \quad (A.106)$$

where

$$\mathscr{T}_{F\mu\nu} = -\, g_{\mu\nu}\, \mathscr{L}_F + \partial\mathscr{L}_F/\partial(\partial\phi/\partial X_\mu)\, \partial\phi/\partial X^\nu \quad (A.107)$$

after some manipulations. Since δX^ν is an arbitrary function of y the differential conservation law follows:

$$\partial/\partial X_\mu\, \mathscr{T}_{F\mu\nu} = 0 \quad (A.108)$$

Eq. A.108 implies the energy-momentum vector

$$P_{F\beta} = \int d^3X\, \mathscr{T}_{F0\beta} \tag{A.109}$$

is conserved:

$$\partial/\partial X^0\, P_{F\beta} = 0 \tag{A.110}$$

The Hamiltonian density (eq. A.83) is

$$\mathscr{H}_F = \mathscr{T}_{F0\beta} \tag{A.111}$$

Thus the field energy

$$H_F = P_{F0} = \int d^3X\, \mathscr{T}_{F00} \tag{A.112}$$

is conserved with respect to the "time" X^0. Later we will see that H_F is trivially conserved in the Coulomb gauge of X_μ. (We will also establish an electromagnetic-like quantum field theory for X_μ with gauge invariance.) In other gauges the conservation of H_F is not trivial.

Separable Lagrangians and Angular Momentum Conservation

We can also verify Lorentz invariance and obtain the form of the conserved angular momentum for a separable Lagrangian by considering the effect of an infinitesimal Lorentz transformation. We will consider the case of a scalar field ϕ.

Under an infinitesimal Lorentz transformation as specified by eqs. A.60a – A.60d the separable Lagrangian changes by

$$\delta\mathscr{L}_s = \epsilon^{\mu\nu}\, y_\nu\, \partial\mathscr{L}_s/\partial y^\mu \tag{A.113}$$

which can also be expressed as

$$\delta\mathscr{L}_s = \partial\mathscr{L}_s/\partial\phi\,\delta\phi + \partial\mathscr{L}_s/\partial(\partial\phi/\partial X^\mu)\,\delta(\partial\phi/\partial X^\mu) + \partial\mathscr{L}_s/\partial X^\mu\,\delta X^\mu + \\ + [\partial\mathscr{L}_s/\partial(\partial X^\mu/\partial y^\nu)]\,\delta(\partial X^\mu/\partial y^\nu) \tag{A.114}$$

Combining eqs. A.113 and A.114 leads to:

$$\epsilon_{\mu\nu}\, \partial/\partial y^\sigma \mathscr{M}_s^{\ \sigma\mu\nu} = 0 \tag{A.115}$$

where

$$\mathscr{M}_s^{\ \sigma\mu\nu} = J\, \mathscr{M}_F^{\ \sigma\mu\nu} + \mathscr{M}_C^{\ \sigma\mu\nu} + \mathscr{M}_M^{\ \sigma\mu\nu} \tag{A.116}$$

$$\mathscr{M}_F^{\ \sigma\mu\nu} = (g^{\mu\sigma}y^\nu - g^{\nu\sigma}y^\mu)\mathscr{L}_F + \partial\mathscr{L}_F/\partial(\partial\phi/\partial y_\sigma)\,(y^\mu\partial\phi/\partial y_\nu - y^\nu\partial\phi/\partial y_\mu) \tag{A.117}$$

$$\mathscr{M}_C^{\ \sigma\mu\nu} = (g^{\mu\sigma}y^\nu - g^{\nu\sigma}y^\mu)\mathscr{L}_C \ +$$

$$+ \partial\mathscr{L}_C/\partial(\partial X^\delta/\partial y^\sigma)(g^{\delta\nu}X^\mu - g^{\delta\mu}X^\nu + y^\mu\,\partial X^\delta/\partial y_\nu - y^\nu\,\partial X^\delta/\partial y_\mu) \tag{A.118}$$

$$\mathscr{M}_M{}^{\sigma\mu\nu} = \mathscr{L}_F\,\partial J/\partial(\partial X^\delta/\partial y^\sigma)(g^{\delta\nu}X^\mu - g^{\delta\mu}X^\nu + y^\mu\,\partial X^\delta/\partial y_\nu - y^\nu\,\partial X^\delta/\partial y_\mu) \tag{A.119}$$

where the third term originates in the dependence of J on derivatives of X^μ. Eq. A.117 was obtained in part by using the identity:

$$\partial\mathscr{L}/\partial(\partial\phi/\partial y^\sigma) = \partial\mathscr{L}/\partial(\partial\phi/\partial X^\alpha)\,\partial y^\sigma/\partial X^\alpha \tag{A.120}$$

where $\mathscr{L}$ and ϕ have the form specified in eq. A.15.

The conserved angular momentum is:

$$M_s^{\mu\nu} = \int dy\,\mathscr{M}_s^{0\mu\nu} \tag{A.121}$$

with

$$\partial M_s^{\mu\nu}/\partial y^0 = 0 \tag{A.122}$$

Angular Momentum and $\mathscr{L}_F$

An alternate conserved angular momentum can be obtained by considering the "field" part of the Lagrangian $\mathscr{L}_F$ under an infinitesimal Lorentz transformation ($\epsilon_{\mu\nu} = -\epsilon_{\nu\mu}$):

$$X'_\mu = X_\mu + \delta X_\mu \tag{A.123a}$$

$$\delta\phi = \phi(X'(y)) - \phi(X(y))$$

$$= \delta X^\mu\,\partial\phi/\partial X^\mu \tag{A.123b}$$

$$\delta X^\mu = S^\mu{}_a X^a(y) - X^\mu(y) \tag{A.123c}$$

$$= \epsilon^\mu{}_a X^a(y) \tag{A.123d}$$

where $S^\mu{}_a$ is the Lorentz transformation matrix for a vector. (If X^μ is a gauge field then an additional operator gauge term would have to be added to eq. A.123d.)

The Lagrangian changes by

$$\delta\mathscr{L}_F = \epsilon^{\mu\nu}X_\nu\,\partial\mathscr{L}_F/\partial X^\mu \tag{A.124}$$

under an infinitesimal Lorentz transformation. The change can also be expressed as:

$$\delta\mathscr{L}_F = \partial\mathscr{L}_F/\partial\phi\,\delta\phi + \partial\mathscr{L}_F/\partial(\partial\phi/\partial X^\mu)\,\delta(\partial\phi/\partial X^\mu) \quad (A.125)$$

Combining eqs. A.124 and A.125 leads to:

$$\epsilon_{\mu\nu}\partial/\partial X^\sigma\,\mathscr{M}_{FX}{}^{\sigma\mu\nu} = 0 \quad (A.126)$$

where

$$\mathscr{M}_{FX}{}^{\sigma\mu\nu} = (g^{\mu\sigma}X^\nu - g^{\nu\sigma}X^\mu)\mathscr{L}_F + \partial\mathscr{L}_F/\partial(\partial\phi/\partial X^\sigma)\,(X^\mu\partial\phi/\partial X_\nu - X^\nu\partial\phi/\partial X_\mu) \quad (A.127)$$

The conserved angular momentum associated with the X coordinates is:

$$M_{FX}{}^{\mu\nu} = \int d^3X\,\mathscr{M}_{FX}{}^{0\mu\nu} \quad (A.128)$$

with

$$\partial M_{FX}{}^{\mu\nu}/\partial X^0 = 0 \quad (A.129)$$

The angular momentum density can be written in the familiar form:

$$\mathscr{M}_{FX}{}^{\sigma\mu\nu} = X^\mu\,\mathscr{T}_F^{\sigma\nu} - X^\nu\,\mathscr{T}_F^{\sigma\mu} \quad (A.130)$$

using eq. A.107.

Separable Lagrangians and Internal Symmetries

We will now consider the case of a set of scalar fields ϕ_r in a separable Lagrangian with an internal symmetry under a local transformation

$$\phi_r(X) \to \phi_r(X) - i\epsilon\lambda_{rs}\,\phi_s(X) \quad (A.131)$$

If the Lagrangian is invariant under this transformation, then

$$\delta\mathscr{L}_S \equiv \delta\mathscr{L}_F = 0 = \partial\mathscr{L}_F/\partial\phi_r\,\delta\phi_r + \partial\mathscr{L}_F/\partial(\partial\phi_r/\partial X^\alpha)\,\delta(\partial\phi_r/\partial X^\alpha) \quad (A.132)$$

Using the equation of motion eq. A.31, which is satisfied by all components ϕ_r, we obtain a conserved current:

$$\mathscr{J}^\nu = -i\,\partial\mathscr{L}_F/\partial(\partial\phi_r/\partial X^\nu)\,\lambda_{rs}\,\phi_s \quad (A.133)$$

satisfying

$$\partial\mathscr{J}^\nu/\partial X^\nu = 0 \quad (A.134)$$

The conserved charge is

$$Q = \int d^3X \, \mathcal{J}^0 \qquad (A.135)$$

$$\partial Q/\partial X^0 = 0 \qquad (A.136)$$

We note eq. A.71 provides a corresponding conservation law for the y coordinate system.

Appendix B. Invariance of Two-Tier Quantum Field Theory under Special Relativity

Invariance of Two-Tier QFT under Special Relativistic Transformations

Turning now to the issue of invariance under special relativistic transformations we begin by noting that the transverse gauge of the Y^a field is not manifestly relativistic. In addition Two-Tier Feynman propagators calculated in the transverse gauge are also not manifestly relativistic.

The situation is similar to the case of the electromagnetic field yet differs because the Y^a field plays the role in coordinates in the Lagrangian. We will now show that the manifestly Lorentz invariant Two-Tier *Lorentz gauge formulation* is equivalent to the transverse gauge formulation. The classical Lorentz gauge condition:

$$\partial Y^\mu/\partial y^\mu = 0 \tag{B.1}$$

is too stringent to make into an operator relation. We will therefore define the Lorentz gauge formulation of the Y^a quantization by implementing a condition on the space of physical states.

We begin with the covariant equal time commutation relations:

$$[Y^\mu(\mathbf{y}, y^0), Y^\nu(\mathbf{y}', y^0)] = [\pi^\mu(\mathbf{y}, y^0), \pi^\nu(\mathbf{y}', y^0)] = 0 \tag{B.2}$$

$$[\pi^\mu(\mathbf{y}, y^0), Y^\nu(\mathbf{y}', y^0)] = i\eta^{\mu\nu}\delta^3(\mathbf{y} - \mathbf{y}') \tag{B.3}$$

where

$$\pi^\mu = \partial\mathscr{L}_C / \partial Y_\mu' \tag{B.4}$$

$$Y^{\mu\prime} = \partial Y^\mu/\partial y^0 \tag{B.5}$$

The Fourier expansion of Y^μ is:

$$Y^\mu(y) = \int d^3k\ N_0(k)[\ a^\mu(k)\ e^{-ik\cdot y} + a^{\mu\dagger}(k)\ e^{ik\cdot y}] \tag{B.6}$$

where

$$N_0(k) = [(2\pi)^3 2\omega_k]^{-1/2} \tag{B.7}$$

and

$$\omega_k = (\mathbf{k}^2)^{1/2} = k^0 \tag{B.8}$$

with $k^\mu k_\mu = 0$.

The commutation relations of the Fourier coefficient operators are:

$$[a^\mu(k), a^{\nu\dagger}(k')] = -\eta^{\mu\nu}\delta^3(\mathbf{k} - \mathbf{k}') \tag{B.9}$$

$$[a^{\mu\dagger}(k), a^{\nu\dagger}(k')] = [a^{\mu}(k), a^{\nu}(k')] = 0 \tag{B.10}$$

It will be convenient to divide the Y field into positive and negative frequency parts:

$$Y^{\mu+}(y) = \int d^3k\, N_0(k) a^{\mu}(k)\, e^{-ik\cdot y} \tag{B.11}$$

and

$$Y^{\mu-}(y) = \int d^3k\, N_0(k)\, a^{\mu\dagger}(k)\, e^{ik\cdot y} \tag{B.12}$$

We define

$$a^4(k) = i\, a^0(k) \tag{B.13}$$

and

$$a^{4\dagger}(k) = i\, a^{0\dagger}(k) \tag{B.14}$$

with the resulting commutation relations:

$$[a^{\mu}(k), a^{\nu\dagger}(k')] = \delta^{\mu\nu}\delta^3(\mathbf{k} - \mathbf{k}') \tag{B.15}$$

for $\mu,\nu = 1, 2, 3, 4$.

Having redefined the operators we then follow the familiar Gupta-Bleuler procedure and introduce an indefinite Dirac metric η with $\eta^2 = 1$ and $\eta^\dagger = \eta$ that will enable us to avoid negative probabilities when inner products are calculated. In particular we note,

$$\eta a^4(k) = -a^4(k)\eta \tag{B.16}$$
$$\eta a^i(k) = a^i(k)\eta \tag{B.17}$$

for i = 1, 2, 3. Thus

$$\eta = (-1)^{n_4} \tag{B.18}$$

where n_4 is the number of time-like Y^4 "particles". Let us now define a particle state Φ_{n_4} with n_4 time-like Y^4 "particles." Then the time-like raising and lowering operators change the number of particles in a state:

$$a^4(k)\, \Phi_{n_4+1} = (n_4 + 1)^{1/2}\, \Phi_{n_4} \tag{B.19a}$$

and

$$a^{4\dagger}(k)\, \Phi_{n_4} = (n_4 + 1)^{1/2}\, \Phi_{n_4+1} \tag{B.19b}$$

We now chose the coordinate system so the z direction is the direction of "propagation." (A general specification of the coordinate system does not change the result.) Consequently, with the **k** direction fixed, we can write

$$\partial Y^1/\partial y^1 + \partial Y^2/\partial y^2 = 0 \tag{B.20}$$

and the Lorentz gauge condition, which becomes a condition on the physical states,[116] denoted Φ_L , can be written as:

$$(a^3(k) + i\, a^4(k))\Phi_L = 0 \tag{B.21}$$

for all k.

Any physical state Φ_L can be written as a superposition of states with sharp Y^3 and Y^4 particle numbers $\Phi_{n,m}$:

$$\Phi_{Ln} = \sum^{n} c_m\, \Phi_{n,n-m} \tag{B.22}$$

where the number of transverse Y "particles" is fixed. In fact, due to our unitarity requirement the number of transverse Y "particles" is always fixed to zero in physical states.

Defining the Lorentz operator

$$L = \partial Y^\mu/\partial y^\mu = L_+ + L_- \tag{B.23}$$

where we separate L into positive and negative frequency parts as in eqs. B.11 and B.12 we see that the Lorentz gauge condition (eq. B.21) becomes

$$L_+\Phi_L = 0 \tag{B.24}$$

with Hermitian conjugate (We reserve the † superscript for later use: $\Phi_L^\dagger = \Phi_L{}^*\eta$.)

$$\Phi_L{}^* L_+{}^\dagger = \Phi_L{}^* \eta L_- = 0 \tag{B.25}$$

which is obtained by multiplying the Hermitian conjugate of eq. B.24 on the right by η and using $L_+{}^\dagger = \eta L_- \eta$ and $\eta^2 = 1$. Therefore the inner product equals zero

$$(\Phi_L{}^* \eta L \Phi_L) = 0 \tag{B.26}$$

as does

$$(\Phi_L{}^* \eta L^2 \Phi_L) = 0 \tag{B.27}$$

which shows the square of the Lorentz condition also vanishes between physical states.

Now, the consideration of the relation between matrix elements in the covariant Lorentz gauge and the transverse gauge requires us to inquire more deeply as to the nature of the physical states defined above. We note that it is easy to show

---

116 Similar approaches to defining the set of physical states appear in string and superstring theories. See, for example, Bailin and Love (1994) pp. 160 –167 and pp. 186 – 190.

$$(\Phi_{Ln}{}^* \, \eta\Phi_{Lm}) = 0 \tag{B.28}$$

if n or m is not equal to zero. In the case n = m = 0 the normalization can be chosen such that

$$(\Phi_{L0}{}^* \, \eta\Phi_{L0}) = 1 \tag{B.29}$$

Thus the states Φ_{Lm} are zero norm states for $m \neq 0$ and orthogonal to the vacuum state Φ_{L0} which happens to be the vacuum state $|0>$ used in our discussions in the earlier chapters. The state $|0>$ contains no transverse, longitudinal or time-like Y "particles."

The zero norm states containing superpositions of longitudinal and time-like Y "particles" are needed due to their role in supporting the gauge invariance of the theory.

Demonstration that the Expectation Values of Y Field Products in the Lorentz Gauge Equal their Value in the Transverse Gauge

We will now show that

$$(\Phi_L{}^* \, \eta B\Phi_L) = (\Phi_{tr}{}^* \, \eta B\Phi_{tr}) = (\Phi_{tr}{}^* \, B_{tr}\Phi_{tr}) \tag{B.30}$$

$$(\Phi_{L0}{}^* \, \eta B\Phi_{L0}) = (\Phi_{tr0}{}^* \, \eta B\Phi_{tr0}) = (\Phi_{tr0}{}^* \, B_{tr}\Phi_{tr0}) \equiv <0|B_{tr}|0> \tag{B.31}$$

where $|0>$ and $<0|$ are the vacua used in previous sections, where Φ_{L0} and $\Phi_{tr0} = |0>$ are the Lorentz gauge and transverse gauge vacuum states respectively, and where

$$B = \sum_\lambda C(\partial/\partial y_1{}^\mu, \partial/\partial y_2{}^\nu, \ldots, \partial/\partial y_m{}^\varrho) Y_{\lambda_1}(y_1)\, Y_{\lambda_2}(y_2) \ldots Y_{\lambda_n}(y_n) \tag{B.32}$$

and

$$B_{tr} = \sum_\lambda C(\partial/\partial y_1{}^\mu, \partial/\partial y_2{}^\nu, \ldots, \partial/\partial y_m{}^\varrho) Y_{\lambda_1}{}^{tr}(y_1)\, Y_{\lambda_2}{}^{tr}(y_2) \ldots Y_{\lambda_n}{}^{tr}(y_n) \tag{B.33}$$

with C a polynomial.

We now write eqs. B.11 and B.12 in the form

$$Y^{\mu+}(y) = \int d^3k\, N_0(k) e^{-ik\cdot y} \{\sum_{\lambda=1}^{2} \varepsilon^\mu(k, \lambda)\, a(k,\lambda) + (k^\mu/|\mathbf{k}| - \varepsilon_0{}^\mu) a_3(k) +$$
$$+ \varepsilon_0{}^\mu[a_3(k) - a_0(k)]\} \tag{B.34}$$

and $Y^{\mu-}(y)$ in a corresponding form. Then

$$Y^{\mu+}(y)\Phi_L = \{Y^{\mu+tr}(y) + \partial\Lambda/\partial y_\mu + \eta^{\mu 0}\, D^+(y)\}\Phi_L \tag{B.35}$$

where $\partial\Lambda/\partial y_\mu$ can be eliminated with a gauge transformation and where $D^+(y)\Phi_L = 0$. As a result

$$Y^{\mu+}(y)\Phi_L = Y^{\mu+tr}(y)\Phi_L \quad \text{and} \quad \Phi_L{}^\dagger Y^{\mu-}(y) = \Phi_L{}^\dagger Y^{\mu-tr}(y) \tag{B.36}$$

where $\Phi_L{}^\dagger = \Phi_L{}^*\eta$. Consequently

$$(\Phi_L{}^\dagger Y^\mu(y)\, \Phi_L) = (\Phi_L{}^\dagger Y^{\mu\, tr}(y)\, \Phi_L) \tag{B.37}$$

for all physical Lorentz gauge states Φ_L. More generally

$$(\Phi_L{}^\dagger B\, \Phi_L) = (\Phi_{tr}{}^\dagger B\, \Phi_{tr}) = (\Phi_{tr}{}^\dagger\, B_{tr}\, \Phi_{tr}) \equiv \langle 0|B_{tr}|0\rangle \tag{B.38}$$

with B and B_{tr} given by eqs. B.32 and B.33, thus proving eqs. B.30 and B.31. The proof of eq. B.38 is directly based on the representation of $Y^{\mu\pm}(y)$ in eq. B.34, and

$$[D^+(k), Y^{\mu\pm}(y)] = 0 \tag{B.39}$$

for all k and y with $D^+(k) = a_3(k) - a_0(k)$, and

$$[D^+(k), D^-(k')] = 0 \tag{B.40}$$

for all k and k' with $D^-(k) = a_3{}^\dagger(k) - a_0{}^\dagger(k)$, and

$$D^-(k)\Phi_{L0} = 0 \quad \text{and} \quad \Phi_{L0}{}^\dagger D^+(k) = 0 \tag{B.41}$$

where Φ_{L0} is the Lorentz gauge vacuum state. As a result of these relations the only surviving terms in B in eqs. B.30 and B.31 are the commutators of $Y^{\mu-tr}$ and $Y^{\nu+tr}$ with the consequence

$$(\Phi_L{}^*\, \eta B\Phi_L) = (\Phi_{tr}{}^*\, \eta B\Phi_{tr}) = (\Phi_{tr}{}^*\, B_{tr}\Phi_{tr}) \tag{B.30}$$

Therefore we have shown the expectation value of a product of Y fields in the Lorentz covariant Lorentz gauge is equal to the expectation value of the corresponding product of transverse gauge Y fields, thus establishing the Lorentz invariance of Two-Tier quantum field theories.

Lorentz Invariant S Matrix Elements

Having established the Lorentz invariance of Two-Tier quantum field theories we now describe a calculation procedure that transforms seemingly non-covariant amplitudes into manifestly covariant amplitudes.

Consider an interaction with a certain number of incoming particles and (possibly) a different number of outgoing particles. Assume a calculation of an S matrix element is performed according to the rules in Blaha (2003), and the present work, in the Y transverse gauge and in the center of mass of the incoming particles. The total momentum in the center of mass is $P^\mu = (P^0, \mathbf{0})$. The S matrix amplitude thus calculated will have a non-manifestly covariant form in the center of mass frame:

$$S_{ab} = S_{ab}(p^0_1, p^0_2, \ldots, p^0_n, |\mathbf{p}_1|, |\mathbf{p}_2|, \ldots |\mathbf{p}_n|, \mathbf{p}_1\cdot\mathbf{p}_2, \ldots \mathbf{p}_i\cdot\mathbf{p}_j, \ldots) \qquad \text{(B.42)}$$

It can easily be rewritten in covariant form using the total momentum P^μ:

$$p^0_i = P^\mu p_{i\mu} / (P^\nu P_\nu)^{1/2} \qquad \text{(B.43)}$$

$$|\mathbf{p}_i| = [\,(P^\mu p_{i\mu})^2/(P^\nu P_\nu) - p_i^\mu p_{i\mu}\,]^{1/2} \qquad \text{(B.44)}$$

$$\mathbf{p}_i\cdot\mathbf{p}_j = (P^\mu p_{i\mu} P^\alpha p_{j\alpha})/(P^\nu P_\nu) - p_i^\mu p_{j\mu} \qquad \text{(B.45)}$$

for i, j = 1, 2, …, n. After substitutions are made we obtain a completely covariant form:

$$S_{ab} = S_{ab}(P^\nu P_\nu, P^\alpha p_{1\alpha}, P^\beta p_{2\beta}, \ldots, P^\kappa p_{n\kappa}, p_1^\rho p_{2\rho}, \ldots, p_i^\sigma p_{j\sigma}, \ldots) \qquad \text{(B.46)}$$

Lastly we note that a gauge transformation can always be made after a Lorentz transformation in the y coordinates that restores the transverse gauge of the Y field.

Dirac Metric Operator

The metric operator η introduced earlier

$$\eta = (-1)^{n_4} \qquad \text{(B.47)}$$

can be combined with the metric operator V (eq. 5.15) introduced in earlier chapters for the transverse gauge

$$V = \exp[-i\pi \sum_{\lambda=1}^{2} \int d^3k\, a^\dagger(k, \lambda) a(k, \lambda)] \qquad \text{(B.48)}$$

with the property

$$V\, Y^j(y)\, V^{-1} = -Y^j(y) \qquad \text{(B.49)}$$

for j = 1, 2, 3.

The result is the invariant metric operator for the Lorentz gauge:

$$V_L = \exp[-i\pi \int d^3k\, a^{\mu\dagger}(k) a_\mu(k)] \qquad \text{(B.50)}$$

which could be used in inner products such as

$$(\Phi_{L0}{}^{*}\, V_L B\Phi_{L0}) = (\Phi_{tr0}{}^{*}\, V_L B\Phi_{tr0}) = (\Phi_{tr0}{}^{*}\, VB_{tr}\Phi_{tr0}) \equiv \langle 0|VB_{tr}|0\rangle \tag{B.51}$$

that occur in the evaluation of S matrix elements.

We note the following properties of V_L:

$$V_L{}^{\dagger} = V_L{}^{-1} = V_L \tag{B.52}$$

$$V_L{}^{2} = I \tag{B.53}$$

Appendix C. CQ Mechanics – PseudoQuantum Mechanics

The Quantum Mechanics version of GiFT is CQ Mechanics – a Quantum Mechanics that also embodies a classical mechanics limit. Its book appears below.

CQ Mechanics

A Unification of Quantum & Classical Mechanics, Quantum/Semi-Classical Entanglement, Quantum/Classical Path Integrals, Quantum/Classical Chaos

Stephen Blaha, Ph.D.*

Blaha Research

* sblaha777@yahoo.com

Pingree-Hill Publishing
P. O. Box 368
Auburn, NH 03032 USA

This book is printed on acid free paper.

C.1 Why Quantum Theory?

A question that is not often considered in these days is the reason that Nature 'chose' to be quantum rather than based on classical, deterministic mechanics. In our Theory of Everything presented in Blaha (2015c) and subsequent books in 2015 and 2016 we assumed that al Natural phenomena were ultimately based on quantum field theory.

There is a more fundamental assumption that we could posit that leads to quantum field theory and then to quantum mechanics (which is based on quantum field theory.[117]) If we assume the following postulate:

All entities in the universe are composed of discrete particles, that are integer countable, and all interactions can ultimately be reduced to the interactions of these particles.

Then, when we define field theories, they must be quantum field theories – describe particles (quanta) – and thus the field theories must be second quantized. *Particles are integer countable*[118] whether free or in perturbation theory interaction terms. And the interactions of these theories must be based on the exchange of particles although the particles have a 'cloud' of virtual particles surrounding them. The quantum mechanics of the particles constituting atoms (matter) then follows as a consequence of quantum field theory.

Classical mechanics then becomes an approximation to quantum mechanics under certain conditions that turn out to be the common conditions of everyday experience.

In basing the origin of quantum theory on the particulate nature of the entities in the universe we assume that particles exist, and can be defined, in our mostly flat space-time (which itself is generated by amalgamations of graviton particles). We also assume that the particle concept can be extended to unusual space-time coordinates such as non-static space-times. However it became apparent many years ago when accelerating coordinate systems and other non-static[119] coordinate systems were considered, that the definition of particles in quantum field theory is problematic.[120]

Sections 2 and 3 describe the correct definition of particles in quantum field theory. The correct definition of bosons furnishes a basis for a better definition of the

---

[117] Heitler (1954) shows how the Heisenberg Uncertainty Principle is a consequence of quantum field theory using an example from Quantum Electrodynamics. The Uncertainty Principle and the Correspondence Principle lead directly to quantum mechanics. Quantum mechanics is thus a consequence of quantum field theory – not an independent fundamental theory.

[118] Theories with continuous matter have not been shown to exist experimentally.

[119] A non-static coordinate system mixes space and time coordinates.

[120] S. Blaha, Il Nuovo Cimento **49 A**, 35 (1979). ___, **49 A**, 113 (1979), which appear in appendices A and B and is discussed in the following chapters, describe how to define particles in any coordinate system. The particle definition issued is discussed in papers which they reference.

Higgs Mechanism. The necessity of higher derivative theories of Gravity and the Strong Interactions[121] to obtain explicit color confinement and to reconcile gravity theory with data on the rotation of stars around galactic centers leads to an extension of the definition of particles to have principal value propagators and thus gives *particulate* action-at-a-distance.[122]

A further issue, that emerges in perturbation theory calculations in quantum field theories, leads to the introduction of a vector field as part of each propagator that eliminates the point-like nature of particle interactions in the high energy (short distance) limit in favor of 'fuzzy' interacting particles. This extension of quantum field theory is called Two-Tier quantum field theory.[123] It is required for the Theory of Everything since a conventional renormalization procedure is not known to exist – and is not likely to exist.

The combination of features that we have developed enables us to create a divergence-free[124] Theory of Everything where particles can be defined in any static or non-static coordinate system and where bosons, neglecting interactions, are stable against decay to negative energy states.[125]

The formalism that we present can be applied to quantum and classical dynamics. We define a quantum-classical formalism, that we call *CQMechanics*, that has a fully quantum sector, a classical sector, and an intermediate sector bridging the quantum and classical sectors.

We then proceed to develop the harmonic oscillator theory within this framework. Subsequently we discuss a generalized Feynman path integral formalism, a generalized Schrödinger equation, a generalized Boltzmann equation, the Fokker-Planck equation, a generalized approach to quantum and classical chaos, and to quantum entanglement as well as semi-quantum entanglement. Our formalism applies to both Quantum Field Theory and Quantum Mechanics as well as the path integrals, the Fokker-Planck equation and the Boltzmann equation.

C.2 Boson Particle Formulation

There are two issues confronting the usual approach to the quantization of boson fields that require resolution through a 'new' quantization procedure for boson particles. One problem is the need to quantize boson fields in unconventional coordinate systems such as accelerating coordinate systems and coordinate systems defined for highly curved space-time. The other major problem is the need to quantize boson fields in such a way that bosons of negative energy have a physical interpretation.[126]

---

[121] Unified theory: See Blaha (2016e).
[122] See appendices A and B.
[123] See Blaha (2005a).
[124] There are no divergences in perturbation theory calculations and no need for renormalization programs to remove divergences. Physical quantities do get renormalized by finite amounts.
[125] Negative energy boson states are equivalent to classical fields.
[126] There is no Pauli Exclusion Principle for bosons. Negative energy fermions 'fill' their Dirac negative energy sea due to the limitation of fermions to one fermion per state imposed by the Pauli Exclusion Principle.

In this section we will define a new quantization procedure for bosons that will eliminate both of these problems. In section 3 we will describe the analogous quantization procedure for fermions that will enable us to create well-defined Dirac field particle states in any coordinate system. (A filled Dirac sea of negative energy fermions will exist in this formulation as it does in the conventional formulation.)

In the case of both bosons and fermions we will see that the flat space-time, static coordinate systems that are normally used will remain valid special case approximations to the new formulations of quantum field theory.

Having resolved these problems for quantum field theories of bosons and fermions we will point out in section 5 that 'ordinary' quantum mechanics also has a problem with quantization in unconventional coordinate systems. There are also difficulties in the transition between classical and quantum 'analogues.' For example the transition from the classical Boltzmann equation to a quantum version is uncertain.

Using a framework analogous to our 'new' approach to the quantum field theory of bosons and fermions we will establish a generalization of quantum mechanics that contains both quantum mechanics and classical mechanics, and an intermediate mixed form. This generalization supports a smooth transition between classical mechanics and quantum mechanics. With this generalization we will be able to examine the transition from quantum to classical mechanics in detail without recourse to methods such as expansions in Planck's constant $\hbar$.

C.2.1 Quantization of Boson Fields in Unconventional (Static and Non-Static) Coordinate Systems

The problems associated with the definition of asymptotic particle states in arbitrary coordinate systems have been pointed out by numerous authors.[127] Our 1978 paper (Appendix A) resolves this problem with a consistent procedure for the local definition of asymptotic boson particle states in any coordinate system, which may or may not have a time-like Killing vector.

The general procedure is described in section 2 in Appendix A starting with eq. 6. The boson particle interpretation is described in section 4. In this section we wish to bring out salient details of the procedure which we will be relevant for our consideration of the generalization of quantum mechanics that we will consider later.

The first distinctive feature of this form of boson second quantization is the use of two fields to define a second quantized boson theory. The use of *two* fields enables us to define states which correspond to quantum field particles in any coordinate system. Further they give us the scope to define, not only quantum field particle states, but also classical boson field states. States, which are composites of both classical fields and quantum particles, can also be defined.

In our formulation[128] the simplest Lagrangian density for a generic massless, scalar Klein-Gordon particle is:

$$\mathcal{L} = \partial\varphi_1/\partial x_\mu \partial\varphi_2/\partial x^\mu \tag{C.2.1}$$

---

[127] See appendix A, which contains our 1978 paper Il Nuovo Cimento **49 A**, 35, for references.

[128] In earlier books we have called this approach to second quantization *PseudoQuantum field theory*.

with Hamiltonian density

$$\mathcal{H} = \pi_1 \pi_2 + \partial\varphi_1/\partial x_i \partial\varphi_2/\partial x^i \qquad \text{(C.2.2)}$$

where i labels spatial coordinates, and $\pi_1 = \partial\varphi_2/\partial t$ and $\pi_2 = \partial\varphi_1/\partial t$. Eqs. C.2.1 and C.2.2 are without a potential or mass term. Eq. 6 and the discussion following it, in appendix A describe the massive boson case.

The fields can be Fourier expanded in terms of creation and annihilation operators:

$$\varphi_i(\mathbf{x}, t) = \int d^3k\, [a_i(k)f_k(x) + a_i^\dagger(k)f_k^*(x)] \qquad \text{(C.2.3)}$$

for i = 1, 2 where

$$f_k(x) = e^{-ik\bullet x}/(2\omega_k(2\pi)^3)^{½}$$

with $\omega_k = |\mathbf{k}|$ in the massless case and $\omega_k = (|\mathbf{k}|^2 + m^2)^{½\, i}$ for a massive boson.

The creation and annihilation operators satisfy the commutation relations:

$$[a_i(k), a_j^\dagger(k')] = (1 - \delta_{ij})\delta^3(\mathbf{k} - \mathbf{k'}) \qquad \text{(C.2.4)}$$
$$[a_i(k), a_j(k')] = 0$$
$$[a_i^\dagger(k), a_j^\dagger(k')] = 0$$

for i, j = 1, 2. The vacuum state |0> satisfies

$$a_1(k)|0> = a_1^\dagger(k)|0> = 0 \qquad \text{(C.2.5)}$$
$$a_2(k)|0> \neq 0 \qquad a_2^\dagger(k)|0> \neq 0 \qquad \text{(C.2.6)}$$

The dual vacuum state satisfies

$$<0|a_2(k) = <0|a_2^\dagger(k) = 0 \qquad \text{(C.2.7)}$$
$$<0|a_1(k) \neq 0 \qquad <0|a_1^\dagger(k) \neq 0 \qquad \text{(C.2.8)}$$

Positive energy single particle *ket* states are defined using $a_2^\dagger(k)$ while negative energy ket states are defined using $a_2(k)$. Positive energy single particle *bra* states are defined using $a_1(k)$ while negative energy bra states are defined using $a_1^\dagger(k)$.

C.2.1.1 Transformations to Other (Possibly Non-Static) Coordinate Systems

The preceding discussion applies directly in the rectangular coordinates with which we are familiar. In eqs. 2 – 4, and their discussion, in appendix E we show that the definition of boson field orthonormal sets according to a different definition of positive frequency is related to the definition above in eq. C.2.3 by a local Bogoliubov transformation. The definition of particle states and vacua are different in general. However, as eqs. 15 – 31 (Appendix A) show, we can define the transformation to preserve the invariance of the particle number operator and thus make the theory under a different definition of positive frequency fully unitarily equivalent to the original theory. Thus the particle interpretation of states is preserved.

The general form of the Bogoliubov transformation (eq. 23) is

$$\begin{aligned} a_i(k, \lambda_1(k), \lambda_2(k)) &= B(\lambda_1(k), \lambda_2(k))a_i(k)B^{-1}(\lambda_1(k), \lambda_2(k)) \\ &= \exp(i\lambda_1(k))\cosh(\lambda_2(k))a_i(k) + \exp(-i\lambda_1(k))\sinh(\lambda_2(k))a_i^\dagger(k) \end{aligned} \tag{23}$$

with $B(\lambda_1(k), \lambda_2(k))$ given by eq. 24 and

$$B^{-1}(\lambda_1(k), \lambda_2(k)) = B^\dagger(\lambda_1(k), \lambda_2(k)) \tag{C.2.9}$$

where † indicates Hermitian conjugate. The text following eq. 23 provide the definition of bra and ket states, inner products, the energy-momentum tensor, equal-time commutation relations, the Green's functions, and the perturbation theory of the 'new' formalism.

Appendix A shows the general form of transformations between type '1' and type '2' creation and annihilation operators in this excerpt:[129]

> The equal-time commutation relations, and the self-adjointness of H and φ_2 place six constraints on the constants C_{ij} and $\tilde{C}_{ij}$ in eqs. (15) and (16). After some algebra we find that we are able to express the field operators in the form
>
> $$\varphi_1(x) = \int d^3k \left[\left(\frac{\cos(\theta_1-\theta_2)}{\sin\theta_1} A_{1k} + \frac{\sin(\theta_1-\theta_2)}{\sin\theta_1} A_{2k}\right) f_k(x) + \left(\frac{\cos(\theta_1-\theta_2)}{\cos\theta_1} A_{1k}^\dagger - \frac{\sin(\theta_1-\theta_2)}{\cos\theta_1} A_{2k}^\dagger\right) f_k^*(x)\right], \tag{40}$$
>
> $$\varphi_2(x) = \int d^3k \left[(\cos\theta_2 A_{2k} + \sin\theta_2 A_{1k}) f_k(x) + (\sin\theta_2 A_{2k}^\dagger - \cos\theta_2 A_{1k}^\dagger) f_k^*(x)\right], \tag{41}$$
>
> where θ_1 and θ_2 are arbitrary constants which fix the boundary conditions of the Green's functions. (They are *not* related to the Bogoliubov transformations

Thus PseudoQuantized Field Theory resolves the particle interpretation ambiguities of second quantization in non-static coordinate systems through Bogoliubov rotations.

C.2.1.2 Negative Energy Bosons

Traditional boson second quantization has the problem of the absence of a barrier to the decay of positive energy states to negative energy states since the Pauli Exclusion Principle does not apply to bosons. This problem has been masked ('overcome') by a clever choice of boundary conditions that are embodied in the creation/annihilation momentum space operator conditions:

---

[129] Excerpt used with the kind permission of Il Nuovo Cimento A.

$$a|0> = 0 \qquad \text{Conventional Approach} \qquad (C.2.10)$$
$$a^{\dagger}|0> \neq 0$$

In this conventional approach the creation of negative energy boson states is eliminated *ab initio* by these conditions. Yet boson quantum fields still have a conceptual physical cloud hanging over them that spin ½ fields do not. A spin ½ particle cannot transition to negative energy because there is a filled sea of negative energy particles. No additional particles can fall into the sea due to the Pauli Exclusion Principle that forbids two fermions with the same 4-momentum and quantum numbers.

In the case of scalar particles the Pauli Exclusion Principle does not apply and so, *physically*, a *filled* negative energy sea of bosons is not possible and positive energy bosons should be able to transition to negative energy states. This problem was "resolved" by the above definition of boson vacuums to exclude transitions to negative energy. But the rationale for the definition is lacking. Dirac was once asked about this issue many years ago. He said he had a solution to the problem. However he did not present it – presumably in keeping with his well-known taciturn nature. So the issue remained an open question.

In this book and earlier work[130] we showed that a more physically satisfactory method exists for avoiding the negative energy state problem. This method relies on the use of a larger Fock space in which *negative energy states (or partially negative energy states) are interpreted as states containing classical fields or a mix of classical fields and individual boson particles.* This approach resolves the negative energy boson issue and provides a common framework for boson particles and classical boson fields.

The issue of the spontaneous decay of a positive energy boson into a negative energy state still seems to exist. However all known fundamental scalar bosons are Higgs bosons that have a vacuum expectation value and a 'heavy' quantum field part of positive energy that immediately decays into other particles such as a pair of photons. The decay of a positive energy boson to a negative energy state is precluded by a separation of the formalism into separate positive energy and negative energy sectors as shown in section 4.5 below.

C.2.2 Classical Field States for Bosons

Classical c-number boson fields exist in our PseudoQuantum Field Theory. A classical c-number field has the form

$$\Phi(\mathbf{x}, t) = \int d^3k\, [\alpha(k)f_k(x) + \alpha^*(k)f_k{}^*(x)] \qquad (C.2.11)$$

A corresponding classical state is a coherent state with the form

$$|\, \Phi, \Pi> = C \exp\{\int d^3k\, [\alpha(k)a_2{}^{\dagger}(k) + \alpha^*(k)a_2(k)]\}|0> \qquad (C.2.12)$$

and correspondingly for $\Pi(x)$ where C is a normalization constant.

---

[130] See Appendix 2-A and references therein.

The defining properties of a classical field state are:

$$\varphi_1(x)|\Phi, \Pi\rangle = \Phi(x)|\Phi, \Pi\rangle \qquad (C.2.13)$$
$$\pi_1(x)|\Phi, \Pi\rangle = \Pi(x)|\Phi, \Pi\rangle$$

where $\Phi(x)$ and $\Pi(x)$ are sharp on the states and where $\varphi_i(x)$ is given by eq. C.2.3.

Additional details on coherent states, which differ somewhat from conventional coherent states such as those of Kibble[131] and others, can be found in Appendix C.

C.2.3 The Enigma of Higgs Particles and the Higgs Mechanism

Our PseudoQuantum Field Theory is ideally suited for describing Higgs Mechanism phenomena. In our previous work on the Standard Model, and its generalization to The Unified SuperStandard Theory described in a series of books entitled *Physics is Logic*, we showed that the fermion spectrum results from Complex Special Relativity, the gauge interactions result from the Reality group, the fermion generations result from the Generation group, the layers of fermions result from the U(4) Layer group, and from the combination with Complex General Relativity in our Theory of Everything. Higgs particles and the Higgs Mechanism were inserted *ad hoc* to generate particle masses and symmetry breaking effects.

The apparent recent discovery of Higgs particles at CERN seems to solidify the existence of the Higgs sector of the Standard Model and of our Unified SuperStandard Theory as described in earlier volumes of *Physics is Logic*.[132]

But whence arises Higgs particles? There does not appear to be a more fundamental cause than the need for particle masses obtained through symmetry breaking. And so the Higgs sector was an expedient mechanism. With our method of avoiding divergences in perturbation theory using Two-Tier quantum field theory the need for the Higgs Mechanism appears to have disappeared with the former need for a symmetry breaking mechanism to generate particle masses. The ElectroWeak sector has no divergences in our approach and thus does not need the renormalization program previously developed that was based on symmetry breaking using the Higgs Mechanism.

In considering the Higgs Mechanism a number of peculiarities appear that diminishes its attractiveness:

1. As remarked above, it is selective in the sense that some gauge fields have associated Higgs particles and utilize the Higgs Mechanism, and some gauge fields do not have associated Higgs particles. In particular, the ElectroWeak gauge fields, the Generation group gauge fields, the Layer group gauge fields, and the complex gravitation fields have associated Higgs particles. The strong interaction (gluon) gauge fields do not.

---

[131] T. W. B. Kibble, Jour. Math. Phys. **2**, 212 (1961).
[132] Blaha (2015a) and (2015b).

2. The conventional Higgs potentials have a quadratic mass term of the "wrong" sign plus a quartic interaction term, which together, generate non-zero vacuum expectation values. They obviously accomplish their goal. But the source of these potentials, and why they have their form, is unknown. One suspects a fundamental principle should be operative here.

3. One can imagine creating a Higgs microscope at some super-accelerator. Using this microscope in the presence of a (classical) condensate could enable the Uncertainty Principle to be violated. This possibility, in the case of a microscope using electromagnetic fields, was the source of a heuristic argument for the need to quantize the electromagnetic field.[133]

4. The standard formulation of the Higgs Mechanism uses classical fields under the assumption that a path integral formulation justifies their use. While this may be true, the path integral formulation relies on implicit, unstated boundary conditions that obscure the physics of the quantum field theoretic nature of the mechanism. A direct quantum field theoretic study of the Higgs Mechanism is needed and would further elucidate its character. It is possible, and it has been shown in our earlier books, that the apparently "true" mechanism described below reveals a number of important new results in a properly formulated version of the Higgs Mechanism.

C.2.4 "True" Origin of an Acceptable Mass Creation Mechanism

In this section we are using *PseudoQuantization*[134] and *PseudoQuantum field theory.* It combines both quantum and classical fields within the same framework. In this extended theory vacuum expectation values appear as coherent ground states that are strictly classical in nature.

This section is based on our 1978 paper that appeared in the peer-reviewed journal *Physical Review D.* The paper is reproduced in appendix C for the reader's convenience.

We suggest the reader skim or read the paper before proceeding. The paper also presents a new formulation of Quantum Theory that incorporates both quantum and classical mechanics within one framework that is of interest in its own right. See section 4 for details. Recently, experimenters have been investigating the possibility of macroscopic and other strange quantum phenomena. The new formulation is ideally suited for tracing the transition from a quantum to a classical regime. For example, it is applicable to "large n atoms" where the outermost electrons approach classical behavior with an almost continuous energy spectrum.

---

[133] Heitler (1954) p. 86 provides a good discussion of the need to quantize the electromagnetic field.
[134] This new formalism was first described in S. Blaha, Phys. Rev. D**17**, 994 (1978).

C.2.5 Higgs-Like Vacuum Expectation Value Generation of Masses

The Higgs Mechanism is based on the appearance of non-zero, c-number vacuum expectation values for Higgs fields due to potential terms directly appearing in Lagrangians.

C.2.5.1 PseudoQuantization of Higgs Particles

We will now consider the PseudoQuantization of a scalar particle using two fields in a manner shown earlier. It will become a "Higgs" particle with a non-zero vacuum expectation value.

Using the formalism described earlier we define $\varphi_1(x)$ and $\varphi_2(x)$[135] for a generic boson suppressing any internal symmetry indices for simplicity. We define a "vacuum state" containing a coherent superposition that satisfies

$$\varphi_1(x)|\Phi, \Pi> = \Phi|\Phi, \Pi> \qquad (C.2.14)$$

where Φ is a constant. Evaluating a fermion interaction term we find a mass term emerges[136]

$$\bar{\psi}(\varphi_1 + \varphi_2)\psi \rightarrow \bar{\psi}(\Phi + \varphi_2)\psi \qquad (C.2.15)$$

It can also generate a mass for an interaction with a gauge field of the form

$$A^\mu(\varphi_1 + \varphi_2)^2 A_\mu \rightarrow A^\mu(\Phi + \varphi_2)^2 A_\mu \qquad (C.2.16)$$

for ElectroWeak and other gauge fields. The φ_2 term leads to the production of Higgs particles in interactions. (The production of Higgs particles that decay into ElectroWeak gauge particles has recently been found at CERN.)

The present formalism provides a clean way to separate the vacuum expectation value of a scalar particle from its quantum field part in contrast to the conventional Higgs Mechanism where one has to separate a Higgs field into parts manually.

To obtain both the vacuum expectation value and the interaction with the quantum part of the PseudoQuantum fields we choose to always specify interactions with fermions and gauge fields using $\varphi = \varphi_1 + \varphi_2$ as seen above.

It appears that our formulation of the mass generation mechanism sheds significant light on the reason for the special prominence of inertial frames. Consider massive scalars.[137] Eq. 6 in appendix A describes a massive scalar particle. If the scalar is massive, then the rest frame particle "vacuum" coherent state below yields a non-zero expectation value Φ:

$$|\Phi, \Pi> = C\exp\{[(2\pi)^3 m/2]^{1/2}\Phi[a_2^\dagger(\mathbf{0},m) + a_2(\mathbf{0},m)]\}|0> \qquad (C.2.17)$$

---

[135] The subscripts on the fields are not gauge symmetry indices but simply identifiers distinguishing the fields from one another.

[136] When matrix elements with a "vacuum state" are calculated.

[137] Experiments at CERN have apparently discovered a Higgs particle with a 125 GeV/c mass.

where m is a generic mass. (We note that the conventional Higgs Mechanism also has mass terms.) *Thus our PseudoQuantum formalism allows us to define coherent "vacuum" states that lead to particle masses and Higgs particles.*

C.2.6 PseudoQuantized Non-Abelian Fields

The previous sections have considered scalar boson field theory. PseudoQuantum Field Theory also applies to non-Abelian fields. See appendix B, Blaha (2016e) and earlier papers by the author.[138]

C.3 Fermion Quantization

Fermion field quantization is problematic in unconventional coordinate systems such as accelerating coordinate systems and coordinate systems defined for highly curved space-time.

In this section we will define a PseudoQuantization procedure for fermions that supports second quantization in non-static and unusual non-rectangular coordinate systems.

Having resolved these problems for both bosons and fermions using PseudoQuantum field theory we will see in section 4 that 'ordinary' quantum mechanics also has a problem with quantization in unconventional coordinate systems. It also has difficulties in the transition between classical and quantum mechanics. For example the transition from the classical Boltzmann equation to a quantum version is uncertain.

Using a framework analogous to our PseudoQuantum field theory formalisms for bosons and fermions we will establish a generalization of quantum mechanics, PseudoQuantization Mechanics, which contains both quantum mechanics and classical mechanics, and intermediate mixed mechanic states. This generalization supports a smooth transition between classical mechanics and quantum mechanics. With this generalization we will be able to examine the transition from quantum to classical mechanics in detail without recourse to methods such as expansions in Planck's constant $\hbar$.

The fermion PseudoQuantization procedure is described in section 3 in Appendix A to which the reader is referred. It is similar to boson PseudoQuantization in that it requires two fermion fields for each fermion particle. Eq. 61 and the following discussion in appendix A show a simple illustrative canonical Lagrangian formulation of fermion PseudoQuantization including Fourier expansions, equal time commutation relations, and creation and annihilation operators. Eq. 69 shows the general form of Fourier expansions while eqs. 77 and 78 show the form restricted by anti-commutation relations and adjointness of the Hamiltonian to have rotations of creation and annihilation operators: b_{1k} and b_{2k}, and d_{1k} and d_{2k} in a manner similar to the analogous boson case in eqs. 40 and 41.

Thus boson and fermion PseudoQuantization field theory both have pairs of fields associated with each particle and utilize rotations to implement unitary equivalence between static and non-static coordinate systems.

---

138 S. Blaha, Phys. Rev. **D10**, 4268 (1974); Phys. Rev. **D11**, 2921 (1975).

C.4 PseudoQuantization Mechanics – Joint Quantum-Classical Mechanics Formalism

Having established the need for paired fields for bosons and fermions using PseudoQuantization field theory to achieve unitary equivalence of quantization in both static and non-static coordinate systems we now turn to the case of Quantum Mechanics. Here again we find there is a problem associated with the transformations between coordinate systems in certain cases. There is also the problem in determining the transition between quantum mechanical entities and their classical equivalents.

These problems are analogous to those described in previous sections for second quantization. Since quantum mechanics is derived from quantum field theory it is reasonable to suspect that the resolution of quantum mechanics problems will ultimately be found in an analogue of PseudoQuantization which we earlier saw resolved quantum field theory problems.

This section will present PseudoQuantization Mechanics,[139] which contains both fully quantum, and fully classical, sectors as well as an intermediate sector that provides a transition between the quantum and classical regimes. With this formalism we can overcome coordinate system issues as well as the challenge of the correspondence between classical and quantum physics.

C.4.1 Coordinate Systems Problems of Quantum Mechanics

Problems exist in coordinate transformations in certain quantum mechanical situations which are 'fixed' through the use of 'recipes' that patch over the difficulties. One example is the change of coordinates in path integrals.[140] Gutzwiller (1990) points out that there is no simple rule for general canonical transformations of coordinates in path integrals. Given the central role of path integrals in quantum theory the difficulties of canonical transformations in path integrals is of concern. Other problems with canonical transformations in quantum mechanics are also discussed in Gutzwiller (1990).

We shall develop the PseudoQuantization formalism with a view towards facilitating canonical transformations in quantum mechanical studies as well as elucidating the transition from the quantum to the classical regimes.

C.4.2 PseudoQuantization Harmonic Oscillator

The harmonic oscillator plays a central role in classical and quantum mechanics due to its appearance in a variety of physical problems. In this section we will describe the PseudoQuantum formulation of the one-dimensional simple harmonic oscillator as a prelude to the general PseudoQuantum description.

Appendix C contains a paper on the PseudoQuantum harmonic oscillator.[141] In this section we will describe it, in part, with some changes, as a formalism that

---

[139] Much of this chapter was presented in S. Blaha, Phys. Rev. D**17**, 994 (1978) which is reprinted in appendix C. See also S. Blaha, Phys. Rev. D**10**, 4268 (July, 1974) and Phys. Rev. D**11**, 2921 (1974). See appendix D.

[140] See pp. 202-3 in Gutzwiller (1990).

[141] See its description in appendix C – section II of S. Blaha, Phys Rev **D17**, 994 (1978). Excerpts used with the kind permission of Physical Review D.

embodies both classical and quantum sectors, and provides a graceful transition between classical and quantum harmonic oscillator dynamics. Thus we will have an example of a new approach to understanding the classical-quantum transition. In later sections we will apply this approach to physical phenomena where the transition from a classical description to a quantum equivalent is problematic.

We begin with (appendix C) two commuting variables x_1 and p_1, which we augment with two new variables x_2 and p_2, defined by

$$x_i = (m\omega/\hbar)^{-1/2} Q_i \qquad (C.4.1)$$
$$p_i = (m\omega\hbar)^{1/2} P_i$$

for i, j = 1, 2 where

$$P_2 = -i\, d/dQ_1 \qquad (C.4.2)$$
$$Q_2 = i\, d/dP_1$$

with the commutation relations:

$$[Q_i, P_j] = i(1 - \delta_{ij}) \qquad (C.4.3)$$

for i, j = 1, 2.

Next we define raising and lowering operators

$$a_i = 2^{-1/2}(Q_i + iP_i) \qquad (C.4.4)$$
$$a_i^\dagger = 2^{-1/2}(Q_i - iP_i)$$
$$Q_i = (a_i + a_i^\dagger)/\sqrt{2}$$
$$P_i = (a_i - a_i^\dagger)/(\sqrt{2}i)$$

with

$$[a_i, a_j^\dagger] = (1 - \delta_{ij}) \qquad (C.4.5)$$
$$[a_i, a_j] = 0$$
$$[a_i^\dagger, a_j^\dagger] = 0$$

for i, j = 1, 2.

We now define an alternate set of raising and lowering operators that will use an angle θ to provide a continuous transition from classical to quantum (and vice versa)[142]

$$b_1 = Q_1\cos\theta + iP_2\sin\theta \qquad (C.4.6)$$
$$b_2 = -Q_2\sin\theta + iP_1\cos\theta$$
$$b_1^\dagger = Q_1\cos\theta - iP_2\sin\theta \qquad (C.4.7)$$
$$b_2^\dagger = -Q_2\sin\theta - iP_1\cos\theta$$

Their commutation relations are

$$[b_1, b_1^\dagger] = \sin(2\theta) \qquad (C.4.8a)$$
$$[b_2, b_2^\dagger] = -\sin(2\theta)$$
$$[b_1, b_2^\dagger] = [b_2, b_1^\dagger] = 0$$
$$[b_1, b_2] = [b_1^\dagger, b_2^\dagger] = 0$$

142 This definition differs from that appearing in appendix C.

The PseudoQuantum Hamiltonian[143] is

$$\begin{aligned}\hat{H} &= p_1p_2/m + m\omega^2x_1x_2 \\ &= ½\omega(\{a_1, a_2^\dagger\} + \{a_2, a_1^\dagger\}) \\ &= \omega(P_1P_2 + Q_1Q_2)\end{aligned} \quad (C.4.8b)$$

In terms of the original P and Q variables we find

$$\begin{aligned}Q_1 &= (b_1 + b_1^\dagger)/(2\cos\theta) \\ Q_2 &= -(b_2 + b_2^\dagger)/(2\sin\theta) \\ P_1 &= -i(b_2 - b_2^\dagger)/(2\cos\theta) \\ P_2 &= -i\sin(\theta)(b_1 - b_1^\dagger)/(2\sin\theta)\end{aligned} \quad (C.4.9)$$

C.4.2.1 Dirac Metric Operator ζ Transforming From Classical to Quantum Oscillator

At this point we define 'number' states with a_2 and $a_2^\dagger$:

$$|n_+, n_-\rangle = (a_2^\dagger)^{n_+}(a_2)^{n_-}|0,0\rangle \quad (C.4.10)$$

where

$$\hat{H}|n_+, n_-\rangle = \omega(n_+ - n_-)|n_+, n_-\rangle \quad (C.4.11)$$

In view of the commutation relations we wish to transform eq. C.4.10 to

$$|n_+, n_-\rangle = (b_1^\dagger)^{n_+}(b_2^\dagger)^{n_-}|0,0\rangle \quad (C.4.12)$$

where the vacuum in eqs. C.4.10 and C.4.12 will be seen to be the same:

$$\begin{aligned}a_1^\dagger|0,0\rangle &= a_1|0,0\rangle = 0 \\ b_1|0,0\rangle &= b_2|0,0\rangle = 0\end{aligned} \quad (C.4.13)$$

We define a 'Dirac-like' metric operator ζ. It satisfies

$$\begin{aligned}\zeta^{-1}a_2^\dagger\zeta &= b_1^\dagger \\ \zeta^{-1}a_2\zeta &= b_2^\dagger\end{aligned} \quad (C.4.14)$$

We provisionally define

$$\zeta = \exp(aP_1Q_1 + bP_1^2 + cQ_2P_1 + dP_2P_1 + eQ_1^2 + fQ_2Q_1 + gP_2Q_1) \quad (C.4.15)$$

Eqs. C.4.4 and C.4.14 imply the values of the constants in eq. 4.15 so that

$$\zeta = \exp[(-i\cos\theta\, P_1Q_1 - (\cos\theta)/2\, P_1^2 + i\sin\theta\, Q_2P_1 - \sin\theta\, P_2P_1 - (\cos\theta)/2\, Q_1^2 - \sin\theta\, Q_2Q_1 + i\sin\theta\, P_2Q_1)/\sqrt{2}] \quad (C.4.16)$$

$$= \exp[(\cos\theta\, (a_1^{\dagger 2} - a_1^2)/2 - (\cos\theta)/4\, (a_1 - a_1^\dagger)^2 + (\sin\theta)/2\, (a_2 + a_2^\dagger)(a_1 - a_1^\dagger) +$$

[143] Eqs. 3, 12, 21 in Appendix C with ω made explicit.

$$+ (\sin\theta)/2\, (a_2 - a_2^\dagger)(a_1 - a_1^\dagger) - (\cos\theta)/4\, (a_1 + a_1^\dagger)^2 - (\sin\theta)/2\, (a_2 + a_2^\dagger)(a_1 + a_1^\dagger) +$$
$$+ (\sin\theta)/2\, (a_2 - a_2^\dagger)(a_1 + a_1^\dagger))/\surd 2]$$

$$\zeta^{-1} = \exp[(\cos\theta\, (a_1^{\dagger 2} - a_1^2)/2 - (\cos\theta)/4\, (a_1 - a_1^\dagger)^2 + (\sin\theta)/2\, (a_2 + a_2^\dagger)(a_1 - a_1^\dagger) + \quad (C.4.17)$$
$$+ (\sin\theta)/2\, (a_2 - a_2^\dagger)(a_1 - a_1^\dagger) - (\cos\theta)/4\, (a_1 + a_1^\dagger)^2 - (\sin\theta)/2\, (a_2 + a_2^\dagger)(a_1 + a_1^\dagger) +$$
$$+ (\sin\theta)/2\, (a_2 - a_2^\dagger)(a_1 + a_1^\dagger))/\surd 2]$$

We note the ground state (vacuum) explicitly satisfies:

$$|0,0> = \zeta^{-1}|0,0> \quad (C.4.18)$$

by eq. C.4.13.

We also note

$$\zeta^{-1}[a_2, a_1^\dagger]\zeta = 1$$

and

$$\zeta^{-1}[a_1, a_2^\dagger]\zeta = 1$$

imply

$$\zeta^{-1}a_1\zeta = b_1/\sin(2\theta) \quad (C.4.19)$$
$$\zeta^{-1}a_1^\dagger\zeta = -b_2/\sin(2\theta)$$

using eq. C.4.8.

The transformed Hamiltonian H can be expressed as

$$\hat{H}_\zeta = \zeta^{-1}\hat{H}\zeta = ½\omega(\{b_1, b_1^\dagger\} - \{b_2, b_2^\dagger\})/ \sin(2\theta) \quad (C.4.20)$$

C.4.2.2 Classical, Intermediate, and Quantum Wave Functions

The classical $(n_+, n_-)^{th}$ coordinate space wave function (eq. 28 appendix A) has the form:

$$\Psi_{n_+,n_-}(x_1, p_1, x_2, p_2, \theta) = (n_+!n_-!)^{-½}<x_1, p_1, x_2, p_2|(a_2^\dagger)^{n_+}(a_2)^{n_-}|0, 0> \quad (C.4.21)$$
$$= (n_+!n_-!)^{-½}<x_1, p_1, x_2, p_2|\zeta\zeta^{-1}(a_2^\dagger)^{n_+}\zeta\zeta^{-1}(a_2)^{n_-}\zeta\zeta^{-1}|0, 0>$$
$$= (n_+!n_-!)^{-½}<x_1, p_1, x_2, p_2|\zeta^\dagger b_1^{\dagger n_+} b_2^{\dagger n_-}|0,0>$$

using the conventional normalization of states with the form $|n> = (n_-!)^{-½}a^{\dagger n}|0>$.

Next we note

$$<x_1, p_1, x_2, p_2|\zeta^\dagger = <x_1, p_1, x_2, p_2| \quad (C.4.22)$$

similarly to eq. C.4.18.

$$b_1^\dagger = Q_1\cos\theta - iP_2\sin\theta \equiv \cos\theta\, Q_1 - \sin\theta\, d/dQ_1 = \cos\theta\, \eta_1 - \sin\theta\, \partial/\partial\eta_1 \quad (C.4.23)$$
$$b_2^\dagger = -Q_2\sin\theta - iP_1\cos\theta \equiv -i(\sin\theta\, d/dP_1 + \cos\theta\, P_1) = -i(\sin\theta\, \partial/\partial\eta_2 + \cos\theta\, \eta_2)$$
$$= \sin\theta\, \partial/\partial\eta_3 - \cos\theta\, \eta_3$$

where $\eta_3 = i\eta_2 = iQ_2$ and $\eta_1 = Q_1$.

Eq. C.4.21 can be expressed as:

$$\Psi_{n_+,n_-}(x_1, p_1, x_2, p_2, \theta) = (n_+!n_-!)^{-1/2}(-1)^{n_-}[\cos\theta\ \eta_1 - \sin\theta\ \partial/\partial\eta_1]^{n_+} \cdot$$
$$\cdot [\cos\theta\ \eta_3 - \sin\theta\ \partial/\partial\eta_3]^{n_-}<x_1, p_1, x_2, p_2|0,0> \qquad (C.4.24)$$

The determination of

$$\Psi_{0,0} \equiv <x_1, p_1, x_2, p_2|0,0>$$

begins with noting

$$<x_1, p_1, x_2, p_2|b_1|0,0> = 0$$

or

$$(\cos\theta\ \eta_1 + \sin\theta\ \partial/\partial\eta_1)\Psi_{0,0} = 0$$

and

$$<x_1, p_1, x_2, p_2|b_2|0,0> = 0$$

or

$$(\cos\theta\ \eta_3 + \sin\theta\ \partial/\partial\eta_3)\Psi_{0,0} = 0$$

These conditions require

$$\Psi_{0,0} = C \exp[-\tfrac{1}{2}\cot\theta(\eta_1^2 + \eta_3^2)] \qquad (C.4.25)$$

where the normalization $C = [m\omega\cot\theta/(i\pi\hbar)]^{1/2}$ is determined by

$$1 = C^2 \int dx_1 dx_2 \exp[-\tfrac{1}{2}\cot\theta(\eta_1^2 + \eta_3^2)] \qquad (C.4.26)$$

Then eq. C.4.24 becomes

$$\Psi_{n_+,n_-}(x_1, p_1, x_2, p_2, \theta) = (n_+!n_-!)^{-1/2}[m\omega\cot\theta/(i\pi\hbar)]^{1/2}(-1)^{n_-}[\cos\theta\ \eta_1 - \sin\theta\ \partial/\partial\eta_1]^{n_+} \cdot$$
$$\cdot [\cos\theta\ \eta_3 - \sin\theta\ \partial/\partial\eta_3]^{n_-}\exp[-\tfrac{1}{2}\cot\theta(\eta_1^2 + \eta_3^2)]$$
$$(C.4.27)$$

$$= (n_+!n_-!)^{-1/2}[m\omega\cot\theta/(i\pi\hbar)]^{1/2}(-1)^{n_-}[\cos\theta\ \eta_1 - \sin\theta\ \partial/\partial\eta_1]^{n_+} \cdot$$
$$\cdot [i\cos\theta\ \eta_2 + i\sin\theta\ \partial/\partial\eta_2]^{n_-}\exp[-\tfrac{1}{2}\cot\theta(\eta_1^2 - \eta_2^2)]$$

Note that eq. C.4.27 contains a product of Hermite polynomials if $\theta = \pi/4$. It is not surprising that we obtain quantum harmonic oscillator factors in the wave function since, as eq. C.4.8a shows the b operators have conventional quantum oscillator commutation relations for $\theta = \pi/4$ – thus this value of θ corresponds to the quantum case. We note that Hermite polynomials $H_n(\eta)$ are generated by

$$(\eta - \partial/\partial\eta)^n \exp(-\tfrac{1}{2}\eta^2) = \exp(-\tfrac{1}{2}\eta^2)H_n(\eta) \qquad (C.4.28)$$

We can generalize Hermite polynomials for other values of θ with

$$H_n(\eta, \theta) = \exp[+\tfrac{1}{2}\cot\theta\ \eta^2]\ [\cos\theta\ \eta - \sin\theta\ \partial/\partial\eta]^n\exp[-\tfrac{1}{2}\cot\theta\ \eta^2] \qquad (C.4.29)$$

Then eq. C.4.27 can be expressed by

$$\Psi_{n_+,n_-}(x_1, p_1, x_2, p_2, \theta) = (n_+!n_-!)^{-½}[m\omega\cot\theta/(i\pi\hbar)]^{½}(-1)^{n_-}H_{n+}(\eta_1,\theta)H_{n-}(\eta_3,\theta)\cdot$$
$$\cdot\exp[-½\cot\theta(\eta_1^{\,2} + \eta_3^{\,2})] \qquad \text{(C.4.30)}$$

$$= (n_+!n_-!)^{-½}[m\omega\cot\theta/(i\pi\hbar)]^{½}(-1)^{n_-}H_{n+}(\eta_1,\theta)H_{n-}(i\eta_2,\theta)\exp[-½\cot\theta(\eta_1^{\,2} - \eta_2^{\,2})]$$
$$= (-1)^{n_-}\Psi_{n_+}(\eta_1, \theta)\Psi_{n-}(\eta_3, \theta)$$
$$= (-1)^{n_-}\Psi_{n_+}((m\omega)^{½}\, x_1, \theta)\Psi_{n-}(i(m\omega)^{½}\, x_2, \theta)$$

where

$$\Psi_n(\eta, \theta) = (n!)^{-½}[m\omega\cot\theta/(i\pi\hbar)]^{1/4}H_n(\eta,\theta)\, \exp[-½\cot\theta\eta^2] \qquad \text{(C.4.30a)}$$

At $\theta = \pi/4$ the wave function factorizes into a harmonic oscillator wave function times an inverted harmonic oscillator wave function:

$$\Psi_{n_+,n-}(x_1, p_1, x_2, p_2, \theta{=}\pi/4) = (n_+!n_-!)^{-½}[m\omega/(i\pi\hbar)]^{½}(-1)^{n_-}2^{-(n_+ + n_-)/2}H_{n+}(\eta_1)H_{n-}(\eta_3)\cdot$$
$$\cdot\exp[-½(\eta_1^{\,2}+\eta_3^{\,2})] \qquad \text{(C.4.30b)}$$

$$= (-1)^{n_-}2^{-(n_+ + n_-)/2}\Psi_{n_+}((m\omega)^{½}\, x_1)\Psi_{n-}(i(m\omega)^{½}\, x_2)$$

where $H_n(\eta)$ is a Hermite polynomial of degree n.

For $\theta = 0$ we find the b commutation relations (eq. 8a) are zero indicating that the wave function is classical in nature. In this case, simply substituting $\theta = 0$ would cause eq. C.4.30 to 'blow up.' However for certain values of n+ and n– the limit $\theta \to 0$ yields a physically interesting result – a wave function that is a delta function similar to that appearing in eq. 43 in appendix C.

Consider first the case n+ = 1 and n– = 0. Then

$$\Psi_{1,0}(x_1, p_1, x_2, p_2, \theta) = [m\omega\cot\theta/(i\pi\hbar)]^{½}[\cos\theta\, \eta_1 \; - \sin\theta\, \partial/\partial\eta_1]\exp[-½\cot\theta(\eta_1^{\,2} + \eta_3^{\,2})]$$

As $\theta \to 0$, and for $\eta_3 = 0$, we find

$$\Psi_{1,0}(x_1, p_1, x_2, p_2, \theta{\to}0) \to [m\omega/(i\hbar)]^{½}\eta_1(\pi\sin\theta)^{-½}\exp[-½\eta_1^{\,2}/\sin\theta]$$
$$\to [m\omega/(i\hbar)]^{½}\eta_1(2\pi\sin\theta)^{-½}\exp[-\eta_1^{\,2}/(2\sin\theta)]$$
$$\to [m\omega/(i\hbar)]^{½}\eta_1\delta(\eta_1) = 0 \qquad \text{(C.4.31)}$$

Now consider the case n+ = 0 and n– = 0:

$$\Psi_{0,0}(x_1, p_1, x_2, p_2, \theta) = [m\omega\cot\theta/(i\pi\hbar)]^{½}\exp[-½\cot\theta(\eta_1^{\,2} + \eta_3^{\,2})]$$

As $\theta \to 0$, and for $\eta_3 = 0$, we find

$$\Psi_{0,0}(x_1, p_1, x_2, p_2, \theta{\to}0) \to [m\omega/(i\hbar)]^{½}(\pi\sin\theta)^{-½}\exp[-½\cot\theta\eta_1^{\,2}/\sin\theta]$$
$$\to [m\omega/(i\hbar)]^{½}\eta_1(2\pi\sin\theta)^{-½}\exp[-\eta_1^{\,2}/(2\sin\theta)]$$
$$\to [m\omega/(i\hbar)]^{½}\delta(\eta_1) = i^{-½}\delta(x_1) \neq 0 \qquad \text{(C.4.32)}$$

using

$$\delta(\eta) = \lim_{\varepsilon\to 0} (\pi\varepsilon)^{-½}\exp[-\eta^2/\varepsilon] \qquad \text{(C.4.33)}$$

Thus the Gaussian factor combined with the preceding $(2\pi\sin\theta)^{-½}$ grows to a delta-function wave function. *Wave functions corresponding to higher values of n+ and n– go to zero in the limit $\theta \rightarrow 0$. Only* $\Psi_{0,0}(x_1, p_1, x_2, p_2, \theta\rightarrow 0)$ *is non-zero.*

The introduction of the time dependence and a shift of the location of the minimum of the harmonic oscillator potential to x_0 would lead to a wave function such as:

$$\Psi_{0,0}(x_1, p_1, x_2, p_2, \theta\rightarrow 0) = i^{-½}\delta(x_1 - x_0\sin(\omega t)) \qquad \text{(C.4.34)}$$

A similar behavior may be seen in the case $\eta_1 = 0$. Then we find a wave function with a factor of $\delta(x_2)$.

Lastly, the case of $\theta \rightarrow \pi/2$ is of interest. Eq. 4.30 yields

$$\Psi_{n_+,n_-}(x_1, x_2, \theta\rightarrow\pi/2) = (n_+!n_-!)^{-½}[m\omega\cos\theta/(i\pi\hbar)]^{½}[-\partial/\partial\eta_1]^{n_+}[\partial/\partial\eta_3]^{n_-}\cdot$$
$$\cdot \exp[-½\cos\theta(\eta_1^{\ 2}+\eta_3^{\ 2})]|_{\theta\rightarrow\pi/2}$$
$$= 0 \qquad \text{(C.4.35)}$$

Figuratively speaking, the wave function progresses from one non-zero 'classical' wave function at $\theta = 0$, to a quantum mechanical wave function at $\theta = \pi/4$, to a zero value wave function at $\theta = \pi/2$. Thus one might say "The good Lord by giving us a quantum universe put us in a position halfway between nothingness and classical mechanics." By implementing Quantum theory we get Second Quantization of particle fields, and thereby, integer countability of particle numbers – a distinct simplification in Nature.

C.4.2.3 Energy Eigenvalues

From eq. C.4.8b, C.4.10, and C.4.11 we see that eq. C.4.11 for the state

$$|n_+, n_-\rangle = (a_2^{\dagger})^{n_+}(a_2)^{n_-}|0, 0\rangle$$

shows the energy of the wave function (eqs. C.4.27 and C.4.30) to be

$$E_{n+,n-} = (n_+ - n_-)\hbar\omega = [n_+ + ½ - (n_- + ½)]\hbar\omega \qquad \text{(C.4.36)}$$

Eq. C.4.27 satisfies the PseudoQuantized Schrödinger equation:

$$\hat{H}\Psi_n(x_1, p_1, x_2, p_2, \theta, t) = i\partial\Psi_n(x_1, p_1, x_2, p_2, \theta, t)\partial t \qquad \text{(C.4.37)}$$

C.4.3 Wave Function as a Function of Position and Momentum

We can define the wave function in terms of x_1 and p_1 with a Fourier transform:

$$\Psi_{n+,n-}(x_1, p_1, \theta) = \int dx_2\, e^{-ip_1x_2}\, \Psi_{n+,n-}(x_1, x_2, \theta) \qquad \text{(C.4.38)}$$

$\Psi_{n+,n-}(x_1, p_1, \theta=\pi/4)$, is a wave function for the combined 'normal', and inverted, harmonic oscillators. Thus the full PseudoQuantum theory enables us to define a wave function that is a function of both position and momentum without inconsistency. We discuss this topic in more detail later when we compare it to the Wigner distribution function.

C.4.4 Intermediate Classical-Quantum Wave Functions

For other values of θ in eq. C.4.27 and C.4.30 we obtain wave functions that are intermediate between quantum and classical operator wave functions. Later we will find it of interest to trace the evolution of a classical wave function to a quantum wave function and vice versa in the general case of non-harmonic oscillator dynamics.

It is interesting to note the dependence of the energy level spacing on the angle θ. The transformed energy (eq. C.4.14 implements the transformation to the b operators) has the form:

$$\hat{H}_\zeta = \zeta^{-1}\hat{H}\zeta = \tfrac{1}{2}\omega(\{b_1, b_1^\dagger\} - \{b_2, b_2^\dagger\})/\sin(2\theta) \qquad \text{(C.4.20)}$$

If either n_+ and n_- change by one unit, then the energy changes by

$$\Delta E = \tfrac{1}{2}\omega/\sin(2\theta)$$

For $\theta = \pi/4$ (the quantum case)

$$\Delta E = \tfrac{1}{2}\omega \qquad \text{(C.4.39)}$$

For $\theta = \pi/8$ (the quantum approaching classical case)

$$\Delta E = \tfrac{1}{2}\omega/0.383 = 1.307\omega \qquad \text{(C.4.40)}$$

As $\theta \to 0$ (the classical case)

$$\Delta E \to \infty \qquad \text{(C.4.41)}$$

Thus 'higher' (lower) energy states beyond the $n_+ = 0$ and $n_- = 0$ state are inaccessible energy-wise. This corresponds to our above finding that only the wave function $\Psi_{0,0}$ is non-zero in the classical limit.

C.5 General Formalism for a PseudoQuantized System

The basic procedure of our PseudoQuantization Formalism are described in our paper Phys. Rev **D17**, 994 (1978) reprinted in appendix C.[144] The relevant excerpt is

[144] Excerpt used with the kind permission of Physical Review D.

We shall now briefly outline the procedure for embedding a classical-mechanical system in a quantum system.[6] Consider a classical Hamiltonian system with one degree of freedom, and commuting canonical variables, x_1 and p_1, which have the equations of motion

$$\dot{x}_1 = -i[x_1, \hat{H}] , \tag{1}$$

$$\dot{p}_1 = -i[p_1, \hat{H}] , \tag{2}$$

where defining

$$\hat{H} = -i\left(\frac{\partial H(x_1, p_1)}{\partial p_1}\frac{\partial}{\partial x_1} - \frac{\partial H(x_1, p_1)}{\partial x_1}\frac{\partial}{\partial p_1}\right) \tag{3}$$

allows us to write Hamilton's equations in commutator form. With Sudarshan[6] we define

$$x_2 = i\frac{\partial}{\partial p_1} \tag{4}$$

and

$$p_2 = -i\frac{\partial}{\partial x_1} \tag{5}$$

so that

$$[x_1, x_2] = [p_1, p_2] = 0 , \tag{6}$$

$$[x_1, p_2] = [x_2, p_1] = i , \tag{7}$$

and $\hat{H}$ can now be taken to be the operator

$$\hat{H} = \frac{\partial H(x_1, p_1)}{\partial p_1}p_2 + \frac{\partial H(x_1, p_1)}{\partial x_1}x_2 . \tag{8}$$

It is now apparent that we can take the above quantities and equations of motion to describe a quantum mechanical system with two degrees of freedom in the "coordinate" representation where the "coordinates" are (x_1, p_1) and the canonical momenta are $\Pi = (p_2, -x_2)$. As we will see below the linearity of $\hat{H}$ in the momenta is crucial for the maintenance of the classical character of x_1 and p_1, and for the observability of the phase-space trajectory. Since we choose to identify the physical observables with the commutative algebra of the coordinate operators, x_1 and p_1, we are led to impose the superselection condition that the momenta, Π, are unobservable. As a result the Hamiltonian and other generators of canonical transformations, which are all linear in the momenta, are also unobservable. However, in each case there is an associated dynamical quantity which is observable.

The required unobservability of the momenta restricts the form of the interaction between a classical-made-quantum system and an inherently quantum system to

$$H_{\rm int} = \Phi_1 x_2 + \Phi_2 p_2 + X , \qquad (9)$$

where Φ_1, Φ_2, and X are functions of x_1, p_1, and the quantum system variables. The commutation relations of these functions are also constrained[6] by the superselection rule and the commutativity of the classical variables, x_1 and p_1, and their time derivatives. In the next section we will study the simple harmonic oscillator in order to exemplify the quantum-mechanical case described above and also for direct use in the field-theoretic generalizations of subsequent sections.

Based on the above discussion we assume that we start with a conventional Hamiltonian that we express as

$$H = H(x_1, p_1) = \tfrac{1}{2}p_1^2 + V(x_1) \qquad \text{(C.5.1)}$$

Introducing x_2 and p_2, as in eqs. 4 and 5 above, we can generalize H to a PseudoQuantum Hamiltonian $\hat{H}$:

$$\hat{H} = p_1 p_2 + x_2 \partial V/\partial x_1 \qquad \text{(C.5.2)}$$

where V is a function of x_1.

We can introduce raising and lowering operators a_i and $a_i^\dagger$ using the procedure of section C.4. Then we can proceed as in the harmonic oscillator case to calculate wave functions. In the next section we apply this procedure to the Boltzmann equation, which has some similarity to the Schrödinger equation.

C.6 PseudoQuantization of the Boltzmann Equation

The Boltzmann equation is a classical dynamics equation that describes the dynamics of a multi-particle system with interactions. The equivalent quantum formulation is not known. However Wigner and others have proposed possible quantum equivalents that of some of the expected features of the quantum Boltzmann function. In this section we will follow a procedure similar to that of section 4 for the Vlasov approximation. For special cases of the collision term of the Boltzmann equation we will obtain quantum equivalents.

C.6.1 Non-Relativistic Boltzmann Equation

The non-relativistic Boltzmann equation for identical particles of one chemical species is

$$[\mathbf{p}\cdot\nabla/m + F\cdot\partial/\partial\mathbf{p}]f = -\partial f/\partial t + (\partial f/\partial t)_{coll}$$

where f(**r**, **p**, t) is Boltzmann's probability density function. It has often been remarked that this Boltzmann equation strongly resembles the Schrödinger equation.

C.6.2 PseudoQuantum Form of the Boltzmann Equation

We can make the case that it even more strongly resembles the PseudoQuantized Schrödinger equation (see eq. C.5.2) by defining

$$-i[\mathbf{p}_1\cdot\mathbf{p}_2/m - \mathbf{x}_2\cdot F]f = -\partial f/\partial t + (\partial f/\partial t)_{coll} \qquad (C.6.1)$$

where $\mathbf{F} = \mathbf{F}(x_1, t)$ and

$$\mathbf{p}_2 = i\nabla \qquad (C.6.2)$$
$$x_2 = -i\partial/\partial\mathbf{p}_1$$

Comparing eq. C.6.1 with section C.4, we find we can define

$$\hat{H} = \mathbf{p}_1\cdot\mathbf{p}_2/m + x_2\cdot\partial V/\partial\mathbf{x}_1 \qquad (C.6.3)$$

where

$$\partial V/\partial\mathbf{x}_1 = \mathbf{F}(x_1, t)$$

Given the close similarity of the Boltzmann equation and the Schrödinger equation it is sensible to treat the solution of the equation as a 'wave function' that initially represents a classical state such as the classical harmonic oscillator wave function that we saw in section C.4. Then we will define the Boltzmann distribution in terms of the wave function solution.

The PseudoQuantized Boltzmann wave equation is

$$\hat{H}\psi = -i\partial\psi/\partial t + i(\partial\psi/\partial t)_{coll} \qquad (C.6.4)$$

where

$$\psi = \psi(\mathbf{x}_1, \mathbf{x}_2, \theta)$$

with θ defined later in specific cases. The value of θ determines whether ψ is classical, quantum, or in an intermediate state.

In a manner somewhat analogous to that of Wigner[145] we define a Boltzmann distribution with

$$f_q(\mathbf{r}_1, \mathbf{p}_1, t, \theta) = \int d^3r_2\, \psi(\mathbf{r}_1, \mathbf{r}_2, t, \theta)\psi^\dagger(\mathbf{r}_1, \mathbf{r}_2, t, \theta) \exp(-2i\mathbf{r}_2\cdot\mathbf{p}_1/\hbar) \qquad \text{(C.6.5)}$$

in three spatial dimensions where we use the suffix 'q' of f_q to signify the PseudoQuantum Boltzmann probability density function $f_q(\mathbf{r}, \mathbf{p}, t)$. The function f_q can be classical, quantum, or intermediate between classical and quantum depending on the value of θ. We will consider examples that illustrate the dependence of f_q on θ. Later we will also see that our definition of f_q eliminates the problems of the Wigner density function.

C.6.3 PseudoQuantum Form of the Vlasov Equation

The collision-less Boltzmann equation is called the Vlasov equation. It is of interest because of the difficulties associated with solving the full Boltzmann equation. Its PseudoQuantized equivalent is

$$\hat{H}\psi = -i\partial\psi/\partial t \qquad \text{(C.6.6)}$$

This equation has 'only' the difficulty of its solution for the various forces $\mathbf{F}(x_1, t)$.

We note that the one-dimensional version of eq. C.6.6 where $V = \frac{1}{2} x_1^2$ is solved in section C.6.4.

C.6.4 PseudoQuantum Vlasov Equation Solution for a Three-dimensional Harmonic Oscillator Force with θ = π/4

The choice of θ = π/4 gives 'quantum' harmonic oscillator solutions consisting of a harmonic oscillator factor and an inverted harmonic oscillator factor.

The three-dimensional harmonic oscillator PseudoQuantum 'Hamiltonian' equation is

$$(\mathbf{p}_1\cdot\mathbf{p}_2/m + \mathbf{x}_2\cdot\mathbf{x}_1)\psi = -i\partial\psi/\partial t \qquad \text{(C.6.7)}$$

or

$$(\mathbf{p}_1\cdot\mathbf{p}_2/(2m') + \mathbf{x}_2\cdot\mathbf{x}_1)\,\psi = -i\partial\psi/\partial t \qquad \text{(C.6.8)}$$

This equation is fully separable[146] for θ = π/4 in rectangular coordinates which we label x, y, and z. The solution is a product of one-dimensional PseudoQuantum harmonic oscillator wave function factors of the form of C.4.30b:

---

[145] E. P. Wigner, Phys. Rev. **40**, 749 (1932).

$$\psi_{n+,n-}(\mathbf{r}_1, \mathbf{r}_2, t, \pi/4) = \Psi_{nx+,nx-}(x_{1x},p_{1x},x_{2x},p_{2x},t,\theta)\ \Psi_{ny+,ny-}(x_{1y},p_{1y},x_{2y},p_{2y},t,\theta)\ \Psi_{nz+,nz-}(x_{1z},p_{1z},x_{2z},p_{2z},t,\theta) \quad \text{(C.6.9)}$$

$$= (-1)^{n_{x-}+n_{y-}+n_{z-}}2^{-(n_{x+}+n_{x-}+n_{y+}+n_{y-}+n_{z+}+n_{z-})/2}\Psi_{nx+}((m\omega)^{1/2}x_1)\Psi_{nx-}(i(m\omega)^{1/2}x_2)\Psi_{ny+}((m\omega)^{1/2}y_1)\cdot$$
$$\cdot\Psi_{ny-}(i(m\omega)^{1/2}y_2)\Psi_{nz+}((m\omega)^{1/2}z_1)\Psi_{nz-}(i(m\omega)^{1/2}z_2)$$

$$= A\Psi_{nx+}((m\omega)^{1/2}x_1)\Psi_{ny+}((m\omega)^{1/2}y_1)\Psi_{nz+}((m\omega)^{1/2}z_1)\Psi_{nx-}(i(m\omega)^{1/2}x_2)\Psi_{ny-}(i(m\omega)^{1/2}y_2)\cdot$$
$$\cdot\Psi_{nz-}(i(m\omega)^{1/2}z_2)$$
$$= A\Psi_1(\mathbf{r}_1)\Psi_2(\mathbf{r}_2) \quad \text{(C.6.9a)}$$

with the time dependence not displayed and where

$$A = (-1)^{n_{x-}+n_{y-}+n_{z-}}2^{-(n_{x+}+n_{x-}+n_{y+}+n_{y-}+n_{z+}+n_{z-})/2} \quad \text{(C.6.9c)}$$

using eq. C.4.30b.

The energy, which is constant since the Hamiltonian is not explicitly time dependent, is

$$E_{n+,n-} = (n_+ - n_-)\hbar\omega \quad \text{(C.6.10)}$$

with

$$n_+ = n_{x+} + n_{y+} + n_{z+} \quad \text{(C.6.11)}$$
$$n_- = n_{x-} + n_{y-} + n_{z-}$$

Following Wigner, a Fourier transform for $\theta = \pi/4$ of eq. C.6.9a factors gives a *quantum* Boltzmann density function:

$$f_q(\mathbf{q}, \mathbf{p}_1, t, \pi/4) = \int d^3Q\ \Psi(\mathbf{q} - \mathbf{Q})\Psi^\dagger(\mathbf{q} + \mathbf{Q}) \exp(-2iQ\cdot\mathbf{p}_1/\hbar)$$

$$= \int d^3Q\ \Psi_1(\mathbf{q} - \mathbf{Q})\Psi_1^\dagger(\mathbf{q} + \mathbf{Q})\Psi_2(\mathbf{q} - \mathbf{Q})\Psi_2^\dagger(\mathbf{q} + \mathbf{Q}) \exp(-2iQ\cdot\mathbf{p}_1/\hbar) \quad \text{(C.6.12)}$$

where we let $\mathbf{r}_1 = \mathbf{q} - \mathbf{Q}$ and $\mathbf{r}_2 = \mathbf{q} + \mathbf{Q}$.

If we define the Fourier transform of a wave function $\Psi(\mathbf{r})$ by

$$\Phi(\mathbf{p}) = (2\pi\hbar)^{-3/2} \int d^3r\ \Psi(\mathbf{r}) \exp(-i\mathbf{r}\cdot\mathbf{p}/\hbar)$$

then

$$f_{qp}(\mathbf{p}_1, t, \pi/4) = \int d^3q\ f_q(\mathbf{q}, \mathbf{p}_1, t, \pi/4) = \int d^3p\ \Phi(\mathbf{p})\Phi^\dagger(\mathbf{p}) \quad \text{(C.6.13)}$$

yields a projection of the phase space distribution into momentum space. In addition

$$f_{qp}(\mathbf{p}_1, t, \pi/4) = \Psi(\mathbf{q})\Psi^\dagger(\mathbf{q}) \quad \text{(C.6.14)}$$

---

[146] For other values of θ the solutions of the equation do not separate type '1' coordinates from type '2' coordinates. See eq. E.4.30 for the general case.

yields a projection of the phase space distribution into coordinate space.

Thus $f_q(\mathbf{q}, \mathbf{p}_1, t, \pi/4)$ can be interpreted as the quantum equivalent of the (classical) Boltzmann distribution. *PseudoQuantization gives us 2n variables just as there are 2n variables in phase space.*

C.6.5 PseudoQuantum Vlasov Equation Solution for a Three-dimensional Harmonic Oscillator Force for Arbitrary θ

The general representation of our PseudoQuantized Vlasov equation solution for any value of θ is given by eq. C.4.30. The 3-dimensional Vlasov representation is

$$\psi_{n_+,n_-}(\mathbf{r}_1, \mathbf{r}_2, t, \theta) = \psi_{n_{x+},n_{x-}}(x_{1x},p_{1x},x_{2x},p_{2x},t,\theta)\psi_{n_{y+},n_{y-}}(x_{1y},p_{1y},x_{2y},p_{2y},t,\theta)\psi_{n_{z+},n_{z-}}(x_{1z},p_{1z},x_{2z},p_{2z},t,\theta) \tag{C.6.15}$$

$$= (-1)^{n_{x-}+n_{y-}+n_{z-}}2^{-(n_{x+}+n_{x-}+n_{y+}+n_{y-}+n_{z+}+n_{z-})/2}\Psi_{n_{x+}}((m\omega)^{1/2}x_1, t,\theta)\Psi_{n_{x-}}(i(m\omega)^{1/2}x_2, t,\theta)\cdot$$
$$\cdot\Psi_{n_{y+}}((m\omega)^{1/2}y_1, t,\theta)\Psi_{n_{y-}}(i(m\omega)^{1/2}y_2,t,\theta)\Psi_{n_{z+}}((m\omega)^{1/2}z_1, t,\theta)\Psi_{n_{z-}}(i(m\omega)^{1/2}z_2, t,\theta)$$

$$= A\Psi_{n_{x+}}((m\omega)^{1/2}x_1, t, \theta)\Psi_{n_{y+}}((m\omega)^{1/2}y_1, t, \theta)\Psi_{n_{z+}}((m\omega)^{1/2}z_1, t, \theta)\Psi_{n_{x-}}(i(m\omega)^{1/2}x_2, t, \theta)\cdot$$
$$\cdot\Psi_{n_{y-}}(i(m\omega)^{1/2}y_2, t, \theta)\,\Psi_{n_{z-}}(i(m\omega)^{1/2}z_2, t, \theta)$$
$$= A\Psi_1(\mathbf{r}_1, t, \theta)\Psi_2(\mathbf{r}_2, t, \theta)$$

Following similar steps as in the previous section we find

$$f_q(\mathbf{q}, \mathbf{p}_1, t, \theta) = \int d^3Q\, \psi_{n_+,n_-}(\mathbf{q} - \mathbf{Q}, t, \theta)\, \psi_{n_+,n_-}{}^{\dagger}(\mathbf{q} + \mathbf{Q}, t, \theta) \exp(-2iQ\cdot\mathbf{p}_1/\hbar)$$

$$= A^2 \int d^3Q\, \Psi_1(\mathbf{q} - \mathbf{Q}, t, \theta)\Psi_1{}^{\dagger}(\mathbf{q} + \mathbf{Q}, t, \theta)\Psi_2(\mathbf{q} - \mathbf{Q}, t, \theta)\Psi_2{}^{\dagger}(\mathbf{q} + \mathbf{Q}, t, \theta)\exp(-2iQ\cdot\mathbf{p}_1/\hbar) \tag{C.6.16}$$

where we let $\mathbf{r}_1 = \mathbf{q} - \mathbf{Q}$ and $\mathbf{r}_2 = \mathbf{q} + \mathbf{Q}$.

Following similar steps we can again obtain eqs. C.6.13 and C.6.14 and establish a connection between phase space, and momentum and coordinate space projections.

If we define the Fourier transform of a wave function $\Psi(\mathbf{r})$ by

$$\Phi(\mathbf{p}, \theta) = (2\pi\hbar)^{-3/2} \int d^3r\, \Psi(\mathbf{r}, \theta) \exp(-i\mathbf{r}\cdot\mathbf{p}/\hbar)$$

then

$$f_{qp}(\mathbf{p}_1, t, \theta) = \int d^3q\, f_q(\mathbf{q}, \mathbf{p}_1, t, \theta) = \int d^3p\, \Phi(\mathbf{p}, \theta)\Phi^{\dagger}(\mathbf{p}, \theta) \tag{C.6.17}$$

yields a projection of the phase space distribution into momentum space. In addition

$$f_{qq}(\mathbf{p}_1, t, \theta) = \Psi(\mathbf{q}, \theta)\Psi^{\dagger}(\mathbf{q}, \theta) \tag{C.6.18}$$

yields a projection of the phase space distribution into coordinate space.

Thus $f_q(\mathbf{q}, \mathbf{p}_1, t, \theta)$ can be interpreted as the quantum equivalent of the (classical) Boltzmann distribution. PseudoQuantization gives us 2n variables just as there are 2n variables in phase space.

C.6.6 PseudoQuantum Vlasov Equation Solution for a Three-dimensional Harmonic Oscillator Force for θ = 0 – The Classical Case

In the $\theta = 0$ case, which is the classical mechanics limit, we find that eq. C.4.34 gives a precise expression for the Vlasov Boltzmann equation solution of eq. C.6.16. We note that only the 0-0 wave functions are non-zero. In one dimension we have:

$$\psi = \Psi_{0,0}(x_1, p_1, x_2, p_2, \theta \to 0) = i^{-1/2}\delta(x_1 - x_0\sin(\omega t)) \qquad \text{(C.4.34)}$$

The 3-dimensional case (eq. C.6.16) gives

$$f_q(\mathbf{q}, \mathbf{p}_1, t, \theta{=}0) = \int d^3Q\, \delta^3(\mathbf{q} - \mathbf{Q} - \mathbf{x}_0\sin(\omega t))\delta^3(\mathbf{q} + \mathbf{Q} - \mathbf{x}_0\sin(\omega t))\exp(-2iQ\cdot\mathbf{p}_1/\hbar)$$
$$= \delta^3(\mathbf{q} - \mathbf{x}_0\sin(\omega t)) \qquad \text{(C.6.19)}$$

a classical solution specifying the classical harmonic oscillator trajectory. We note that all other solutions (for other values of n_+ and n_-) are 'pushed' to $E = \infty$ according to eq. C.4.41.

We note $f_q(\mathbf{q}, \mathbf{p}_1, t, \theta{=}0)$ is positive definite as a probability should be. The integral

$$\int d^3q\, f_q(\mathbf{q}, \mathbf{p}_1, t, \theta{=}0) = 1$$

shows the sum of the probabilities of the normalized Boltzmann distribution in coordinate space is unity.

We thus have achieve a quantum-classical Boltzmann distribution in phase space in both coordinates and momenta using PseudoQuantization where the number of phase space parameters is 2n = 6 in this case—unlike the case of the Wigner density alternative.

C.6.7 Comparison to the Wigner Density Function

The Wigner density function in n dimensions is defined as:

$$\Psi(p, q) = \int d^nQ\, \psi(q - Q)\, \psi^\dagger(q + Q)\, \exp(-2ipQ/\hbar) \qquad \text{(C.6.20)}$$

where $\psi(q)$ of the wave function of the system. The interpretation of $\Psi(p, q)$ as the quantum probability in phase space corresponds to $f(p, q)$ – the classical Boltzmann distribution. It is often interpreted in that manner.

However, several concerns are usually expressed about this interpretation:

1. Although real-valued $\Psi(p, q)$ can have a negative value making a probability interpretation problematic.

2. $\Psi(p, q)$ appears to depend on 2n values. However the wave function $\psi(q)$, upon which it is defined, only depends on n variables. Thus the domains of each function are different and $\Psi(p, q)$ can only be viewed as dependent on n variables.

Wigner attempted to overcome these objections by using the quantum mechanics density matrix $\rho(q, q')$ in an attempt to reflect the usual situation that a quantum system is in a mixed state consisting of a superposition of orthogonal states $\psi_k(q)$ with a probability of $\rho_k \geq 0$ with the $\Sigma\, \rho_k = 1$. The density matrix for this case is defined to be

$$\rho(q, q') = \Sigma\, \rho_k\, \psi_k(q)\psi_k^{\dagger}(q') \qquad \text{(C.6.21)}$$

Using the density matrix the Wigner distribution now is

$$\Psi(p, q) = \int d^nQ\; \rho(q - Q, q + Q) \exp(-2ipQ/\hbar) \qquad \text{(C.6.22)}$$

The new form of the Wigner distribution is a function of 2n variables and $\Psi(p, q)$ is positive definite. However, the density matrix (eq. C.6.14) has all eigenvalues between 0 and 1 and a trace equal to one. These properties are not shared by every potential Boltzmann probability f(p, q, t). Thus the representation is limited to 'special cases.'

Our form of the *quantum* Boltzmann probability distribution is $f_q(\mathbf{r}_1, \mathbf{p}_1, t, \theta)$ which we have shown overcomes the redundancy of variables in the Wigner quantum generalization of the Boltzmann distribution and gives a sensible result in the classical limit.[147]

C.6.8 PseudoQuantum Form of the BGK Approximation to the Boltzmann Equation

The BKG approximation to the Boltzmann equation is

$$[\mathbf{p}\cdot\nabla/m + F\cdot\partial/\partial\mathbf{p}]f = -\partial f/\partial t + \upsilon(f_0 - f) \qquad \text{(C.6.23)}$$

where f_0 is the local Maxwell distribution $f_0 = f_0(\mathbf{r}, \mathbf{p})$ and υ is the molecular collision frequency. This model of the collision term due to Bhatnagar, Gross, and Crook[148] has been a much studied approximation.

Before introducing the PseudoQuantum form of the BKG approximation we use the local Maxwell-Boltzmann distribution to re-express the BKG approximation in the form

$$f_0 = n[m/(2\pi kT)]^{3/2} \exp[-m(p - p_0)^2/(2kT)] \qquad \text{(C.6.24)}$$

---

[147] Our PseudoQuantum equivalent density has the same form as the Wigner density (eq. E.6.21).

[148] P. L. Bhatnagar, E. P. Gross, and M. Crook, Phys. Rev. **94**, 511 (1954).

where n is the particle density (assumed constant at equilibrium), k is Boltzmann's constant, T is the temperature, p_0 is the average momentum, and m is the mass of a particle. Inserting f_0 in eq. C.6.23 and letting

$$f = f_0 g \qquad \text{(C.6.25)}$$

we obtain

$$[\mathbf{p}\cdot\nabla/m + \mathbf{F}\cdot\partial/\partial\mathbf{p} - (m/kT)\mathbf{F}\cdot(\mathbf{p} - \mathbf{p}_0) + \upsilon]g = -\partial g/\partial t + \upsilon \qquad \text{(C.6.26)}$$

The PseudoQuantized equivalent of the expanded BKG approximation (eq. C.6.26) is

$$[\mathbf{p}_1\cdot\mathbf{p}_2/m - \mathbf{x}_2\cdot F(\mathbf{x}_1) + (m/kT)\mathbf{F}(\mathbf{x}_1)\cdot(\mathbf{p}_2 + i\mathbf{p}_0) + i\upsilon]g = -i\partial g/\partial t + i\upsilon \qquad \text{(C.6.27)}$$

with the Maxwell-Boltzmann distribution term acting as a 'driving force.'

Given a force F we can proceed to PseudoQuantize using operators that are similar to those of eqs. C.4.1 – C.4.7 but adapted to the force and the Maxwell-Boltzmann distribution 'driving force.' We will not consider BKG examples in this book although a harmonic driving force $\mathbf{F}(\mathbf{x}_1) = -m\omega^2\mathbf{x}_1$ is an interesting case to consider.

C.6.9 Relativistic Boltzmann Equation

The Boltzmann equation is non-relativistic and is appropriate in systems that are at rest or moving at non-relativistic velocities. If a system is traveling at relativistic velocities then the relativistic Boltzmann equation must be used. In this section we first generalize the Boltzmann equation to its special relativistic form by making all terms covariant. Then, when we 'go to' a rest frame, the relativistic equation becomes the non-relativistic Boltzmann equation.

C.6.9.1 Relativistic Generalization of the Boltzmann Equation

The relativistic form of the non-relativistic Boltzmann equation of section C.6.1 is

$$[\mathbf{p}^\mu\nabla_\mu/m + F^\mu\partial/\partial\mathbf{p}^\mu]f = (\partial f/\partial t)_{\text{collRelativistic}} \qquad \text{(C.6.28)}$$

where we use indices to transform vectors into 4-vectors: the momentum, derivative operators and the force become Lorentz 4-vectors. The collision term must now be in a relativistic form.

C.6.9.2 PseudoQuantized Relativistic Boltzmann Equation

The PseudoQuantum form of the Boltzmann equation is discussed earlier:

$$-i[\mathbf{p}_1\cdot\mathbf{p}_2/m - \mathbf{x}_2\cdot F]f = -\partial f/\partial t + (\partial f/\partial t)_{\text{coll}} \qquad \text{(C.6.1)}$$

$$\mathbf{p}_2 = i\nabla \qquad \text{(C.6.2)}$$

$$x_2 = -i\partial/\partial\mathbf{p}_1$$

We make it relativistic in a manner similar to the approach in subsection C.6.9.1. The result is the relativistic PseudoQuantum Boltzmann equation:

$$[p_1{}^{\mu}p_{2\mu} - mx_2{}^{\mu}F_{\mu}(x_1{}^{\alpha})]f = im(\partial f/\partial t)_{collRelativistic} \qquad (C.6.29)$$

where the collision term is relativistic. This formalism, superficially, has two times. However when the wave functions are calculated only one time $x_1{}^0$ is relevant as the calculations in section 4 in the Vlasov approximation suggest.

C.6.10 Quantum and Classical Entropy

The von Neumann entropy for a system described by a density matrix ρ is defined as

$$S = -\,\mathrm{tr}[\rho\ln\rho] \qquad (C.6.30)$$

Using eigenvectors |n> the density matrix can be expressed as

$$\rho = \sum_i \eta_i|i\rangle\langle i| \qquad (C.6.31)$$

Then ρ can be expressed in the information theory Shannon formulation of entropy:

$$S = -\sum_i \eta_i \ln \eta_i \qquad (C.6.32)$$

If we use the harmonic oscillator development of section 4 we can express the von Neumann entropy in a form which ranges from classical to quantum as a function of the angle θ. If we define the harmonic oscillator states

$$|\,n+, n-\rangle = b_1{}^{\dagger n_+} b_2{}^{\dagger n_-}|0,0\rangle$$

as in section 4 where

$$b_1{}^{\dagger} = Q_1\cos\theta - iP_2\sin\theta \qquad (C.4.23)$$
$$b_2{}^{\dagger} = -Q_2\sin\theta - iP_1\cos\theta$$

then the density matrix is

$$\rho(\theta) = \sum_{n+,n-} |n+, n-\rangle\langle n+, n-| = \sum_{n+,n-} \eta_{n+,n-}\, b_1{}^{\dagger}(\theta)^{n_+} b_2{}^{\dagger}(\theta)^{n_-}|0,0\rangle\langle 0,0|b_1(\theta)^{n_+} b_2(\theta)^{n_-} \qquad (C.6.33)$$

In the quantum limit where θ = π/4 we see

$$\rho(\pi/4) = \sum_{n+,n-} (\eta_{n+,n-}/2^{n+\,+\,n-})(Q_1 - iP_2)^{n_+}(Q_2 + iP_1)^{n_-}|0,0\rangle\langle 0,0|(Q_1 + iP_2)^{n_+}(Q_2 - iP_1)^{n_-} \qquad (C.6.34)$$

yielding a quantum density matrix and thus a von Neumann quantum entropy.

In the classical limit where $\theta = 0$ we see

$$\rho(0) = \sum_{n+,n-} (\eta_{n+,n-}/2^{n+ + n-})Q_1^{n_+}P_1^{n_-}|0,0><0,0|Q_1^{n_+}P_1^{n_-} = \rho(Q_1, P_1) \quad (C.6.35)$$

yielding a purely classical function of Q_1 and P_1 as the density matrix. The von Neumann entropy's classical limit in this case is the classical phase space quantity

$$S = -\{\Sigma(\eta_{n+,n-}/2^{n+ + n-})Q_1^{2n_+}P_1^{2n_-}\}\ln\{\Sigma(\eta_{n+,n-}/2^{n+ + n-})Q_1^{2n_+}P_1^{2n_-}\} \quad (C.6.36)$$
$$= S(Q_1, P_1)$$

since Q_1 and P_1 commute.

C.7 PseudoQuantum Path Integral Formulation

The path integral formulation of quantum mechanics (and also of quantum field theory) plays an important role in the understanding of quantum physics. One of its major issues is the transition from a quantum mechanical framework to a classical mechanical framework. We will examine this issue from the point of view of a PseudoQuantum path integral formulation. Earlier we have seen that we can embody both quantum and classical mechanics phenomena within the PseudoQuantum framework and 'rotate' between quantum and classical mechanics solutions.

In order to establish a PseudoQuantum path integral formalism we must first generate a Lagrangian from a PseudoQuantum Hamiltonian. In section C.5 we described the general formalism for deriving a PseudoQuantum Hamiltonian from a classical Hamiltonian. We now construct the PseudoQuantum Lagrangian. Starting with the equations in section C.5:

$$x_2 = id/dp_1$$
$$p_2 = -id/dx_1$$

$$\hat{H}(x_1, p_1, x_2, p_2) = \partial H(x_1, p_1)/\partial p_1\, p_2 + \partial H(x_1, p_1)/\partial x_1\, x_2$$

we define the velocities[149]

$$x'_1 = \partial\hat{H}(x_1, p_1, x_2, p_2)/\partial p_1 = \partial^2 H(x_1, p_1)/\partial p_1^2 p_2 + \partial^2 H(x_1, p_1)/\partial x_1 \partial p_1\, x_2 \quad (C.7.1)$$

$$x'_2 = \partial\hat{H}(x_1, p_1, x_2, p_2)/\partial p_2 = \partial H(x_1, p_1)/\partial p_1|_{p_2 = p_1} \quad (C.7.2)$$

The Lagrangian L is constructed in the canonical way using Legendre transformations

$$L = p_1 x'_1 + p_2\, x'_2 - \hat{H}(x_1, p_1, x_2, p_2) \quad (C.7.3)$$

$$= p_1[\partial^2 H(x_1, p_1)/\partial p_1^2 p_2 + \partial^2 H(x_1, p_1)/\partial x_1 \partial p_1\, x_2] - \partial H(x_1, p_1)/\partial x_1\, x_2$$

---

[149] The velocity x'_2 is a defined quantity, which is defined in a manner consistent with the definition of x'_1.

$$= \partial^2H(x_1, p_1)/\partial p_1^{\ 2}\, p_1p_2 + \partial^2H(x_1, p_1)/\partial x_1\partial p_1\, p_1x_2 - \partial H(x_1, p_1)/\partial x_1\, x_2 \qquad \text{(C.7.3a)}$$

where p_1 and p_2 are extracted from eqs. C.7.1 and C.7.2.

We now consider the example of a harmonic oscillator where

$$H = p^2/(2m) + ½\, m\omega^2x^2$$
$$\hat{H} = p_1p_2/m + m\omega^2x_1x_2 \qquad \text{(C.4.8b)}$$

Substituting in eq. C.7.3 we find

$$\begin{aligned} L &= \partial^2H(x_1, p_1)/\partial p_1^{\ 2}\, p_1p_2 + \partial^2H(x_1, p_1)/\partial x_1\partial p_1\, p_1x_2 - \partial H(x_1, p_1)/\partial x_1\, x_2 \\ &= p_1p_2/m - m\omega^2x_1x_2 \\ &= mx'_1x'_2 - m\omega^2x_1x_2 \end{aligned} \qquad \text{(C.7.4)}$$

with p_1 and p_2 determined, and replaced, as functions of x'_1 and x'_2 by eqs. C.7.1 and C.7.2.

The Lagrange equations of motion determined for i = 1, 2 are:

$$d/dt\,(\partial L/\partial x'_i) - \partial L/\partial x_i = 0 \qquad \text{(C.7.5)}$$

$$mx''_2 + m\omega^2x_2 = 0 \qquad \text{(C.7.6)}$$
$$mx''_1 + m\omega^2x_1 = 0 \qquad \text{(C.7.7)}$$

C.7.1 Feynman Path Integral formulation

The propagator K(x – y, t) for the Feynman path integral formulation has the form:

$$K(x - y, T) = A \lim_{n\to\infty} \int\!\!\int\!\!\int\!\!\int_{-\infty}^{+\infty} \ldots \int dx_0dx_1\ldots\, dx_n \exp[i/\hbar \int_t^{t+T} L(x, v, t_a)\, dt_a] \qquad \text{(C.7.8)}$$

where A is a constant, and the integral over the dx's ranges from –∞ to ∞.

C.7.1.1 Conventional Formulation – Free Particle Case

In the one dimensional free particle case the path integral is a product of n infinitesimal paths of time interval ε:

$$K(x - y, T) = \prod_n G_\varepsilon \qquad \text{(C.7.9)}$$

where T = nδ. Using ~ to denote proportionality up to a constant we find the Fourier transform of an interval of path

$$\begin{aligned} G_\delta &= \int dx\, e^{-ipx} \exp[-ix^2/(2\varepsilon)] \\ &\sim \exp[-p^2/(2\varepsilon)] \end{aligned} \qquad \text{(C.7.10)}$$

Then the product of the incremental factors that total to time T give

$$K(p, T) \sim \exp[-iTp^2/2] \qquad (C.7.11)$$

A Fourier transformation yields the free particle propagator

$$K(x - y, T) \sim \int dp\, e^{-ip(x - y)} \exp[-iTp^2/2]$$
$$\sim \exp[-i(x - y)^2/T] \qquad (C.7.12)$$

where we normalize the propagator to unity

$$\int dy\, K(x - y, T) = 1 \qquad (C.7.13)$$

C.7.1.2 PeseudoQuantum Formulation – Free Particle Case

We will now develop the PseudoQuantum path integral formalism for the case of a free particle. The form of the path integral now is

$$K(x - y, T) =$$
$$= A \lim_{n \to \infty} \iiiint_{-\infty}^{+\infty} \ldots \int dx_{10} dx_{11} \ldots dx_{1n}\, dx_{20} dx_{21} \ldots dx_{2n} \exp[i/\hbar \int_t^{t+T} L(x_1, x_2, x'_1, x'_2, v, t_a)\, dt_a] \qquad (C.7.14)$$

where A is a constant, and all integrals over the dx's ranges from $-\infty$ to ∞. Note that we use two sets of coordinates and momenta.

Following a similar path to subsection C.7.1.1 we first we determine the Fourier transform on the path integral for an infinitesimal time interval ε:

$$G_\varepsilon = \iint dx_1 dx_2 \exp[-ip_1x_1 - ip_2x_2] \exp[-imx_1x_2/\varepsilon]$$
$$\sim \exp[-i\varepsilon p_1p_2/m] \qquad (C.7.15)$$

Upon combining the intervals to a total time T we obtain the Fourier transform of the total path integral

$$K(p, T) \sim \exp[-iTp_1p_2/m] \qquad (C.7.16)$$

which yields the spatial path integral

$$K(x_1 - y_1, x_2 - y_2, T) \sim \int d\, p_1 dp_2 \exp[ip_1(x_1 - y_1) + ip_2(x_2 - y_2)] \exp[-iTp_1p_2/m]$$

$$\sim \exp[-im(x_1 - y_1)(x_2 - y_2)/T] \qquad (C.7.17)$$

which we normalize to unity

$$\int dy_1 dy_2\, K(x_1 - y_1, x_2 - y_2, T) = 1 \qquad (C.7.18)$$

with the result

$$K(x_1 - y_1, x_2 - y_2, T) = (m/T)\exp[-im(x_1 - y_1)(x_2 - y_2)/T] \quad (C.7.19)$$

If we now use the path integral on a free particle wave function we see that it displaces the wave function by the time T:

$$\Psi_0(x_1, x_2, t) = \exp[-ip_1x_1 - ip_2x_2 - iEt] \qquad \text{where } E = p_1p_2/m \quad (C.7.20)$$

$$\Psi(y_1, y_2, t + T) = \iint dx_1dx_2\, K(x_1 - y_1, x_2 - y_2, T)\Psi_0(x_1, x_2, t) \quad (C.7.21)$$
$$= \exp[-ip_1y_1 - ip_2y_2 - iE(t + T)]$$
$$= \Psi_0(y_1, y_2, t + T) \quad (C.7.22)$$

C.7.1.3 Introducing the Rotation Between the Quantum and Classical Cases of the Path Integral

We begin by expressing the coordinates in terms of new 'rotated' coordinates:

$$\begin{aligned} x_1 &= u_1 \cos\theta + u_2 \sin\theta \quad (C.7.23)\\ x_2 &= -u_1 \sin\theta + u_2 \cos\theta \\ p_1 &= -p_{u1} \sin\theta + p_{u2} \cos\theta \\ p_2 &= p_{u1} \cos\theta + p_{u2} \sin\theta \end{aligned}$$

Then the quantities of interest are now expressed as

$$\Psi_0(x_1, x_2, t) = \exp[-ip_1x_1 - ip_2x_2 - iEt] \qquad \text{where } E = p_1p_2/m \quad (C.7.24)$$

$$= \Psi_0(u_1, u_2, t) = \exp\{-i[(p_{u2}u_2 - p_{u1}u_1)\sin(2\theta) + (p_{u2}u_1 + p_{u1}u_2)\cos(2\theta)] - iEt\} \quad (C.7.25)$$

where the energy E now haves the form

$$E = (-p_{u1} \sin\theta + p_{u2} \cos\theta)(p_{u1} \cos\theta + p_{u2} \sin\theta)/m$$
$$= [(p_{u2}^2 - p_{u1}^2)\sin(2\theta)]/(2m) + [p_{u2}p_{u1} \cos(2\theta)]/m \quad (C.7.26)$$

We will now examine the two special cases: $\theta = 0$ corresponding to classical mechanics and $\theta = \pi/4$ corresponding to quantum mechanics.

<u>$\theta = 0$</u>

The wave equation in this case is

$$\Psi_0(u_1, u_2, t) = \exp\{-i(p_{u2}u_1 + p_{u1}u_2) - iEt\} \quad (C.7.27)$$

with

$$E = p_{u2}p_{u1}/m$$

The wave function has a 'classical' form as we showed in eq. 47 in appendix C.

$\theta = \pi/4$

The wave equation in this case is

$$\Psi_0(u_1, u_2, t) = \exp\{-i(p_{u2}u_2 - p_{u1}u_1) - iEt\} \quad \text{(C.7.28)}$$

with

$$E = (p_{u2}^2 - p_{u1}^2)/(2m)$$

This wave function has a quantum form with a positive energy part and a negative energy part:

C.7.1.4 Free Path Integral with Rotation Between Classical and Quantum Mechanics

The incremental path integral factor, expressed in terms of u_1 and u_2, and then Fourier transformed is:

$$G_\varepsilon(p_{u1}, p_{u2}) = \iint du_1 du_2 \exp\{-i[(p_{u2}u_2 - p_{u1}u_1)\sin(2\theta) + (p_{u2}u_1 + p_{u1}u_2)\cos(2\theta)]\} \cdot$$
$$\cdot \exp\{-im[(u_2^2 - u_1^2)\sin(2\theta)/2 + u_2u_1\cos(2\theta)]\}/\varepsilon\} \quad \text{(C.7.29)}$$
$$\sim \exp\{-i\varepsilon[(p_{u2}^2 - p_{u1}^2)\sin(2\theta)/2 + p_{u1}p_{u2}\cos(2\theta)]/m\}$$

yielding the cumulative product for the time interval T

$$K(p_{u1}, p_{u2}, T) \sim \exp[-iT[(p_{u2}^2 - p_{u1}^2)\sin(2\theta)/2 + p_{u1}p_{u2}\cos(2\theta)]/m] \quad \text{(C.7.30)}$$

Upon Fourier transforming to coordinate space we find

$$K(u_1 - v_1, u_2 - v_2, T) \sim \int dp_{u1}dp_{u2} \exp\{i[(p_{u2}w_2 - p_{u1}w_1)\sin(2\theta) + (p_{u2}w_1 + p_{u1}w_2)\cos(2\theta)]\}K(p_{u1}, p_{u2}, T)$$

$$\sim \exp[im[(w_2^2 - w_1^2)\sin(2\theta)/2 + w_2w_1\cos(2\theta)]/T] \quad \text{(C.7.31)}$$

where

$$w_i = u_i - v_i$$

The special cases of interest are:

$\theta = 0$

$$K(u_1 - v_1, u_2 - v_2, T) \sim \exp[imw_2w_1]/T] \quad \text{(C.7.32)}$$

This gives the *Classical* path integral without the use of any limiting or approximation procedure such as one often finds in the literature.

$\theta = \pi/4$

This case yields the familiar free particle path integral – but with a part for a positive energy particle and a part for a negative energy particle. The negative energy part can be removed easily yielding the conventional free particle quantum path integral.

$$K(u_1 - v_1, u_2 - v_2, T) \sim \exp[im(w_2^2 - w_1^2)/(2T)] \quad \text{(C.7.33)}$$

C.7.2 General PseudoQuantum Formulation

The propagator K(x − y, t) for the conventional Feynman path integral formulation has the form:

$$K(x - y, T) = A \lim_{n\to\infty} \int\!\!\int\!\!\int\!\!\int_{-\infty}^{+\infty} \ldots \int dx_0 dx_1 \ldots dx_n \exp[i/\hbar \int_t^{t+T} L(x, v, t_a)\, dt_a] \quad \text{(C.7.34)}$$

where A is a constant, and the integral over the dx's ranges from −∞ to ∞.

The PseudoQuantum form of the path integral formalism is based on the PseudoQuantum Lagrangian

$$L(x_1, v_1, x_2, v_2) = \partial^2H(x_1, p_1)/\partial p_1^2\, p_1p_2 + \partial^2H(x_1, p_1)/\partial x_1\partial p_1\, p_1x_2 - \partial H(x_1, p_1)/\partial x_1\, x_2 \quad \text{(C.7.35)}$$

$$K(x_1 - y_1, x_2 - y_2, T) = A \lim_{n\to\infty} \int\!\!\int\!\!\int\!\!\int_{-\infty}^{+\infty} \ldots \int dx_{10}dx_{11}\ldots dx_{1n}\, dx_{20}dx_{21}\ldots dx_{2n} \exp[i/\hbar \int_t^{t+T} L(x_1, v_1, x_2, v_2)\, dt_a] \quad \text{(C.7.36)}$$

From subsection C.7.1.2, where

$$G_\varepsilon = \int\!\!\int dx_1dx_2 \exp[-ip_1x_1 - ip_2x_2] \exp[-imx_1x_2/\varepsilon]$$
$$\sim \exp[-i\varepsilon p_1p_2/m] \quad \text{(C.7.37)}$$

we can perform the x_2 integration in G_ε using

$$x'_2 \cong x_2(t + \varepsilon) - x_{2(}(t)\, /\varepsilon \quad \text{(C.7.38)}$$

if

$$p_2 = mx'_2$$

Then

$$G_\varepsilon = \int\!\!\int dx_1dx_2 \exp[-ip_1x_1 - ip_2x_2] \exp\{-i\varepsilon[\partial^2H(x_1, p_1)/\partial p_1^2\, p_1mx_2/\varepsilon + \partial^2H(x_1, p_1)/\partial x_1\partial p_1\, p_1x_2 - \partial H(x_1, p_1)/\partial x_1\, x_2]]$$

$$= \int dx_1 \exp[-ip_1x_1]\, \delta(p_2 - (m\partial^2H(x_1, p_1)/\partial p_1^2\, p_1 + \varepsilon\partial^2H(x_1, p_1)/\partial x_1\partial p_1\, p_1 - \varepsilon\partial H(x_1, p_1)/\partial x_1))$$

$$= \int dx_1 \exp[-ip_1x_1]\, \delta(p_2 - (m\partial^2H(x_1, p_1)/\partial p_1^2\, p_1 + \varepsilon\partial^2H(x_1, p_1)/\partial x_1\partial p_1\, p_1 - \varepsilon\partial H(x_1, p_1)/\partial x_1)) \quad \text{(C.7.39)}$$

C.7.2.1 PseudoQuantum Wave Functions and Schrödinger equation

In the presence of a potential, the path integral formulation leads us to transform it into a Schrödinger equation. The incremental time displacement form of the wave equation is

$$\Psi(y_{1k+1}, y_{2k+1}, t + \varepsilon) = \int\!\!\int dx_{1k}dx_{2k} \exp\{(i\varepsilon/\hbar)[m(x_{1k+1} - x_{1k})(x_{2k+1} - y_{2k})/\varepsilon^2 -$$
$$- x_{2k+1}(\partial H(x_1, p_1)/\partial x_1)|_{x_1=x_{1k+1}}]\}\Psi_0(x_{1k}, x_{2k}, t) \quad \text{(C.7.40)}$$

$$= \iint dx_{1k}dx_{2k} \exp\{(i\varepsilon/\hbar)[m(x_{1k+1} - x_{1k})(x_{2k+1} - y_{2k})/\varepsilon^2 - - x_{2k+1}(\partial V(x)/\partial x)|_{x=x_{1k+1}}]\}\Psi_0(x_{1k}, x_{2k}, t)$$

for a potential V(x). In the case of the harmonic oscillator the PseudoQuantum potential term is

$$x_{2k+1}\partial V(x)/\partial x)|_{x=x_{1k+1}} = m\omega^2 x_{1k+1}x_{2k+1} \quad (C.7.41)$$

The PseudoQuantum Schrödinger equation that results is

$$i\hbar\partial\Psi(x_1, x_2, t)/\partial t = (-\hbar^2/m)\partial^2\Psi(x_1, x_2, t)/\partial x_1\partial x_2 + V(x_1, x_2)\Psi(x_1, x_2, t) \quad (C.7.42)$$

C.7.2.2 Decomposition of PseudoQuantum Schrödinger Equation into Quantum and Classical Parts

The Schrödinger equation can be decomposed into a quantum and a classical part. Starting from eq. C.7.42:

$$i\hbar\partial\Psi(x_1, x_2, t)/\partial t = (-\hbar^2/m)\partial^2\Psi(x_1, x_2, t)/\partial x_1\partial x_2 + V(x_1, x_2)\Psi(x_1, x_2, t) \quad (C.7.43)$$

we find

$$i\hbar\partial\Psi(u_1, u_2, t, \theta)/\partial t = (-\hbar^2/m)[\sin(2\theta)\,(\partial^2/\partial u_2^2 - \partial^2/\partial u_1^2)\,/2 + \cos(2\theta)\partial^2/\partial u_1\partial u_2]\cdot \cdot\Psi(u_1, u_2, t, \theta) + V(u_1 \cos\theta + u_2 \sin\theta, -u_1 \sin\theta + u_2 \cos\theta)\Psi(u_1, u_2, t, \theta) \quad (C.7.44)$$

using the relation to u_1 and u_2:

$$x_1 = u_1 \cos\theta + u_2 \sin\theta$$
$$x_2 = -u_1 \sin\theta + u_2 \cos\theta$$

Then

$$\Psi(u_1, u_2, t, \theta) = \exp\{-i[(p_{u2}u_2 - p_{u1}u_1)\sin(2\theta) + (p_{u2}u_1 + p_{u1}u_2)\cos(2\theta)] - iEt\} \quad (C.7.45)$$

and the energy is

$$E = [(p_{u2}^2 - p_{u1}^2)\sin(2\theta)]/(2m) + [p_{u2}p_{u1}\cos(2\theta)]/m \quad (C.7.46)$$

Again there are two cases of interest:

<u>The Quantum part $\theta = \pi/4$</u>

$$i\hbar\partial\Psi(u_1, u_2, t, \theta = \pi/4)/\partial t = (-\hbar^2/m)[(\partial^2/\partial u_2^2 - \partial^2/\partial u_1^2)\,/2]\Psi(u_1, u_2, t, \theta = \pi/4) + + V((u_1 + u_2)/\sqrt{2}, (-u_1 + u_2)/\sqrt{2})\Psi(u_1, u_2, t, \theta = \pi/4) \quad (C.7.47)$$

In the quantum free particle case

$$\Psi(u_1, u_2, t, \theta = \pi/4) = \exp\{-i(p_{u2}u_2 - p_{u1}u_1) - iEt\} \quad (C.7.48)$$

where

$$E = (p_{u2}^2 - p_{u1}^2)/(2m)$$

<u>The Classical part θ = 0</u>

$$i\hbar\partial\Psi(u_1, u_2, t, \theta = 0)/\partial t = (-\hbar^2/m)\partial^2/\partial u_1\partial u_2\, \Psi(u_1, u_2, t, \theta = 0) + V(u_1, u_2)\Psi(u_1, u_2, t, \theta = 0)$$

In the classical free particle case

$$\Psi(u_1, u_2, t, \theta = 0) = \exp\{-i(p_{u2}u_1 + p_{u1}u_2) - iEt\} \quad (C.7.49)$$

where

$$E = p_{u2}p_{u1}/m$$

C.7.3 Classical Part of Free Particle PseudoQuantum Feyman Path Propagator

The classical part of the free particle PseudoQuantum path integral is described as follows. First the time incremental factor is

$$G_\varepsilon\,(p_{u1}, p_{u2}) = \iint du_1du_2\, \exp\{-i[(p_{u2}u_2 - p_{u1}u_1)\sin(2\theta) + (p_{u2}u_1 + p_{u1}u_2)\cos(2\theta)]\} \cdot \cdot \exp\{-im[(u_2^2 - u_1^2)\sin(2\theta)/2 + u_2u_1\cos(2\theta)]\}/\varepsilon\}$$

$$\sim \exp\{-i\varepsilon[(p_{u2}^2 - p_{u1}^2)\sin(2\theta)/2 + p_{u1}p_{u2}\cos(2\theta)]/m\} \quad (C.7.50)$$

The product of the incremental terms is

$$K(p_{u1}, p_{u2}, T) \sim \exp[-iT[(p_{u2}^2 - p_{u1}^2)\sin(2\theta)/2 + p_{u1}p_{u2}\cos(2\theta)]/m] \quad (C.7.51)$$

which yields the *classical* path integral

$$K(u_1 - v_1, u_2 - v_2, T) \sim \int dp_{u1}dp_{u2}\, \exp\{i[(p_{u2}w_2 - p_{u1}w_1)\sin(2\theta) + (p_{u2}w_1 + p_{u1}w_2)\cos(2\theta)]\}K(p_{u1}, p_{u2}, T)$$
$$\sim \exp[im[(w_2^2 - w_1^2)\sin(2\theta)/2 + w_2w_1\cos(2\theta)]/T] \quad (C.7.52)$$

where

$$w_i = u_i - v_i$$

<u>For θ = 0 the classical path integral steps are:</u>

$$G_\varepsilon\,(p_{u1}, p_{u2}) \sim \exp\{-i\varepsilon p_{u1}p_{u2}/m\} \quad (C.7.53)$$

$$K(p_{u1}, p_{u2}, T) \sim \exp[-iTp_{u1}p_{u2}/m] \quad (C.7.54)$$

$$K(u_1 - v_1, u_2 - v_2, T) \sim \exp[im(u_1 - v_1)(u_2 - v_2)/T] \quad (C.7.55)$$

This gives us a classical path integral formulation that avoids approximation techniques which have hitherto been used. The above development can be rewritten in terms of the x and p variables.

C.7.4 Fokker-Planck Equation

The Feynman path integral formulation and equations can be transformed into similar forms by letting $i\hbar$ be changed to a positive constant. The Fokker-Planck equation gives the probability density of a particle velocity's time evolution under the impact of forces. This equation is also known as the Smoluchowski equation—named after its originator.

In its formulation a variable x_2 is introduced as the 'response' variable to the 'primary' variable x_1.The Fokker-Planck equation for the probability density is

$$p(x_1', t + \varepsilon) = (1/2\pi i) \int_{-\infty}^{\infty} dx_1 \int_{-i\infty}^{i\infty} dx_2 \exp\{\varepsilon[-x_2(x_1' - x_1)/\varepsilon + x_2 D_1(x_1, t) + x_2^2 D_2 (x_1, t)]\} p(x_1, t) \tag{C.7.56}$$

The origin of x_2 in our formalism, and in the Fokker-Planck equation, are very different. The Fokker-Planck equation has the Lagrangian:

$$L = \int dt\, [x_2 D_1(x_1, t) + x_2^2 D_2 (x_1, t) - x_2\, \partial x_1/\partial t]$$

A comparison of this formulation with the PseudoQuantum path integral formulation shows an apparently remarkable similarity. Thus one might view the Fokker-Planck equation as a precursor of the PseudoQuantum formulation. The papers appearing in the appendices of this book show that the origin of our formalism and the Fokker-Planck formalism are very different.

C.8 The Transition Between Classical and Quantum Chaos

Chaos has become an increasingly important field of activity. While classical chaos has been the more studied aspect of chaos there has been an increasingly larger interest in quantum chaos. In this section we wish to show that the PseudoQuantum formalism appears to be a useful means of relating classical chaos to quantum chaos for many systems. It makes it possible to trace the transition from a classical chaos situation to quantum chaos. It also offers the possibility to determine the quantum analogue of a classical chaos phenomenon.

Given a quantum system it is often difficult to determine whether it has a chaotic regime. Frequently extensive numerical analysis is needed for this determination.

In this section we show that the PseudoQuantum formalism enables us to determine a classical Hamiltonian from a quantum formalism where chaos is known to occur based on extensive numerical investigations.

A well-studied[150,151] Hamiltonian for a quantum theory, known to have chaotic regions, is

$$H = (p_x^2 + p_y^2/2 + x^2y^2 + \beta(x^4 + y^4)/4 \qquad \text{(C.8.1)}$$

Creating the equivalent PseudoQuantum Hamiltonian we obtain

$$\hat{H} = p_{x1}p_{x2} + p_{y1}p_{y2} + y_1^2x_1x_2 + x_1^2y_1y_2 + \beta(x_1^3x_2 + y_1^3y_2) \qquad \text{(C.8.2)}$$

Introducing new variables we can develop a form of eq. C.8.2 that allows us to trace the transition from a quantum theory to a classical theory, which also should have chaotic regimes.

$$x_1 = u_{x1}\cos\theta + u_{x2}\sin\theta \qquad \text{(C.8.3)}$$
$$x_2 = -u_{x1}\sin\theta + u_{x2}\cos\theta$$

$$p_{x1} = p_{ux1}\cos\theta + p_{ux2}\sin\theta$$
$$p_{x2} = -p_{ux1}\sin\theta + p_{ux2}\cos\theta$$
$$y_1 = u_{y1}\cos\theta + u_{y2}\sin\theta$$
$$y_2 = -u_{y1}\sin\theta + u_{y2}\cos\theta$$

$$p_{y1} = p_{uy1}\cos\theta + p_{uy2}\sin\theta$$
$$p_{y2} = -p_{uy1}\sin\theta + p_{uy2}\cos\theta$$

Then we obtain the PseudoQuantum Hamiltonian

$$\begin{aligned}\hat{H}(\theta) &= p_{x1}p_{x2} + p_{y1}p_{y2} + y_1^2x_1x_2 + x_1^2y_1y_2 + \beta(x_1^3x_2 + y_1^3y_2) \\ &= (p_{ux2}^2 - p_{ux1}^2 + p_{uy2}^2 - p_{uy1}^2)\sin(2\theta)/2 + (p_{uy1}p_{uy2} + p_{ux1}p_{ux2})\cos(2\theta) + \\ &\quad + (u_{y1}\cos\theta + u_{y2}\sin\theta)^2[(u_{x2}^2 - u_{x1}^2)\sin(2\theta)/2 + u_{x1}u_{x2}\cos(2\theta)] + \\ &\quad + (u_{x1}\cos\theta + u_{x2}\sin\theta)^2[(u_{y2}^2 - u_{y1}^2)\sin(2\theta)/2 + u_{y1}u_{y2}\cos(2\theta)] + \\ &\quad + \beta\{(u_{x1}\cos\theta + u_{x2}\sin\theta)^2[(u_{x2}^2 - u_{x1}^2)\sin(2\theta)/2 + u_{x1}u_{x2}\cos(2\theta)] + \\ &\quad + (u_{y1}\cos\theta + u_{y2}\sin\theta)^2[(u_{y2}^2 - u_{y1}^2)\sin(2\theta)/2 + u_{y1}u_{y2}\cos(2\theta)]\}\end{aligned} \qquad \text{(C.8.4)}$$

The classical Hamiltonian that results from this analysis is

$$\hat{H}(0) = (p_{uy1}p_{uy2} + p_{ux1}p_{ux2}) + u_{y1}^2u_{x1}u_{x2} + u_{x1}^2u_{y1}u_{y2} + \beta\{u_{x1}^2u_{x1}u_{x2} + u_{y1}^2u_{y1}u_{y2}\} \qquad \text{(C.8.5)}$$

---

150 The Hamiltonian model above has been studied by: Y. Y. Bai, G. Hose, K. Stefański, and H. S. Taylor, Phys. Rev. **A31**, 2821 (1985), R. L. Waterland, J.-M. Yuan, C. C. Martens, R. E. Gillilan, and W. P. Reinhardt, Phys. Rev. Lett. **61**, 2733 (1988, and other papers..

151 Another much studied model—the 2-Dimensional stadium, Quantum billiard ball model has classical chaotic dynamics, and a quantum approximation that is chaotic: S. W. McDonald and A. N. Kaufman, Phys. Rev. **A37**, 3067 (1988); _______, Phys. Rev. Lett., **42**, 1189 (1979), and other papers.

The Quantum Hamiltonian that emerges is

$$\hat{H}(\pi/4) = (p_{ux2}^2 - p_{ux1}^2 + p_{uy2}^2 - p_{uy1}^2)/2 + u_{y2}^2(u_{x2}^2 - u_{x1}^2)/2 + u_{x2}^2(u_{y2}^2 - u_{y1}^2)/2 + \\ + \beta\{u_{x2}^2(u_{x2}^2 - u_{x1}^2)/2 + u_{y2}^2(u_{y2}^2 - u_{y1}^2)/2\} \qquad (C.8.6)$$

While these Hamiltonians require numerical analysis to understand their chaotic features, they offer the possibilities of comparative studies of quantum and classical chaos.

The study of other models of a similar character would appear to be of importance in elucidating quantum chaos.

C.9 The Transition Between Classical & Quantum Entanglement Dynamics

Quantum Entanglement has become of great importance judging from its increasing number of papers. It offers the possibility of new forms of communication that might be of great value in interstellar communication should Mankind reach the stars.

In this section we will study a prototype example of quantum entanglement with a view towards investigating the transition from quantum entanglement through 'semi-classical' quantum entanglement to a 'classical' limit.

The example that we consider will use a combination of a positive energy single particle state entangled with a negative energy particle state to simulate the more commonly studied case of entangled spins. The particles, in a superposed state, are assumed to separate, and one particle will be measured thus determining the state of the other particle due to entanglement. We define the 'entangled' NOON-type state:

$$\Psi = (|n+ = 1, n- = 0> + |\, n+ = 0, n- = 1>)/\sqrt{2} \qquad (C.9.1)$$

where one pure state $|n+ = 1, n- = 0>$ is a one particle state of positive energy and the other state $|\, n+ = 0, n- = 1>$ has a one negative energy particle.

We define projection operators of the form:[152]

$$\rho(\theta) = ||n+(\theta),n-(\theta)><n+(\theta),n-(\theta)| \qquad (C.9.2)$$

using an angle θ as we have done previously to specify the quantum-classical content of the projection.

More generally we will define harmonic oscillator states:

$$|\, n+, n-> = b_1^{\dagger n_+} b_2^{\dagger n_-}|0,0> \qquad (C.9.3)$$

with a density operator

$$\rho(\theta) = \sum_{n+,n-} |n+, n-><n+, n-| = \sum_{n+,n-} b_1^\dagger(\theta)^{n_+} b_2^\dagger(\theta)^{n_-}|0,0><0,0|b_1(\theta)^{n_+} b_2(\theta)^{n_-} \qquad (C.9.4)$$

[152] We will be using the harmonic oscillator formalism of E.4 although the conclusions will be more far reaching.

We will consider the particular projection:

$$P(\theta) = |1(\theta),0><1(\theta),0| \quad (C.9.5)$$

which we will apply to Ψ:

$$P(\theta)\Psi = |1(\theta), 0>/\sqrt{2} = \sin(2\theta)b_1^\dagger(\theta)|0, 0>/\sqrt{2} = \sin(2\theta)(Q_1\cos\theta - iP_2\sin\theta)|0, 0>/\sqrt{2} \quad (C.9.6)$$

Using the familiar relations:

$$b_1 = Q_1\cos\theta + iP_2\sin\theta \quad (C.4.6)$$
$$b_2 = -Q_2\sin\theta + iP_1\cos\theta$$
$$b_1^\dagger = Q_1\cos\theta - iP_2\sin\theta \quad (C.4.7)$$
$$b_2^\dagger = -Q_2\sin\theta - iP_1\cos\theta$$

with commutation relations

$$[b_1, b_1^\dagger] = \sin(2\theta) \quad (C.4.8a)$$
$$[b_2, b_2^\dagger] = -\sin(2\theta)$$
$$[b_1, b_2^\dagger] = [b_2, b_1^\dagger] = 0$$
$$[b_1, b_2] = [b_1^\dagger, b_2^\dagger] = 0$$

We find

$$\Psi' = P(\theta)\Psi \quad (C.9.7)$$

has the following forms for $\theta = \pi/4$ and $\theta=0$:

$\underline{\theta = \pi/4}$

$$\Psi' = (Q_1 - iP_2)|0, 0>/2 \quad (C.9.8)$$

gives a quantum state.

$\underline{\theta = 0}$

$$\Psi' = 0 \quad (C.9.9)$$

We note that we showed in section C.4 that 'classical' states containing particles have infinite energy—this above result, $\Psi' = 0$, reflects the infinite energy of states containing one or more particles. The value of $\Psi' = 0$ leaves the positivity or negativity of the other particle undetermined since entanglement is a purely quantum phenomena.

Other values of θ

$$\Psi' = \sin(2\theta)(Q_1\cos\theta - iP_2\sin\theta)|0, 0>/\sqrt{2} \quad (C.9.10)$$
$$= \sin(2\theta)b_1^\dagger(\theta)|0, 0>/\sqrt{2} = \sin(2\theta)|1(\theta),0>/\sqrt{2}$$

An intermediate result occurs but the other entangled separated particle state's energy is determined to be negative.

C.10 PseudoQuantum Transition Between Quantum and [153]Classical Dynamics

The preceding sections have shown that the use of the PseudoQuantum framework, which contains a purely quantum sector (albeit with both positive and negative energy parts that are separable), a purely classical sector, and an intermediate sector that is partly quantum and partly classical, enables us to

1. Relate the corresponding quantum and classical dynamics of a physical phenomenon.

2. Study the transition from quantum to classical behavior without recourse to approximations or limits such as $\hbar \to 0$.

3. Determine the classical equivalent of a quantum dynamical system.

4. Determine the quantum equivalent of a classical dynamical system.

These advantages appear to be fairly general in nature—as evidenced by the harmonic oscillator case studied in section C.4—since the harmonic oscillator plays such a prominent role in many physical situations.

We conclude that the PseudoQuantum formalism, which is not only relevant for quantum-classical mechanics dynamics, but is also of importance in Quantum Field Theories as shown in our earlier sections and in our papers in the appendices. It has also recently been used in a new GraviStrong unified theory that relates quark confinement to deviations from Newtonian gravitation at galactic distance scales using a canonical PseudoQuantum formulation of a higher derivative Quantum Field Theory. (See Blaha (2016e).)

---

[153] Gutzwiller (1990) points out the use of harmonic oscillator wave functions in several studies of the quantum-classical connection.

Appendix D. PseudoQuantum Theory

PseudoQuantum Field Theory (and its Quantum Mechanics analogue CQ Mechanics[154]) originates in the need to second quantize in unusual coordinate systems, and in curved space-time coordinate systems. The papers in Appendix E provide a detailed introduction to PseudoQuantum Field Theory to which the reader is referred.

In this appendix[155] we point out its advantages in a variety of field theory contexts that are relevant for the Unified SuperStandard Theory. The advantages of PseudoQuantum Field Theory are:

1. Quantization in any coordinate system in flat or curved space-times with an invariant definition of asymptotic particle states. An n particle asymptotic state in one coordinate system is a unitarily equivalent n particle asymptotic state in any other coordinate system. Therefore particle number is invariant under change of coordinate system. This is important for the Unified SuperStandard Theory in curved space-times. It is also important for quantization in higher dimensional Euclidean spaces such as the Megaverse. The method was developed in the late 1970's by the author to provide a quantization procedure which supports a unique particle interpretation of states in arbitrary non-static space-times where no global time-like coordinate (Killing vector) exists. PseudoQuantum Field Theory which we developed in a series of books[156] also can be formulated in the Megaverse. Thus we can use it in the Megaverse to implement the Higgs Mechanism to generate particle masses and symmetry breaking.

2. PseudoQuantum Field Theory enables one to define Higgs particle dynamics in such a way that a non-zero vacuum expectation value cleanly separates from the quantum field part of the Higgs fields. This technique can be used in symmetry breaking mechanisms, mass generation, and possible generation of coupling constants as vacuum expectation values.

3. It supports the canonical definition of higher derivative field theories through the use of the Ostrogradski bootstrap. See Appendix B where a fourth order theory of the Strong interaction is defined that has color

---

[154] See Appendix C for details. CQ Mechanics encompasses both classical mechanics and quantum mechanics, and provides a method of rotating between them. It has applications to transitions between Quantum/Semi-Classical Entanglement, and Quantum/Classical Path Integrals, and Quantum/Classical Chaos.

[155] This appendix is chapter 10 in Blaha (2018e).

[156] See Blaha (2017b) for the discussion of the PseudoQuantum field theory formalism for Higgs particles in our Extended Standard Model. See chapter 20 of Blaha (2017b), and earlier books, for a more detailed view than that presented here.

confinement and a linear r potential. The potential part of this theory was used by the Cornell group to calculate the Charmonium spectrum. (See Blaha (2017b) for details.)

An associated advantage of using PseudoQuantum Field Theory is that it provides for retarded propagators and an Arrow of Time.

D.1 General Case of PseudoQuantization in Differing Coordinate Systems

Appendix A describes the PseudoQuantization procedure that relates second quantizations in differing coordinate systems. We can epitomize the general concept in the following short example.

Consider the case of a scalar particle in D space-time dimensions that we second quantize in coordinate system denoted 1 with coordinates x based on a time-like Killing vector

$$\varphi(x) = \sum_\alpha [\chi_\alpha(x)A_\alpha + \chi_\alpha{}^*(x)A_\alpha{}^\dagger] \qquad \text{(D.1)}$$

where the $\chi_\alpha(x)$ are positive frequency with respect to a definition of positive frequency within a universe – following the notation of Appendix A.

Consider now the second quantization of the particle field in a second coordinate system denoted 2 with coordinates y based on a different time-like Megaverse Killing vector

$$\varphi(y) = \sum_\beta [\psi_\beta(y)b_\beta + \psi_\beta{}^*(y)b_\beta{}^\dagger] \qquad \text{(D.2)}$$

where the $\psi_\beta(y)$ are positive frequency with respect to 2's definition of positive frequency.

Comparing above definitions we see the difference in the definition of the coordinates used in the field expansions as well as the implicit difference in the definitions of positive frequency. To relate the quantizations to each other, we must use the relation between the x and y coordinates:

$$y_i = f_i(x)$$

or, in vector form,

$$y = f(x)$$

for i = 1, 2, … , D. Thus

$$\varphi(f(x)) = \sum_\beta [\psi_\beta(f(x))b_\beta + \psi_\beta{}^*(f(x))b_\beta{}^\dagger] \qquad \text{(D.3)}$$

Inverting the above equations to obtain the relation of the Fourier coefficient operators we see:

$$A_\alpha = \sum_\beta [C_{\alpha\beta}\, b_\beta + C'_{\alpha\beta}\, b_\beta{}^\dagger]$$

where $C_{\alpha\beta}$ and $C'_{\alpha\beta}$ are c-number functions of α and β:

$$C_{\alpha\beta} = (\chi_\alpha(x), \varphi(f(x))$$
$$C'_{\alpha\beta} = (\chi_\alpha{}^*(x), \varphi(f(x))) \tag{D.4}$$

The above equations imply an N particle state in one coordinate system will appear as a superposition of states of various numbers of particles in the other coordinate system IF the standard quantum field theory formulation is used.

TO REMEDY this situation – which we take to be unphysical – we must reformulate quantum field theory using the PseudoQuantum formulation presented Appendix A. The scalar particle case is discussed in Appendix A between eqs. 6 – 31, to which the reader is referred.

The conclusions of that section and the sections following it, in Appendix A are:

1. One can define corresponding unitarily equivalent particle states in two quantizations with invariant particle numbers.

2. The Fourier coefficient operators of the two quantizations are related by Bogoliubov transformations and are unitarily equivalent.

3. The group of the local Bogoliubov transformations is an infinite tensor product of $SU_{1,1}$ groups.

4. The vacua of the particle are invariant under Bogoliubov transformations that relate the Megaverse and the universe quantizations.

5. Unitarily equivalent perturbation theories of both quantizations can be defined.

We now consider the case of Two-Tier PseudoQuantization, and then turn to various applications of PseudoQuantization.

D.2 Two-Tier PseudoQuantum Field Theory

The combination of the Two-Tier procedure with the PseudoQuantiztion procedure leads to a somewhat more complicated situation. In principle, both are required for a Unified SuperStandard Theory in any coordinate system in flat or curved space-times in any number of dimensions. However their direct combination is both complicated and unphysical.

The main purpose of PseudoQuantization is to have particle number invariance under a change of coordinate system. Two-Tier Field Theory 'cloaks' each particle in infinite 'clouds' of Y^μ quanta as Fig. 9.1 illustrates. We define PseudoQuantization as implementing particle number invariance for 'bare' particles without their clouds of Y^μ quanta. Thus an asymptotic particle state of n particles (neglecting its Y^μ quanta cloud) remains a unitarily equivalent n particle state (neglecting its Y^{μ} quanta cloud) under a change of coordinate system.

To implement this concept we first define quantization of a particle in coordinate systems without Two-Tier quanta. We then 'dress' the quantization by replacing the coordinates y^{μ} in each coordinate system with the corresponding Two-Tier coordinates:

$$y^{\mu} \rightarrow X^{\mu}(y) = y^{\mu} + i\, Y^{\mu}(y)/M_c^2 \qquad \text{(D.5)}$$

It appears the most convenient gauge in each coordinate system is the Lorentz gauge:

$$\partial Y^{\mu}/\partial y^{\mu} = 0 \qquad \text{(D.6)}$$

We now briefly consider the case of a scalar particle PseudoQuantization. This case is considered in more detail in Appendix A. Following Appendix A we must introduce two fields $\varphi_1(y)$ and $\varphi_2(y)$ with the free fields' Lagrangian

$$\mathscr{L}(y) = \partial^{\mu}\varphi_1\partial_{\mu}\varphi_2 - \tfrac{1}{2}\,\partial^{\mu}\varphi_1\partial_{\mu}\varphi_1 - m^2\varphi_1\varphi_2 + \tfrac{1}{2}\, m^2\, \varphi_1^{\,2} \qquad \text{(D.7)}$$

in a coordinate system with coordinates y. Then following the steps indicated in Appendix A from eq. 7 onward we arrive at a PseudoQuantum formulation in the coordinate system with coordinates y that is unitarily equivalent to that of a different coordinate system defined a similar manner.

From eq. 43 onwards we can replace the c-number coordinates x and y with Two-Tier coordinates of the form

$$X^{\mu}(y) = y^{\mu} + i\, Y^{\mu}(y)/M_c^2$$

and proceed to calculate propagators and perturbation theory diagrams.

Thus we have a straight-forward procedure to unite the PseudoQuantum formalism with Two-Tier coordinates to obtain finite perturbation theory results with unitary equivalence to quantization in other coordinate systems in both flat and curved space-times.

The use of two fields per particle of PseudoQuantum field theory will be seen to part of the applications consider in the remainder of this subsection. We will put aside the consideration of quantizations in other coordinate systems in what follows to keep the presentation as simple as possible.

D.3 PseudoQuantum Higgs Scalar Particle Field Theory in D-dimensional Space-Time

D.3.1 The Enigma of Higgs Particles and the Higgs Mechanism

In our previous work on the Standard Model, and its generalization to The Unified SuperStandard Theory described in a series of books entitled *Physics is Logic*

..., we showed that the fermion spectrum results from Complex Special Relativity, the gauge interactions result from the Reality group, the fermion generations result from the Generation group, and the Theory of Everything results from a combination with Complex General Relativity. The Higgs particles and the Higgs Mechanism were inserted to generate particle masses and symmetry breaking effects.

Whence come Higgs particles? A more fundamental cause has not been suggested until our analysis, which is presented here in chapter 11. So the Higgs sector appeared to be an expedient mechanism to insert much needed symmetry breaking and masses into the theory.

There are a number of peculiarities in the implementation of the Higgs Mechanism:

1. First, it is selective in the sense that some gauge fields have associated Higgs particles and utilize the Higgs Mechanism, and some gauge fields do not have associated Higgs particles. In particular, the ElectroWeak gauge fields, the Generation group gauge fields, the Layer group fields, and the complex gravity Species gauge fields have associated Higgs particles. The strong interaction (gluon) gauge fields do not.[157]

2. The Higgs potentials have a quadratic mass term of the "wrong" sign plus a quartic interaction term, which together, generate non-zero vacuum expectation values. They obviously accomplish their goal. But the source of these potentials, and why they have the same form, is unknown. One expects a fundamental principle should be operative here.

3. One can imagine creating a Higgs microscope at some super-accelerator. Using this microscope in the presence of a (classical) condensate could enable the Uncertainty Principle to be violated. This possibility, in the case of a microscope using electromagnetic fields, was the source of a heuristic argument for the need to quantize the electromagnetic field.[158]

4. The formulation of the Higgs Mechanism uses classical fields under the assumption that a path integral formulation justifies their use. While this may be true, the path integral formulation relies on implicit, unstated boundary conditions that obscure the physics of the quantum field theoretic nature of the mechanism. A direct quantum field theoretic study of the Higgs Mechanism is needed and would further elucidate its character.

5. Scalar fields have a cloud hanging over them that spin ½ fields do not. A spin ½ particle cannot transition to negative energy because there is a filled sea of negative energy particles. No additional particles can fall into the sea

---

[157] See section 4.2.8 for an explanation.

[158] Heitler (1954) p. 86 provides a good discussion of the need to quantize the electromagnetic field.

due to the Pauli Exclusion Principle that forbids two fermions with the same 4-momentum and quantum numbers. In the case of scalar particles the Pauli Exclusion Principle does not apply and so a *filled* negative energy sea of scalar particles is not possible and positive energy scalar particles can transition to negative energy without hindrance. This problem has been "resolved" by an appropriate definition of the scalar particle vacuum to exclude transitions to negative energy. But the rationale for the definition is lacking. Dirac was asked about this issue many years ago. He said he had a solution to the problem. However he did not present it – in keeping with his well-known taciturn nature. So the issue remains an open question.

For the above reasons we will show that a more satisfactory method of achieving the goals of mass generation and symmetry breaking exists.[159] This method relies on a larger Fock space that enables the appearance of a vacuum expectation value for Higgs particles to be understood within a truly quantum framework. More importantly, this method is a consequence the PseudoQuantization procedure described above that enables unitarily equivalent quantizations in different coordinate systems. So a profound fundamental justification for our Higgs boson formulation exists. One major consequence of this approach is the appearance of a local Arrow of Time – a concept that has been a subject of interest for over one hundred years. Another consequence is a rationale for ElectroWeak Higgs bosons and for their absence for the strong (gluon) interaction.

D.3.2 PseudoQuantization of Scalar Particles

We now consider the PseudoQuantization[160] of a scalar particle field that will become a Higgs particle with a non-zero vacuum expectation value.[161] We begin by defining two fields that correspond to the scalar particle: $\varphi_1(x)$ and $\varphi_2(x)$.[162] These fields will be assumed to have the equal time commutators

$$
\begin{aligned}
[\varphi_i(x), \pi_j(y)] &= i(1 - \delta_{ij})\delta^3(\mathbf{x} - \mathbf{y}) \qquad \text{(D.8)} \\
[\varphi_i(x), \varphi_j(y)] &= 0 \\
[\pi_i(x), \pi_j(y)] &= 0
\end{aligned}
$$

[159] In the Extended Standard Model of Blaha (2015a) we have shown that the basic particles have a mass, the Landauer mass, so that the theory is symmetry violating from the very start. We have also shown that our Two-Tier formalism for quantum field theories always yields finite results in perturbation theory calculations – making the renormalization approach of t'Hooft and others, which relied on initially massless gauge fields, unnecessary.

[160] PseudoQuantization in a D-dimensional space-time is described in Blaha (2017c). This discussion is relevant to PseudoQuantization in the Megaverse, or in other universes.

[161] Much of this section appears in Blaha (2016c), and earlier books, as well as in S. Blaha, Phys. Rev. **D17**, 994 (1978). The case of fermion PseudoQuantization is also discussed in Appendix A – S. Blaha, Il Nuovo Cimento **49A**, 35 (1979).

[162] The subscripts on the fields are not gauge symmetry indices but simply identifiers distinguishing the fields from each other.

where δ_{ij} is the Kronecker δ and where $\pi_i(x)$ is the canonically conjugate momentum to $\varphi_i(x)$. The fields $\varphi_1(x)$ and $\pi_1(y)$ will be observable classical fields. The fields $\varphi_2(x)$ and $\pi_2(y)$ will not be observables so that $\varphi_1(x)$ and $\pi_1(y)$ can both be sharp on the set of physical states.

We now specify the Lagrangian density for a scalar Klein-Gordon particle:

$$\mathcal{L} = \partial\varphi_1/\partial x_\mu \partial\varphi_2/\partial x^\mu \tag{D.9}$$

with Hamiltonian density

$$\mathcal{H} = \pi_1\ \pi_2\ + \partial\varphi_1/\partial x_i \partial\varphi_2/\partial x^i$$

where i labels spatial coordinates, and $\pi_1 = \partial\varphi_2/\partial t$ and $\pi_2 = \partial\varphi_1/\partial t$. The Lagrangian $\mathcal{L}$ is without a potential or mass term.

The Lagrangian and Hamiltonian for a massive scalar particle in this formalism are

$$\mathcal{L} = \partial\varphi_1/\partial x_\mu \partial\varphi_2/\partial x^\mu - m^2\ \varphi_1\varphi_2 \tag{D.10}$$

with Hamiltonian density

$$\mathcal{H} = \pi_1\ \pi_2\ + \partial\varphi_1/\partial x_i \partial\varphi_2/\partial x^i + m^2\ \varphi_1\varphi_2$$

The fields can be Fourier expanded in terms of creation and annihilation operators:

$$\varphi_i(\mathbf{x}, t) = \int d^3k\ [a_i(k)f_k(x) +\ a_i^\dagger(k)f_k{}^*(x)] \tag{D.11}$$

for i = 1, 2 where

$$f_k(x) = e^{-ik\bullet x}\ /(2\omega_k(2\pi)^3)^{1/2}$$

with $\omega_k = |\mathbf{k}|$.

The creation and annihilation operators satisfy the commutation relations:

$$[a_i(k), a_j^\dagger(k')] = (1 - \delta_{ij})\delta^3(\mathbf{k} - \mathbf{k'})$$
$$[a_i(k), a_j(k')] = 0$$
$$[a_i^\dagger(k), a_j^\dagger(k')] = 0$$

for i, j = 1, 2.

In this formulation the defining properties of a physical state are:

$$\varphi_1(x)|\Phi, \Pi> = \Phi(x)|\Phi, \Pi> \tag{D.12}$$
$$\pi_1(x)|\Phi, \Pi> = \Pi(x)|\Phi, \Pi>$$

where $\Phi(x)$ and $\Pi(x)$ are sharp on the states and thus classical fields with

$$\Phi(\mathbf{x}, t) = \int d^3k\ [\alpha(k)f_k(x) +\ \alpha^*(k)f_k{}^*(x)] \tag{D.13}$$

and correspondingly for $\Pi(x)$.

D.3.3 Vacuum States for Scalar (Higgs) Particles with Non-Zero Vacuum Expectation Values

When we implement the mass mechanism, Φ is constant. We can define a set of states

$$a_1(k)|\alpha> = \alpha(k)|\alpha>$$
$$a_1^\dagger(k)|\alpha> = \alpha^*(k)|\alpha>$$

and correspondingly a set of coherent states

$$|\alpha> = C\exp\{\int d^3k\, [\alpha(k)a_2^\dagger(k) + \alpha^*(k)a_2(k)]\}|0> \quad \text{(D.14)}$$

where C is a normalization constant and where the vacuum state |0> satisfies

$$a_1(k)|0> = a_1^\dagger(k)|0> = 0 \quad \text{(D.15)}$$

$$a_2(k)|0> \neq 0 \qquad a_2^\dagger(k)|0> \neq 0$$

The dual vacuum state satisfies

$$<0|a_2(k) = <0|a_2^\dagger(k) = 0$$

$$<0|a_1(k) \neq 0 \qquad <0|a_1^\dagger(k) \neq 0$$

With this coherent state formalism, which gives purely classical fields and yet also has quantum fields through the use of φ_2 and its creation and annihilation operators, we now have the machinery to define a mass mechanism without the introduction of a potential whose origin can only be described as dubious.

For we can define a coherent state for some k as

$$|\Phi, \Pi> = C\exp\{[(2\pi)^3\omega_k/2]^{½}\Phi[a_2^\dagger(k) + a_2(k)]\}|0> \quad \text{(D.16)}$$

where C is a normalization constant, that yields a non-zero vacuum expectation value:

$$\varphi_1(x)|\Phi, \Pi> = \Phi|\, \Phi, \Pi> \quad \text{(D.17)}$$

where Φ is a constant. Evaluating a fermion interaction term we find a mass term emerges[163]

$$\psi\, (\varphi_1 + \varphi_2)\psi \;\rightarrow \bar{\psi}(\Phi + \varphi_2)\psi \quad \text{(D.18)}$$

It generates a mass for an interaction with a gauge field of the form

[163] When matrix elements with a "vacuum state" are taken.

$$A^{\mu}(\varphi_1 + \varphi_2)^2 A_{\mu} \rightarrow A^{\mu}(\Phi + \varphi_2)^2 A_{\mu} \quad \text{(D.19)}$$

It also yields a quantum field theoretic interaction that would result in the production of ElectroWeak particles from these scalar fields. The production of Higgs particles that decay into ElectroWeak gauge particles has recently been found at CERN.

The present formalism provides a clean way to separate the vacuum expectation value of a scalar particle from its quantum field part in contrast to the Higgs Mechanism where one has to separate a Higgs field into parts manually.

D.3.4 Interpretation of Negative Energy Scalar Particle States

As we noted earlier, scalar particle physics has the problem of no barrier to the decay of positive energy states to negative energy states due to the absence of a Pauli Exclusion Principle for bosons. The PseudoQuantization procedure that we developed in 1978 and describe here allows negative energy states as one would physically expect and raises the possibility of disastrous particle decays to negative energy. The above equations show that negative energy states are possible in this theory.

However they also show that combined positive and negative energy boson states can be interpreted as classical field states. In addition, the ability of any number of boson particles to have the same 4-momentum and quantum numbers shows that a *macroscopic* classical scalar field state can be constructed.

Thus we can view states containing negative energy particles as classical field states and thus solve[164] *the issue of interpreting negative energy particle states – a more satisfactory approach than the standard quantization procedure does – with due respect to Professor Dirac.*

We note that macroscopic many particle fermion states can only have one particle in any mode unlike bosons. Therefore we cannot use this formalism to create macroscopic classical fermion field states.[165] And the filled Dirac Sea of negative energy fermions precludes the transition of a positive energy Dirac fermion to a negative energy state. *Thus there is a certain complementarity between fermions that cannot become classical fields but have a filled sea precluding decays to negative energy states, and bosons that can become classical fields but support decays to negative energy states.*

D.3.5 Contrast with Conventional Second Quantization of Scalar Particles

The PseudoQuantization procedure followed here uses different boundary conditions than the usual scalar particle quantization procedure. The essence of the difference is embodied in a comparison of the definition of the vacuum above and the definition of the conventional second quantized field vacuum:

$$a|0\rangle = 0 \qquad \text{Conventional Approach}$$

---

[164] Also a boson that has no interactions cannot transition from to a positive energy state to a negative energy state due to conservation of energy.

[165] However we can create PseudoQuantum fermion states. See S. Blaha, Phys. Rev. **D17**, 994 (1978) (reproduced in Appendix I) and references therein to earlier papers by the author.

$$a^{\dagger}|0> \neq 0$$

In the conventional approach the creation of negative energy boson states is eliminated *ab initio* whereas in our approach it is allowed in order to support classical field states with non-zero vacuum expectation values that are a form of classical field. While one cannot discredit the conventional choice for conventional scalar fields, one can see that our approach yields a physically more important result – particularly for Higgs fields – because it leads to an Arrow of Time *locally* – an important feature of physical phenomena that has been a subject of much discussion and dispute. One can say that the conventional approach sweeps the issue "under the rug" rather than seeking a deeper justification – differing from Dirac's implied notion that the issue merited attention. We will discuss the "Arrow of Time" within the framework of our PseudoQuantization approach later.

D.3.6 Why Inertial Reference Frames are Special

The great physicists of the early 20th century raised numerous questions about Special Relativity after Einstein and Poincarè's discovery. Prominent among them was the question of why inertial reference frames are of especial importance in Special Relativity, and afterwards in General Relativity.

It appears that our formulation of the mass generation mechanism sheds significant light on the reason for the special prominence of inertial frames. Earlier we considered the case of a massless PseudoQuantized scalar. We now consider massive scalars since experiments at CERN have apparently discovered a Higgs particle with a 125 GeV/c mass. The above equations describe a massive scalar particle. If the scalar is massive, then the "vacuum" state that yields a non-zero expectation value must change to

$$|\Phi, \Pi> = C\exp\{(2\pi)^3 m/2]^{\frac{1}{2}}[a_2^{\dagger}(\mathbf{0},m) + a_2(\mathbf{0},m)]\}|0> \qquad \text{(D.20)}$$

to have operators for a particle of mass m in its rest frame. Then, having established this preferred frame for a Higgs particle, in The Unified SuperStandard Theory, and requiring that invariant intervals

$$ds^2 = dt^2 - d\mathbf{x}^2 \quad \text{(in rectangular coordinates)}$$

are unchanged by a (complex or real) Lorentz transformation, we find that inertial reference frames are singled out as "special" in the sense that they are the only accessible reference frames that can be generated by a Lorentz boost/transformation from the Higgs particle rest frame. *The Higgs particle vacuum state singles out the class of inertial reference frames.*

Thus Higgs particles play a central role in establishing the basis of physical reality.

D.3.7 PseudoQuantization Reveals More Physical Consequences than the Higgs Mechanism of Scalar Particles

Earlier we pointed out that our PseudoQuantization theory of Higgs particles reveals more physical consequences than the conventional approach, which implements the Higgs Mechanism by simply using a potential term that has a minimum at a non-zero vacuum expectation value. This section shows the major results of a properly implemented mechanism. We find a better explanation of the negative energy state problem of boson field theories. We find a local arrow of time that explains the direction of time that we, and all of nature, experiences. We find the reason why inertial reference frames have a special physical significance – a result long sought by physicists.

In addition we will see in chapter 11 that real gauge fields should have an associated Higgs particle, while necessarily complex gauge fields (the Strong interaction gauge field in The Unified SuperStandard Theory) do not have an associated gauge field. These results correspond to experimental reality.

D.3.8 The T Invariance Issues of Our PseudoQuantized Scalar Particle Theory

The PseudoQuantized scalar particle Hamiltonian equations are invariant under time reversal $t \rightarrow t' = -t$. The 'new' vacuum states defined above break the time reversal invariance of the theory resulting in retarded particle propagators.

The Hamiltonian equations

$$[H, \varphi_1(\mathbf{x}, t)] = -i\partial\varphi_1/\partial t \qquad \text{(D.21)}$$
$$[H, \varphi_2(\mathbf{x}, t)] = -i\partial\varphi_2/\partial t$$

are invariant under time reversal. If we define a time reversal operator transformation U then the time reversed equations are

$$[UHU^{-1}, \varphi_1(\mathbf{x}, -t)] = +i\partial\varphi_1(\mathbf{x}, -t)/\partial(-t)$$
$$[UHU^{-1}, \varphi_2(\mathbf{x}, -t)] = +i\partial\varphi_2(\mathbf{x}, -t)/\partial\,(-t)$$

The operator U, which is unitary, transforms H into –H. This operation is legal because the Hamiltonian – in this case – is not positive definite and admits negative energy states.[166] Thus

$$[H, \varphi_1(\mathbf{x}, -t)] = -i\partial\varphi_1(\mathbf{x}, -t)/\partial\,(-t)$$
$$[H, \varphi_2(\mathbf{x}, -t)] = -i\partial\varphi_2(\mathbf{x}, -t)/\partial\,(-t)$$

and the time reversal invariance of the equations of motion is established for this case.

Time reversal invariance is broken by our choice of vacuum states. This choice is necessary to obtain classical field states as we showed earlier. A demonstration of the

166 Unlike the usual case of second quantized Klein-Gordon quantum field theory.

time reversal symmetry breaking is presented later where we show theory has retarded propagators for particle propagation to and from asymptotic states.

Within the interaction region the particle propagators are the sum of retarded and advanced parts that combine to yield principle value propagators – not Feynman propagators. Many years ago Feynman and Wheeler championed principle value propagators for electrodynamics to obtain an action-at-a distance theory of Quantum Electrodynamics. While their theory, and ours, differ from the standard quantum field theory approach there is no reason to view them as faulty, or having serious physical defects. The only question is whether nature chooses conventional quantum field theory or PseudoQuantized quantum field theory. In our case the need for a classical scalar particle non-zero vacuum expectation value strongly motivates our choice of psedoquantized Higgs particles.

D.3.9 Retarded Propagators for Our Quantized Higgs Particles

In the previous section we pointed out that our PseudoQuantization Higgs theory has an arrow of time due to is boundary conditions as expressed by its definition of the vacuum state and it's dual. In this section we will show that the theory uses retarded propagators for propagation to and from the interaction region to asymptotic in-states and out-states. Within an interaction region the theory uses half-retarded – half-advanced propagators. We discuss aspects of the perturbation theory and propagators of our scalar particles in this chapter.

First we note that in-states at $t = -\infty$ are composed of superpositions of $a_2(k)$ and $a_2^\dagger(k)$ creation and annihilation operators:

$$a_2(k)|0> \neq 0 \qquad\qquad a_2^\dagger(k)|0> \neq 0$$

while the out-states composed of superpositions of $a_1(k)$ and $a_1^\dagger(k)$ creation and annihilation operators:

$$<0|a_1(k) \neq 0 \qquad\qquad <0|a_1^\dagger(k) \neq 0$$

Consequently when in-state particles (x_1) propagate into the interaction region (x_2) the relevant propagators are retarded propagators with the form

$$\begin{aligned} G_{in}(x_2, x_1) &= <0|T(\varphi_{1\,in}(x_2), \varphi_{2\,in}(x_1))|0> \\ &= \theta(x_{20} - x_{10})<0|[\varphi_{1\,in}(x_2), \varphi_{2\,in}(x_1)]\,|0> \end{aligned} \tag{D.22}$$

This is a manifestly retarded propagator. The choice of vacuums clearly results in a time asymmetry giving a retarded propagation reflecting the familiar Arrow of Time.

A similar situation prevails for propagation to out-states (x_3) from the interaction (x_2) region:

$$\begin{aligned} G_{out}(x_3, x_2) &= <0|T(\varphi_{1\,out}(x_3), \varphi_{2\,out}(x_2))|0> \\ &= \theta(x_{30} - x_{20})<0|[\varphi_{1\,out}(x_3), \varphi_{2\,out}(x_2)]\,|0> \end{aligned} \tag{D.23}$$

Within the interaction region the Higgs particles have principle value propagators.

Thus we find PseudoQuantized Higgs particles embody a local Arrow of Time. The locality of the Arrow of Time is embodied in all the particles that interact with the Higgs particle. Since the mass of *every* particle – bosons and fermions – has a Higgs contribution, and thus *every* particle interacts with the Higgs particles, the Arrow of Time permeates The Unified SuperStandard Theory as well as the more familiar Standard Model known from experiment.

D.3.10 The Local Arrow of Time

In the *Physics is Logic* series of monographs we saw that complex coordinates led to the form of the fermion spectrum, that the mapping of complex coordinates to real-valued coordinates yielded the Reality group and The Unified SuperStandard Theory gauge interactions, that Complex General Relativity led to Higgs particles that were directly united with elementary particle masses and gave us the equality of inertial mass and gravitational mass. Later we will see the reduction of complex gauge fields to real gauge fields explains the appearance of Higgs fields in The Unified SuperStandard Theory.

The PseudoQuantization procedure leads to retarded Higgs field propagators and thence to a *local* arrow of time. Many arguments have been put forward over the past hundred plus years for the Arrow of Time. Many arguments based on Statistical Mechanics, Entropy, and Boltzmann's statistical atomic theory have suggested the Arrow of Time is a global statistical consequence. This view seems to contradict the results of elementary particle experiments where a *local* Arrow of Time is evident.

Our rationale for the Arrow of Time begins with retarded Higgs fields. Then we note that Higgs field quantum interactions appear for all fermions and gauge particles. Thus all particle interactions are imbued with an Arrow of Time. Particles united to form macroscopic matter inherit their combined Arrows of Time producing the global Arrow of Time we experience.

Thus our PseudoQuantization approach offers a more satisfactory solution of the origin of the Arrow of Time.

It is remarkable that complex quantities – coordinates and fields – through the Higgs phenomena that we have considered, lead to the equality of inertial mass and gravitational mass, and an Arrow of Time. This unity of mass and time phenomena may reflect the deeper fact that we can have no practical Arrow of Time if all particles were massless, for particle dynamics at light speed would then be pointless. This view has been expressed by DeWitt, Unruh, and others who have pointed out that, physically, time is meaningful and measurable only if masses exist; the larger the mass, the more accurate the time measurement in principle.[167]

D.3.11 Space-Time Dependent Particle Masses

It is possible that the ultimate Unified SuperStandard Theory has masses that evolve with time and may also be spatially varying – different values in different parts

---

[167] No mass, no clock; no clock, no physical time. See Blaha (2015a) pp. 368-371 for a discussion including comments by DeWitt and Unruh.

of the universe. Presently there is no decisive evidence for this possibility although astrophysical studies continue. In this section we will describe the mechanism for space-time dependent masses.

Consider a classical field (time and spatially varying):

$$\Phi(\mathbf{x}, t) = \int d^3k\, [\alpha(k)f_k(x) + \alpha^*(k)f_k{}^*(x)] \tag{D.24}$$

If we define the coherent vacuum state

$$|\alpha\rangle = C \exp\{\int d^3k\, [\alpha(k)a_2{}^\dagger(k) + \alpha^*(k)a_2(k)]\}|0\rangle \tag{D.25}$$

then

$$\varphi_1(x)|\Phi, \Pi\rangle = \Phi(x)|\Phi, \Pi\rangle$$
$$\pi_1(x)|\Phi, \Pi\rangle = \Pi(x)|\Phi, \Pi\rangle$$

where

$$\varphi_i(\mathbf{x}, t) = \int d^3k\, [a_i(k)f_k(x) + a_i{}^\dagger(k)f_k{}^*(x)] \tag{D.26}$$

for i = 1, 2 and where

$$f_k(x) = e^{-ik\bullet x}/(2\omega_k(2\pi)^3)^{1/2}$$

with ω_k equal to the energy.

D.3.12 Inertial Mass Equals Gravitational Mass

From the days of Newton through Einstein[168] to the present the equality of gravitational mass and inertial mass has been a topic of interest. Mach, who played an important role, in this ongoing discussion, thought distant masses in the universe were the source of the equality. However the origin of the equality, which has been shown experimentally to very high accuracy, remained uncertain until the *Physics is Logic* series of books, in which we showed the interconnection of the Unified SuperStandard Theory and Complex Gravitation via Higgs generated masses that united gravitational and inertial mass.

In Blaha (2016h) we showed that a Complex General Relativity transformation can be factored into the product of a complex-valued transformation and a real-valued General Coordinate transformation. The set of complex valued transformations form a U(4) group that we called the General Coordinate Reality group. Later we will define the Internal Symmetry Species Group as the corresponding analogue. The Species Group has gauge fields that undergo spontaneous symmetry breaking and generate contributions to all fermion masses.

Since fermion field masses are now sums of ElectroWeak Higgs contributions, Generation group Higgs contributions, Layer group Higgs contributions and Species group contributions, and since the gravitational Higgs fields appear in all fermion

[168] For example, Einstein and Grossman in 1913 stated, "The theory herein described originates in the conviction that the proportionality between the inertial and gravitational mass of a body is an exact law of nature that must be expressed as a foundation principle of theoretical physics."

masses, the equality of inertial and gravitational mass is proven. The gravitational Higgs particles' equations depend, in part, on the gravitational field by Blaha (2016h) and so set the mass scale of gravitational mass, and thereby of all Higgs mass contributions. They set the scale of inertial masses equal to the scale of gravitational masses. ***Since an expression cannot mix mass scales, the gravitational mass scale must be the same as the inertial mass scale. Inertial Mass equals gravitational mass.***

We have established the equality of inertial and gravitational mass at the short distance quantum level. In our view, this explanation is far more satisfying than basing the equality on a combination of large distance phenomena and quantum phenomena. As Einstein and Weyl have pointed out, all fundamental physics phenomena should be based on a local theory. Complex Gravity as we have constructed it, combined with the Unified SuperStandard Theory, furnishes a completely local basic Theory of Everything.

The equation above contains a coherent state $|\alpha>$ for a time and spatially varying mass. The above equations can be generalized to the case of multiple space-time varying masses.[169]

$$|\Phi_1,\Phi_2, \ldots ,\Phi_n;\Pi_1,\Pi_2, \ldots ,\Pi_n> = C \prod_{i=1}^{n} \exp\left\{\int d^3k \, [\alpha_i(k)a_{2i}^\dagger(k) + \alpha_i^*(k)a_{2i}(k)]\right\}|0> \quad (D.27)$$

Then all n mass vacuum expectation values are space-time dependent:

$$\varphi_{1i}(x) \,|\, \Phi_1, \Phi_2, \ldots , \Phi_n; \Pi_1, \Pi_2, \ldots , \Pi_n> = \Phi_i(x) \,|\, \Phi_1, \Phi_2, \ldots , \Phi_n; \Pi_1, \Pi_2, \ldots , \Pi_n> \quad (D.28)$$

Thus our formalism can accommodate space-time varying masses should they be found in the Cosmos.

D.3.13 Benefits of the PseudoQuantization Method

In this book and earlier work we showed that a more physically satisfactory method for avoiding the negative energy state problem exists. This method relies on the use of a larger Fock space in which negative energy states (or partially negative energy states) are interpreted as states containing classical fields or a mix of classical fields and individual boson particles. This approach resolves the negative energy boson issue and provides a common framework for boson particles and classical boson fields.

One consequence of the PseudoQuantization method is that it enables the appearance of a vacuum expectation value for Higgs particles (a constant classical field) to be understood within a truly quantum framework. Another major consequence of this approach is the appearance of a *local* Arrow of Time due to the Higgs mass generation mechanism – a concept that has been a subject of interest for over one hundred years. A macroscopic arrow of time is often described as a statistical result. But our approach yields an arrow of time at the single particle level.

---

[169] The "vacuum" state |0> also implicitly has factors for the vacuum expectation values used for fields that give masses to fermions and vector bosons as described in Blaha (2016h).

The conventional approach to boson field quantization sweeps these issues "under the rug" rather than seeking a deeper justification. It differs from Dirac's implied notion that the issue merited attention.

Another important consequence of the PseudoQuantization method is that it singles out inertial reference frames when applied to the case of Higgs particles.

Yet another more subtle consequence of boson PseudoQuantization is that it provides a rationale/explanation for the presence of ElectroWeak Higgs bosons, *and for their absence for the strong (gluon) interactions. The question of why there are no strong interaction Higgs bosons has not been previously considered to the best of this author's knowledge.*

Appendix E. Some PseudoQuantum Papers by the Author

These papers are reprinted with the kind permission of Physical Review D and Il Nuovo Cimento..

S. Blaha, “The Local Definition of Asymptotic Particle States”, IL Nuovo Cimento **49A**, 35 (1979).

S. Blaha, “New Framework for Gauge Field Theories”, IL Nuovo Cimento **49A**, 113 (1979).

S. Blaha, Phys. Rev. **D17**, 994 (1978).

IL NUOVO CIMENTO VOL. 49 A, N. 1 1 Gennaio 1979

The Local Definition of Asymptotic Particle States (*).

S. BLAHA

Physics Department, Williams College - Williamstown, Ma. 01267

(ricevuto il 28 Luglio 1978)

Summary. — A generalization of quantum field theory is described which has a unique particle interpretation even in space-times where no global timelike co-ordinate exists. The formulation is described in detail for the case of scalar bosons and spin-one-half fermions in flat space-time. We show that it is possible to construct a model in our approach which is physically equivalent to any given model in the usual formulation. In addition, a new class of models can be constructed which are not possible in the usual formulation. This class includes quantum action-at-a-distance models which can be used to develop models with higher-derivative field equations which are unitary. Our formulation allows some latitude in the choice of boundary conditions, so that one can opt for a continuum of possible Green's functions ranging from Feynman propagators to principal-value propagators (half advanced-half retarded).

1. – Introduction.

Our experience in flat space-time has fostered the opinion that a given action leads to a unique quantum field theory upon implementation of the canonical quantization procedure. This is apparently not true in general. A given action corresponds to an infinity of physically inequivalent quantum field theories in nonstatic space-times where no timelike Killing vector exists [1,2]. The origin of this plurality of quantum theories can be seen in free field the-

(*) Supported in part by grants from the National Science Foundation, and Research Corporation.

(1) S. A. FULLING: *Phys. Rev. D*, **7**, 2850 (1973); C. SOMMERFIELD: *Ann. Phys.*, **84**, 285 (1974).

(2) B. DEWITT: *Phys. Rep.*, **19**, 295 (1975).

ories (cf. FULLING (1)). The usual quantization procedure is based on a definition of positive frequency which selects an acceptable complete orthonormal set of field equation solutions to use in field quantization. In nonstatic space-times no unique criterion exists for defining positive frequency. As a result, there is no restriction on the choice of complete orthonormal set of field equation solutions used to Fourier-expand fields. Having different choices leads to unitarily (and physically) inequivalent representations of the field algebra. The set of physical particle states in one quantization is generally not unitarily related to the set of physical states in another quantization (1).

The absence of a criterion to select the « correct » quantum-field theory in the usual formulation has led us to consider a generalization of quantum field theory. In this generalization we introduce extra degrees of freedom in such a way that quantizations based on differing definitions of positive frequency are unitarily equivalent. Thus for a given action there is one resulting quantum field theory up to unitary equivalence.

In particular the physical particle states of different quantizations are related by a unitary transformation. Since the particle number operator is invariant under this transformation, a N-particle state in one quantization is a superposition of N-particle states in any other quantization. This is made possible by a local definition of particle states in the Fourier-transformed space (momentum space in the case of flat space-time).

It is important to note that the plethora of inequivalent quantizations in the usual formulation is faced by *one* observer. It is not a question of quantizations in different co-ordinate systems corresponding to different observers. The differences in the quantizations of two relatively accelerating observers, for example, are physically real and, in fact, also exist within the framework of our formulation. Relatively accelerating observers will, in general, « see » different numbers of particles.

Sections **2** and **3** contain our formulation of a free-scalar-boson field theory and a free spin-one-half fermion field theory in flat space-time. Significant differences exist between our formulation and the usual formulation. However, models exist in our formulation which make predictions which are identical to those of conventional field theory models, *e.g.*, quantum electrodynamics. Models also exist in our formulation which are completely outside the framework of the usual formulation. For example, a choice of boundary conditions is possible in our formulation which allows for virtual particles to propagate via non-Feynman propagators. In general, our particle propagator has the form

$$G = \sin^2\theta\, G_F + \cos^2\theta\, C G_F^* C^{-1}\,, \tag{1}$$

where θ is an arbitrary angle, G_F the usual Feynman propagator with G_F^* its complex conjugate, and where C is the relevant charge conjugation matrix. G is a Feynman propagator if $\theta = \pi/2$.

If $\theta = \pi/4$ then G is a principal-value propagator (half advanced-half retarded). This type of Green's function has appeared in classical action-at-a-distance theories. Our formulation thus encompases quantum action-at-a-distance theories. The use of principal-value propagators allows a substantial enlargement of the class of unitary, renormalizable field theory models. For example, models with higher-derivative field equations cannot simultaneously satisfy the requirements of positive probabilities and unitarity, if Feynman propagators are used. But if principal-value propagators are used, both requirements can be consistently satisfied (3). This has allowed us to previously construct a unitary, higher-derivative non-Abelian model of the strong interactions with a manifest linear potential and quark confinement (4). Of course, the use of principal-value propagators leads to a different type of analytic structure for amplitudes. We take the view that analyticity is an experimental question rather than a fundamental requirement on field theory (5). It is amusing to note that confinement of color in this model serves to sharply dampen if not eliminate the potential nonanalyticity.

We will discuss our formulation of non-Abelian field theories in detail in a subsequent paper (6).

2. – Boson quantization.

In flat space-time a timelike co-ordinate exists and as a result the Hamiltonian occupies a privileged position in defining positive frequency. In a non-static space-time, with no global timelike co-ordinate, no corresponding operator exists and the definition of « positive frequency » appears to be arbitrary. Consider a free scalar-field theory in such a situation. The field equation has an infinite number of possible complete orthonormal sets of solutions which span the space of solutions. Consider two possible sets: $\{\chi_\alpha, \chi_\alpha^*\}$ and $\{\psi_\beta, \psi_\beta^*\}$, where the χ_α are positive frequency with respect to one definition of positive frequency, and ψ_β are positive frequency with respect to a different definition. Then mode expansions of the scalar field

$$\varphi(x) = \sum_\alpha [\chi_\alpha(x) A_\alpha + \chi_\alpha^*(x) A_\alpha^\dagger], \tag{2}$$

$$\varphi(x) = \sum_\beta [\psi_\beta(x) b_\beta + \psi_\beta^*(x) b_\beta^\dagger], \tag{3}$$

(3) S. Blaha: *Phys. Rev. D*, **10**, 4268 (1974).
(4) S. Blaha: *Phys. Rev. D*, **11**, 2921 (1975).
(5) R. E. Cutkowsky, P. V. Landshoff, D. I. Olive and J. C. Polkinghorne: *Nucl. Phys.*, **12** B, 281 (1969); T. D. Lee: in *Quanta-Essays in Theoretical Physics Dedicated to Gregor Wentzel*, edited by P. G. O. Freund, C. J. Goebel and Y. Nambu (Chicago, Ill., 1970); H. Rechenberg and E. C. G. Sudarsham: *Nuovo Cimento*, **14** A, 299 (1973).
(6) S. Blaha: *Nuovo Cimento*, **49** A, 58 (1978).

can be inverted to relate the Fourier coefficient operators

$$A_\alpha = \sum_\beta [C_{\alpha\beta} b_\beta + \bar{C}_{\alpha\beta} b_\beta^\dagger], \tag{4}$$

where $C_{\alpha\beta}$ and $\bar{C}_{\alpha\beta}$ are c-number functions of α and β. Equation (4) shows that A_α is related to b_β and $b_\beta^\dagger$ through a local Bogoliubov transformation. As a result, the quantizations are, in general, not unitarily equivalent, have different vacua, and different particle interpretations [1,2]. The basis of this difficulty is the noncommutativity of Fourier coefficient operators and their Hermitian conjugates

$$[b_\beta, b_{\beta'}^\dagger] = \delta_{\beta\beta'}. \tag{5}$$

We shall propose a generalization of quantum field theory in which (in the free-field case) the Fourier coefficient operators and their Hermitian conjugates commute. In order to maintain the quantum character of the theory a supplementary field and the corresponding Fourier coefficient operators will be introduced. We shall confine our discussion to flat space-time in this section, and in sect. **3** which deals with free spin-one-half fermion quantization. In sect. **4** we discuss the particle interpretation of the formulation in nonstatic space-time.

Let us provisionally introduce the Lagrangian

$$\mathscr{L} = \partial_\mu\varphi_1\partial^\mu\varphi_2 - \tfrac{1}{2}\partial_\mu\varphi_1\partial^\mu\varphi_1 - m^2\varphi_1\varphi_2 + \tfrac{1}{2}m^2\varphi_1^2. \tag{6}$$

Following canonical procedures, we obtain the field equations

$$(\Box + m^2)\varphi_i = 0, \tag{7}$$

for $i = 1, 2$ and the canonical momenta

$$\pi_1 = \dot{\varphi}_2 - \dot{\varphi}_1 \tag{8}$$

and

$$\pi_2 = \dot{\varphi}_1, \tag{9}$$

which are taken to satisfy the canonical equal-time commutation relations

$$[\varphi_i(x), \pi_j(y)] = i\delta_{ij}\delta^3(\boldsymbol{x} - \boldsymbol{y}), \tag{10}$$

$$[\varphi_i(x), \varphi_j(y)] = [\pi_i(x), \pi_j(y)] = 0, \tag{11}$$

for $i, j = 1, 2$. Equations (9) and (10) imply

$$[\varphi_1(x), \dot{\varphi}_1(y)] = 0 \,, \tag{12}$$

$$[\varphi_1(x), \dot{\varphi}_2(y)] = i\delta^3(\boldsymbol{x} - \boldsymbol{y}) \,, \tag{13}$$

$$[\varphi_2(x), \dot{\varphi}_2(y)] = i\delta^3(\boldsymbol{x} - \boldsymbol{y}) \,, \tag{14}$$

at equal times. The most general form for the mode expansion of the fields is

$$\varphi_1(x) = \int \mathrm{d}^3k\, [(C_{11}A_{1k} + C_{12}A_{2k})f_k(x) + (\tilde{C}_{11}A^\dagger_{1k} + \tilde{C}_{12}A^\dagger_{2k})f^*_k(x)] \,, \tag{15}$$

$$\varphi_2(x) = \int \mathrm{d}^3k\, [(C_{21}A_{1k} + C_{22}A_{2k})f_k(x) + (\tilde{C}_{21}A^\dagger_{1k} + \tilde{C}_{22}A^\dagger_{2k})f^*_k(x)] \,, \tag{16}$$

where $(2\pi)^{\frac{3}{2}}(2\omega_k)^{\frac{1}{2}}f_k(x) = \exp[-ik\cdot x]$ and where C_{ij} and $\tilde{C}_{ij}$ are a set of constants. In view of the afore-mentioned difficulties stemming from the noncommutativity of a Fourier-coefficient operator and its Hermitian conjugate, we are led to impose the commutation relations

$$[A_{ik}, A_{jk'}] = [A^\dagger_{ik}, A^\dagger_{jk'}] = 0 \tag{17}$$

and

$$[A_{ik}, A^\dagger_{jk'}] = (1 - \delta_{ij})\,\delta^3(\boldsymbol{k} - \boldsymbol{k}') \,, \tag{18}$$

for $i, j = 1, 2$. We define two vacua (which are in fact related) $|0\rangle_1$ and $|0\rangle_2$ by

$$A_{1k}|0\rangle_2 = A^\dagger_{1k}|0\rangle_2 = 0 \,, \tag{19}$$

$$A_{2k}|0\rangle_1 = A^\dagger_{2k}|0\rangle_1 = 0 \tag{20}$$

with

$$A_{2k}|0\rangle_2 \neq 0 \,, \qquad A^\dagger_{2k}|0\rangle_2 \neq 0 \,, \tag{21}$$

and

$$A_{1k}|0\rangle_1 \neq 0 \,, \qquad A^\dagger_{1k}|0\rangle_1 \neq 0 \,, \tag{22}$$

for all k. These definitions are motivated by the need for vacua which would be invariant under Bogoliubov transformations—a necessary requirement if the difficulties of particle interpretation caused by relations of the form of eq. (4) are to be avoided. Let us define the local Bogoliubov transformation

$$A_{ik}(\lambda_1, \lambda_2) \equiv B_{\lambda_1\lambda_2}A_{ik}B^{-1}_{\lambda_1\lambda_2} = \\ = \exp[i\lambda_1]\cosh\lambda_2 A_{ik} + \exp[-i\lambda_1]\sinh\lambda_2 A^\dagger_{ik} \,, \tag{23}$$

where λ_1 and λ_2 are functions of the momentum k. The operator B has the form

$$B_{\lambda_1\lambda_2} = \exp\left[2i\int d^3k\lambda_1(k)\Gamma_{3k}\right]\exp\left[2i\int d^3k\lambda_2(k)\Gamma_{2k}\right], \tag{24}$$

where

$$\Gamma_{3k} = (A^\dagger_{2k}A_{1k} + A_{2k}A^\dagger_{1k})/2\,, \tag{25}$$

$$\Gamma_{2k} = i(A^\dagger_{2k}A^\dagger_{1k} - A_{2k}A_{1k})/2\,. \tag{26}$$

If we also define

$$\Gamma_{1k} = -(A^\dagger_{2k}A^\dagger_{1k} + A_{2k}A_{1k})/2\,, \tag{27}$$

then these operators satisfy the commutation relations of a $SU_{1,1}$ algebra:

$$[\Gamma_{1k}, \Gamma_{2k'}] = -i\delta_{kk'}\Gamma_{3k}\,, \quad [\Gamma_{2k}, \Gamma_{3k'}] = i\delta_{kk'}\Gamma_{1k}\,, \quad [\Gamma_{3k}, \Gamma_{1k'}] = i\delta_{kk'}\Gamma_{2k}\,. \tag{28}$$

Thus the group of local Bogoliubov transformations is an infinite tensor product of $SU_{1,1}$ groups. It should be noted that $|0\rangle_2$ and $|0\rangle_1$ are invariant under this group. The equations of motion and equal-time commutation relations are also invariant under this group. These properties will enable us to show the uniqueness of the particle interpretation of our formulation in nonstatic space-time in sect. **4**. The Casimir operator for the k-th $SU_{1,1}$ algebra,

$$\Gamma^2_k = \Gamma^2_{3k} - \Gamma^2_{1k} - \Gamma^2_{2k}\,, \tag{29}$$

$$\Gamma^2_k = N_k(N_k + 2)\,, \tag{30}$$

allows us to identify the particle number operator (cf. sect. **4** below for its derivation)

$$N = \int d^3kN_k = \int d^3k[A^\dagger_{2k}A_{1k} - A_{2k}A^\dagger_{1k}]\,, \tag{31}$$

which is left invariant by the Bogoliubov transformations.

If we compare our formulation to the usual one at this stage, we see that the enlargement of the field algebra has allowed us to define a group of local Bogoliubov transformations which is unitary and leaves the vacuum invariant —two properties not possible in the usual approach.

We now define inner products in our formalism. The structure of the commutation relations eq. (17) and (18) together with the nature of the defined vacua suggest that kets can be taken to have the form

$$|\alpha\rangle = A^\dagger_{2k_1}A^\dagger_{2k_2}\ldots A_{2q_1}A_{2q_2}\ldots|0\rangle_2 \tag{32}$$

and that bras should have the form

$$\langle\alpha| = {}_1\langle 0|A_{1k_1}A_{1k_2}\ldots A^\dagger_{1q_1}A^\dagger_{1q_2}\ldots\,. \tag{33}$$

(We could have chosen to construct kets using $|0\rangle_1$ and bras using ${}_2\langle 0|$ with no change in consequences.) The form of the commutation relations (which are used to reduce the inner products to a multiple of ${}_1\langle 0|0\rangle_2$) imply that the dual of the ket space is not its Hermitian conjugate. In our case the algebra reduces inner products to ${}_1\langle 0|0\rangle_2$ which we define to be unity.

We can relate the dual of a ket to its Hermitian conjugate through the introduction of a Dirac metric operator (8). We define the operator, γ, by

$$\gamma^{-1} A_{1k} \gamma = A_{2k} , \qquad \gamma^{-1} A_{2k} \gamma = A_{1k} , \tag{34}$$

$$\gamma |0\rangle_1 = |0\rangle_2 . \tag{35}$$

We find

$$\gamma = \exp\left[-\frac{i\pi}{2}\int \mathrm{d}^3k \, [A^\dagger_{2k} A_{2k} + A^\dagger_{1k} A_{1k} - A^\dagger_{2k} A_{1k} - A_{2k} A^\dagger_{1k}]\right] , \tag{36}$$

which implies that γ satisfies the necessary conditions for a metric operator, $\gamma = \gamma^\dagger = \gamma^{-1}$. The norm of a state $|\alpha\rangle$ can thus be defined by

$$(|\alpha\rangle)^\dagger \gamma |\alpha\rangle \tag{37a}$$

and inner products will generally have the form

$$(|\beta\rangle)^\dagger \gamma |\alpha\rangle . \tag{37b}$$

The adjoint operator is defined by

$$A^* = \gamma^{-1} A^\dagger \gamma . \tag{38}$$

Physical observables must be self-adjoint, $A^* = A$. Self-adjoint operators play the same role as Hermitian operators do in the usual formulation. In particular the Hamiltonian must be self-adjoint, if we are to have conservation of norm. Because φ_2 satisfies a Jordan-Pauli commutation relation, we shall introduce interactions in our model using only $\varphi_2(x)$. As a result φ_2 must also be self-adjoint if the Hamiltonian is to be self-adjoint.

---

(7) Earlier two field formalisms have been considered by G. MIE: *Ann. Phys. Lpz.*, **37**, 511 (1912); P. A. M. DIRAC: *Comm. Dublin Inst. Advanced Studies*, **180** A, 1 (1942); W. PAULI: *Rev. Mod. Phys.*, **15**, 175 (1943); M. FROISSART: *Suppl. Nuovo Cimento*, **14**, 197 (1959); T. D. LEE and G. C. WICK: *Phys. Rev. D*, **2**, 1033 (1970). Our motivation and formulation differ substantially from them. Ref. (3,4) above do describe models which can be directly incorporated within the framework of our formulation.

(8) W. PAULI: *Rev. Mod. Phys.*, **15**, 175 (1943).

The energy-momentum tensor is defined by

$$T^{\mu\nu} = -g^{\mu\nu}\mathscr{L} + \frac{\delta\mathscr{L}}{\delta\partial_\mu\varphi_1}\partial^\nu\varphi_1 + \frac{\delta\mathscr{L}}{\delta\partial_\mu\varphi_2}\partial^\mu\varphi_2 \tag{39}$$

with the Hamiltonian given by the 0-0 component. It is easy to verify that the requirements of Poincaré invariance, and the Schwinger commutation relations for $T^{\mu\nu}$ are met.

The equal-time commutation relations, and the self-adjointness of H and φ_2 place six constraints on the constants C_{ij} and $\tilde{C}_{ij}$ in eqs. (15) and (16). After some algebra we find that we are able to express the field operators in the form

$$\varphi_1(x) = \int \mathrm{d}^3k \left[\left(\frac{\cos(\theta_1-\theta_2)}{\sin\theta_1}A_{1k} + \frac{\sin(\theta_1-\theta_2)}{\sin\theta_1}A_{2k}\right)f_k(x) + \right.$$
$$\left. + \left(\frac{\cos(\theta_1-\theta_2)}{\cos\theta_1}A^\dagger_{1k} - \frac{\sin(\theta_1-\theta_2)}{\cos\theta_1}A^\dagger_{2k}\right)f^*_k(x)\right], \tag{40}$$

$$\varphi_2(x) = \int \mathrm{d}^3k\,[(\cos\theta_2 A_{2k} + \sin\theta_2 A_{1k})f_k(x) + (\sin\theta_2 A^\dagger_{2k} - \cos\theta_2 A^\dagger_{1k})f^*_k(x)], \tag{41}$$

where θ_1 and θ_2 are arbitrary constants which fix the boundary conditions of the Green's functions. (They are *not* related to the Bogoliubov transformations defined above.) We also find

$$H = \int \mathrm{d}^3k\,\omega_k(A^\dagger_{2k}A_{1k} + A_{2k}A^\dagger_{1k}) = 2\int \mathrm{d}^3k\,\omega_k\Gamma_{3k} \tag{42}$$

in the free-field case independent of θ_1 and θ_2.

The theory is not invariant under Bogoliubov transformations due to their noncommutativity with H. This is consonant with the absence of any evidence in nature for such an invariance (and related constants of motion). The point of our formulation is to ensure that representations of the field algebra and dynamics, which are related to each other by Bogoliubov transformations, are unitarily equivalent. In the case of flat space-time the unitary equivalence is a moot point, since a unique generator of the dynamics, the Hamiltonian, is apparent. In nonstatic space-times, where no unique generator of the dynamical motion is determined, the unitary equivalence is necessary in order to have an unambiguous quantum field theory (given the action).

Different choices for the generator of the dynamics lead to representations which can be related by Bogoliubov transformations. These representations are unitarily equivalent in our formulation, but not equivalent in the usual formulation. We return to this issue in sect. **4**.

The role of θ_1 and θ_2 is evident in the Green's functions. As usual we define the Green's functions as the vacuum expectation values of the time-ordered product of the field operators:

$$iG_{ij}(x-y) = {}_1\langle 0|T(\varphi_i(x)\varphi_j(y))|0\rangle_2 . \tag{43}$$

Equation (41) implies

$$G_{22}(x-y) = \sin^2\theta_2 G_F(x-y) + \cos^2\theta_2 G_F^*(x-y) , \tag{44}$$

where $G_F(x-y)$ is the usual Feynman propagator. G_{12} and G_{11} also depend on θ_1 and θ_2, but their precise expressions will not be of use in our presentation.

We shall now show that a model exists within our formulation which is physically equivalent to any conventional scalar quantum field theory with interaction $\mathscr{L}_I(\varphi)$. Our model Lagrangian is given by eq. (6) plus the interaction Lagrangian $\mathscr{L}_I(\varphi_2)$, where $\mathscr{L}_I(\varphi_2)$ is the same function of φ_2 as $\mathscr{L}_I(\varphi)$ is of φ. In order to have Feynman propagators, it is necessary to choose the boundary condition $\theta_2 = \pi/2$. We shall demonstrate that an asymptotic state exists in our formulation which corresponds to any asymptotic state of the usual formulation, and then show that S-matrix elements between corresponding states in the two models are equal in any order of perturbation theory.

The construction of asymptotic fields and states in our model is based on the renormalized quadratic part of the Lagrangian (eq. (6)). Therefore the previous development of this section can be used if appropriate subscripts « in » or « out » are appended to the operators. In particular, since $\theta_2 = \pi/2$, we have

$$\varphi_{2\text{in}}(x) = \int d^3k\,[f_k(x)A_{1k\text{in}} + f_k^*(x)A^\dagger_{2k\text{in}}] \tag{45a}$$

by eq. (41). We shall express the in-field operator of the usual formulation by

$$\varphi_{\text{in}}(x) = \int d^3k\,[f_k(x)A_{k\text{in}} + f_k^*(x)A^\dagger_{k\text{in}}] . \tag{45b}$$

Note that the form of $\varphi_{2\text{in}}$ and φ_{in} is identical except for the subscripts « 1 » and « 2 » on the operators. In addition, the commutation relations of the field operators and the Fourier-coefficient operators are also identical except for numerical subscripts. Furthermore, the application of the field operators to the vacua is also identical in effect (except for subscripts)

$$\begin{cases} \varphi_{\text{in}}(x)|0\rangle = \varphi^{(-)}_{\text{in}}(x)|0\rangle , & \varphi_{2\text{in}}(x)|0\rangle_2 = \varphi^{(-)}_{2\text{in}}(x)|0\rangle_2 , \\ \langle 0|\varphi_{\text{in}}(x) = \langle 0|\varphi^{(+)}_{\text{in}}(x) , & {}_1\langle 0|\varphi_{2\text{in}}(x) = {}_1\langle 0|\varphi^{(+)}_{2\text{in}}(x) , \end{cases} \tag{45c}$$

where the superscript « + » labels positive-frequency parts of the field operator and « − » labels negative-frequency parts. This close parallel in properties between our model and the model of the usual formulation implies the identity

$$\langle 0|\mathscr{P}(\varphi_{\text{in}})|0\rangle = {}_1\langle 0|\mathscr{P}(\varphi_{2\text{in}})|0\rangle_2 , \tag{45d}$$

where $\mathscr{P}(\varphi_{\text{in}})$ is any polynomial in the field φ_{in}. Later we shall use this identity to demonstrate the equality of the S-matrices in our model and the given model of the usual formulation. (Note that a straightforward application of eq. (45*d*) implies that the propagator G_{22} in our formulation equals the time-ordered propagator of the usual formulation.)

We now state the rule associating asymptotic particle states in our formulation with those of the usual formulation: given an in or out ket of the usual formulation, the corresponding ket in our formulation is obtained by appending the subscript « 2 » to every Fourier-coefficient operator (and to the vacuum) (*e.g.* $A^{\dagger}_{k\text{in}}|0\rangle \Leftrightarrow A^{\dagger}_{2k\text{in}}|0\rangle_2$). Given an in or out bra of the usual formulation, the corresponding bra in our formulation is obtained by appending « 1 » to each Fourier-coefficient operator (and to the vacuum) (*e.g.* $\langle 0|A_{k\text{in}} \Leftrightarrow$ $\Leftrightarrow {}_1\langle 0|A_{1k\text{in}}$). It is easily seen that energy-momentum eigenstates in the usual formulation correspond to energy-momentum eigenstates in our formulation. Thus we have identified the set of physical states in our model and find a detailed correspondence to those of the usual formulation.

The development of the perturbation theory of our model is completely analogous to the usual development. The S-matrix relates in and out fields: $\varphi_{2\text{in}}(x) = S\varphi_{2\text{out}}(x)S^{-1}$. LSZ reduction formulae are derived in the same manner as in the usual formulation. We find the reduction formula for a particle from an in-state and from an out-state to be, respectively,

$$(46)\qquad \begin{cases} \langle \beta\,\text{out}|\alpha p\,\text{in}\rangle = \\ \qquad = \langle \beta - p\,\text{out}|\alpha\,\text{in}\rangle + \dfrac{i}{\sqrt{Z}}\displaystyle\int d^4x\, f_p(x)(\overrightarrow{\Box + m^2})\langle \beta\,\text{out}|\varphi_2(x)|\alpha\,\text{in}\rangle\,, \\ \langle \beta k\,\text{out}|\alpha\,\text{in}\rangle = \\ \qquad = \langle \beta\,\text{out}|\alpha - k\,\text{in}\rangle + \dfrac{i}{\sqrt{Z}}\displaystyle\int d^4x\, f_k^*(x)(\overrightarrow{\Box + m^2})\langle \beta\,\text{out}|\varphi_2(x)|\alpha\,\text{in}\rangle\,, \end{cases}$$

where $\varphi_2(x)$ is the interacting field and where we use the notation of ref. (9). The reduction of several particles leads to expressions which are identical to corresponding expressions of the usual model if the subscript « 2 » is appended to $\varphi(x)$.

Just as in the conventional model, we can formally develop a perturbation theory based on the U-matrix. The U-matrix relates the interacting and asymptotic field operator

$$(47a)\qquad \varphi_2(\boldsymbol{x}, t) = U^{-1}(t)\varphi_{2\text{in}}(\boldsymbol{x}, t)U(t)$$

---

(9) We follow the conventions and notation of J. D. BJORKEN and S. D. DRELL: *Relativistic Quantum Fields* (New York, N. Y., 1965).

and is easily shown to satisfy the differential equation

$$(47b)\qquad i\frac{\partial U}{\partial t}=-\left[\int \mathrm{d}^3x\,\mathscr{L}_1(\varphi_{2\mathrm{in}})\right]U\,.$$

Defining $U(t,t')=U(t)U^{-1}(t')$ and solving eq. (47b) gives

$$(48)\qquad U(t,t')=T\exp\left[i\int_{t'}^{t}\mathrm{d}^4x\,\mathscr{L}_1(\varphi_{2\mathrm{in}})\right].$$

The LSZ procedure defined above reduces the calculation of S-matrix elements to the evaluation of time-ordered products of the interacting fields, ${}_1\langle 0|T(\varphi_2(x_1)\varphi_2(x_2)\dots\varphi_2(x_N))|0\rangle_2$. The U-matrix can then be used to reduce this quantity to the ratio of matrix elements involving only in-fields:

$$(49)\qquad \frac{{}_1\langle 0|T(\varphi_{2\mathrm{in}}(x_1)\dots\varphi_{2\mathrm{in}}(x_N)\exp[i\int\mathrm{d}^4x\,\mathscr{L}_1(\varphi_{2\mathrm{in}})])|0\rangle_2}{{}_1\langle 0|T(\exp[i\int\mathrm{d}^4x\,\mathscr{L}_1(\varphi_{2\mathrm{in}})])|0\rangle_2}\,.$$

Expanding to any order in the interaction in eq. (49) gives matrix elements of polynomials in $\varphi_{2\mathrm{in}}$ which are equal—term by term—to corresponding matrix elements of the perturbation theory of the model of the conventional formulation by eq. (45d). Thus S-matrix elements between corresponding states are identically equal in the conventional model and our corresponding model.

It should be noted that only a subset of the possible asymptotic states in our model are identified as physical particle states which correspond to states in the usual model. The operator $A_{2k\mathrm{in}}$ can also be used to create in-kets (and $A^\dagger_{1k\mathrm{out}}$ to create out bras), but the S-matrix elements between physical kets and any ket (or bra) in which these operators appear is zero. (This follows from the fact that $[\mathscr{L}_1(\varphi_{2\mathrm{in}}),A_{2k\mathrm{in}}]=[\mathscr{L}_1(\varphi_{2\mathrm{in}}),A^\dagger_{1k\mathrm{in}}]=0$ and ${}_1\langle 0|A_{2k\mathrm{in}}=0=A^\dagger_{1k\mathrm{in}}|0\rangle_2$.) Thus the S-matrix is block diagonal in our model. The part of it corresponding to the physical state sector is identical to the S-matrix of the given model of the conventional formulation.

The expression for the vacuum expectation value from which S-matrix elements may be calculated, eq. (49), can be used to show the unitary equivalence of representations which are related by a Bogoliubov transformation. Suppose that we had not used the representation of eq. (45a), but instead the Bogoliubov-transformed representation

$$(50)\qquad \varphi^{\mathrm{B}}_{2\mathrm{in}}(x)=\int\mathrm{d}^3k\,[f_k(x)(A_{1k\mathrm{in}}\cosh\lambda+A^\dagger_{1k\mathrm{in}}\sinh\lambda)+ + f^*_k(x)(A^\dagger_{2k\mathrm{in}}\cosh\lambda+A_{2k\mathrm{in}}\sinh\lambda)]=B_{0\lambda}\varphi_{2\mathrm{in}}(x)B^{-1}_{0\lambda}\,.$$

The canonical nature of the transformation guarantees that the canonical commutation relations will be maintained. If we follow the development of the perturbation theory given by eqs. (45)-(49) with φ^{B}_{2in} replacing φ_{2in} and $\varphi^{B}_{2} = U^{-1}(t)\varphi^{B}_{2in}U(t)$ replacing φ_2, then we find that S-matrix elements are calculated from vacuum expectation values involving only φ^{B}_{2in} fields:

$$(51)\qquad \frac{{}_1\langle 0|T(\varphi^{B}_{2in}(x_1)\ldots\varphi^{B}_{2in}(x_N)\exp[i\int d^4x\,\mathscr{L}_1(\varphi^{B}_{2in})])|0\rangle_2}{{}_1\langle 0|T(\exp[i\int d^4x\,\mathscr{L}_1(\varphi^{B}_{2in})])|0\rangle_2}\,.$$

Since $B^{-1}_{0\lambda}|0\rangle_2 = |0\rangle_2$ and ${}_1\langle 0|B_{0\lambda} = {}_1\langle 0|$ we find that eq. (51) is equal to eq. (49). Thus the unitary equivalence of representations of the quantum field theory differing by a Bogoliubov transformation is demonstrated. (One can formally define Bogoliubov transformations for interacting fields $B^{int} = UBU^{-1}$, but B^{int} is not unitary due to the well-known difficulties of the U-matrix in the conventional formulation which are also present in our formulation. We circumvent this problem by working with the definition of the S-matrix in terms of vacuum expectation values of asymptotic in-fields, where the unitary equivalence under Bogoliubov transformation can be unambiguously shown to hold.)

A comparison of our formulation with the usual formulation shows a certain similarity of form at the Lagrangian level if our Lagrangian is put in the form

$$(52)\qquad \mathscr{L} = -\frac{1}{2}\partial_\mu(\varphi_1-\varphi_2)\partial^\mu(\varphi_1-\varphi_2) + \partial_\mu\varphi_2\partial^\mu\varphi_2 + \\ + \mathscr{L}_1(\varphi_2) + \frac{m^2}{2}(\varphi_1-\varphi_2)^2 - \frac{m^2}{2}\varphi_2^2\,.$$

In the usual approach $\varphi_3 = \varphi_1 - \varphi_2$ is an ignorable field and it would not have been surprising that we found equal S-matrix elements above. However, our formulation differs from the usual formulation in two respects—first, the field operators are both expanded in type «1» and «2» Fourier coefficient operators and, more importantly, the vacuum is defined in a way which correlates the φ_3 and φ_2 sectors. In the $\theta_2 = \pi/2$ case the first difference can be eliminated by a relabeling of Fourier-coefficient operators. However, for other values of θ_2 both differences are present and lead to a very different theory from the usual formulation. While it is clear that the correlation between the φ_3 and φ_2 sectors can be implemented in free field theory, one might ask if this remains true in the interacting case. Certainly the correlation can be implemented in the asymptotic fields and states, since that is free field theory. One can also *formally* implement the correlation for interacting fields through eq. (47*a*). But the implementation of the correlation in the interacting case is actually based on the reduction of the S-matrix element to the vacuum expectation value of products of asymptotic fields. Since the correlation can be maintained for the asymptotic fields and states, the physical quantities of the models, S-matrix elements, embody the correlation. (The value of the correlation we introduce is twofold: first, it is necessary in order to obtain the unitary equivalence of Bogoliubov rotated representations, and secondly, it widens the range of allowed flat–space-time quantum field theories to include those with principal-value propagators [6].)

We conclude this section with a brief discussion of our formulation of the charged-scalar-particle case. In the usual approach, the free-charged-scalar-particle Lagrangian may be expressed in terms of complex fields, $\varphi(x)$ and $\varphi^*(x)$ or in terms of two real fields $\varphi_a(x)$ and $\varphi_b(x)$ with

$$\varphi(x) = [\varphi_a(x) + i\varphi_b(x)]/\sqrt{2}\,. \tag{53}$$

If we follow the same procedure as above for the real fields, double their number and use the Lagrangian form of eq. (6), we are led to the complex field expression of the Lagrangian:

$$\mathscr{L} = \partial_\mu\tilde{\varphi}_2\partial^\mu\varphi_1 + \partial_\mu\varphi_2\partial^\mu\tilde{\varphi}_1 - \partial_\mu\tilde{\varphi}_1\partial^\mu\varphi_1 - m_2\tilde{\varphi}_2\varphi_1 - m^2\varphi_2\tilde{\varphi}_1 + m^2\tilde{\varphi}_1\varphi_1\,, \tag{54}$$

where

$$\varphi_i(x) = [\varphi_{ia}(x) + i\varphi_{ib}(x)]/\sqrt{2} \tag{55}$$

and

$$\tilde{\varphi}_i(x) = [\varphi_{ia}(x) - i\varphi_{ib}(x)]/\sqrt{2}\,. \tag{56}$$

We require φ_{ia} and φ_{ib} to embody the same boundary conditions, so that the expansion of φ_{ia} and φ_{ib} utilizes the same constants, c_{ij} and $\tilde{c}_{ij}$. Consequently

$$\varphi_i(x) = \int d^3k[(c_{i1}A_{+1k} + c_{12}A_{+2k})f_k + (\tilde{c}_{i1}A^\dagger_{-1k} + \tilde{c}_{i2}A^\dagger_{-2k})f^*_k]\,, \tag{57}$$

$$\tilde{\varphi}_i(x) = \int d^3k[(c_{i1}A_{-1k} + c_{i2}A_{-2k})f_k + (\tilde{c}_{i1}A^\dagger_{+1k} + \tilde{c}_{i2}A^\dagger_{+2k})f^*_k]\,, \tag{58}$$

where φ_i and $\tilde{\varphi}_i$ are related via the charge conjugation operator [9]. Following the quantization pattern discussed above, with only minor changes due to the presence of two types of Fourier coefficient operators: positive charge, A_{+ik}, and negative charge, A_{-ik}, leads eventually to the following Green's function:

$$G_{22}(x-y) = G_F(x-y)\sin^2\theta_2 + G^*_F(x-y)\cos^2\theta_2\,. \tag{59}$$

Note that it has the same form as eq. (1). (Lagrangian interaction terms are expressed solely in terms of φ_2 and $\tilde{\varphi}_2$.) As a result we require $\gamma\tilde{\varphi}_2\gamma^{-1} = \varphi_2^\dagger$, the Hermitian conjugate of φ_2, where γ is the metric operator so that

$$C\varphi_2C^{-1} = \gamma\varphi_2^\dagger\gamma^{-1}\,, \tag{60}$$

where C is the charge conjugation operator.

3. – Fermion quantization.

In this section we describe our formulation of spin-one-half fermion quantum field theory. Again we are motivated by the need for a unique particle interpretation in nonstatic space-time. The formulation for fermions has close similarities to boson quantization.

Two fields are needed to describe a spin-one-half particle. The Lagrangian is

$$\mathscr{L} = \tilde{\psi}_2\gamma^0(i\nabla - m)\psi_1 + \tilde{\psi}_1\gamma^0(i\nabla - m)\psi_2 - \tilde{\psi}_1\gamma^0(i\nabla - m)\psi_1 . \tag{61}$$

We follow the conventions and notation of ref. (9). The fields $\tilde{\psi}_i$ will be related to the transpose of the charge conjugate field via

$$\tilde{\psi}_i = \psi_i^{cT}\gamma^0 C^T \tag{62}$$

for $i = 1, 2$. (In the usual formulation $\psi^\dagger = \tilde{\psi}$ would hold.) The equations of motion are

$$(i\nabla - m)\psi_i = 0 , \qquad \tilde{\psi}_i(i\overleftarrow{\nabla} - m) = 0 , \tag{63}$$

for $i = 1, 2$. The momentum conjugate to ψ_1 is

$$\pi_1 = i(\tilde{\psi}_2 - \tilde{\psi}_1) \tag{64}$$

and the conjugate to ψ_2 is

$$\pi_2 = i\tilde{\psi}_1 . \tag{65}$$

The canonical equal-time anticommutation relations imply

$$\{\psi_{1\alpha}(x), \tilde{\psi}_{1\beta}(y)\} = 0 , \tag{66}$$

$$\{\psi_{1\alpha}(x), \tilde{\psi}_{2\beta}(y)\} = \delta_{\alpha\beta}\delta^3(\boldsymbol{x} - \boldsymbol{y}) \tag{67}$$

and

$$\{\psi_{2\alpha}(x), \tilde{\psi}_{2\beta}(y)\} = \delta_{\alpha\beta}\delta^3(\boldsymbol{x} - \boldsymbol{y}) . \tag{68}$$

The most general form for the mode of expansion of the fields is (9)

$$\psi_i = \sqrt{2m}\sum_s\int d^3k[(c_{i1}b_{1ks} + c_{i2}b_{2ks})f_k(x)u_{ks} + (\tilde{c}_{i1}d^\dagger_{1ks} + \tilde{c}_{i2}d^\dagger_{2ks})f^*_k(x)v_{ks}] . \tag{69}$$

Just as in the charged scalar case, we develop our formulation in such a way that the even and odd charge conjugation combinations, $\psi_i \pm \psi_i^c$, implement the same boundary conditions. Therefore

$$\tilde{\psi}_i = \sqrt{2m}\sum_s \int d^3k[(c_{i1}d_{1ks} + c_{i2}d_{2ks})f_k(x)v_{ks}^\dagger + (\tilde{c}_{i1}b_{1ks}^\dagger + \tilde{c}_{i2}b_{2ks}^\dagger)f_k^*(x)u_{ks}^\dagger]. \tag{70}$$

The nonzero Fourier-coefficient anti-commutation relations are

$$\{d_{iks}, d_{jk's'}^\dagger\} = \{b_{iks}, b_{jk's'}^\dagger\} = (1-\delta_{ij})\delta_{ss'}\delta^3(\boldsymbol{k}-\boldsymbol{k}') \tag{71}$$

for $i, j = 1, 2$. The definition of states and inner products mirror the boson case. The vacua are defined by

$$b_{1ks}|0\rangle_2 = b_{1ks}^\dagger|0\rangle_2 = d_{1ks}|0\rangle_2 = d_{1ks}^\dagger|0\rangle_2 = 0 \tag{72}$$

and

$$b_{2ks}|0\rangle_1 = b_{2ks}^\dagger|0\rangle_1 = d_{2ks}|0\rangle_1 = d_{2ks}^\dagger|0\rangle_1 = 0 \tag{73}$$

and are related by a metric operator η:

$$\eta|0\rangle_1 = |0\rangle_2 \tag{74}$$

which satisfies $\eta = \eta^\dagger = \eta^{-1}$. We conventionally choose to construct kets from $|0\rangle_2$ and define their dual as their Hermitian conjugate multiplied by the metric operator. Thus inner products have the form

$$\langle\alpha|\beta\rangle = (|\alpha\rangle)^\dagger\eta|\beta\rangle. \tag{75}$$

Physical observables must be self-adjoint, $A = A^* = \eta^{-1}A^\dagger\eta$, in order to have real eigenvalues. The Hamiltonian must be self-adjoint in order to conserve the norm. In view of eq. (68) we only use ψ_2 and $\tilde{\psi}_2$ in interaction terms and therefore require $\tilde{\psi}_2 = \eta^{-1}\psi_2^\dagger\eta$, so that

$$C\psi_2C^{-1} = \eta^{-1}C\bar{\psi}_2^T\eta, \tag{76}$$

which bears comparison with eq. (15.112) of ref. (9) and also eq. (60) above. The equal-time anticommutation relations, eq. (76), and the adjointness of H restrict the constants c_{ij} and $\tilde{c}_{ij}$ so that

$$\psi_1 = \sqrt{2m}\sum_s \int d^3k[(\cos(\theta_1-\theta_2)b_{1ks} + \sin(\theta_1-\theta_2)b_{2ks})f_k(x)u_{ks}/\sin\theta_1 + \\ + (\cos(\theta_1-\theta_2)d_{1ks}^\dagger - \sin(\theta_1-\theta_2)d_{2ks}^\dagger)f_k^*(x)v_{ks}/\cos\theta_1] \tag{77}$$

and

$$\psi_2 = \sqrt{2m}\sum_s \int d^3k[(\sin\theta_2 b_{1ks} + \cos\theta_2 b_{2ks})f_k(x)u_{ks} + \\ + (\cos\theta_2 d_{1ks}^\dagger + \sin\theta_2 d_{2ks}^\dagger)f_k^*(x)v_{ks}]. \tag{78}$$

The Hamiltonian is

$$(79)\qquad H = \sum_s \int d^3k\,\omega_k (b^\dagger_{2ks} b_{1ks} - b_{2ks} b^\dagger_{1ks} + d^\dagger_{2ks} d_{1ks} - d_{2ks} d^\dagger_{1ks})\,.$$

In contrast to the usual formulation, we see that our Hamiltonian does not have an infinite vacuum energy with respect to $|0\rangle_2$. It is not positive definite, but we will be able to develop a unitary S-matrix theory in the space of positive-energy asymptotic states, if we choose $\theta_2 = \pi/2$. This is evident from an examination of the Green's function

$$(80)\qquad S_{22}(x-y) = -i\,{}_1\langle 0|T(\psi_2(x)\,\tilde{\psi}_2(y)\gamma_0)|0\rangle_2 =$$

$$(81)\qquad = \sin^2\theta_2 S_F(x-y) + \cos^2\theta_2\, C\gamma^0 S^*_F (C\gamma^0)^{-1}\,,$$

which gives $S_{22} = S_F$, the usual Feynman propagator, if $\theta_2 = \pi/2$. As in the boson case, we introduce interactions only through type-2 operators, and use type-2 operators to LSZ reduce in and out particles. The result is a perturbation theory which, in the Fermion sector, only involves time-ordered products of ψ_2 and $\tilde{\psi}_2$. Thus we can establish a model in our formulation which is equivalent, so far as S-matrix elements are concerned, to any given model of the usual formulation. In particular, our model electrodynamics has the Lagrangian

$$(82)\qquad \mathscr{L} = \tfrac{1}{2} F^1_{\mu\nu} F^{2\mu\nu} + \tfrac{1}{4} F^1_{\mu\nu} F^{1\mu\nu} + \tilde{\psi}_2 \gamma^0 (i\nabla - m)\psi_1 + \\ + \tilde{\psi}_1 \gamma^0 (i\nabla - m)\psi_2 - \tilde{\psi}_1 \gamma^0 (i\nabla - m)\psi_1 - e_0 \tilde{\psi}_2 \gamma^0 \hat{A}_2 \psi_2$$

with

$$(83)\qquad F^i_{\mu\nu} = \partial_\nu A_{i\mu} - \partial_\mu A_{i\nu}$$

for $i = 1, 2$. While we will discuss this model more fully elsewhere (6) two things are worth noting. First the interaction is expressed solely in terms of fields of type 2—both for fermions and the photon. Following our quantization procedure leads to S-matrix expressions which are term-by-term equal to corresponding expressions in QED. Secondly, the model is gauge invariant. The gauge transformation is

$$(84)\qquad \psi_2 \to \exp[i\Lambda]\psi_2\,,$$

$$(85)\qquad \tilde{\psi}_2 \to \exp[-i\Lambda]\tilde{\psi}_2\,,$$

$$(86)\qquad A_{2\mu} \to A_{2\mu} - \frac{1}{e}\partial_\mu \Lambda\,,$$

$$(87)\qquad \psi_1 \to \psi_1 + (\exp[i\Lambda] - 1)\psi_2\,,$$

$$(88)\qquad \tilde{\psi}_1 \to \tilde{\psi}_1 + (\exp[-i\Lambda] - 1)\tilde{\psi}_2\,,$$

and its associated conserved current is

$$J_\mu = -i\frac{\delta\mathscr{L}}{\delta\partial^\mu\psi_1}\psi_2 - i\frac{\delta\mathscr{L}}{\delta\partial^\mu\psi_2}\psi_2\,, \tag{89}$$

$$J_\mu = \tilde{\psi}_2\gamma^0\gamma_\mu\psi_2\,. \tag{90}$$

We close our discussion of fermions by considering the case $\theta_2 = \pi/4$ which, by eq. (81), gives the principal-value propagator

$$S_{22}(x-y) = \int\frac{d^4k}{(2\pi)^4}\exp[-ik\cdot(x-y)](k+m)\frac{P}{k^2-m^2}\,, \tag{91}$$

where

$$\frac{P}{k^2-m^2} = \frac{1}{2}\left(\frac{1}{k^2-m^2+i\varepsilon}+\frac{1}{k^2-m^2-i\varepsilon}\right). \tag{92}$$

Thus a quantum action-at-a-distance model of fermions can be constructed within the framework of our formulation.

4. – Particle interpretation.

In this section we shall show that the particle interpretation of our formulation of quantum field theory is well defined for the case of a free scalar particle in a nonstatic space-time where no global timelike co-ordinate exists. We assume an action of the form

$$S = \int d^4x\,[\varphi_2 D\varphi_1 - \tfrac{1}{2}\varphi_1 D\varphi_1] + (\text{surface terms})\,, \tag{93}$$

which under the variation of S gives the field equations

$$D\varphi_1 = D\varphi_2 = 0\,. \tag{94}$$

The self-adjointness of D implies

$$\int_V [f^* Dg - (Df)^* g]\,d^4x = \int_{\Sigma_v} f^*\overleftrightarrow{D}^\mu g\,d\Sigma_\mu\,, \tag{95}$$

where Σ_v is the surface bounding V, $d\Sigma_\mu$ is an outward directed surface element of Σ_v and D^μ is a two-edged vector differential operator. If Σ is a spacelike complete Cauchy hypersurface for the field equations (we assume they exist), then an inner product for complex solutions of the field equations can be defined by

$$(v_1, v_2) = i\int_{\Sigma} v_1^* \overleftrightarrow{D}^\mu v_2 \, d\Sigma_\mu . \tag{96}$$

We now choose an arbitrary complete orthonormal set of pairs of complex conjugate solutions of eq. (94), $\{V_\alpha, V_\alpha^*\}$, satisfying

$$(V_\alpha, V_{\alpha'}) = -(V_\alpha^*, V_{\alpha'}^*) = \delta_{\alpha\alpha'} , \tag{97}$$

$$(V_\alpha, V_{\alpha'}^*) = 0 , \tag{98}$$

and use them in the mode expansion of the field operators

$$\varphi_i = \sum_\alpha [(c_{i1}A_{1\alpha} + c_{i2}A_{2\alpha}) V_\alpha + (\tilde{c}_{i1}A_{1\alpha}^\dagger + \tilde{c}_{i2}A_{2\alpha}^\dagger) V_\alpha^*] , \tag{99}$$

where c_{ij} and $\tilde{c}_{ij}$ are real c-numbers. The Fourier coefficient operators satisfy

$$[A_{i\alpha}, A_{j\alpha'}^\dagger] = (1 - \delta_{ij})\delta_{\alpha\alpha'} \tag{100}$$

with all other commutators equal to zero. The commutativity of Fourier-coefficient operators and their Hermitian conjugate allows us to define the vacua $|0\rangle_1$ and $|0\rangle_2$ by

$$A_{1\alpha}|0\rangle_2 = A_{1\alpha}^\dagger|0\rangle_2 = A_{2\alpha}|0\rangle_1 = A_{2\alpha}^\dagger|0\rangle_1 = 0 . \tag{101}$$

We choose to construct states from $|0\rangle_2$. The one-particle ket corresponding to the Fourier transform variable α is

$$|\alpha\rangle = -(v_\alpha^*, \varphi_2)|0\rangle_2/\tilde{c}_{22} \tag{102}$$

and the one-particle bra dual to it is

$$\langle\alpha| = (|\alpha\rangle)^\dagger \gamma , \tag{103}$$

where γ is a metric operator satisfying $\gamma = \gamma^{-1} = \gamma^\dagger$ and

$$\gamma^{-1} A_{i\alpha} \gamma = \varepsilon_{ij} A_{j\alpha} \tag{104}$$

with $\varepsilon_{11} = \varepsilon_{22} = 0$ and $\varepsilon_{12} = \varepsilon_{21} = 1$. The further development of this quantization proceeds along the lines of sect. **2**. In particular φ_2 is self-adjoint.

We now introduce a quantization of the particle described by S which parallels the above development in every detail except that a different complete orthonormal set of field equation solutions $\{W_\beta, W_\beta^*\}$ is used in the mode expansion of the fields

$$\varphi_i = \sum_\beta [(c_{i1} A_{1\beta} + c_{i2} A_{2\beta}) W_\beta + (\bar{c}_{i1} A^\dagger_{1\beta} + \bar{c}_{i2} A^\dagger_{2\beta}) W^*_\beta] . \tag{105}$$

The question arises: how are we to relate the two quantizations? In the usual formulation only one answer is apparent—the field operators are to be identified [1,2], since they are uniquely determined by the field equations and the canonical commutation relations. But in the present case, the field operators are not uniquely determined, so that the identification of the fields in eq. (105) and (99) is not required. The relation between the quantizations must obviously be well defined (in the sense that every operator and state in one quantization can be uniquely expressed in terms of operators and states of the other representations). More importantly, it must only relate operators whose properties are fixed by the field equation and the canonical commutation relations; and whose expectation values are uniquely specified by purely geometrical restrictions on their support and do not embody a definition of positive frequency. In our formalism, the operators which satisfy these requirements are linear combinations of

$$\varphi^{\mathrm{II}}_{iv} = \sum_\alpha (A_{i\alpha} V_\alpha + A^\dagger_{i\alpha} V^*_\alpha) , \tag{106}$$

$$\varphi^{\mathrm{II}}_{iw} = \sum_\beta (A_{i\beta} W_\beta + A^\dagger_{i\beta} W^*_\beta) , \tag{107}$$

for $i = 1, 2$ in the respective quantizations we can restrict the discussion to these quantities. In particular, the vacuum expectation value

$${}_1\langle 0|\varphi^{\mathrm{II}}_1(x)\varphi^{\mathrm{II}}_2(y)|0\rangle_2 = \tfrac{1}{2}\,{}_1\langle 0|[\varphi^{\mathrm{II}}_1(x), \varphi^{\mathrm{II}}_2(y)]|0\rangle_2 , \tag{108}$$

$${}_1\langle 0|\varphi^{\mathrm{II}}_1(x)\varphi^{\mathrm{II}}_2(y)|0\rangle_2 = \frac{i}{2}\Delta(x-y) , \tag{109}$$

where $\Delta(x-y)$, the commutator function, has the well-defined geometrical property that it vanishes at spacelike distances.

Identifying $\varphi^{\mathrm{II}}_{iw}$ with $\varphi^{\mathrm{II}}_{iv}$, for $i = 1, 2$, leads to the relations

$$A_{i\beta} = (W_\beta, \varphi^{\mathrm{II}}_i) = \sum_\alpha [(W_\beta, V_\alpha) A_{i\alpha} + (W_\beta, V^*_\alpha) A^\dagger_{i\alpha}] , \tag{110}$$

for $i = 1, 2$ plus Hermitian-conjugate expressions. The form of the inner products on the right-hand side of eq. (110) is determined by requiring that the definition of positive frequency implicit in the separation of the orthonormal set $\{W_\beta, W^*_\beta\}$ into complex conjugate pairs of solutions can also be implemented by linear combinations of V_α and V^*_α. Specifically, we assume that a complete orthonormal set of pairs of complex conjugate functions, $\{V_\alpha, V^*_\alpha\}$, exists which satisfies

(111) $$V_\alpha = c_1 \tilde{V}_\alpha + c_2 \tilde{V}^*_\alpha ,$$

(112) $$V^*_\alpha = c^*_1 \tilde{V}^*_\alpha + c^*_2 \tilde{V}_\alpha ,$$

for all α, where c_1 and c_2 are c-number functions of α only with $|c_1| > |c_2|$, and where

(113) $$(W_\beta, \tilde{V}^*_\alpha) = 0 ,$$

(114) $$(W^*_\beta, \tilde{V}_\alpha) = 0 ,$$

for all β. The orthogonality conditions imply

(115) $$c_1 = \exp[i\lambda_1] \cosh \lambda_2 ,$$

(116) $$c_2 = \exp[i\lambda_1] \sinh \lambda_2 ,$$

where λ_1 and λ_2 are solely functions of α. The substitution of eqs. (111) and (112) in eq. (110) and use of eqs. (113) and (114) gives

(117) $$A_{i\beta} = \sum_\alpha (W_\beta, \tilde{V}_\alpha)[\exp[i\lambda_1] \cosh \lambda_2 A_{i\alpha} + \exp[-i\lambda_1] \sinh \lambda_2 A^\dagger_{i\alpha}]$$

for $i = 1, 2$. Note that the bracketed term on the right-hand side of the equation has the same form as the Bogoliubov rotated Fourier-coefficient operator given in eq. (23). In the present case we can rewrite eq. (117) in the form

(118) $$A_{i\beta} = \sum_\alpha (W_\beta, \tilde{V}_\alpha) B_{\lambda_1\lambda_2} A_{i\alpha} B^{-1}_{\lambda_1\lambda_2}$$

with

(119) $$B_{\lambda_1\lambda_2} = \exp\left[2i \sum_\alpha \lambda_1(\alpha) \Gamma_{3\alpha}\right] \exp\left[2i \sum_\alpha \lambda_2(\alpha) \Gamma_{2\alpha}\right],$$

where $\Gamma_{3\alpha}$ and $\Gamma_{2\alpha}$ are obtained from eqs. (25) and (26) by replacing the subscripts k with α.

The particle interpretations of the two quantizations will now be shown to be identical. First we note that the vacuum $|0\rangle_2$ of the « α » quantization is invariant under $B_{\lambda_1\lambda_2}$, so that it may be taken to be identical with the $|0\rangle_2$ vacumm of the « β » quantization. Next we note that the canonical commutation relations and the vacuum expectation value of any product of field operators are invariant under B:

(120) $${}_1\langle 0|\varphi_{i_1}(x_1)\varphi_{i_2}(x_2) \ldots |0\rangle_2 = {}_1\langle 0|B^{-1}\varphi_{i_1}(x_1)\varphi_{i_2}(x_2) \ldots B|0\rangle_2 .$$

This implies that we could replace $A_{i\alpha}$ with $B_{\lambda_1\lambda_2} A_{i\alpha} B^{-1}_{\lambda_1\lambda_2}$ in the mode expansions, eq. (107), with no change in physical consequences. In particular, this applies to the definition of particle kets. Equation (102) becomes

$$(121)\qquad |\alpha\rangle = (\exp[-i\lambda_1]\cosh\lambda_2 A^\dagger_{2\alpha} + \exp[i\lambda_1]\sinh\lambda_2 A_{2\alpha})|0\rangle_2 .$$

Consequently, $A^\dagger_{2\beta}|0\rangle_2$ is a superposition of one-particle states in the « α » quantization. In general, the N-particle state in the « β » quantization is a superposition of N-particle states in the « α » quantization.

The invariance of particle number under Bogoliubov transformations is reflected in the relation between the particle number operator,

$$(122)\qquad N = \sum_\alpha (A^\dagger_{2\alpha} A_{1\alpha} - A_{2\alpha} A^\dagger_{1\alpha}) ,$$

which is invariant under Bogoliubov transformations, and related to the Casimir operator of the Bogoliubov group (cf. eqs. (29)-(31)). Our identification of N as the particle number operator is based, as most charge and number operators are, on an invariance of the action under a global change of phase of fields. In our case we note that the action of eq. (93) is invariant under the infinitesimal phase change

$$(123)\qquad \varphi_1 \to \varphi_1 + i\varepsilon\varphi_1 , \qquad \varphi_2 \to \varphi_2 + i\varepsilon(\varphi_1 - \varphi_2) .$$

The corresponding conserved-number operator is given by

$$(124)\qquad N = i\int_{\Sigma_\nu} \varphi_1 \overleftrightarrow{D}^\mu \varphi_2 \, d\Sigma_\mu .$$

Because φ_1 and φ_2 implicitly embody a definition of positive frequency, we are led to replace them with Hermitian operators:

$$(125)\qquad N = i\int_{\Sigma_\nu} \varphi^{\rm H}_1 D^\mu \varphi^{\rm H}_2 \, d\Sigma_\mu$$

with $\varphi^{\rm H}_i$ given in eq. (106). Equation (125) can be evaluated by using eq. (96) and (97) to give eq. (122). Thus our definition of number operator is physically motivated. It is also consistent with our expectations of a number operator.

We shall now summarize our picture of second quantization in curved space-time where no global timelike Killing vector is present. Consider a complete spacelike Cauchy hypersurface. At each point on the surface there is a local timelike direction. There is, in general, a class of operators, which will locally generate a displacement in the timelike direction, but which globally generate very different motions. Due to the absence of a global timelike Killing vector, no member of the class of potential generators of the dynamics is physically selected as the generator of the dynamics. One is free to choose any member as the generator of the dynamics locally. Each choice implies a different definition of positive and negative frequency when field operators are represented by Fourier expansions.

In the usual formulation of quantum field theory any choice of generator of the dynamics (and thus Fourier representation of field operators) is unitarily inequivalent to any other choice in general. As a result each choice gives a *different physical theory* and the second quantization of a theory is not unique. Practically, this means that 1) a one-particle state in one quantization is a many-particle state in any other quantization (particle number is ambiguous), 2) in general one can construct a one-particle state which is an eigenstate of a generator in one representation, but one cannot construct a one-particle state in another representation which is an eigenstate of the same generator (the space of states is different), and 3) (if interactions are introduced) the S-matrix differs from quantization to quantization. Obviously, there are only two acceptable alternatives in this situation; either some new principle selects one representation as the correct physical representation, or a modification of quantum field theory is necessary. In the absence of a new physical principle, we have formulated a modification of quantum field theory.

Our formulation allows one to quantize a field theory with any of the potential generators of the dynamics and yet to have a physically unique theory. Different quantizations can be related by Bogoliubov transformations and, in our formulation, are unitarily equivalent. Consequently, particle number is invariant—N-particle states in one representation are superpositions of N-particle states in any other representation; the set of states in one representation is unitarily equivalent to the set of states in any other representation; and the S-matrix is uniquely determined in the case of an interacting theory (the proof is analogous to that of the flat space-time case discussed in sect. **2**). Our formulation associates a unique physical theory with any given action. In a sense, it implements an equivalence principle in the space of solutions to the field equations—any complete orthonormal set of solutions to the field equations can be used in the Fourier expansion of field operators and a unique physical theory results (cf. eqs. (111)-(114)).

In conclusion, we note that the problem we have addressed relates to one observer and the ambiguities of conventional quantum field theory he must face. Different observers in relatively accelerating frames will not see the same number of particles in our formulation. Neither is particle creation near black holes precluded in our formulation.

* * *

I am grateful to the Aspen Center for Physics for its hospitality while part of this work was being done, and to M. A. B. BEG, S. MANDELSTAM and D. PARK for stimulating conversations.

● RIASSUNTO (*)

Si descrive una generalizzazione della teoria quantistica dei campi che ha un'unica interpretazione particellare — anche negli spazio-tempo in cui non esiste alcuna coordinata globale di tipo tempo. La formulazione è descritta in dettaglio per i casi di bosoni scalari e di fermioni con spin ½ nello spazio-tempo piatto. Si mostra che è possibile costruire un modello nel nostro approccio che è fisicamente equivalente a un qualsiasi modello nella solita formulazione. Inoltre, si può costruire una nuova classe di modelli che non sono possibili nella solita formulazione. Questa classe comprende modelli quantici di azione a distanza che possono essere usati per sviluppare modelli con equazioni di campo a derivata più alta che sono unitarie. La nostra formulazione permette ampia scelta delle condizioni limite, cosicché si può optare per un continuo di possibili funzioni di Green che vanno dai propagatori di Feynman a propagatori del valore principale (mezzo avanzato-mezzo ritardato).

(*) *Traduzione a cura della Redazione.*

Локальное определение асимптотических состояний частиц.

Резюме (*). — Описывается обобщение квантовой теории поля, которая имеет единую частичную интерпретацию — даже в пространстве и времени, где не существует глобальной времениподобной координаты. Подробно описывается формулировка для случаев скалярных бозонов и фермионов со спином половина в плоском пространстве-времени. Мы показываем, что имеется возможность сконструировать модель в нашем подходе, которая физически эквивалентна любой заданной модели в обычной формулировке. Кроме того, может быть сконструирован новый класс моделей, которые являются невозможными в обычной формулировке. Этот класс включает модели квантового действия на расстоянии, которые могут быть использованы для развития моделей с полевыми уравнениями с высшими производными, которые являются унитарными. Наша формулировка допускает некоторую свободу в выборе граничных условий, так что имеется возможность выбрать континуум возможных гриновских функций, от фейнмановских пропагаторов до пропагаторов главных значений (наполовину опережающая - наполовину запаздывающая).

(*) *Переведено редакцией.*

IL NUOVO CIMENTO VOL. 49 A, N. 1 1 Gennaio 1979

New Framework for Gauge Field Theories (*).

S. BLAHA

Physics Department, Williams College - Williamstown, Ma. 01267

(ricevuto l'8 Agosto 1978)

Summary. — We formulate gauge theories within the framework of a generalization of quantum field theory. In particular, we discuss models of electrodynamics and of Yang-Mills theories, a model of the strong interaction with higher-order derivatives and quark confinement and a renormalizable model of pure quantum gravity with Einstein Lagrangian. In the case of electrodynamics we show that two models are possible: one with predictions which are identical to QED and one which is a quantum action-at-a-distance model of electrodynamics. In the case of Yang-Mills theories we can construct a model which is identical in predictions to any conventional model, or a quantum action-at-a-distance model. In the second case it is possible to eliminate all loops of Yang-Mills particles (in all gauges) in a manner consistent with unitarity. A variation of Yang-Mills models exists in our formulation which has higher-order derivative field equations. It is unitary and has positive probabilities. It can be used to construct a model of the strong interactions which has a linear potential and manifest quark confinement. Finally we show how to construct an action-at-a-distance model of pure quantum gravity (whose classical limit is the dynamics of the Einstein Lagrangian) coupled to an external classical source. The model is trivially renormalizable.

1. – Introduction.

Because of the absence of an acceptable physical interpretation of conventional quantum field theory in the case of curved nonstatic space-time, we

(*) Supported in part by Research Corporation and the National Science Foundation.

recently developed a modified formulation of quantum field theory (¹). We showed it had a unique physical particle interpretation in nonstatic space-time. We described the flat–space-time formulation for the cases of scalar particles and spin–one-half fermions. In both cases we found that it was possible to construct a model which was equivalent in predictions to any model of the usual formulation. However, it was also possible to construct models which had no analogue in the usual formulation. In these models the quantum exchange of a « particle » did not take place via a Feynman propagator. Rather, the propagator G could be chosen to be any of a continuum of possibilities ranging from the Feynman propagator to the principal-value (half advanced-half retarded) propagator. The choice is parametrized by an angle θ, whose specification is equivalent to a choice of boundary condition

$$G(x-y) = \sin^2\theta\, G_{\mathrm{F}}(x-y) + \cos^2\theta\, C G_{\mathrm{F}}^* C^{-1}, \tag{1}$$

where G_{F} is the usual Feynman propagator with G_{F}^* its complex conjugate and C is an appropriate charge conjugation matrix.

The new degree of freedom, represented by θ, leads to new possibilities for the formulation of flat–space-time quantum field theory models which will be especially evident in the case of gauge theories.

In sect. **2** we shall explore models of quantum electrodynamics which occur within the framework of our formulation. We shall show that one model is completely equivalent to QED in its predictions. In addition, we shall show that a quantum action-at-a-distance electrodynamics is also possible.

In sect. **3** we describe model Yang-Mills theories and show that a model equivalent to any conventional model can be formulated as well as a quantum action-at-a-distance model. In the second case all non-Abelian boson loops can be eliminated (in all gauges) in a manner which is consistent with unitarity.

In sect. **4** we describe a non-Abelian model of quark confinement based on higher-derivative field equations for which a unitary physical S-matrix is obtained by the choice of principal-value propagators. This demonstrates the utility of principal-value propagators for widening the class of unitary quantum field theories to include higher-derivative theories.

In sect. **5** we describe a model of pure quantum gravity based on the Einstein Lagrangian which is trivially renormalizable, if gravitons propagate via principal-value propagators. (No graviton loops.) This illustrates the potential of the principal-value propagator to ameliorate renormalization problems by allowing one to limit the self-interactions of quantum fields.

In view of the unfamiliar features of principal-value propagators, we shall

---

(¹) S. BLAHA: *Nuovo Cimento*, **49** A, 35 (1978).

devote the remainder of this section to a discussion of their properties in relation to causality and analyticity.

Among the first appearances of principal-value, or half advanced-half retarded, propagators was in Feynman's space-time approach to quantum electrodynamics [2], where he observed that one is naturally led to such a propagator for the photon, if one considers a quantum one-photon exchange between two electrons wherein the photons propagate, as in classical electrodynamics, via retarded propagators. Since one cannot distinguish between the process where a photon propagates from electron A to B and the process where the photon propagates from B to A for short time intervals one must sum the two amplitudes and a principal-value propagator results. This observation illustrates the absence of a direct relation between the propagator of the classical field and the propagator of the corresponding quantum field. In particular, it is perfectly possible for a quantum field to have a principal-value propagator, and yet have the classical field propagate via a retarded propagator. This can be understood within the framework of the absorber model of Feynman and Wheeler [3]. For a quantum process, where a finite number of quanta is exchanged, the effective propagator of the quanta cannot have this aspect of its character changed. But in a classical situation, where infinite numbers of quanta are involved, the effective propagator of the quanta can be changed through interaction with the absorber in the manner outlined in ref. [3]. We therefore conclude that the propagation of quanta via principal-value propagators does not imply that macroscopic causality (as represented by retarded propagators) is lost. This remark is relevant to our models of the strong interaction and quantum gravity discussed later.

So far as microscopic causality is concerned, it will be seen that principal-value propagators are completely consistent with the vanishing of commutators of field operators for spacelike distances. Thus neither macroscopic nor microscopic causality is necessarily inconsistent with the use of quantum principal-value propagators.

The use of principal-value propagators does lead to a different analytic structure of S-matrix amplitudes. Amplitudes are now piecewise analytic. However, several authors [4] have shown that such amplitudes are not necessarily inconsistent with experiment—even for the case of the pion-nucleon dispersion relations. In particular, considering the absence of any direct relation between the analytic properties of quark-quark scattering amplitudes,

---

(2) R. Feynman: *Phys. Rev.* **76**, 769 (1949), footnote (7).

(3) J. Wheeler and R. Feynman: *Rev. Mod. Phys.*, **17**, 157 (1945); **21**, 425 (1949); F. Hoyle and J. Narlikar: *Ann. of Phys.*, **54**, 207 (1969); **62**, 44 (1971).

(4) R. E. Cutkowsky, P. V. Landshoff, D. I. Olive and J. C. Polkinghorne: *Nucl. Phys.*, **12** B, 281 (1969); T. D. Lee and G. C. Wick: *Phys. Rev. D*, **2**, 1033 (1970); M. Gundzik and E. C. G. Sudarshan: *Phys. Rev. D*, **6**, 798 (1972); H. Rechenberg and E. C. G. Sudarshan: *Nuovo Cimento*, **14** A, 299 (1973).

and the analytic properties of the scattering amplitudes of their bound states one cannot rule out the possibility of principal-value propagators for color gluons. In the case of quantum gravity, another potential application of principal-value propagators, the analytic structure of scattering amplitudes due to graviton exchange is unknown and likely to remain so. In the absence of such information, the renormalizability of pure quantum gravity, and the compatibility with Mach's principle, resulting from the use of principal-value propagators for gravitons is encouraging.

Perhaps the most important question facing field theory models with principal-value propagators is unitarity. The appendix contains a detailed discussion of this issue. The physical S-matrix is shown to be unitary. In addition the value of using infinite-momentum frame variables to compute S-matrix elements is pointed out.

2. – Model electrodynamics.

In this section we shall describe a model electrodynamics which is identical to quantum electrodynamics in its predictions. We shall also discuss a model for quantum action-at-a-distance electrodynamics.

The Lagrangian density of our models (¹) is

$$\mathscr{L} = -\tfrac{1}{2} F^1_{\mu\nu} F^{2\mu\nu} - \tfrac{1}{4} F^1_{\mu\nu} F^{1\mu\nu} + \tilde{\psi}_2 \gamma^0 (i(\gamma\cdot\nabla) - e_0(\gamma\cdot A_2) - m)\psi_2 + (\tilde{\psi}_1 - \tilde{\psi}_2)\gamma^0(i(\gamma\cdot\nabla) - m)(\psi_1 - \psi_2)\,, \tag{2}$$

where we have introduced two electromagnetic fields, $A_{1\mu}$ and $A_{2\mu}$, so that

$$F^i_{\mu\nu} = \partial_\nu A_{i\mu} - \partial_\mu A_{i\nu}\,, \tag{3}$$

for $i = 1, 2$, and two fields, ψ_1 and ψ_2 for electrons. The field $\tilde{\psi}_i$ is related to the transpose of the charge conjugate of ψ_i by (¹)

$$\tilde{\psi}_i = \psi_i^{cT} \gamma^0 C^T\,, \tag{4}$$

where C is a charge conjugation matrix. We follow the notation and conventions of ref. (⁵). The free field theory for fermions has been developed in detail in ref. (¹).

(⁵) J. Bjorken and S. Drell: *Relativistic Quantum Fields* (New York, N. Y., 1965).

Before discussing the free field theory of the photons we note the invariance of $\mathscr{L}$ under a restricted gauge transformation of the second kind,

$$(5)\qquad \psi_2 \to \psi_2 \exp[i\Lambda],$$

$$(6)\qquad \tilde{\psi}_2 \to \tilde{\psi}_2 \exp[-i\Lambda],$$

$$(7)\qquad e_0 A_{2\mu} \to e_0 A_{2\mu} - \partial_\mu \Lambda,$$

$$(8)\qquad \psi_1 \to \psi_1 + [\exp[i\Lambda] - 1]\psi_2,$$

$$(9)\qquad \tilde{\psi}_1 \to \tilde{\psi}_1 + [\exp[-i\Lambda] - 1]\tilde{\psi}_2.$$

The associated conserved current is

$$(10)\qquad J_\mu = \tilde{\psi}_2 \gamma^0 \gamma_\mu \psi_2.$$

It should be noted that the Lagrangian is also invariant under an independent gauge transformation, $A_{1\mu} \to A_{1\mu} - \partial_\mu \Lambda_1$.

Upon varying the action associated with $\mathscr{L}$, we find the field equations

$$(11)\qquad \partial^\mu F^1_{\mu\nu} = \partial^\mu F^2_{\mu\nu},$$

$$(12)\qquad \partial^\nu F^2_{\mu\nu} = e_0 \tilde{\psi}_2 \gamma^0 \gamma_\nu \psi_2,$$

$$(13)\qquad (i(\gamma\cdot\nabla) - m)\psi_1 = (i(\gamma\cdot\nabla) - m)\psi_2,$$

$$(14)\qquad (i(\gamma\cdot\nabla) - e_0(\gamma\cdot A_2) - m)\psi_2 = 0.$$

Equations (12) and (14) demonstrate the equivalence of our model to the usual model of electrodynamics at the level of c-number fields.

We now turn to the case of free photons. From the Lagrangian we can identify the canonical momentum conjugate to $A_{1\mu}$ as

$$(15)\qquad \pi_{1\mu} = F^2_{0\mu} - F^1_{0\mu},$$

while the momentum conjugate to $A_{2\mu}$ is

$$(16)\qquad \pi_{2\mu} = F^1_{0\mu}.$$

Since the free electromagnetic Lagrangian is invariant under independent gauge transformations of $A_{1\mu}$ and $A_{2\mu}$ the choice of gauges for $A_{1\mu}$ and $A_{2\mu}$ are also independent. We shall second quantize the fields in the joint Coulomb gauge

$$(17)\qquad \vec{\nabla}\cdot\vec{A}_1 = \vec{\nabla}\cdot\vec{A}_2 = 0.$$

The resulting equal-time commutation relations are

$$[F^1_{0i}(\vec{x}, t), A_{1j}(\vec{y}, t)] = 0 , \tag{18}$$

$$[F^1_{0i}(\vec{x}, t), A_{2j}(\vec{y}, t)] = i\delta^{tr}_{ij}(\vec{x} - \vec{y}) , \tag{19}$$

$$[F^2_{0i}(\vec{x}, t), A_{2j}(\vec{y}, t)] = i\delta^{tr}_{ij}(\vec{x} - \vec{y}) , \tag{20}$$

$$[F^2_{0i}(\vec{x}, t), A_{1j}(\vec{y}, t)] = i\delta^{tr}_{ij}(\vec{x} - \vec{y}) , \tag{21}$$

with

$$\delta^{tr}_{ij}(\vec{x} - \vec{y}) = \int \frac{d^3k}{(2\pi)^3} \exp[-i\vec{k}\cdot(\vec{x} - \vec{y})](\delta_{ij} - k_i k_j/|\vec{k}|^2) . \tag{22}$$

The field operator expansions are completely analogous to the expansion of a free scalar field discussed in ref. [1]. The major difference is the necessary presence of polarization vectors in the electromagnetic-field expansions

$$\vec{A}_1(x) = \int d^3k \sum_{\lambda=1}^{2} \vec{\varepsilon}(k, \lambda)[f_k(x)(\cos(\theta_1 - \theta_2)a_{1k\lambda} + \sin(\theta_1 - \theta_2)a_{2k\lambda})/\sin\theta_1 + \\ + f^*_k(x)(\cos(\theta_1 - \theta_2)a^\dagger_{1k\lambda} - \sin(\theta_1 - \theta_2)a^\dagger_{2k\lambda})/\cos\theta_1] \tag{23}$$

and

$$\vec{A}_2(x) = \int d^3k \sum_{\lambda=1}^{2} \vec{\varepsilon}(k, \lambda)[f_k(x)(\cos\theta_2 a_{2k\lambda} + \sin\theta_2 a_{1k\lambda}) + \\ + f^*_k(x)(\sin\theta_2 a^\dagger_{2k\lambda} + \cos\theta_2 a^\dagger_{1k\lambda})] , \tag{24}$$

where θ_1 and θ_2 are arbitrary angles. As discussed in ref. [1] the choice of these angles represents a choice of boundary conditions for the free-field Green's functions. If the Fourier coefficient operators satisfy

$$[a_{ik\lambda}, a_{jk'\lambda'}] = [a^\dagger_{ik\lambda}, a^\dagger_{jk'\lambda'}] = 0 , \tag{25}$$

$$[a_{ik\lambda}, a^\dagger_{jk'\lambda'}] = (1 - \delta_{ij})\delta_{\lambda\lambda'}\delta^3(\vec{k} - \vec{k}') , \tag{26}$$

for $i, j = 1, 2$, then the equal-time commutation relations will be satisfied for arbitrary θ_1 and θ_2. In addition the Hamiltonian which is the 0-0 component of the energy-momentum tensor derived from the Lagrangian, has the form

$$H = \int d^3k \sum_{\lambda=1}^{2} \omega_k(a^\dagger_{2k\lambda}a_{1k\lambda} + a_{2k\lambda}a^\dagger_{1k\lambda}) , \tag{27}$$

independent of θ_1 and θ_2. There is an analogy [6] between the quantization

[6] S. BLAHA: *Phys. Rev. D*, **17**, 994 (1978).

of the mode amplitudes, eqs. (25) and (26), and the co-ordinates of an assembly of one-dimensional harmonic oscillators just as in the second quantization of conventional field theory. We shall define photon vacua in the present case in a manner consistent with the procedure of ref. (1) and in analogy with the simple harmonic-oscillator model of ref. (6):

$$a_{1k\lambda}|0\rangle_2 = a^\dagger_{1k\lambda}|0\rangle_2 = 0\,, \tag{28}$$

$$a_{2k\lambda}|0\rangle_2 \neq 0\,, \qquad a^\dagger_{2k\lambda}|0\rangle_2 \neq 0\,, \tag{29}$$

for one vacuum, $|0\rangle_2$, and

$$a_{2k\lambda}|0\rangle_1 = a^\dagger_{2k\lambda}|0\rangle_1 = 0\,, \tag{30}$$

$$a_{1k\lambda}|0\rangle_1 \neq 0\,, \qquad a^\dagger_{1k\lambda}|0\rangle_1 \neq 0\,, \tag{31}$$

for the other vacuum, $|0\rangle_1$, for all k and λ. The vacua are related to each other by a Dirac-metric operator in a manner familiar from the scalar case. The metric operator γ is necessary in view of the commutation relations of the Fourier-coefficient operators, eqs. (25) and (26). A one-photon ket is defined by

$$|k\lambda\rangle = a^\dagger_{2k\lambda}|0\rangle_2\,, \tag{32}$$

while the dual one-photon bra is defined by

$$\langle k\lambda| = {}_1\langle 0|a_{1k\lambda}\,, \tag{33}$$

so that

$$\langle k'\lambda'|k\lambda\rangle = \delta_{\lambda\lambda'}\,\delta^3(\vec{k}-\vec{k}')\,, \tag{34}$$

by means of eq. (26) and ${}_1\langle 0|0\rangle_2 = 1$. The metric operator is introduced in order to relate the dual of a ket to its Hermitian conjugate:

$$\langle k\lambda| = (|k\lambda\rangle)^\dagger\gamma\,. \tag{35}$$

Consequently the metric operator is defined to have the properties

$$\gamma|0\rangle_1 = |0\rangle_2\,, \tag{36}$$

$$\gamma^{-1}a_{1k\lambda}\gamma = a_{2k\lambda}\,, \tag{37}$$

$$\gamma^{-1}a_{2k\lambda}\gamma = a_{1k\lambda} \tag{38}$$

with $\gamma = \gamma^\dagger = \gamma^{-1}$. It has a form similar to the metric operator of the scalar case which is given in eq. (36) of ref. (1). The definition of many-photon bras and kets is a direct generalization of eqs. (32), (33) and (35).

The time-ordered propagator which will later be of use in the development of perturbation theory is

$$G^{22}_{\mu\nu}(x-y) = -i\,{}_1\langle 0|T(A_{2\mu}(x)A_{2\nu}(y))|0\rangle_2 = \sin^2\theta_2\, G_{F\mu\nu}(x-y) + \cos^2\theta_2\, G^*_{F\mu\nu}(x-y)\,, \tag{39}$$

where $G_{F\mu\nu}(x-y)$ is the usual Feynman propagator for photons, while G^*_F is its complex conjugate. Another nonzero free-field propagator, $G^{12}_{\mu\nu}$ is defined as the vacuum expectation value of the time-ordered product of $A_{1\mu}(x)$ and $A_{2\nu}(y)$. It depends on both θ_1 and θ_2. It does not appear in perturbation theory.

Equation (39) shows that $G^{22}_{\mu\nu}$ is a Feynman propagator if $\theta_2 = \pi/2$, and a principal-value propagator if $\theta_2 = \pi/4$. Thus

$$G^{22}_{\mu\nu}(x-y)\big|_{\theta_2=\pi/4} = g_{\mu\nu}\int\frac{d^4k}{(2\pi)^4}\exp[-ik\cdot(x-y)]\,\frac{P}{k^2}\,, \tag{40}$$

where

$$\frac{P}{k^2} = \frac{1}{2}\left(\frac{1}{k^2+i\varepsilon} + \frac{1}{k^2-i\varepsilon}\right), \tag{41}$$

in the Feynman gauge. This Green's function has previously appeared in action-at-a-distance formulations (3) of classical electrodynamics. The generality of our formulation of quantum field theory allows it to appear as a special case. This does not seem to be possible within the framework of the conventional formulation (7). If we choose $\theta_2 = \pi/4$ and proceed to develop perturbation theory for the full interacting theory then we obtain a quantum-action-at-a-distance theory. We obtain quantum exchange of energy and momentum to which we associate lines in Feynman diagrams. The only difference is in the analytic structure associated with the « epsilontics » of the pole locations. Scattering amplitudes are piecewise analytic in this case (4). More importantly, quanta associated with the electromagnetic field do not appear in asymptotic states, if unitarity is to be maintained. Thus the electromagnetic field does not have its own degrees of freedom, but is constrained to have its source in the charged matter fields which are present. The physical expectations of a quantum action-at-a-distance electrodynamics are therefore satisfied. Except for the absence of « photons » from asymptotic states, and the principal-value nature of the electromagnetic-field propagator, the perturbation theory rules, diagrams and combinatorics are identical to those of QED, although they must be unitarized using the method described in the appendix.

---

(7) R. FEYNMAN: *Science*, **153**, 699 (1966).

We now turn to the Feynman case ($\theta_2 = \pi/2$). The field $\vec{A}_2$ takes the form

$$\vec{A}_2 = \int \mathrm{d}^3k \sum_{\lambda=1}^{2} \vec{\varepsilon}(k, \lambda)[f_k(x)a_{1k\lambda} + f_k^*(x)a^\dagger_{2k\lambda}]\,, \tag{42}$$

in the Coulomb gauge. Note that the form of eq. (42) and the commutators, eq. (25) and (26) imply the free-field identity

$$ {}_1\langle 0|A_{2i_1}(x_1)A_{2i_2}(x_2)\dots A_{2i_n}(x_n)|0\rangle_2 = \langle 0|A_{i_1}(x_1)A_{i_2}(x_2)\dots A_{i_n}(x_n)|0\rangle \tag{43}$$

for all spatial indices $i_1, i_2, \dots, i_n$, and all n for which the right-hand side is the vacuum expectation value of n free fields in the usual formulation. This identity is the basis of our claim that the Feynman case of our formulation is *completely* equivalent to QED in its predictions. An important part of establishing the correspondence is the LSZ reduction formulae for photons: for an in photon $(k\lambda)$ we have

$$\begin{aligned} \langle\beta\ \mathrm{out}|\alpha(k\lambda)\ \mathrm{in}\rangle &= \langle\beta\ \mathrm{out}|a^\dagger_{2k\lambda\,\mathrm{in}}|\alpha\ \mathrm{in}\rangle = \\ &= \langle\beta-(k\lambda)\ \mathrm{out}|\alpha\ \mathrm{in}\rangle - \frac{i}{\sqrt{Z_3}}\int \mathrm{d}^4x\, A^\mu_{k\lambda}(x)\overleftarrow{\Box}\langle\beta\ \mathrm{out}|A_{2\mu}(x)|\alpha\ \mathrm{in}\rangle \end{aligned} \tag{44}$$

and for an out photon $(k\lambda)$ we have

$$\begin{aligned} \langle\beta(k\lambda)\ \mathrm{out}|\alpha\ \mathrm{in}\rangle &= \langle\beta\ \mathrm{out}|a_{1k\lambda\,\mathrm{out}}|\alpha\ \mathrm{in}\rangle = \\ &= \langle\beta\ \mathrm{out}|\alpha-(k\lambda)\ \mathrm{in}\rangle - \frac{i}{\sqrt{Z_3}}\int \mathrm{d}^4x\langle\beta\ \mathrm{out}|A_{2\mu}(x)|\alpha\ \mathrm{in}\rangle\overleftarrow{\Box}A^{\mu*}_{k\lambda}(x) \end{aligned} \tag{45}$$

in the notation of ref. [5].

The second quantization of the free-electron field is described in detail in ref. [1]. We assume an in and out electron field formalism is developed along those lines. Our formulation of spin–one-half fermion theory also allows one to choose Feynman propagators or principal-value propagators for fermions. In order to establish a model equivalent to QED we choose Feynman propagators for the electrons.

The perturbation theory for our model is based on the interaction Lagrangian

$$\mathscr{L}_{\mathrm{int}} = -e_0\,\tilde{\psi}_2\gamma^0(\gamma\cdot A_2)\psi_2\,. \tag{46}$$

We shall need the LSZ reduction formulae for electrons

$$\begin{aligned} \langle\beta\ \mathrm{out}|(ps)\alpha\ \mathrm{in}\rangle &= \langle\beta-(ps)\ \mathrm{out}|\alpha\ \mathrm{in}\rangle - \\ &- \frac{i}{\sqrt{Z_2}}\int \mathrm{d}^4x\langle\beta\ \mathrm{out}|\tilde{\psi}_2(x)\gamma^0|\alpha\ \mathrm{in}\rangle(-i\overleftarrow{\nabla} - m)\,U_{ps}(x) \end{aligned} \tag{47}$$

and

$$(48)\qquad \langle\beta(ps)\,\text{out}|\alpha\,\text{in}\rangle = \langle\beta\,\text{out}|\alpha - (ps)\,\text{in}\rangle - \frac{i}{\sqrt{Z_2}}\int d^4x\,\overline{U}_{ps}(x)(i\overrightarrow{\nabla} - m)\langle\beta\,\text{out}|\psi_2(x)|\alpha\,\text{in}\rangle$$

in the notation of ref. ([5]). To remove a positron from an in state, we use

$$(49)\qquad \langle\beta\,\text{out}|(ps)\alpha\,\text{in}\rangle = \langle\beta - (ps)\,\text{out}|\alpha\,\text{in}\rangle + \frac{i}{\sqrt{Z_2}}\int d^4x\,\overline{V}_{\overline{ps}}(x)(i\overrightarrow{\nabla} - m)\langle\beta\,\text{out}|\psi_2(x)|\alpha\,\text{in}\rangle$$

and, to remove a positron from an out state, we use

$$(50)\qquad \langle\beta(ps)\,\text{out}|\alpha\,\text{in}\rangle = \langle\beta\,\text{out}|\alpha - (ps)\,\text{in}\rangle + \frac{i}{\sqrt{Z_2}}\int d^4x\langle\beta\,\text{out}|\tilde{\psi}_2\gamma^0|\alpha\,\text{in}\rangle(-i\overleftarrow{\nabla} - m)\,V_{\overline{ps}}(x)\,.$$

In view of the interaction Lagrangian and eqs. (47)-(50) it is clear that only time-ordered products of $\psi_{2\text{in}}$ and $\tilde{\psi}_{2\text{in}}$ appear in the perturbation theory expansion. The Wick theorem, which applies in the present case, reduces all vacuum expectation values of time-ordered products of $\psi_{2\text{in}}$ and $\tilde{\psi}_{2\text{in}}$ to products of the time-ordered two-point function

$$(51)\qquad iS(x-y) = {}_1\langle 0|T(\psi_2(x)\,\tilde{\psi}_2(y)\gamma^0)|0\rangle_2$$

$$(52)\qquad = i\int\frac{d^4p}{(2\pi)^4}\exp[-ip\cdot(x-y)]\frac{p + m}{p^2 - m^2 + i\varepsilon}\,,$$

where eqs. (51) is in the notation of ref. ([1]).

The above considerations and the familiar development* of the S-matrix as an expansion in vacuum expectation values of time-ordered products of in-field operators lead us to assert that our model electrodynamics is identical to QED in its predictions. In particular

$$(53)\qquad {}_1\langle 0|T\Big(A_{2\mu_1\text{in}}(x_1)A_{2\mu_2\text{in}}(x_2)\ldots\psi_{2\text{in}}(y_1)\psi_{2\text{in}}(y_2)\ldots\tilde{\psi}_{2\text{in}}(z_1)\tilde{\psi}_{2\text{in}}(z_2)\ldots$$

$$\ldots\exp\left[i\int_{-\infty}^{\infty}dt\,L_{\text{int}}\right]\Big)|0\rangle_2 =$$

$$= \langle 0|T\Big(A_{\mu_1\text{in}}(x_1)A_{\mu_2\text{in}}(x_2)\ldots\psi_{\text{in}}(y_1)\psi_{\text{in}}(y_2)\ldots\bar{\psi}_{\text{in}}(z_1)\bar{\psi}_{\text{in}}(z_2)\ldots\exp\left[i\int_{-\infty}^{\infty}dt\,L_{\text{int}}^{\text{QED}}\right]\Big)|0\rangle\,,$$

where the right-hand side is the corresponding time-ordered product of the conventional formulation of QED with

$$L^{\text{QED}}_{\text{int}} = -i : \int \mathrm{d}^3 x e_0 \bar{\varphi}_{\text{in}} (\gamma \cdot A_{\text{in}}) \varphi_{\text{in}} : . \tag{54}$$

Thus, the perturbation theory of our model electrodynamics is term by term equivalent to QED—the equivalent QED term being obtained by ignoring the subscripts 1 and 2 on fields and vacua, and letting $\tilde{\varphi}_2 \gamma^0 \to \bar{\varphi}$. Feynman diagrams can be appropriated without modification for use in our formulation. We conclude with a quote from Feynman (7): « It always seems odd to me that the fundamental laws of physics can appear in so many different forms that are not apparently identical at first.... Theories of the known which are described by different physical ideas may be equivalent in all their predictions and hence scientifically indistinguishable. However they are not psychologically identical when one is trying to move from that base into the unknown. For different views suggest different modifications ... » In that spirit we turn to the investigation of non-Abelian theories.

3. – Yang-Mills models.

The models we have hitherto investigated are members of a class whose Lagrangians have the form

$$\mathscr{L} = \mathscr{L}_0(\varphi_2) + \mathscr{L}_1(\varphi_1 - \varphi_2) , \tag{55}$$

where $\mathscr{L}_0$ has a form which is essentially identical to a Lagrangian of the conventional formulation. For example, in the case of electrodynamics the Lagrangian of eq. (2) can be put in the form of eq. (55), if we let

$$\mathscr{L}_0(A_{2\mu}, \varphi_2) = -\tfrac{1}{4} F^2_{\mu\nu} F^{2\mu\nu} + \tilde{\varphi}_2 \gamma^0 \big(i(\gamma \cdot \nabla) - e_0(\gamma \cdot A_2) - m\big) \varphi_2 \tag{56}$$

and

$$\begin{aligned} \mathscr{L}_1(A_{1\mu} - A_{2\mu}, \varphi_1 - \varphi_2) = \\ = -\tfrac{1}{4}(F^1_{\mu\nu} - F^2_{\mu\nu})^2 + (\tilde{\varphi}_2 - \tilde{\varphi}_1)\gamma^0 \big(i(\gamma \cdot \nabla) - m\big)(\varphi_1 - \varphi_2) . \end{aligned} \tag{57}$$

At the level of c-number fields it is clear from eqs. (56) and (57) that $A_{2\mu}$ and φ_2 reproduce the usual electrodynamics. At the level of quantum fields one can opt to have a model reproducing the conventional quantum theory—we call it the Feynman case. It maintains the implied decoupling of $A_{1\mu} - A_{2\mu}$ and $\varphi_1 - \varphi_2$ from $A_{2\mu}$ and φ_2. On the other hand, we can take advantage of the larger space of states in the larger-manifestly-indefinite-metric theory and couple the fields together by taking advantage of degrees of freedom in the

Fourier expansion of asymptotic fields and the definition of the vacuum. Since we « perturb around » the asymptotic fields, perturbation theory will reflect this coupling. Our principal-value or quantum action-at-a-distance models are based on this possibility.

We shall use the form of eq. (55) as an ansatz to generate Yang-Mills models in our formulation. This leads to the gauge field Lagrangian

$$\mathcal{L} = -\tfrac{1}{4}\boldsymbol{F}_{2\mu\nu}\cdot\boldsymbol{F}_2^{\mu\nu} + \tfrac{1}{4}(\boldsymbol{G}_{1\mu\nu} - \boldsymbol{G}_{2\mu\nu})\cdot(\boldsymbol{G}_1^{\mu\nu} - \boldsymbol{G}_2^{\mu\nu}), \tag{58}$$

where

$$\boldsymbol{F}_{2\mu\nu} = \partial_\mu \boldsymbol{A}_{2\nu} - \partial_\nu \boldsymbol{A}_{2\mu} + g\boldsymbol{A}_{2\mu}\times\boldsymbol{A}_{2\nu} \tag{59}$$

and

$$\boldsymbol{G}_{i\mu\nu} = \partial_\mu \boldsymbol{A}_{i\nu} - \partial_\nu \boldsymbol{A}_{i\mu} \tag{60}$$

for $i = 1, 2$. Under a local gauge transformation, $S = S(x)$,

$$A_{2\mu} \to A'_{2\mu} = S^{-1}A_{2\mu}S + \frac{i}{g}S^{-1}\partial_\mu S, \tag{61}$$

$$A_{1\mu} \to A'_{1\mu} = A_{1\mu} + A'_{2\mu} - A_{2\mu}, \tag{62}$$

the Lagrangian $\mathcal{L}$ is invariant, where $A_{i\mu} = \boldsymbol{A}_{i\mu}\cdot\boldsymbol{T}$ for $i = 1, 2$ with $\boldsymbol{T}$ a vector composed of matrices representing generators of the gauge group. There is an additional invariance under the local transformation, $\boldsymbol{A}_{1\mu} \to \boldsymbol{A}_{1\mu} + \partial_\mu\boldsymbol{\Lambda}$, which is a straightforward generalization of the Abelian gauge transformation.

The field equations derived from eq. (58), upon varying the action with respect to $\boldsymbol{A}_{1\mu}$ and $\boldsymbol{A}_{2\mu}$, are

$$\partial^\mu(\boldsymbol{G}_{1\mu\nu} - \boldsymbol{G}_{2\mu\nu}) = 0 \tag{63}$$

and

$$(\partial^\mu + g\boldsymbol{A}_2^\mu\times)\boldsymbol{F}_{2\mu\nu} = 0. \tag{64}$$

The canonical momentum conjugate to $\boldsymbol{A}_{2\mu}$ is

$$\boldsymbol{\Pi}_{2\mu} = \boldsymbol{F}_{2\mu0} + \boldsymbol{G}_{1\mu0} - \boldsymbol{G}_{2\mu0} \tag{65}$$

and the momentum conjugate to $\boldsymbol{A}_{1\mu}$ is

$$\boldsymbol{\Pi}_{1\mu} = \boldsymbol{G}_{2\mu0} - \boldsymbol{G}_{1\mu0}. \tag{66}$$

We choose the Coulomb gauge to implement the field quantization. Due to the form of the Lagrangian, we can treat $\boldsymbol{A}_{2\mu}$ and $\boldsymbol{A}_{3\mu} = \boldsymbol{A}_{1\mu} - \boldsymbol{A}_{2\mu}$ as independent fields. Choosing the Coulomb gauge for $A_{2\mu}$, $\vec{\nabla}\cdot\vec{A}_2 = 0$, we can isolate the independent field quantities and establish their equal-time commutation relations in the well-known manner (*). We also can choose to work in the Coulomb gauge of $A_{3\mu}$, $\vec{\nabla}\cdot\vec{A}_3 = 0$ which is equivalent to $\vec{\nabla}\cdot\vec{A}_1 = 0$. This possibility follows from the invariance of $\mathscr{L}$ under the local « Abelian » transformation of $A_{3\mu}$ or $A_{1\mu}$ mentioned above. The resulting equal-time commutation relations are equivalent to

$$(67)\qquad [G^{\mathrm{T}}_{\alpha 0 i a}(x), A_{\beta j b}(y)] = i(1 - \delta_{\alpha 1}\delta_{\beta 1})\,\delta_{ab}\,\Delta^{\mathrm{tr}}_{ij}(x - y)\,,$$

where $\alpha, \beta = 1, 2$; i and j are spatial- and a and b are internal-symmetry indices. G^{T} is the transverse part of G. One can now proceed to linearize the field equations. The Fourier expansions of the solutions of the linearized equations have the form of eqs. (23) and (24), if an internal-symmetry index is appended to the Fourier component operators. The discussion then reduces to the Abelian case considered in the last section. In particular the Green's function of the linearized model which is relevant to perturbation theory is

$$(68)\qquad G^{22}_{ab\mu\nu}(x - y) = -i\,{}_1\langle 0|T\big(A_{2\mu a}(x)\,A_{2\nu b}(y)\big)|0\rangle_2\,.$$

It can be expressed as a linear combination of a Feynman propagator and its complex conjugate as in eq. (39). The arbitrary angle can be chosen to give Feynman propagators or principal-value propagators.

The perturbation theory is based on the linearized model. In the Feynman case the LSZ reduction of asymptotic non-Abelian quanta is essentially given in eqs. (44) and (45), if appropriate internal-symmetry indices are introduced. Only fields of « type 2 » are introduced by the LSZ reduction. Furthermore, the cubic and quartic terms of the Lagrangian « interaction » term also involve only fields of type 2. If we restrict all couplings to other fields—such as fermions—to involve only $A_{2\mu}$, then only time-ordered products of $A_{2\mu}$ appear in the perturbation theory expansion of the S-matrix. The consequence of this development is that one has constructed a model which makes predictions which are identical to a conventional formulation based on a Lagrangian of the usual type

$$(69)\qquad \mathscr{L} = -\tfrac{1}{4}\boldsymbol{F}_{\mu\nu}\cdot\boldsymbol{F}^{\mu\nu}\,.$$

This can be verified in a path integral approach based on the Lagrangian of eq. (58).

The point of our formulation is that one can make other choices—such as the principal-value propagator choice. (Note that the path integral formulation

(*) E. Abers and B. W. Lee: *Phys. Rep.*, **9** C, 1 (1973).

only has a formal validity in these cases, since the Wick rotation to Euclidean co-ordinates is highly nontrivial.) The choice of principal-value propagators has several important consequences.

In the appendix we show how to calculate the unitary physical S-matrix in this case. We show that loops of principal-value propagators do not contribute to the absorptive parts of S-matrix elements. As a result Faddeev-Popov ghost loops are not needed to maintain unitarity (in any gauge). We also point out a radical possibility: a unitary S-matrix can be defined in which all non-Abelian boson loops are excluded. Thus the non-Abelian boson sector consists solely of tree diagrams. This is a quantum analogue of the absence of self-interactions in classical action-at-a-distance theories.

4. – Non-Abelain model of the strong interactions.

Several years ago a non-Abelian model of the strong interaction was constructed (9) which had manifest quark confinement and a linear potential. The model was based on higher-order field equations. Such equations lead to well-known unitarity problems, if the usual quantization program is implemented. The author chose to avoid the unitarity problems through an *ad hoc* quantization procedure which gave principal-value propagators. The result was a model having unconventional analyticity properties. Despite the successful confinement scheme in this model, there was some reason for uneasiness, due to the unusual form of the Lagrangian (where two sets of fields were associated with the gluons) and the *ad hoc* method of quantization.

The new framework we have developed in ref. (1) and this paper eliminates these sources of concern. Our formulation is based on fundamental ground—the need for an acceptable physical particle interpretation of quantum field theory in flat and curved space-time. We shall see that the two-field Lagrangian of the strong-interaction model lies within the general framework we have established. The principal-value propagators, required to have a unitary model, will also be obtained in a straightforward manner.

Since the strong-interaction model is described in some detail in ref. (9), we shall only outline the aspects necessary to connect that model with the present work.

The strong-interaction Lagrangian departs slightly from the ansatz of eq. (55) (mainly in the replacement $\boldsymbol{G}^1_{\mu\nu}\cdot\boldsymbol{G}^{1\mu\nu}\to\Lambda^2\boldsymbol{A}_{1\mu}\cdot\boldsymbol{A}_1^\mu$):

$$\mathscr{L} = -\tfrac{1}{2}\boldsymbol{F}^1_{\mu\nu}\cdot\boldsymbol{F}^{2\mu\nu} - \tfrac{1}{2}\Lambda^2\boldsymbol{A}_{1\mu}\cdot\boldsymbol{A}_1^\mu + \bar{\psi}_2\gamma^0\big(i(\gamma\cdot\nabla) + g(\gamma\cdot A_2) - m\big)\psi_2 + (\bar{\psi}_1 - \bar{\psi}_2)\gamma^0\big(i(\gamma\cdot\nabla) - m\big)(\psi_1 - \psi_2)\,. \tag{70}$$

(9) S. BLAHA: *Phys. Rev. D*, **10**, 4268 (1974); **11**, 2921 (1975). It should be noted that the claim that loops of principal-value propagators are identically zero is not true. In the case of one-loop diagrams only the real part is zero.

where ψ_1 and ψ_2 are the quark fields and

$$F^2_{\mu\nu} = \partial_\mu \boldsymbol{A}_{2\nu} - \partial_\nu \boldsymbol{A}_{2\mu} + g\boldsymbol{A}_{2\mu} \times \boldsymbol{A}_{2\nu} \tag{71}$$

and

$$\boldsymbol{F}^1_{\mu\nu} = D_\mu \boldsymbol{A}_{1\nu} - D_\nu \boldsymbol{A}_{1\mu} \tag{72}$$

with the covariant derivative defined by

$$D_\mu = \partial_\mu + g\boldsymbol{A}_{2\mu} \times . \tag{73}$$

The Lagrangian is invariant under the local color SU_3 gauge transformation

$$\psi_2 \to S^{-1}\psi_2 , \tag{74}$$

$$\psi_1 \to \psi_1 + (S^{-1} - I)\psi_2 , \tag{75}$$

$$A_{1\mu} \to S^{-1}A_{1\mu}S , \tag{76}$$

$$A_{2\mu} \to S^{-1}A_{2\mu}S + \frac{i}{g}S^{-1}\partial_\mu S , \tag{77}$$

with $A_{i\mu} = \boldsymbol{A}_{i\mu}\cdot\boldsymbol{T}$, where $\boldsymbol{T}$ is a vector composed of matrices representing SU_3 generators. The equations of motion and the Coulomb gauge quantization are discussed in detail in ref. (9). (The subscripts, 1 and 2, on the gluon fields are reversed in ref. (9) relative to the present development.)

The essential feature of the model can be seen in the linearized field equations

$$\Box \boldsymbol{A}_{2\mu} = -\Lambda^2 \boldsymbol{A}_{1\mu} , \tag{78}$$

$$\Box \boldsymbol{A}_{1\mu} = -g\boldsymbol{J}_\mu , \tag{79}$$

where $\boldsymbol{J}_\nu$ is the color quark current. Equations (78) and (79) show an apparent mass term, $\Lambda^2(\boldsymbol{A}_{1\mu})^2$, actually leads to a higher-order derivative field equation. The implications of this feature for quark confinement are discussed in ref. (9). The discussion is based on the assumption that all gluon propagators are principal-value propagators. We shall now show that the propagators for color gluons in the linearized model can be principal-value propagators.

We work in the Coulomb gauge, $\vec{\nabla}\cdot\vec{\boldsymbol{A}}_2 = 0$. A mode expansion of the gluon fields must be consistent with the free linearized field equations and the equal-time commutation relations,

$$[F^{\lambda T}_{0ia}(x), A_{\beta jb}(y)] = i\delta_{ab}(1 - \delta_{\lambda\beta})\delta^{tr}_{ij}(\vec{x} - \vec{y}) \tag{80}$$

(cf. eqs. (65) and (66) of ref. (9)). These conditions and the requirement of a unitary theory imply the Fourier expansions

(81) $$\vec{A}_{1a}(x) = \int \frac{\mathrm{d}^3k}{\sqrt{2}} \sum_{\lambda=1}^{2} \vec{\varepsilon}(k, \lambda)[(a_{1k\lambda a} + a_{2k\lambda a}) f_k(x) + (a^\dagger_{1k\lambda a} + a^\dagger_{2k\lambda a}) f_k^*(x)],$$

(82) $$\vec{A}_{2a}(x) = \int \frac{\mathrm{d}^3k}{\sqrt{2}} \sum_{\lambda=1}^{2} \vec{\varepsilon}(k, \lambda)[(a_{2k\lambda a} - a_{1k\lambda a}) f_k(x) + (a^\dagger_{2k\lambda a} - a^\dagger_{1k\lambda a}) f_k^*(x)] +$$

$$+ \Lambda^2 \int \frac{\mathrm{d}^4k}{\sqrt{2}} 2\omega_k \theta(k_0) \delta'(k^2) \sum_{\lambda=1}^{2} \vec{\varepsilon}(k, \lambda)[(a_{1k\lambda a} + a_{2k\lambda a}) f_k(x) + (a^\dagger_{1k\lambda a} - a^\dagger_{2k\lambda a}) f_k^*(x)],$$

where $\delta'(k^2) = \partial\delta(k^2)/\partial k^2$ and $\omega_k = |\vec{k}|$. From eqs. (81) and (82) we can determine the Green's functions of the linearized model:

(83) $$iG^{12}_{\mu\nu ab}(x - y) = {}_1\langle 0|T(A_{1\mu a}(x) A_{2\nu b}(y))|0\rangle_2 =$$

$$= -i\delta_{ab} \int \frac{\mathrm{d}^4k}{(2\pi)^4} \exp[-ik\cdot(x - y)] r^{12}_{\mu\nu} P \frac{1}{k^2}$$

and

(84) $$iG^{22}_{\mu\nu ab}(x - y) = {}_1\langle 0|T(A_{2\mu a}(x) A_{2\nu b}(y))|0\rangle_2 =$$

$$= -i\Lambda^2 \delta_{ab} \int \frac{\mathrm{d}^4k}{(2\pi)^4} \exp[-ik\cdot(x - y)] r^{22}_{\mu\nu} P \frac{1}{k^4},$$

where

(85) $$P \frac{1}{k^{2N}} = \frac{1}{2}\left[\frac{1}{(k^2 + i\varepsilon)^N} + \frac{1}{(k^2 - i\varepsilon)^N}\right]$$

and where $r^{12}_{\mu\nu}$ and $r^{22}_{\mu\nu}$ are gauge-dependent tensors. While the path integral formulation is only of formal value for the present model, it can be used to determine the form of the Green's functions for any choice of gauge. It can also be used to generate the perturbation theory rules for the model. The character of the perturbation theory is described in ref. (9). Confinement arises through the Schwinger mechanism in a manner reminiscent of the two-dimensional Schwinger model. The mass scale is set by Λ. The Lagrangian of eq. (70) leads to an interaction between quarks which has the linear potential r as its Coulomb potential in lowest order. The r potential has had notable success in the explication of the charmonium system. A r^{-1} term may also be needed to successfully describe charmonium. Such a term can be introduced in our model by adding another interaction term

(86) $$\mathscr{L}_r = g' \bar{\psi}_2 \gamma^0 (\gamma\cdot A_1) \psi_2$$

to eq. (70). Since $A_{1\mu}$ transforms homogeneously under a gauge transformation, the gauge invariance of the Lagrangian is not altered by the addition of this term.

In conclusion, we have established the basis for a manifestly quark-confining model of the strong interaction within the framework of our formulation of quantum field theory. This example illustrates the value of our formulation in the construction of unitary quantum field theories with higher-order derivative field equations. Another interesting application of this approach would be to quantum gravity which may have its renormalization problems ameliorated by introducing higher-order derivatives. Certain formal similarities, and this possibility, have led the author to propose a unified model of vierbein gravity and the strong interaction (10). The higher-order derivative gravity terms which lead to power counting renormalizability for the quantum gravity sector are linked to the higher-order derivative strong-interaction sector terms which lead to quark confinement. The recent discovery of important spin effects in high-energy p-p scattering (11) is suggestive in view of the presence of strong spin-spin interactions in unified models of the strong-interaction and vierbein gravity.

5. - Model of quantum gravity.

In this section we develop a trivially renormalizable model of quantum gravity (based on the Einstein Lagrangian) which is coupled to an external classical source. In addition to obtaining a physically interesting model, we shall see the utility of principal-value propagators in ameliorating divergence problems. The self-interactions of the « gravitons » will be reduced if principal-value propagators are used. This corresponds to the absence of self-interactions in classical action-at-a-distance models.

The model which we shall develop lies within the framework established above. It can be described as quantum action-at-a-distance gravity. We begin with the Einstein action

$$I = -2\varkappa^{-2}\int \mathrm{d}^4x\, g^{\frac{1}{2}} R \tag{87}$$

with $\varkappa = 32\pi G$, where G is Newton's constant, $g = \det g_{\mu\nu}$, $R = g^{\mu\nu}R^{\lambda}_{\mu\nu\lambda}$ and where

$$R^{\alpha}_{\mu\nu\lambda} = \partial_\nu \Gamma^{\alpha}_{\mu\lambda} + \Gamma^{\alpha}_{\nu\beta}\Gamma^{\beta}_{\mu\lambda} - (\nu \leftrightarrow \lambda) \tag{88}$$

and

$$\Gamma^{\alpha}_{\mu\nu} = \frac{1}{2} g^{\alpha\beta}[\partial_\mu g_{\nu\beta} + \partial_\nu g_{\mu\beta} - \partial_\beta g_{\mu\nu}]\,. \tag{89}$$

---

(10) S. BLAHA: *Lett. Nuovo Cimento*, **18**, 60 (1977).

(11) J. R. O'FALLON, L. G. RATNER, P. F. SCHULTZ, K. ABE, R. C. FERNOW, A. D. KRISCH, T. A. MULERA, A. J. SALTHOUSE, B. SANDLER, K. M. TERWILLIGER, D. G. CRABB and P. H. HANSEN: *Phys. Rev. Lett.*, **39**, 733 (1977).

Our approach is analogous to Feynman's covariant quantization around a flat background field ([12]). We therefore let

$$g_{\mu\nu} = \eta_{\mu\nu} + \varkappa h^2_{\mu\nu}, \tag{90}$$

where $\eta_{\mu\nu} = (-1, 1, 1, 1)$. We then introduce a second tensor field, $h^1_{\mu\nu}$; and choose the action for our model to be (cf. eq. (55))

$$I = \int \mathrm{d}^4x [\tfrac{1}{2}(h^3_{\mu\nu,\sigma})^2 - (h^3_\mu)^2 - \tfrac{1}{2}(h^3_{,\mu})^2 - 2\varkappa^{-2} g^{\frac{1}{2}} R], \tag{91}$$

where $h^3_{\mu\nu} = h^1_{\mu\nu} - h^2_{\mu\nu}$, $h^3_\mu = h^3_{\mu\nu,\nu}$, $h^3_{,\mu} = h_{\nu\nu,\mu} = \partial_\mu h$ and where the last term on the right-hand side of eq. (91) is expanded in $h^2_{\mu\nu}$ by using eq. (90). The gauge invariance of the Lagrangian is maintained by requiring $h^1_{\mu\nu}$ and $h^2_{\mu\nu}$ to satisfy the same transformation law

$$h^i_{\mu\nu} \to h^{i\prime}_{\mu\nu} = h^i_{\mu\nu} - \partial_\nu \varepsilon_\mu - \partial_\mu \varepsilon_\nu, \tag{92}$$

where ε_μ are four small, but otherwise arbitrary, functions of the co-ordinates. We introduce a de Donder harmonic gauge fixing term

$$\mathscr{L}_{\mathrm{D}} = -\frac{1}{2\gamma}\left(h^2_\mu - \frac{1}{2} h^2_{,\mu}\right)^2 + \frac{1}{2\gamma}\left(h^3_\mu - \frac{1}{2} h^3_{,\mu}\right)^2, \tag{93}$$

in order to obtain a regular kinetic matrix from the quadratic terms of the augmented Lagrangian. The quadratic terms are (if we let $\gamma = \frac{1}{2}$)

$$\mathscr{L}_{\text{quad}} = -h^1_{\alpha\beta,\mu} h^2_{\alpha\beta,\mu} + \tfrac{1}{2} h^1_{,\mu} h^2_{,\mu} + \tfrac{1}{2}(h^1_{\alpha\beta,\mu})^2 - \tfrac{1}{4}(h^1_{,\mu})^2 . \tag{94}$$

These terms plus the higher-order « interaction » terms can be introduced into a suitable path integral ([13]) expression and the perturbation theory rules can be developed. The path integral approach is only of formal significance, in general, in our models. It gives the correct algebraic expressions and combinational rules. The nontriviality of Wick rotation to Euclidean co-ordinates in any case, but the Feynman case, precludes attributing anything more than formal value to the path integral approach. Of course, the applicability of the path integral formalism to the Feynman case, and the absence of any difference at the algebraic level (before integrations) between the Feynman and principal-value cases, means that the path integral formalism also describes the combinatorics of the principal-value case.

([12]) R. FEYNMAN: *Acta Phys. Polonica*, **24**, 697 (1963).
([13]) V. N. POPOV and L. FADEEV: Kiev Report. No. ITP 67-36 (1967).

In order to have a nontrivial model in the principal-value case, we shall assume the gravitational field is coupled to an external classical source. Note that covariance requires that only $h^2_{\mu\nu}$ couples to that source. This fact, plus the absence of $h^1_{\mu\nu}$ in higher-order terms, implies that the only free-field propagator necessary to develop perturbation theory is the vacuum expectation value of the time-ordered product of free $h^2_{\mu\nu}$ fields:

$$(95)\qquad iG^{22}_{\mu\nu,\varrho\sigma} = {}_1\langle 0|T\big(h^2_{\mu\nu}(x)h^2_{\varrho\sigma}(y)\big)|0\rangle_2 = \\ = -\frac{i}{2}(\eta_{\mu\varrho}\eta_{\nu\sigma} + \eta_{\mu\sigma}\eta_{\nu\varrho} - \eta_{\mu\nu}\eta_{\varrho\sigma})\int \frac{d^4k}{(2\pi)^4}\,\frac{\exp[-ik\cdot(x-y)]}{k^2}\,.$$

If one chooses the propagator in eq. (95) to be a Feynman propagator, then the usual model of quantum gravity results with its attendant renormalizability problems.

We shall consider the case, in which G^{22} is a principal-value propagator. One can establish this case by following a canonical quantization procedure which is completely analogous to those of models considered earlier. The propagation of « gravitons » by principal-value propagators has important consequences for perturbation theory. All graviton loops can be excluded from the physical S-matrix without leading to unitarity problems (cf. appendix). As a result, only tree diagrams occur and unitarity requires that all gravitons are absorbed—either within trees or on the external classical source. Thus we have a model quantum gravity with no renormalization problem. In the classical limit the model becomes Einstein's classical theory of gravity ([14]). There is, of course, the question of whether the classical limit is a retarded model of gravity or not. The answer lies in cosmology—is the absorber mechanism operative?

An action-at-a distance model of gravity with absorber seems to be very much in the spirit of Mach's principle. The usual approach—where the gravitational field is treated as having its own degrees of freedom—almost invariably leads to the surrepitious introduction of absolute space. The reason is simple. The solution of the field equations for a localized mass distribution is asymptotically flat. The only apparent way to avoid this problem is to say Mach restricts us to the class of closed-universe solutions.

In an action-at-a-distance gravity ([15]) the metric field exists only in the presence of matter, so that any asymptotic flatness of solutions is a mathematical

---

([14]) There is an opinion which is sometimes expressed that a tree diagram model is equivalent to a classical theory. This is not true. The correct view is that the classical limit of a tree diagram model is the corresponding classical theory. Cf. S. DESER's paper in *Quantum Gravity*, edited by C. J. ISHAM, R. PENROSE and D. W. SCIAMA (Oxford, 1975).

([15]) A. WHITEHEAD: *The Principle of Relativity* (Cambridge, 1922).

artifact. The metric properties of space are a consequence of the presence of mass-energy. (Consistent with this view, we note that mass-energy must be present to measure the metric properties of space.)

The ultimate realization of Mach's principle from the present viewpoint requires that inertia be the result of the gravitational effects of distant matter. Preliminary work ([16]) in this direction can be easily incorporated within the framework of an action-at-a-distance model of gravity. In fact, Sciama's model of the inertial effects of the distant stars displays a close analogy to the Feynman-Wheeler absorber model.

In closing we remark that the introduction of quantum matter fields in our action-at-a-distance quantum gravity appears to result in a nonrenormalizable model. It seems that a higher-order derivative Lagrangian of the type mentioned in the last section may be necessary for a fully renormalizable quantum gravity. Of course, that would also be a quantum action-at-a-distance model.

APPENDIX

In this appendix we define a unitary physical S-matrix for action-at-a-distance quantum field theories. The usual formal definition of the S-matrix is unacceptable because it introduces negative-metric states which lead to negative probabilities.

We begin by defining the set of physical asymptotic states to be those states of positive metric which do not contain quanta of action-at-a-distance fields. In an action-at-a-distance electrodynamics the set of physical states includes all electron and positron states not containing « action at a distance » photons. The problem is to define a unitary S-matrix taking « in » physical states to « out » physical states.

We choose to use a variation of Bogoliubov's procedure ([17]) for defining a unitary physical S-matrix as implemented by SUDARSHAN and co-workers ([18]). The starting point is the expansion of the S-matrix in « old fashioned » perturbation theory:

$$S_{\beta\alpha} = \delta_{\beta\alpha} - i(2\pi)^4\delta^4(P_\beta - P_\alpha)T_{\beta\alpha} \tag{A.1}$$

with

$$T_{\beta\alpha} = \langle\beta|H_I|\alpha\rangle + \sum_N \frac{\langle\beta|H_I|N\rangle\langle N|H_I|\alpha\rangle}{E_\alpha - E_N + i\varepsilon} + \dots, \tag{A.2}$$

([16]) D. W. SCIAMA: *Roy. Astron. Soc. Month. Not.*, **113**, 34 (1953).

([17]) N. N. BOGOLIUBOV: *Annales of Invitational Conference on High-Energy Physics*, CERN (1958).

([18]) E. C. G. SUDARSHAN: *Fields and Quanta*, **2**, 175 (1972); C. A. NELSON: Louisiana State University preprint (1972); J. L. RICHARD: *Phys. Rev. D*, **7**, 3617 (1973); C. C. CHIANG: University Göteborg preprint (1972), and references therein.

where H_I is the interaction Hamiltonian. We wish to modify eq. (A.2) so that states with «action-at-a-distance quanta» do not contribute to the absorptive part of amplitudes. This would allow unitarity to be maintained under the restriction of unitarity sums to the set of physical states. SUDARSHAN and co-workers ([18]) have shown that this can be done by taking the energy denominator factor of unphysical states in principal value:

$$(\text{A.3}) \qquad T_{\beta\alpha} = \langle\beta|H_I|\alpha\rangle + \sum_p \frac{\langle\beta|H_I|p\rangle\langle p|H_I|\alpha\rangle}{E_\alpha - E_p + i\varepsilon} + \sum_u \langle\beta|H_I|u\rangle\langle u|H_I|\alpha\rangle P\frac{1}{E_\alpha - E_u} + \dots,$$

where p labels physical states and u labels unphysical states (*i.e.* those containing action-at-a-distance quanta in our case). Equation (A.3) defines the physical T-matrix in our models. It could serve as the basis for calculating S-matrix elements.

However, we would like to re-express this S-matrix in terms of covariant perturbation theory. In order to do this we shall take advantage of Weinberg's study ([19]) of the infinite-momentum frame limit of old-fashioned perturbation theory, and Chang and Ma's realization of infinite-momentum frame results by a change of variable ([20]).

Since the difference between the physical T-matrix and the unmodified T-matrix lies only in the character of the singularity in the energy denominator, the algebraic development of Weinberg, and his power counting arguments, in particular, can be taken over without change to our case. (We also will ignore possible complications due to interchanging the order of integration and the $P \to \infty$ limit. Actually we define our T-matrix to be the result of this procedure.) The result is that Weinberg's rules apply in the present case also —except that Weinberg's rule (d) must be modified, so that an energy denominator is taken in principal value, if the corresponding intermediate state is unphysical—*i.e.* contains action-at-a-distance quanta.

CHANG and MA showed that it was not necessary to go to the infinite-momentum frame limit in order to realize Weinberg's rules. They showed that the expression of the usual Feynman rules in terms of infinite momentum frame variables, $\eta = p^0 + p^3$ and $s = p^0 - p^3$ for each momentum p^μ, led to the same results as Weinberg's rules. This fact can be used to advantage in the case of action-at-a-distance models. S-matrix elements can be calculated in our models from the usual Feynman rules in the following way: i) for a given diagram follow the usual Feynman rules to obtain a S-matrix contribution using infinite-momentum frame variables *à la* Chang and Ma—in particular, associate Feynman propagators with action-at-a-distance particle lines; ii) evaluate the «energy» integrals by complex integration; iii) the result will be a series of terms corresponding to different intermediate states in old-fashioned perturbation theory language. Those denominators corresponding to intermediate states with principal-value quanta should be taken in principal value. Those denominators corresponding to physical intermediate states should remain unmodified.

([19]) S. WEINBERG: *Phys. Rev.*, **150**, 1313 (1966).
([20]) S. J. CHANG and S. K. MA: *Phys. Rev.*, **180**, 1506 (1969).

Let us examine the results of this procedure for some simple cases. First, consider the pole diagram for two-particle-to-two-particle scattering. If the pole corresponds to a principal-value particle the amplitude will be proportional to

$$P\frac{1}{q^2-m^2}, \tag{A.4}$$

using the above rules ([21]). This result also follows from explicit evaluation of the diagram in canonical perturbation theory, using eq. (1) of the text for $\theta = \pi/4$.

Next consider a second-order self-energy correction due to the emission and absorption of a principal-value quantum by a particle (propagating via Feynman propagators). Our modified Feynman rules require us to evaluate an integral

$$I = i\int d^4k \frac{1}{k^2-m^2+i\varepsilon}\,\frac{1}{(p-k)^2-\mu^2+i\varepsilon}, \tag{A.5}$$

using-infinite-momentum frame variables:

$$\eta' = k^0 + k^3 \qquad s' = k^0 - k^3 \qquad \vec{q} = (k^1, k^2) \tag{A.6}$$

and $p = (s, \eta, \vec{0})$. The result is

$$I = \frac{\pi}{2}\int d^2q \int\limits_0^\eta \frac{d\eta'}{\eta'(\eta-\eta')}\,\frac{1}{s-(\vec{q}^{\,2}+\mu^2)/(\eta-\eta')-(\vec{q}^{\,2}+m^2)/\eta'+i\varepsilon} \tag{A.7}$$

which must be modified to

$$I' = \frac{\pi}{2}\int d^2q \int\limits_0^\eta \frac{d\eta'}{\eta'(\eta-\eta')}\,P\frac{1}{s-(\mathbf{q}^2+\mu^2)/(\eta-\eta')-(\mathbf{q}^2+m^2)/\eta'}. \tag{A.8}$$

In this case we see that $2I' = I + I^*$. The propagation of the particle which would ordinarily propagate via Feynman propagators is manifestly different in states where virtual action-at-a-distance quanta are present. This is necessary in order to define a unitary physical S-matrix. Thus the physical S-matrix only agrees with the canonically defined S-matrix for one-principal-value particle intermediate states and states without principal value particles.

If we consider a non-Abelian action-at-a-distance gauge theory coupled to an external classical source, it is possible to define the physical S-matrix in

(21) Compare to eqs. (5) and (6) of ref. (9).

a different manner from the above. Consider the usual definition of the S-matrix and in particular the set of tree diagrams with no external principal-value particle lines. It is easy to verify that the absorptive part of any tree diagram is zero *in the physical region* since

$$\text{Abs}\left[P\frac{1}{k^2-m^2}\right]=0\,. \tag{A.9}$$

As a result the set of tree diagrams defines a unitary (gauge invariant) S-matrix. (This should be contrasted with the usual theory where taking the absorptive part of a tree diagram introduces states containing non-Abelian bosons which in turn introduce loops and Fadeev-Popov ghosts.) In the present case no Fadeev-Popov ghost loops are needed to maintain unitarity ([22]). Thus we wind up with a loopless, tree diagram model. The possibility of limiting the self-interaction of fields in this way allows us to develop a renormalizable model of quantum gravity.

Finally we note that the lack of contributions to unitarity sums from intermediate states containing principal-value particles allows us to avoid the introduction of Fadeev-Popov ghosts in *all* non-Abelian action-at-a-distance models.

([22]) Cf. ref. ([12]) and B. W. LEE and J. ZINN-JUSTIN: *Phys. Rev. D*, **5**, 3121 (1972).

● RIASSUNTO (*)

Si formulano teorie di gauge nel sistema di una generalizzazione della teoria quantistica dei campi. In particolare si discutono modelli di teorie di elettrodinamica e di Yang-Mills, un modello dell'interazione forte con derivate di ordine più alto e confinamento dei quark, e un modello rinormalizzabile di gravità quantistica pura con una Lagrangiana di Einstein. Nel caso dell'elettrodinamica si mostra che due modelli sono possibili: uno con predizioni che sono identiche a QED e uno che è un modello quantistico di azione a distanza dell'elettrodinamica. Nel caso delle teorie di Yang-Mills si può costruire un modello che è identico per quanto riguarda le predizioni a qualsiasi modello convenzionale o modello di azione a distanza. Nel secondo caso è possibile eliminare tutti i cappi di particelle di Yang-Mills (in tutti i gauge) in una maniera consistente con l'unitarietà. Esiste una variazione dei modelli di Yang-Mills nella nostra formulazione che ha equazioni di campo con derivate ad ordine più alto. È unitario ed ha probabilità positive. Può essere usato per costruire un modello d'interazioni forti che ha un potenziale lineare e manifesta confinamento dei quark. Infine si mostra come costruire un modello di azione a distanza della gravità quantistica pura (il cui limite classico è la dinamica della Lagrangiana di Einstein) accoppiato ad una sorgente esterna classica. Il modello è grossolanamente rinormalizzabile.

(*) *Traduzione a cura della Redazione.*

Новый подход к калибровочным теориям поля.

Резюме (*). — Мы формулирусм калибовочные теории в рамках обобщения квантовой теории поля. В частности, мы обсуждаем модели электродинамики и теорий Янга-Миллса, модель сильных взаимодействий с производными высших порядков и удержанием кварков и перенормируемую модель для чистой квантовой гравитации с Лагранжианом Эйнштейна. Мы показываем, что в случае электродинамики возможны две модели: одна модель имеет предсказания, которые идентичны предсказаниям квантовой электродинамики, и другая модель, которая представляет модель элзктродинамики с квантовым действием на расстоянии. В случае теорий Янга-Миллса мы можем сконструировать модель, которая идентична по предсказаниям любой общепринятой модели или модели с квантовым действием на расстоянии. Во втором случае имеется возможность исключить все петли для частиц Янга-Миллса (во всех калибровках). Наша формулировка содержит изменение моделей Янга-Миллса, которое включает уравнения поля с производными высших порядков. Наша модель является унитарной и имеет положительные вероятности. Наша формулировка может быть использована для конструирования модели сильных взаимодействий, которая имеет линейный потенциал и обеспечивает удержание кварков. Мы показываем, как сконструировать модель действия на расстоянии для квантовой гравитации, связанной с внешним классическим источником. Предложенная модель является перенормируемой.

(*) *Переведено редакцией.*

Direttore responsabile: CARLO CASTAGNOLI

Stampato in Bologna dalla Tipografia Compositori coi tipi della Tipografia Monograf
Questo fascicolo è stato licenziato dai torchi il 17-I-1979

PHYSICAL REVIEW D VOLUME 17, NUMBER 4 15 FEBRUARY 1978

Embedding classical fields in quantum field theories

Stephen Blaha*
Physics Department, Syracuse University, Syracuse, New York 13210
(Received 2 August 1976; revised manuscript received 7 November 1977)

We describe a procedure for quantizing a classical field theory which is the field-theoretic analog of Sudarshan's method for embedding a classical-mechanical system in a quantum-mechanical system. The essence of the difference between our quantization procedure and Fock-space quantization lies in the choice of vacuum states. The key to our choice of vacuum is the procedure we outline for constructing Lagrangians which have gradient terms linear in the field variables from classical Lagrangians which have gradient terms which are quadratic in field variables. We apply this procedure to model electrodynamic field theories, Yang-Mills theories, and a vierbein model of gravity. In the case of electrodynamics models we find a formalism with a close similarity to the coherent-soft-photon-state formalism of QED. In addition, photons propagate to $t = +\infty$ via retarded propagators. We also show how to construct a quantum field for action-at-a-distance electrodynamics. In the Yang-Mills case we show that a previously suggested model for quark confinement necessarily has gluons with principal-value propagation which allows the model to be unitary despite the presence of higher-order-derivative field equations. In the vierbein-gravity model we show that our quantization procedure allows us to treat the classical and quantum parts of the metric field in a unified manner. We find a new perturbation scheme for quantum gravity as a result.

I. INTRODUCTION

The relation between classical and quantum systems has been a subject of continuing interest over the years: First, in the original development of quantum mechanics, second, in the study of the classical limit and infrared divergences of quantum-electrodynamic processes,[1,2] and third, in recent attempts to construct strong-interaction models of quark confinement which are for the most part either classical field theory models in search of quantization[3] or quantized gluon models wherein quark confinement is a consequence of infrared behavior.[4,5]

We will describe a new quantization procedure (called pseudoquantization) for field theory which is the analog of Sudarshan's method for embedding a classical-mechanical system in a quantum-mechanical system. It can be used with advantage to either embed a classical field theory in a quantum field theory in such a way as to maintain the classical character of the embedded fields (while studying the interaction between the classical and quantum sectors on essentially the same footing), or to quantize a class of field theories, members of which have been used as models for gravity and as models for the strong interaction with quark confinement.[7-9]

We shall begin (Sec. II) by pseudoquantizing a classical simple harmonic oscillator. This case is of particular importance because of the analogy between the mode amplitudes of a quantum field and the coordinates of a set of simple harmonic oscillators which we will take advantage of in later sections.

In Sec. III we describe the pseudoquantization procedure for field theory. We apply it to electrodynamic models and show that the propagation of photons to $t=+\infty$ is necessarily retarded in this formalism. Further, we display a close analogy between the present formalism and the coherent-soft-photon-state formalism[10] of QED.

In Sec. IV we apply the pseudoquantization procedure to a classical Yang-Mills field. The resulting field theory (with a slight but important modification) has been used as a model for the strong interactions with quark confinement.[7-9] We also apply the pseudoquantization procedure to a vierbein model of gravity and obtain a new perturbation theory for quantum gravity.

In Sec. V we show that principal-value propagators naturally arise in certains sectors of pseudoquantized theories thus verifying an *ad hoc* procedure devised to unitarize a model of quark confinement.[7-9] We also show how to construct a quantum version of action-at-a-distance electrodynamics.

We shall now briefly outline the procedure for embedding a classical-mechanical system in a quantum system.[6] Consider a classical Hamiltonian system with one degree of freedom, and commuting canonical variables, x_1 and p_1, which have the equations of motion

$$\dot{x}_1 = -i[x_1, \hat{H}]\,, \tag{1}$$

$$\dot{p}_1 = -i[p_1, \hat{H}]\,, \tag{2}$$

where defining

$$\hat{H} = -i\left(\frac{\partial H(x_1,p_1)}{\partial p_1}\frac{\partial}{\partial x_1} - \frac{\partial H(x_1,p_1)}{\partial x_1}\frac{\partial}{\partial p_1}\right) \tag{3}$$

allows us to write Hamilton's equations in commutator form. With Sudarshan[6] we define

$$x_2 = i\frac{\partial}{\partial p_1} \tag{4}$$

and

$$p_2 = -i\frac{\partial}{\partial x_1} \tag{5}$$

so that

$$[x_1, x_2] = [p_1, p_2] = 0\,, \tag{6}$$

$$[x_1, p_2] = [x_2, p_1] = i\,, \tag{7}$$

and $\hat{H}$ can now be taken to be the operator

$$\hat{H} = \frac{\partial H(x_1, p_1)}{\partial p_1} p_2 + \frac{\partial H(x_1, p_1)}{\partial x_1} x_2\,. \tag{8}$$

It is now apparent that we can take the above quantities and equations of motion to describe a quantum mechanical system with two degrees of freedom in the "coordinate" representation where the "coordinates" are (x_1, p_1) and the canonical momenta are $\Pi = (p_2, -x_2)$. As we will see below the linearity of $\hat{H}$ in the momenta is crucial for the maintenance of the classical character of x_1 and p_1, and for the observability of the phase-space trajectory. Since we choose to identify the physical observables with the commutative algebra of the coordinate operators, x_1 and p_1, we are led to impose the superselection condition that the momenta, Π, are unobservable. As a result the Hamiltonian and other generators of canonical transformations, which are all linear in the momenta, are also unobservable. However, in each case there is an associated dynamical quantity which is observable.

The required unobservability of the momenta restricts the form of the interaction between a classical-made-quantum system and an inherently quantum system to

$$H_{\text{int}} = \Phi_1 x_2 + \Phi_2 p_2 + X\,, \tag{9}$$

where Φ_1, Φ_2, and X are functions of x_1, p_1, and the quantum system variables. The commutation relations of these functions are also constrained[6] by the superselection rule and the commutativity of the classical variables, x_1 and p_1, and their time derivatives. In the next section we will study the simple harmonic oscillator in order to exemplify the quantum-mechanical case described above and also for direct use in the field-theoretic generalizations of subsequent sections.

II. SIMPLE HARMONIC OSCILLATOR

In this section we discuss the embedding of a classical simple harmonic oscillator in a quantum system. We shall see that the space of states for the indefinite-metric classical-made-quantum system is far larger than the set of states of a classical harmonic oscillator. However, there is a subset of coherent states which may be placed in one-to-one correspondence with the classical harmonic-oscillator states. The classical-made-quantum oscillator is necessarily an indefinite-metric quantum theory for the simple physical reason that the classical bound states cannot have quantized energy levels. Indefinite-metric quantum theories normally have severe problems of physical interpretation. The present work raises the possibility of a partial resolution of some of these problems through a reinterpretation of an indefinite-metric quantum system as a system composed of a classical subsystem interacting with an essentially quantum subsystem of positive metric.

The classical simple harmonic oscillator of frequency ω has the Hamiltonian

$$\mathcal{H} = \frac{1}{2m}(p_1^2 + m^2\omega^2 x_1^2)\,, \tag{10}$$

and the motion is described by

$$x_1 = A\sin(\pi t + \delta)\,, \tag{11}$$

where A and δ are constants. To embed this classical system in a quantum-mechanical system we introduce the variables x_2 and p_2, and, using Eq. (8), obtain the quantum Hamiltonian

$$\hat{H} = \frac{1}{m} p_1 p_2 + m\omega^2 x_1 x_2\,. \tag{12}$$

We eliminate constants by defining (for $i = 1, 2$)

$$x_i = \left(\frac{1}{m\omega}\right)^{1/2} Q_i\,, \tag{13}$$

$$p_i = (m\omega)^{1/2} P_i\,, \tag{14}$$

and

$$\hat{H} = H\omega \tag{15}$$

so that

$$H = P_1 P_2 + Q_1 Q_2\,. \tag{16}$$

The raising and lowering operators are defined by

$$a_j = \frac{1}{\sqrt{2}}(Q_j + iP_j)\,, \tag{17}$$

and

$$a_j^\dagger = \frac{1}{\sqrt{2}}(Q_j - iP_j) \tag{18}$$

for $j = 1, 2$. They have the commutation relations

$$[a_i, a_j] = [a_i^\dagger, a_j^\dagger] = 0\,, \tag{19}$$

$$[a_i, a_j^\dagger] = 1 - \delta_{ij} \tag{20}$$

for $i,j = 1,2$. As a result H is seen to have the form

$$H = \tfrac{1}{2}(a_1 a_2^\dagger + a_2 a_1^\dagger + a_1^\dagger a_2 + a_2^\dagger a_1). \tag{21}$$

The number operators are defined by

$$N_1 = a_2 a_1^\dagger \tag{22}$$

and

$$N_2 = a_2^\dagger a_1 \tag{23}$$

and are not Hermitian. However, their sum is Hermitian and we see that

$$H = N_1 + N_2. \tag{24}$$

The number operators have the following commutation relations with the raising and lowering operators:

$$N_i a_j = a_j(N_i + \delta_{ij} - 1) \tag{25}$$

and

$$N_i a_j^\dagger = a_j^\dagger(N_i - \delta_{ij} + 1) \tag{26}$$

for $i,j = 1,2$.

Up to this point we have maintained a symmetry of the dynamics under the exchange of the subscripts, $1 \longrightarrow 2$. Now we must break that symmetry by choosing a vacuum state which is an eigenstate of Q_1 and P_1 or alternately a_1 and $a_1^\dagger$. The commutativity of Q_1 and P_1 permit this. The observability of Q_1 and P_1 for all time requires it. So we define

$$a_1^\dagger|0\rangle = a_1|0\rangle = 0. \tag{27}$$

As a result $a_2|0\rangle \neq 0$ and $a_2^\dagger|0\rangle \neq 0$. The eigenstates of the number operators are

$$|n_+, n_-\rangle = (a_2^\dagger)^{n_+}(a_2)^{n_-}|0,0\rangle \tag{28}$$

and satisfy

$$N_1|n_+,n_-\rangle = -n_-|n_+,n_-\rangle, \tag{29}$$

$$N_2|n_+,n_-\rangle = n_+|n_+,n_-\rangle, \tag{30}$$

so that

$$H|n_+,n_-\rangle = (n_+ - n_-)|n_+,n_-\rangle. \tag{31}$$

The lack of a lower bound to the energy spectrum is in a sense a problem but a necessary one in that it leads to the possibility of bound states with a continuous energy spectrum—a requirement of a faithful representation of the classical oscillator states. There is a subset of coherent states which can be put in a one-to-one relation with the set of classical oscillator states. The defining property of that subset is that its elements are eigenstates of the operators a_1 and $a_1^\dagger$. If we expand an element of that subset in terms of the number eigenstates

$$|z\rangle = \sum_{n_+,n_-=0}^{\infty} f(z|n_+,n_-)|n_+,n_-\rangle \tag{32}$$

and use

$$a_1^\dagger|n_+,n_-\rangle = -n_-|n_+,n_- - 1\rangle, \tag{33}$$

$$a_1|n_+,n_-\rangle = n_+|n_+ - 1,n_-\rangle \tag{34}$$

to evaluate the eigenvalue equations

$$a_1|z\rangle = iz^*|z\rangle, \tag{35}$$

$$a_1^\dagger|z\rangle = -iz|z\rangle, \tag{36}$$

we find

$$f(z|n_+,n_-) = \frac{C(iz^*)^{n_+}(iz)^{n_-}}{n_+!\,n_-!}, \tag{37}$$

where C is a constant. As a result

$$|z\rangle = C\exp[i(za_2 + z^* a_2^\dagger)]|0,0\rangle. \tag{38}$$

We shall call the $|z\rangle$ states coherent states because of their close formal resemblance to the coherent states used in the study of the classical limit of harmonic oscillators, and of quantum electrodynamics[11] (which were eigenstates of the lowering operator but not of the raising operator).

Since $[H,a_1] = -a_1$, and $[H,a_1^\dagger] = a_1^\dagger$, it is clear that the (x_1,p_1) phase-space trajectory is sharp on the set of coherent $|z\rangle$ states. The classical trajectory represented by the state $|z\rangle$ is easily seen to be

$$x_1 = \left(\frac{2}{m\omega}\right)^{1/2} R\sin(\omega t + \delta) \tag{39}$$

and

$$p_1 = (2m\omega)^{1/2} R\cos(\omega t + \delta), \tag{40}$$

where $z = Re^{i\delta}$. The linearity of H in the "momenta", $\Pi = (p_2, -x_2)$, is crucial for the observability of the phase-space trajectory. In fact, the linearity of all generators of canonical transformations in the momenta is necessary if the canonical transformations are not to take states out of the subset of coherent states.

The superselection rule which follows from the unobservability of the momenta, Π, is best approached by a consideration of the momentum-and coordinate-space representations of the coherent states. In the coordinate-space representation we find that Eqs. (35) and (36) give

$$\left[\left(\frac{m\omega}{2}\right)^{1/2} x_1 + i\left(\frac{1}{2m\omega}\right)^{1/2} p_1\right]\langle x_1 p_1|z\rangle = iz^*\langle x_1 p_1|z\rangle \tag{41}$$

and

$$\left[\left(\frac{m\omega}{2}\right)^{1/2} x_1 - i\left(\frac{1}{2m\omega}\right)^{1/2} p_1\right]\langle x_1 p_1|z\rangle = -iz\langle x_1 p_1|z\rangle, \tag{42}$$

so that

$$\langle x_1 p_1 | z\rangle = \sqrt{2}\,\delta\left(x_1 - \left(\frac{2}{m\omega}\right)^{1/2} \mathrm{Im}z\right) \times \delta(p_1 - (2m\omega)^{1/2}\mathrm{Re}z). \quad (43)$$

We have normalized $\langle x_1 p_1 | z\rangle$ so that

$$\langle z' | z\rangle = \int_{-\infty}^{\infty} dx_1 dp_1 \langle z' | x_1 p_1\rangle\langle x_1 p_1 | z\rangle = \delta(\mathrm{Re}z - \mathrm{Re}z')\delta(\mathrm{Im}z - \mathrm{Im}z'). \quad (44)$$

In momentum space Eqs. (35) and (36) lead to the differential equations

$$\left[\left(\frac{m\omega}{2}\right)^{1/2} i\frac{d}{dp_2} + \left(\frac{1}{2m\omega}\right)^{1/2} \frac{d}{dx_2}\right]\langle x_2 p_2 | z\rangle = iz^*\langle x_2 p_2 | z\rangle \quad (45)$$

and

$$\left[\left(\frac{m\omega}{2}\right)^{1/2} i\frac{d}{dp_2} - \left(\frac{1}{2m\omega}\right)^{1/2} \frac{d}{dx_2}\right]\langle x_2 p_2 | z\rangle = -iz\langle x_2 p_2 | z\rangle . \quad (46)$$

They are easily integrated to give

$$\langle x_2 p_2 | z\rangle = \frac{1}{\sqrt{2\pi}} \exp\left[-ip_2\left(\frac{2}{m\omega}\right)^{1/2} \mathrm{Im}z + ix_2(2m\omega)^{1/2}\mathrm{Re}z\right] \quad (47)$$

with the normalization condition

$$\langle z' | z\rangle = \int_{-\infty}^{\infty} dx_2 dp_2 \langle z' | x_2 p_2\rangle\langle x_2 p_2 | z\rangle = \delta(\mathrm{Re}z - \mathrm{Re}z')\delta(\mathrm{Im}z - \mathrm{Im}z'). \quad (48)$$

The transformation function between the two representations is

$$\langle x_1 p_1 | x_2 p_2\rangle = \frac{1}{2\pi} \exp(+ip_2 x_1 - ip_1 x_2), \quad (49)$$

so that

$$\langle x_1 p_1 | z\rangle = \int_{-\infty}^{\infty} dx_2 dp_2 \langle x_1 p_1 | x_2 p_2\rangle\langle x_2 p_2 | z\rangle . \quad (50)$$

Each coherent state, $|z\rangle$, is a superselection sector in itself. There is no measurable dynamical variable $F = F(a_1, a_1^\dagger)$ which connects different states:

$$\langle z' | F(a_1, a_1^\dagger) | z\rangle = F(iz^*, -iz)\delta^2(z - z'). \quad (51)$$

This reflects the lack of a superposition principle in classical mechanics.

The operator formalism for coherent states is incomplete in that we have not defined an inner product. To remedy this deficiency we define the vacuum dual to $|0,0\rangle$ to satisfy

$$\langle 0,0 | a_2 = \langle 0,0 | a_2^\dagger = 0 \quad (52)$$

with $\langle 0,0|0,0\rangle = 1$. The dual state corresponding to the physical state, z, we define to be

$$\langle z | = \langle 0,0 | \delta(ia_1 + z^*)\delta(ia_1^\dagger - z) \equiv \langle 0,0 | \int_{-\infty}^{\infty} \frac{d\alpha\, d\beta}{(2\pi)^2} \exp[i\alpha(\mathrm{Im}z - 2^{-1/2}Q_1) + i\beta(\mathrm{Re}z - 2^{-1/2}P_1)] \quad (53)$$

so that Eqs. (48) and (51) follow if we choose $C = 1$.

Sometimes the dynamical state of a classical system is incompletely known and one only has a set of probabilities that the system is at a particular phase-space point at $t = 0$. If we let $P(z)$ be the probability that the system is at a phase-space point corresponding to z (as defined above), then using the properties

$$P(z) \geq 0, \quad \int d^2z\, P(z) = 1 \quad (54)$$

one sees that a density operator

$$\rho\delta^2(0) = \int d^2z\, |z\rangle P(z)\langle z| \quad (55)$$

may be defined which satisfies

$$\mathrm{Tr}\rho = 1 \quad (56)$$

and

$$\langle z' | \rho | z'\rangle \equiv \lim_{z'' \to z'} \langle z'' | \rho | z'\rangle = P(z'). \quad (57)$$

The mean value of an observable $A = A(a_1, a_1^\dagger)$ is given by

$$\langle A\rangle = \mathrm{Tr}\rho A = \int d^2z\, A(iz^*, -iz)P(z), \quad (58)$$

and one can develop a formalism similar to the density-matrix formalism of quantum mechanics.

We now turn to a closer investigation of the relation of the pseudoquantum mechanics discussed above and true quantum-mechanical systems. We shall be particularly interested in the relation of the coherent states described above and the coherent states of a quantum-mechanical harmonic oscillator—to which they bear such a remarkable resemblance. We shall see that the pseudoquantum oscillator system is equivalent to an indefinite-metric quantum system composed of a harmonic oscillator (thus the connection to the coherent-state quantum oscillator formalism) and an "inverted" oscillator to be described below.

Let us define the following rotated raising and lowering operators in terms of the operators defined in Eqs. (17) and (18):

$$b_1 = a_1 \cos\theta + a_2 \sin\theta, \quad (59)$$

$$b_2 = -a_1 \sin\theta + a_2 \cos\theta. \quad (60)$$

Their commutation relations are

$$[b_1, b_1^\dagger] = \sin(2\theta), \quad (61)$$

$$[b_2, b_2^\dagger] = -\sin(2\theta), \quad (62)$$

$$[b_2, b_1^\dagger] = [b_1, b_2^\dagger] = \cos(2\theta) \quad (63)$$

with all other commutators equal to zero. The Hamiltonian of Eq. (21) becomes

$$H = \tfrac{1}{2}(\{b_1, b_1^\dagger\} - \{b_2, b_2^\dagger\})\sin(2\theta) + \tfrac{1}{2}(\{a_1, a_2^\dagger\} + \{a_2, a_1^\dagger\})\cos(2\theta), \quad (64)$$

where $\{u, v\} = uv + vu$.

Now θ is an arbitrary angle and it is obvious that choosing $\theta = 0$ gives the commutation relations and Hamiltonian studied above. However, the choice $\theta = \pi/4$ results in a new form for H and the commutation relations, which can be interpreted as a harmonic oscillator (the b_1 and $b_1^\dagger$ sector) and an "inverted" harmonic oscillator (the b_2 and $b_2^\dagger$ sector) where the commutator and b_2 terms in the Hamiltonian have the wrong sign. The commutativity of the oscillator raising and lowering operators with the inverted oscillator raising and lowering operators leads to a simple factorization of the coherent states which lays bare the basic of the close similarity of form for our coherent states and the coherent states of a quantum oscillator[10]:

$$|z\rangle = \frac{1}{\sqrt{2\pi}} \exp\left[\frac{i}{\sqrt{2}}(zb_1 + z^* b_1^\dagger)\right] \times \exp\left[\frac{i}{\sqrt{2}}(zb_2 + z^* b_2^\dagger)\right] |0, 0\rangle, \quad (65)$$

while the coherent state of Ref. 11 has the form

$$|\alpha\rangle = \exp(\alpha b^\dagger - \alpha^* b)|0\rangle, \quad (66)$$

where α is a complex numer and $[b, b^\dagger] = 1$. It should be remembered that our choice of vacuum state such that $a_1|0,0\rangle = a_1^\dagger|0,0\rangle = 0$ obviates a simple direct relationship.

Since we have uncovered an interesting relation between a classical-made-quantum system and a "quantum" system of indefinite metric the possibility of reinterpreting indefinite-metric quantum systems as systems containing classical subsystems naturally arises.

III. EMBEDDING OF CLASSICAL FIELDS

In this section we shall discuss the embedding of a classical field theory in a quantum field theory. We shall study the embedding in detail for a scalar field and then describe the features of a classical-made-quantum electrodynamics which we shall call pseudoquantum electrodynamics for the sake of brevity.

Consider a classical field, $\phi_1(x)$, with canonically conjugate momentum, $\pi_1(x)$, and Hamiltonian equations of motion

$$\frac{d}{dt}\phi_1(x) = \frac{\delta \hat{H}}{\delta \pi_1(x)}, \quad (67)$$

$$\frac{d}{dt}\pi_1(x) = \frac{-\delta H}{\delta \phi_1(x)}, \quad (68)$$

where $\hat{H}$ is the Hamiltonian. We wish to define a "quantum" Hamiltonian, H, which allows us to rewrite Eqs. (67) and (68) in commutator form:

$$\frac{d}{dt}\phi_1(x) = i[H, \phi_1(x)], \quad (69)$$

$$\frac{d}{dt}\pi_1(x) = i[H, \pi_1(x)]. \quad (70)$$

Equations (69) and (70) are satisfied if

$$H = \int d^3x \left[\frac{\delta H}{\delta \pi_1(x)} \frac{1}{i} \frac{\delta}{\delta \phi_1(x)} - \frac{\delta H}{\delta \phi_1(x)} \frac{1}{i} \frac{\delta}{\delta \pi_1(x)}\right]. \quad (71)$$

We now formally define

$$\phi_2(x) = i\frac{\delta}{\delta \pi_1(x)} \quad (72)$$

and

$$\pi_2(x) = -i\frac{\delta}{\delta \phi_1(x)}, \quad (73)$$

so that

$$H = \int d^3x \left[\frac{\delta \hat{H}}{\delta \pi_1(x)} \pi_2(x) + \frac{\delta \hat{H}}{\delta \phi_1(x)} \phi_2(x)\right]. \quad (74)$$

The fields satisfy the equal-time commutation relations

$$[\phi_i(x), \pi_j(y)] = i(1 - \delta_{ij})\delta^3(\vec{x} - \vec{y}), \quad (75)$$

$$[\phi_i(x), \phi_j(y)] = 0, \quad (76)$$

$$[\pi_i(x), \pi_j(y)] = 0, \quad (77)$$

where δ_{ij} is the Kronecker δ.

We note that the linearity of H in ϕ_2 and π_2 is necessary to maintain the classical character of ϕ_1 and π_1. This is best seen by an examination of Eqs. (69) and (70) and the corresponding Hamiltonian equations for ϕ_2 and π_2. (Other generators of canonical transformations are also linear in π_2 and ϕ_2.)

$\phi_2(x)$ and $\pi_2(x)$ will not be observables on the set of physical states, so that $\phi_1(x)$ and $\pi_1(x)$ will both be sharp on the set of physical states and satisfy superselection rules.

If we wish to couple the classical field to a truly quantum system and maintain the classical nature of the field then certain restrictions exist on the form of the total Hamiltonian H_{tot} and on the commutation relations of the various terms occurring in it. First, the coupling must satisfy the requirement that H_{tot} is linear in $\phi_2(x)$ and $\pi_2(x)$. If we denote the quantum fields by ψ and write the general form of the Hamiltonian as

$$H_{tot}=H+H_Q(\psi)+H_{int}, \tag{78}$$

where H is given by Eq. (74), $H_Q(\psi)$ depends only on the quantum fields, ψ, and

$$H_{int}=\int d^3x[\tilde{A}(\phi_1,\pi_1,\psi)\phi_2(x)+\tilde{B}(\phi_1,\pi_1,\psi)\pi_2(x)+\tilde{C}(\phi_1,\pi_1,\psi)], \tag{79}$$

then we can rearrange the Hamiltonian so that

$$H_{tot}=\int d^3x[A(\phi_1,\pi_1,\psi)\phi_2(x)+B(\phi_1,\pi_1,\psi)\pi_2(x)+C(\phi_1,\pi_1,\psi)], \tag{80}$$

where

$$A=\frac{\delta\hat{H}}{\delta\phi_1(x)}+\tilde{A}, \tag{81}$$

$$B=\frac{\delta\hat{H}}{\delta\pi_1(x)}+\tilde{B}, \tag{82}$$

and

$$C=\tilde{C}+\mathcal{H}_Q \tag{83}$$

with $H_Q=\int d^3x\,\mathcal{H}_Q$. An examination of the equations of motion of $\phi_1(x)$, $\pi_1,(x)$, and ψ,

$$\frac{d}{dt}\phi_1=B(\phi_1,\pi_1,\psi), \tag{84}$$

$$\frac{d}{dt}\pi_1=A(\phi_1,\pi_1,\psi), \tag{85}$$

$$\frac{d}{dt}\psi=i[H_{tot},\psi], \tag{86}$$

and the second time derivatives of ϕ_1 and π_1, such as

$$\frac{d^2}{dt^2}\phi_1(x)=i[H,B]=\int d^3y\left(-A\frac{\delta B}{\delta\pi_1(y)}+B\frac{\delta B}{\delta\phi_1(y)}+i\phi_2(y)[A,B]+i\pi_2(y)[B(y),B(x)]+i[C,B]\right), \tag{87}$$

leads us to require the equal-time commutation relations

$$[A(x),A(y)]=[A(x),B(y)]=[B(x),B(y)]=0, \tag{88}$$

where $A(x)=A(\phi_1(x),\pi_1(x),\psi(x))$, etc., so that $\phi_1(x)$ and $\pi_1(x)$ are independent of ϕ_2 and π_2 and hence observable for all time. An examination of higher time derivatives of ϕ_1 and π_1 lead to further restrictions on the equal-time commutation relations of A, B, and C. Examples are

$$[A,[C,B]]=0, \tag{89}$$

$$[B,[C,B]]=0, \tag{90}$$

$$[A,[C,[C,[C,B]]]]=0, \tag{91}$$

etc. A sufficient condition for satisfying all relations of this class consists of having equal-time commutation relations with the form

$$[A,C]=F_1(A,B,\phi_1,\pi_1) \tag{92}$$

and

$$[B,C]=F_2(A,B,\phi_1,\pi_1). \tag{93}$$

Finally, we note that another obvious requirement [cf. Eqs. (84) and (85)] for the observability of ϕ_1 and π_1 is that A and B depend only on an (equal-time) commutative subset of the quantum field variables, ψ.

The above restrictions on the equal-time commutation relations have a direct interpretation in terms of Feynman diagrams for quantum corrections to the classical field behavior. For example, consider the interaction of the classical field sector with a scalar quantum field, ψ, expressed in the interaction

$$H_{int}=g\phi_2(x)\psi^2(x). \tag{94}$$

If $H_Q(\psi)$ is the conventional free Klein-Gordon Hamiltonian, then we find that Eq. (92) is not satisfied so that the Green's function for the classical ϕ_1 field receives quantum corrections from vacuum polarization loops of ψ particles and thus loses its classical character.

We now define a Lagrangian appropriate to our pseudoquantum field theory and then verify the reasonableness of our definition, and the pseudoquantization procedure described above, by studying the equivalent path-integral formulation. The Lagrangian corresponding to the pseudoquantum Hamiltonian, H, is

$$L=\int d^3x(\pi_1\dot{\phi}_2+\pi_2\dot{\phi}_1)-H, \tag{95}$$

where $L=L(\phi_1,\dot{\phi}_1,\phi_2,\dot{\phi}_2)$ and

$$\pi_1=\frac{\delta L}{\delta\dot{\phi}_2}, \tag{96}$$

$$\pi_2 = \frac{\delta L}{\delta \dot{\phi}_1}. \tag{97}$$

The vacuum-vacuum transition amplitude for the field theory corresponding to the H_{tot} of Eq. (78) will be shown to be

$$W = \int \prod_x d\phi_1(x) d\phi_2(x) d\pi_1(x) d\pi_2(x) d\psi(x) \exp(iS), \tag{98}$$

where $S = \int dt\, L_{tot}$ up to external source terms. We begin by considering the vacuum-vacuum transition amplitude corresponding to H_Q,

$$W_Q = \int \prod_x d\psi(x) \exp(iS_Q), \tag{99}$$

where ϕ_1 has the character of an external source.

We can now introduce the classical behavior of the ϕ_1 field through functional δ functions

$$\int \prod_x d\psi(x) d\phi_1(x) d\pi_1(x) \delta(B(\phi_1, \pi_1, \psi) - \dot{\phi}_1) \times \delta(A(\phi_1, \pi_1, \psi) + \dot{\pi}_1) e^{iS_Q}, \tag{100}$$

which can be put in the form

$$\int \prod_x d\phi_1(x) d\pi_1(x) d\phi_2(x) d\pi_2(x) \times \exp\left\{ i \int d^4x [(\dot{\phi}_1 - B)\pi_2 - (\dot{\pi}_1 + A)\phi_2] + iS_Q \right\}. \tag{101}$$

After performing a partial integration on the $\dot{\pi}_1\phi_2$ term and discarding a surface term we see that the definition of L in Eq. (95) is correct and that the vacuum-vacuum transition amplitude is indeed given by Eq. (98).

The restrictions on the commutation relations of the various terms in the H_{tot} [expressed in Eqs. (88)–(93)] translate into the requirement that the "quantum completion"[11] of the ϕ_2 field does not take place, i.e., that all N-point functions of the ϕ_2 field are zero:

$$\frac{\delta^n W}{\delta J_2(x_1) \delta J_2(x_2) \cdots \delta J_2(x_n)} = 0, \tag{102}$$

where J_2 is an external source coupled to ϕ_2.

We now discuss the embedding of a free classical Klein-Gordon field in a quantum field theory. The Lagrangian density is

$$\mathcal{L} = \frac{\partial \phi_1}{\partial x^\mu} \frac{\partial \phi_2}{\partial x_\mu} - m^2 \phi_1 \phi_2, \tag{103}$$

from which one obtains the Euler-Lagrange equations (for $i = 1, 2$)

$$(\Box + m^2)\phi_i(x) = 0. \tag{104}$$

The canonical momenta are (note that π_2 is conjugate to ϕ_1, etc.)

$$\Pi_i = \dot{\phi}_i \tag{105}$$

for $i = 1, 2$ with the equal-time commutation relations given by Eqs. (75)–(77). We expand the fields in Fourier integrals:

$$\phi_1(\vec{x}, t) = \int d^3k [a_1(k) f_k(x) + a_1^\dagger f_k^*(x)] \tag{106}$$

and

$$\phi_2(\vec{x}, t) = \int d^3k [a_2(k) f_k(x) + a_2^\dagger(k) f_k^*(x)], \tag{107}$$

where

$$f_k(x) = (2\pi)^{-3/2} (2\omega_k)^{-1/2} e^{-ik\cdot x} \tag{108}$$

with $\omega_k = (\vec{k}^2 + m^2)^{1/2}$. The Fourier component operators satisfy the commutation relations

$$[a_i(k), a_j^\dagger(k')] = (1 - \delta_{ij}) \delta^3(\vec{k} - \vec{k}') \tag{109}$$

and

$$[a_i(k), a_j(k')] = [a_i^\dagger(k), a_j^\dagger(k')] = 0 \tag{110}$$

for $i, j = 1, 2$.

In terms of the Fourier coefficients

$$H \equiv \int d^3x (\dot{\phi}_1 \dot{\phi}_2 + \vec{\nabla}\phi_1 \cdot \vec{\nabla}\phi_2 + m^2 \phi_1 \phi_2) \tag{111}$$

becomes

$$H = \int d^3k\, \omega_k [\{a_1(k), a_2^\dagger(k)\} + \{a_2(k), a_1^\dagger(k)\}]. \tag{112}$$

The analogy between the mode amplitudes of the fields and the raising and lowering operators of the simple harmonic oscillator has been previously remarked. We can therefore use the considerations of Sec. II to establish the spectrum of physical states. The defining properties of a physical state are that $\phi_1(x)$ and $\pi_1(x)$ are sharp on it for all time:

$$\phi_1(x) |\Phi, \Pi\rangle = \Phi(x) |\Phi, \Pi\rangle \tag{113}$$

and

$$\pi_1(x) |\Phi, \Pi\rangle = \Pi(x) |\Phi, \Pi\rangle, \tag{114}$$

where $\Phi(x)$ and $\Pi(x)$ are c-number functions of x:

$$\Phi(x) = \int d^3k [\alpha(k) f_k(x) + \alpha^*(k) f_k^*(x)] \tag{115}$$

and

$$\Pi(x) = -i \int d^3k\, \omega_k [\alpha(k) f_k(x) - \alpha^*(k) f_k^*(x)] \tag{116}$$

with $\alpha(k)$ a c-number function of k.

As a result we are led to define a set of physical states, $|\alpha\rangle$, which are in one-to-one correspondence with the classical solutions of the Klein-Gordon equation and satisfy

$$a_1(k)|\alpha\rangle = \alpha(k)|\alpha\rangle, \qquad (117)$$

$$a_1^\dagger(k)|\alpha\rangle = \alpha^*(k)|\alpha\rangle. \qquad (118)$$

In analogy with the states of the simple harmonic oscillator (Sec. II) we further define

$$|\alpha\rangle = C\exp\left\{\int d^3k'[\alpha(k')a_2^\dagger(k') - \alpha^*(k')a_2(k')]\right\}|0\rangle, \qquad (119)$$

where the vacuum state, $|0\rangle$, satisfies

$$a_1(k)|0\rangle = a_1^\dagger(k)|0\rangle = 0. \qquad (120)$$

The physical states, $|\alpha\rangle$, lie in a space which is the infinite tensor product of single-mode spaces. While ϕ_1 and π_1 are sharp for all time on the subset of physical states, we see that ϕ_2 and π_2 are not and, in fact, when applied to a physical state map it into an unphysical state. The superselection rules are embodied in

$$\langle\alpha'|\mathcal{O}|\alpha\rangle = \mathcal{O}_\alpha\delta^2(\alpha-\alpha'), \qquad (121)$$

where $\mathcal{O}$ is the operator corresponding to any observable, $\mathcal{O}_\alpha$ is its eigenvalue for the state $|\alpha\rangle$, and $\delta^2(\alpha-\alpha')$ is a functional δ function in the real and imaginary parts of $\alpha-\alpha'$. The functional δ functions have their origin in the definition of the dual set of physical states. We define the dual vacuum state $\langle 0|$ by

$$\langle 0|a_2(k) = 0 \qquad (122a)$$

and

$$\langle 0|a_2^\dagger(k) = 0 \qquad (122b)$$

for all k with $\langle 0|0\rangle = 1$. The dual state corresponding to $\alpha(k)$ we define by

$$\langle\alpha| = \langle 0|\prod_k \delta(\alpha(k)-a_1(k))\delta(\alpha^*(k)-a_1^\dagger(k)) \equiv \langle 0|\delta(\alpha-a_1)\delta(\alpha^*-a_1^\dagger), \qquad (123)$$

so that

$$\langle\alpha'|\alpha\rangle = \delta^2(\alpha'-\alpha) \qquad (124)$$

if $C=1$.

We have now established a procedure for embedding a classical field in a quantum field theory. Given a Lagrangian, L, for a classical field theory describing a field $\phi_1(x)$, the Lagrangian density for the pseudoquantum field theory, $\mathcal{L}_{PQ}$ is

$$\mathcal{L}_{PQ}(\phi_1,\dot\phi_1,\phi_2,\dot\phi_2) = \frac{\delta L}{\delta\phi_1(x)}\phi_2(x) + \frac{\delta L}{\delta\dot\phi_1(x)}\pi_2(x) \qquad (125)$$

up to a divergence with

$$\pi_2(x) = \frac{\delta}{\delta\dot\phi_1(x)}\int d^3x\,\mathcal{L}_{PQ}. \qquad (126)$$

In the case of a classical electromagnetic field interacting with a quantum electron field, one pseudoquantum model, which describes some electromagnetic processes, has the Lagrangian

$$\mathcal{L} = -\tfrac{1}{2}F^1_{\mu\nu}F^2_{\mu\nu} + \bar\psi(i\nabla\!\!\!/ - eA\!\!\!/_1 - m_0)\psi, \qquad (127)$$

where $A^1_\mu(x)$ is the classical electromagnetic field, ψ is the electron field, $A^2_\mu(x)$ is the unobservable auxiliary field, and $F^i_{\mu\nu} = \partial_\nu A^i_\mu - \partial_\nu A^i_\mu$ for $i=1,2$. Although our interpretation of the free electromagnetic part of the Lagrangian, $-\frac{1}{2}F^1_{\mu\nu}F^2_{\mu\nu}$, is new, the actual form of this term appeared some time ago in a generalization of electrodynamics by Mie,[12] and was recently used in an Abelian prototype model for quark confinement.[8] The equations of motion are

$$\partial^\mu F^1_{\mu\nu} = 0, \qquad (128)$$

$$\partial^\mu F^2_{\mu\nu} + eJ_\nu = 0, \qquad (129)$$

and

$$(i\nabla\!\!\!/ - eA\!\!\!/^1 - m)\psi = 0. \qquad (130)$$

The canonical momentum which is conjugate to A^1_μ is

$$\Pi^2_\mu = F^2_{0\mu} \qquad (131)$$

and that conjugate to A^2_μ is

$$\Pi^1_\mu = F^1_{0\mu}. \qquad (132)$$

We take A^1_μ and Π^1_μ to be classical fields which are observable for all time. A^2_μ and Π^2_μ are not observable. Note that $\mathcal{L}$ is invariant under the independent gauge transformations

$$A^1_\mu \to A^1_\mu + \partial_\mu\Lambda^1(x) \qquad (133)$$

and

$$A^2_\mu \to A^2_\mu + \partial_\mu\Lambda^2(x). \qquad (134)$$

Since $\Pi^1_0 = \Pi^2_0 = 0$, it is apparent that A^1_0 and A^2_0 are c numbers. If we chose the Coulomb gauge for A^1_μ,

$$\vec\nabla\cdot\vec{A}^1 = 0, \qquad (135)$$

and for A^2_μ,

$$\vec\nabla\cdot\vec{A}^2 = 0, \qquad (136)$$

then we can establish the equal-time commutation relations

$$[\Pi_i^a(\vec{x},t),A_j^b(\vec{y},t)]=i(1-\delta_{ab})\times\int\frac{d^3k}{(2\pi)^3}e^{i\vec{k}\cdot(\vec{x}-\vec{y})}\left(\delta_{ij}-\frac{k_ik_j}{|\vec{k}|^2}\right)=i(1-\delta_{ab})\delta_{ij}^{tr}(\vec{x}-\vec{y}) \quad (137)$$

for $a,b=1,2$ and $i,j=1,2,3$.

This pseudoquantum field theory describes the dynamics of quantum electron fields interacting with a free, classical electromagnetic field. A typical perturbation theory matrix element would have the form

$$\langle \mathcal{a}',0|T(\bar{\psi}(x)J^{\mu_1}(x_1)A^1_{\mu_1}(x_1)J^{\mu_2}(x_2)A^1_{\mu_2}(x_2)\cdots J^{\mu_n}(x_n)A^1_{\mu_n}(x_n)\psi(y))|\mathcal{a},0\rangle, \quad (138)$$

where $|\mathcal{a},0\rangle$ is the tensor product of an electron vacuum state and an electromagnetic state corresponding to the classical field $\mathcal{a}_\mu(z)$. Because $A^1_\mu(x)$ is sharp on this state, the matrix element becomes

$$\langle 0|T(\bar{\psi}(x)J^{\mu_1}(x_1)\cdots J^{\mu_n}(x_n)\psi(y))|0\rangle\mathcal{a}_{\mu_1}(x_1)\mathcal{a}_{\mu_2}(x_2)\cdots\mathcal{a}_{\mu_n}(x_n) \quad (139)$$

modulo a functional δ function in $\mathcal{a}'-\mathcal{a}$. Thus this model is equivalent to a quantized electron field interacting with an external electromagnetic field.

Another possibility for a model electrodynamics is realized by letting the interaction term in Eq. (127) above be replaced with

$$L_{int}=-e\bar{\psi}A\!\!\!/_2\psi. \quad (140)$$

Because the equivalent of the equal-time commutation relation, Eq. (92), is not true in this model, the A^1_μ field loses its purely classical character due to quantum corrections. However, this model may be of value for the study of the modification of the A^1_μ field resulting from the emission of many soft photons by a current.

Since vacuum polarization effects modify the electromagnetic field in this case we define in-field eigenstates (in the transverse gauge) by

$$\vec{A}^1_{in}|\mathcal{a}\rangle_{in}=\vec{\mathcal{a}}_{in}|\mathcal{a}\rangle_{in}, \quad (141)$$

where

$$|\mathcal{a}\rangle_{in}=\exp\left[\int d^3k\sum_{\lambda=1}^{2}(\alpha(k,\lambda)a_2^\dagger(k,\lambda)-\alpha^*(k,\lambda)a_2(k,\lambda))\right]|0\rangle \quad (142)$$

and

$$\vec{\mathcal{a}}_{in}=\int d^3k\sum_{\lambda=1}^{2}\vec{\epsilon}(k,\lambda)[\alpha(k,\lambda)f_k(x)+\alpha^*(k,\lambda)f_k^*(x)] \quad (143)$$

with

$$\vec{A}^i_{in}=\int d^3k\sum_{\lambda=1}^{2}\vec{\epsilon}(k,\lambda)[a_i(k,\lambda)f_k(x)+a_i^\dagger(k,\lambda)f_k^*(x)] \quad (144)$$

for $i=1,2$. The vacuum state is defined by

$$a_1(k,\lambda)|0\rangle=a_1^\dagger(k,\lambda)|0\rangle=0$$

for all k,λ. The interacting field, $\vec{A}^1$, is apparently not sharp on $|\mathcal{a}\rangle_{in}$ but is sharp on

$$|\mathcal{a}\rangle=U^{-1}(t,-\infty)|\mathcal{a}\rangle_{in}, \quad (145)$$

where

$$U(t,-\infty)=T\left(\exp\left[-i\int_{\infty}^{t}d^4x\,H_{int}(A^2_{in},\psi_{in})\right]\right) \quad (146)$$

because

$$\vec{A}^1(\vec{x},t)=U^{-1}(t,-\infty)\vec{A}^1_{in}(\vec{x},t)U(t,-\infty). \quad (147)$$

With these preliminaries completed, the study of physical processes within the framework of these models is now possible, although we shall not pursue it in this report.

Before turning to a discussion of non-Abelian gauge field theories, it is worth noting that the choice of vacuum state we have made necessitates a redefinition of normal-ordering. By normal-ordering a Lagrangian term we shall mean that the observable fields (to which we have consistently appended the superscript or subscript one) are to be placed to the right, and unobservable fields, labeled by two, are to be placed to the left. Thus Wick's theorem (with our definition of normal-ordering) becomes in the case of two fields

$$\begin{aligned}T(\phi_{1\,in}(x_1)\phi_{2\,in}(x_2))&=:\phi_{1\,in}(x_1)\phi_{2\,in}(x_2):+\langle 0|T(\phi_{1\,in}(x_1)\phi_{2\,in}(x_2))|0\rangle\\&=\phi_{2\,in}(x_2)\phi_{1\,in}(x_1)+\theta(x_{10}-x_{20})[\phi_{1\,in}(x_1),\phi_{2\,in}(x_2)].\end{aligned} \quad (148)$$

Note that the Green's function

$$G(x_1,x_2)=\langle 0|T(\phi_{1\,in}(x_1)\phi_{2\,in}(x_2))|0\rangle \quad (149)$$

is necessarily retarded. From this we can conclude that the models of electrodynamics, which we have considered, naturally embody the observed retarded nature of classical electrodynamics. Another way of stating this result is: If classical electrodynamics is to have a pseudoquantum formulation, its Green's functions are necessarily retarded. The origin of the asymmetry is the definition of the vacuums (which is equivalent to a specification of boundary conditions). Just as in classical electrodynamics retarded propagation is implemented by a choice of boundary conditions which do not require a commitment to any specific cosmological model.

Finally we would like to note that the Lagrangian obtained from adding L_{int} of Eq. (140) to the Lagrangian of Eq. (127) is equivalent to the usual Lagrangian of electrodynamics plus a term describing a massless Abelian gauge field with the wrong sign. (This is seen by defining new fields equal to the sum and difference of A_μ^1 and A_μ^2.) This field theory may be quantized following the procedure we have outlined. A_μ^1 loses its classical character due to quantum corrections.

IV. NON-ABELIAN GAUGE THEORIES

In this section we shall describe the procedure for embedding a classical non-Abelian Yang-Mills field in a quantum field theory. Then we will discuss a vierbein formulation of quantum gravity which could have been interpreted as a pseudoquantum field theory for a classical metric field if it were not for one term in the Lagrangian which makes it a truly quantum field theory. Nevertheless we suggest a new canonical quantization procedure based on our pseudoquantum approach.

Consider a classical Yang-Mills field, $A_\mu^1 = \underline{A}_\mu^1 \cdot \underline{T}$, where the jth component of $\underline{T}$ is a matrix representing a generator of a non-Abelian group G in the defining representation with commutation relations

$$[T_j, T_k] = i t_{jkl} T_l . \tag{150}$$

We can define a pseudoquantum field theory, wherein the classical character of A_μ^1 is maintained, which has the Lagrangian density

$$\begin{aligned}\mathcal{L} = {} & \tfrac{1}{2}\underline{F}^1_{\mu\nu}\cdot\underline{F}^{2\mu\nu} - \tfrac{1}{2}\underline{F}^{2\mu\nu}\cdot(\partial_\mu \underline{A}^1_\nu - \partial_\nu \underline{A}^1_\mu + g\underline{A}^1_\mu \times \underline{A}^1_\nu) \\ & - \tfrac{1}{2}\underline{F}^{1\mu\nu}\cdot(\partial_\mu \underline{A}^2_\nu - \partial_\nu \underline{A}^2_\mu + g\underline{A}^1_\mu \times \underline{A}^2_\nu - g\underline{A}^1_\nu \times \underline{A}^2_\mu) \\ & + \bar{\psi}(i\not{\nabla} + g\not{A}^1 - m)\psi ,\end{aligned} \tag{151}$$

where ψ is a fermion field. The theory is invariant under the local gauge transformation, $S \in G$,

$$\psi' = S^{-1}\psi , \tag{152}$$

$$A_\mu^{1'} = S^{-1} A_\mu^1 S + \frac{i}{g} S^{-1}\partial_\mu S , \tag{153}$$

$$F_{\mu\nu}^{1'} = S^{-1} F_{\mu\nu}^1 S , \tag{154}$$

$$A_\mu^{2'} = S^{-1} A_\mu^2 S , \tag{155}$$

$$F_{\mu\nu}^{2'} = S^{-1} F_{\mu\nu}^2 S . \tag{156}$$

Except for one important term this Lagrangian with its attendant gauge invariance properties has been suggested as a possible model for the quark-confining strong interaction.[8] Since the omitted term has a masslike character $\Lambda^2 \underline{A}_\mu^2 \cdot \underline{A}^{2\mu}$, where Λ has the dimensions of a mass, it is clear that the strong-interaction model's ultraviolet behavior approaches that of the present pseudoquantum theory if the same quantization procedure is followed in both cases. We shall discuss this question further in the next section and show that the *ad hoc* procedure followed in Ref. 8 leads to the same result as the quantization procedure developed in this report.

The Euler-Lagrange equations of motion which are obtained from $\mathcal{L}$ in the canonical manner are

$$\underline{F}^1_{\mu\nu} = \partial_\mu \underline{A}^1_\nu - \partial_\nu \underline{A}^1_\mu + g\,\underline{A}^1_\mu \times \underline{A}^1_\nu , \tag{157}$$

$$\underline{F}^2_{\mu\nu} = \partial_\mu \underline{A}^2_\nu - \partial_\nu \underline{A}^2_\mu + g\underline{A}^1_\mu \times \underline{A}^2_\nu - g\underline{A}^1_\nu \times \underline{A}^2_\mu , \tag{158}$$

$$(\partial_\mu + g\,\underline{A}^1_\mu \times)\underline{F}^{1\mu\nu} = 0 , \tag{159}$$

$$(\partial_\mu + g\underline{A}^1_\mu \times)\underline{F}^{2\mu\nu} + gA^2_\mu \times \underline{F}^{1\mu\nu} + g\underline{J}^\nu = 0 , \tag{160}$$

$$(i\vec{\nabla} + g\not{A}^1 - m)\psi = 0 , \tag{161}$$

with the conservation law

$$(\partial_\nu + g\underline{A}^1_\nu \times)\underline{J}^\nu = 0 . \tag{162}$$

The canonical momentum which is conjugate to $\underline{A}^1_j$ is

$$\underline{\Pi}^2_j = \underline{F}^2_{0j} \tag{163}$$

and the canonical momentum conjugate to $\underline{A}^2_j$ is

$$\underline{\Pi}^1_j = \underline{F}^1_{0j} \tag{164}$$

for $j = 1, 2, 3$. The canonical momentum corresponding to the fields $\underline{A}^i_0$ is zero for $i = 1, 2$. The existence of equations of constraint among the Euler-Lagrange equations implies that not all field components are independent, so that we must isolate the independent components prior to defining the canonical equal-time commutation relations.

Following Ref. 8 we choose to work in the Coulomb gauge, $\nabla_i \underline{A}^1_i = 0$, and define the field variables

$$\underline{A}^2_i = \underline{A}^{2T}_i + \underline{A}^{2L}_i , \tag{165}$$

$$\underline{\Pi}^a_i = \underline{\Pi}^{aT}_i + \underline{\Pi}^{aL}_i , \tag{166}$$

where

$$\nabla_i \cdot \underline{A}^{2T}_i = \nabla_i \cdot \underline{\Pi}^{aT}_i = 0 \tag{167}$$

and $a = 1, 2$. Then the nonzero equal-time commutation relations are

$$[\Pi^{aT}_{ip}(x), A^{bT}_{jq}(y)] = i\delta_{pq}(1 - \delta_{ab})\delta^{\text{tr}}_{ij}(\vec{x} - \vec{y}) , \tag{168}$$

where p and q are internal-symmetry indices, $a, b = 1, 2$, and $i, j = 1, 2, 3$.

While the classical character of A_μ^1 can be maintained with our choice of $\mathcal{L}$, this theory has features due to its non-Abelian nature which make it less trivial and therefore more interesting than the corresponding Abelian theory discussed in the last section. If we follow a procedure similar to that in the Abelian case [Eq. (127)] and introduce a set of states appropriate to the quadratic part of the Lagrangian, then the cubic and quartic Yang-Mills terms in the interaction part of the Lagrangian will act to transform $A_{\text{in}\,\mu}^1$ eigenstates into eigenstates of the interacting field A_μ^1. This is, of course, necessary for the classical Yang-Mills equations of motion to be satisfied. Our formalism, thus, offers a perturbative method for calculating solutions of the classical Yang-Mills equations. In addition, it gives an interesting interpretation to the short-distance behavior of the quark-confining field theory of Ref. 8. At short distances the gluon field A_μ^1 effectively decouples from the quark sector and becomes, in effect, a free field. This type of short-distance behavior is certainly not at odds with the seemingly simple behavior observed in hadron processes at high energy. Therefore, it is possible that pseudoquantum field theory may be relevant to the short-distance behavior of hadron interaction. Certainly, it is interesting that elementary fermions fall into two similar groups: those which appear to be individually observable (leptons) and those which are not individually observable (quarks).

We now turn to a consideration of a vierbein model of gravity which has certain close similarities to the pseudoquantum field theories we have been studying. In Weyl's formulation[13] of the Einstein-Cartan theory of gravity a vierbein field, $l^{\mu a}(x)$, is introduced which is the "square root" of the metric tensor

$$g^{\mu\nu}=\eta_{ab}l^{\mu a}l^{\nu b}, \tag{169}$$

where η_{ab} is the constant metric tensor of special relativity, where Roman indices transform as vectors under the SL(2, C) group of local Lorentz transformations, and where Greek indices transform as vectors under general coordinate transformations. It is useful to introduce the constant Dirac matrices, γ_a and $4S_{ab}=i[\gamma_a,\gamma_b]$. Under an SL(2, C) transformation,

$$S=\exp[iC^{ab}(x)S_{ab}], \tag{170}$$

a spinor, $\psi(x)$, becomes

$$\psi'=S\psi. \tag{171}$$

The local nature of the transformation requires the introduction of a gauge field

$$B_\mu^{ab}=-B_\mu^{ba} \tag{172}$$

which transforms inhomogeneously,

$$B_\mu\to SB_\mu S^{-1}-\frac{i}{g}S\partial_\mu S^{-1}, \tag{173}$$

so that a Lorentz transformation gauge-covariant derivative can be defined

$$\nabla_\mu\psi=(\partial_\mu+igB_\mu)\psi, \tag{174}$$

where $B_\mu=B_\mu^{ab}S_{ab}$ and $g=12\pi G$ where G is Newton's constant. Under a gauge transformation we have

$$l^\mu=l^{\mu a}\gamma_a\to Sl^\mu S^{-1}, \tag{175}$$

so that the gauge-covariant derivative of l^μ is defined to be

$$\nabla_\nu l^\mu=(\partial_\nu+igB_\nu\times)\,l^\mu, \tag{176}$$

where $B_\nu\times l^\mu=[B_\nu,l^\mu]$. The commutator

$$igB_{\mu\nu}=[\partial_\mu+igB_\mu,\partial_\nu+igB_\nu] \tag{177}$$

transforms homogeneously under a gauge transformation

$$B_{\mu\nu}\to SB_{\mu\nu}S^{-1}, \tag{178}$$

and as a second-rank tensor under general coordinate transformations. With these field quantities we are able to construct a Lagrangian $\mathcal{L}_{\text{Weyl}}$ which reduces to the Einstein Lagrangian for gravity when no matter is present,[13]

$$\mathcal{L}=\mathcal{L}_{\text{Weyl}}+\mathcal{L}_{\text{matter}}, \tag{179}$$

where

$$\mathcal{L}_{\text{Weyl}}=\frac{i}{8l}\operatorname{Tr}l^\mu l^\nu B_{\mu\nu} \tag{180}$$

and where, for example, we might let

$$l\,\mathcal{L}_{\text{matter}}=\bar\psi(il^\mu\nabla_\mu+m)\psi \tag{181}$$

with $l=\det(l^{\mu a})$.

We observe that the terms containing derivatives in $\mathcal{L}_{\text{Weyl}}$ are linear in the field B_μ—a suggestive feature in view of our previous discussion. However, the quadratic term in B_μ eliminates the possibility of regarding $\mathcal{L}_{\text{Weyl}}$ as a pseudoquantum field theory for a classical field $l^{\mu a}$. But, regardless of this consideration, the fact that $l^{\mu a}$ is necessarily classical in part leads us to consider quantizing vierbein gravity in a manner which is based on the pseudoquantization procedure described above. Remembering that a successful perturbation theory requires the perturbation to be around known solutions we introduce a quadratic Lagrangian term via

$$\mathcal{L}=\mathcal{L}_0+(\mathcal{L}-\mathcal{L}_0)=\mathcal{L}_0+\mathcal{L}_{\text{int}}, \tag{182}$$

where

$$\mathcal{L}_0=-\tfrac{1}{4}i\operatorname{Tr}(B'_{\mu a}l^\mu\gamma^a+ig[B_a,B_b]\gamma^a\gamma^b) \tag{183}$$

and

$$B'_{\mu a}=\partial_\mu B_a-\partial_a B_\mu . \tag{184}$$

Our plan is to follow the pseudoquantization procedure for the "free" part of the Lagrangian $\mathfrak{L}_0$. Therefore we will (i) choose a particular coordinate system (harmonic coordinates) and a particular gauge, the "Lorentz" gauge, $\partial^\mu B_\mu=0$, (ii) establish equal-time commutation relations, (iii) define a set of eigenstates of $l^{\mu a}$, and (iv) proceed to calculate quantum corrections in perturbation theory.

The equations of motion for the "free" Lagrangian $\mathfrak{L}_0$ are

$$\partial_\mu B_b^{ab}-\partial_b B_\mu^{ab}=0 \tag{185}$$

and

$$\partial_\mu(l^{\mu a}\eta^{\nu b}-l^{\nu a}\eta^{\mu b})+2g(\eta^{\nu a}B_c^{cb}-\eta^{\nu b}B_c^{ca}-\eta^{ac}B_c^{\nu b}+\eta^{bc}B_c^{\nu a})=0 . \tag{186}$$

We work in the gravitational equivalent of the Lorentz gauge of electrodynamics,

$$\partial^\mu B_\mu^{ab}=0 , \tag{187}$$

and choose harmonic coordinates

$$\partial_\mu l^{\mu a}=\tfrac{1}{2}\,\partial^a\eta_{\sigma\tau}l^{\sigma\tau} . \tag{188}$$

The Green's function associated with Eq. (185) is

$$G_{\alpha ef,\rho\sigma}(x,y)=-\tfrac{1}{2}\int\frac{d^4k}{k^2}\,e^{-ik\cdot(x-y)}g_{\alpha ef,\rho\sigma}(k) , \tag{189}$$

where

$$g_{\alpha ef,\rho\sigma}(k)=k_e\left(\eta_{\alpha\rho}\eta_{f\sigma}+\eta_{\alpha\sigma}\eta_{f\rho}-\eta_{\alpha f}\eta_{\rho\sigma}-\frac{k_\alpha k_\rho\eta_{f\sigma}+k_\alpha k_\sigma\eta_{f\rho}}{k^2}\right)-k_f\left(\eta_{\alpha\rho}\eta_{e\sigma}+\eta_{\alpha\sigma}\eta_{e\rho}-\eta_{\alpha e}\eta_{\rho\sigma}-\frac{k_\alpha k_\rho\eta_{e\sigma}+k_\alpha k_\sigma\eta_{e\rho}}{k^2}\right). \tag{190}$$

In order to relate the above Green's function to a time-ordered product of the quantum fields it is first necessary to introduce a set of coherent states, $|L\rangle$, which are eigenstates of $l^{\mu a}$:

$$l^{\mu a}(x)|L\rangle=L^{\mu a}(x)|L\rangle , \tag{191}$$

where $L^{\mu a}(x)$ is a c-number function of x. In particular, we define $|\eta\rangle$ to satisfy

$$l^{\mu a}|\eta\rangle=\eta^{\mu a}|\eta\rangle , \tag{192}$$

where $\eta^{\mu a}$ is the constant Lorentz metric tensor of special relativity. Given a state $|L\rangle$ we define the field

$$l_L^{\mu a}=l^{\mu a}-L^{\mu a} . \tag{193}$$

This field corresponds to the quantum part of $l^{\mu a}$ and when applied to the purely classical state $|L\rangle$ has the eigenvalue zero.

We now make the identification

$$iG_{\alpha ef,\rho\sigma}(x,y)=\langle L|T(B_{\alpha ef}(x),l_{L\rho\sigma}(y))|L\rangle . \tag{194}$$

If we desire to calculate quantum corrections to $l_{\rho\sigma}=\eta_{\rho\sigma}$ we choose $|L\rangle=|\eta\rangle$. (It should be noted that $G_{\alpha ef,\rho\sigma}$ is independent of the choice of $|L\rangle$ as we have defined it.) Because $l_{L\rho\sigma}(y)$ is sharp on $|L\rangle$ we find that the right side of Eq. (194) becomes

$$iG_{\alpha ef,\rho\sigma}(x,y)=\theta(y_0-x_0)[l_{\rho\sigma}(y),B_{\alpha ef}(x)] \tag{195}$$

up to a functional δ function. From the form of $\mathfrak{L}_0$ we see that the commutator is not zero. It is fully determined by an equal-time commutation relation of $l_{\rho\sigma}$ and $B_{\alpha ef}$ (which by the way is the only nonzero equal-time commutator if the canonical procedure is followed), the equations of motion, and the requirement that it be zero at spacelike distances. The "retarded" form of $G_{\alpha ef,\rho\sigma}$ fixes the integration contour around poles in Eq. (192). The other nonzero Green's function in the free Lagrangian model specified by $\mathfrak{L}_0$ is

$$iH^{\mu\nu,\rho\sigma}(x,y)=\langle L|T(l_L^{\mu\nu}(x),l_L^{\rho\sigma}(y))|L\rangle . \tag{196}$$

It is nonzero owing to the presence of the $[B_\mu,B_\nu]$ term in $\mathfrak{L}_0$. We shall show in the next section that it is a principal-value propagator rather than a Feynman propagator. In coordinate space this results in $H^{\mu\nu,\rho\sigma}$ being the sum of the advanced and retarded propagators. As a result our model is equivalent to an action-at-a-distance theory in some sectors.

The classical part of $l_{\mu a}$ is the solution of the classical linearized field equations with appropriate matter sources. The linearized field equations are derived from a Lagrangian consisting of $\mathfrak{L}_0$ plus matter terms. (Note that the form of $\mathfrak{L}_0$ is obtained by substituting $l_{\mu a}=\eta_{\mu a}+h_{\mu a}$ in $\mathfrak{L}_{\text{Weyl}}$, expanding, and keeping quadratic terms.) Thus the class of possible background metrics is restricted.

A simplification occurs in perturbation theory when the classical part of $l_{\mu\alpha}$ is $\eta_{\mu\alpha}$. In this case $(\mathfrak{L}_{\text{Weyl}}-\mathfrak{L}_0)|\eta\rangle=0$ when $\mathfrak{L}_0$ and $\mathfrak{L}_{\text{Weyl}}$ are expressed in terms of asymptotic fields.

V. PRINCIPAL-VALUE PROPAGATORS AND ACTION AT A DISTANCE

In this section we shall show that certain propagators, in field theories where the pseudoquantization procedure has been followed, are principal-value propagators (i.e., the sum of the advanced and retarded Green's functions in coordinate space) rather than Feynman propagators. We also describe a quantum field theory for action-at-a-distance electrodynamics which completes the program initiated by Schwarzschild, Tetrode, and Fokker.[14]

To illustrate the origin of the principal-value propagator we return to the scalar field model of Eq. (103) which described a classical field, $\phi_1(x)$. We introduce an interaction term

$$L_{\text{int}} = -\int d^3z \, \tfrac{1}{2}\lambda^2 [\phi_2(z)]^2 \tag{197}$$

(where λ is a constant), which destroys the purely classical nature of ϕ_1. Suppose we consider the Green's function

$$i\tilde{G}(x, y) = \langle 0 | T(\phi_1(x)\phi_1(y)) | 0\rangle , \tag{198}$$

which would be zero if L_{int} were not present. In terms of in-fields we have

$$i\tilde{G}(x, y) = \left\langle 0 \left| T\left(\phi_{1\text{in}}(x)\phi_{1\text{in}}(y)\exp\left(i\int dt\, L_{\text{int}}\right)\right)\right| 0\right\rangle , \tag{199}$$

where the vacuum states, $|0\rangle$ and $\langle 0|$, are defined as in Eqs. (120) and (122). From the definition of the vacuum we find (dropping "in" labels)

$$i\tilde{G}(x, y) = \frac{-i\lambda^2}{2}\int d^4z \langle 0 | T(\phi_1(x)\phi_1(y)\phi_2{}^2(z)) | 0\rangle , \tag{200}$$

which becomes

$$i\tilde{G}(x, y) = \frac{-i\lambda^2}{2}\epsilon(x_0 - y_0)\frac{\partial}{\partial m^2}\Delta(x - y) \tag{201}$$

with

$$\Delta(x - y) = -i\int \frac{d^4k}{(2\pi)^3}\delta(k^2 - m^2)\epsilon(k_0)e^{-ik\cdot(x-y)} . \tag{202}$$

Using

$$\tfrac{1}{2}\epsilon(x_0 - y_0)\Delta(x - y) = \int \frac{d^4k}{(2\pi)^4} \mathrm{P}\frac{1}{k^2 - m^2} \times e^{-ik\cdot(x-y)} , \tag{203}$$

we see that

$$\tilde{G}(x, y) = -\lambda^2\int \frac{d^4k}{(2\pi)^4}\mathrm{P}\frac{1}{(k^2 - m^2)^2}e^{-ik\cdot(x-y)} , \tag{204}$$

where

$$\mathrm{P}\frac{1}{(k^2 - m^2)^2} \equiv \frac{1}{2}\left[\frac{1}{(k^2 - m^2 + i\epsilon)^2} + \frac{1}{(k^2 - m^2 - i\epsilon)^2}\right]. \tag{205}$$

The form of $\tilde{G}$ is consistent with the equations of motion:

$$(\Box + m^2)\phi_1 + \lambda^2\phi_2 = 0 , \tag{206}$$

$$(\Box + m^2)\phi_2 = \delta^4(x - y) . \tag{207}$$

The appearance of the principal-value dipole propagator rather than the Feynman dipole propagator in Eq. (204) is useful because it eliminates certain unitarity problems associated with indefinite-metric fields. However, depending on the model under consideration, it could lead to difficulties with causality. To illustrate the manner in which unitarity problems are resolved, consider the interaction of the ϕ_1 dipole field with a scalar quantum field ψ with

$$L'_{\text{int}} = g\phi_1(x)[\psi(x)]^2 . \tag{208}$$

Suppose we consider the subset of in and out states containing arbitrary numbers of ψ particles but no ϕ_1 or ϕ_2 particles. These states have positive metric. If one could systematically exclude indefinite-metric ϕ_1 and ϕ_2 particles from physical states one would avoid negative probabilities and other problems. But the sum over states in a unitarity sum would normally include states with ϕ_1 particles if the ϕ_1 field had Feynman propagators. In the case of principal-value propagators, no intermediate states with ϕ_1 particles occur, since the pole term is not present. The interaction mediated by the ϕ_1 field is a form of action at a distance and ϕ_1 is properly described by the phrase adjunct field, coined by Feynman and Wheeler.[14] A more detailed discussion of the unitarity question is given in Refs. 7 and 8. In those articles a dipole gluon model for quark confinement was proposed which introduced principal-value propagators in an *ad hoc* manner to resolve unitarity problems. It was pointed out that causality problems did not necessarily exist in those models because the non-Abelian dipole gluons were confined for the same reason as the quarks so that—at the worst—there would be unobservable causality violations at distances of the order of hadron dimensions.

The pseudoquantization procedure may be used to construct a quantum field-theoretic version of action-at-a-distance electrodynamics. Consider the Lagrangian

$$\mathcal{L} = -\tfrac{1}{2}F^{\mu\nu}(\partial_\nu A_\mu - \partial_\mu A_\nu) + \tfrac{1}{4}F^{\mu\nu}F_{\mu\nu} + \bar{\psi}(i\not{\partial} - e\not{A} - m_0)\psi . \tag{209}$$

We define the momentum

$$\Pi_\mu = \frac{\delta \mathcal{L}}{\delta A^\mu} = F_{0\mu}. \tag{210}$$

Going to the transverse gauge as in Sec. IV, we define the equal-time commutation relation

$$[\Pi_i(\vec{x}, t), A_j(\vec{y}, t)] = i\delta_{ij}^{tr}(\vec{x} - \vec{y}). \tag{211}$$

Suppose we neglect interaction terms in $\mathcal{L}$ for the moment and choose $F_{\mu\nu}$ to be an observable classical field (as it is up to quantum corrections which we neglect) and A_μ to be unobservable (as it is because it is not gauge invariant). Then we follow our pseudoquantization procedure for

$$\mathcal{L}_0 = -\tfrac{1}{2} F^{\mu\nu}(\partial_\nu A_\mu - \partial_\mu A_\nu) + \tfrac{1}{4} F^{\mu\nu} F_{\mu\nu}. \tag{212}$$

In particular, we define a vacuum such that

$$F_{\mu\nu}|0\rangle = 0, \quad A_\mu|0\rangle \neq 0, \tag{213}$$

while

$$\langle 0|A_\mu = 0, \quad \langle 0|F_{\mu\nu} \neq 0. \tag{214}$$

Then

$$iG_{\mu\nu}(x, y) = \langle 0|T(A_\mu(x)A_\nu(y))|0\rangle \tag{215}$$

would be zero were it not for $F_{\mu\nu}F^{\mu\nu}$ in $\mathcal{L}_0$. In terms of appropriate in-fields it becomes

$$2iG_{\mu\nu}(x, y) = \int d^4z\,(\theta(x_0 - y_0)\theta(y_0 - z_0) + \theta(y_0 - x_0)\theta(x_0 - z_0)) \times [A_{\mu\,\text{in}}(x), F_{\alpha\beta\,\text{in}}(z)][A_{\mu\,\text{in}}(y), F_{\text{in}}^{\alpha\beta}(z)]. \tag{216}$$

Note that we are treating $F_{\mu\nu}F^{\mu\nu}$ in $\mathcal{L}_0$ as an interaction term. The structure of $G_{\mu\nu}(x, y)$ is the same as that of Eq. (200) so we can conclude that

$$G_{\mu\nu}(x, y) = -g_{\mu\nu} \int \frac{d^4k}{(2\pi)^4}\, \mathrm{P}\frac{1}{k^2}\, e^{-ik\cdot(x-y)} \tag{217}$$

in the Feynman gauge. Thus the action-at-a-distance interaction follows from the pseudoquantization of electrodynamics. The classical character of $F_{\mu\nu}$ is lost owing to quantum corrections resulting from the presence of $J_\mu A^\mu$ in the Lagrangian.

The example we have just studied has a certain parallel in the vierbein model of gravitation studied in the last section. The forms of the Lagrangian and commutation relations are similar. As a result it is clear that

$$D^{\mu\nu,\lambda\sigma}(x, y) = \left\langle L \left| T\left(l_{L\,\text{in}}^{\mu\nu}(x) l_{L\,\text{in}}^{\lambda\sigma}(y) \int d^4z\, \tilde{\mathcal{L}}_{\text{int}}(z)\right) \right| L \right\rangle \tag{218}$$

with

$$\tilde{\mathcal{L}}_{\text{int}} = \tfrac{1}{4} g \operatorname{Tr}[B_{\mu\,\text{in}}, B_{\nu\,\text{in}}]\gamma^\mu\gamma^\nu \tag{219}$$

is a principal-value propagator. Therefore we have constructed an action-at-a-distance version of quantum gravity. Our motivation was to take account of the classical part of $l^{\mu a}$ in a way which did not divorce it from the quantum part to which it is intimately related.

VI. CONCLUSION

We have seen that an alternative to Fock-space quantization exists for a class of field theories which have Lagrangian gradient terms which are linear in field variables. A method was also proposed for constructing Lagrangians of that type from classical Lagrangians with gradient terms which are quadratic in field variables. To some extent this process has a parallel in the passage from Klein-Gordon field Lagrangians which are quadratic in derivatives to Dirac field Lagrangians which are linear in derivatives.

The quantization procedure we have outlined is canonical so far as the fields are concerned. We do, however, make a choice of vacuum states which differs from the usual choice. As a result we have found free propagators which were either retarded, or half-advanced and half-retarded. The choice of vacuum state does not in itself preclude the appearance of Feynman propagators. If one has a good reason to modify the canonical commutation relations then it is possible to obtain Feynman propagators.[15] The procedure we have outlined has, therefore, a greater generality than the particular class of models studied in the present work. It can enable one to embed a classical field theory in a quantum field theory in such a way as to maintain its classical character. It can also be applied to study classical field theories which obtain quantum corrections. Finally it can be applied in order to obtain a fully second-quantized field theory (cf. Ref. 15).

ACKNOWLEDGMENT

This work was supported in part by the U.S. Energy Research and Development Administration.

*Present address: Physics Department, Williams College, Williamstown, Mass. 01267.

[1]D. R. Yennie, S. C. Frautschi, and H. Suura, Ann. Phys. (N.Y.) 13, 379 (1961).

[2]R. J. Glauber, Phys. Rev. 131, 2766 (1963).

[3]W. A. Bardeen, M. S. Chanowitz, S. D. Drell, M. Weinstein, and T.-M. Yan, Phys. Rev. D 11, 1094 (1975).

[4]J. M. Cornwall and G. Tiktopoulos, Phys. Rev. D 13, 3370 (1976).
[5]S. Blaha, Phys. Lett. 56B, 373 (1975).
[6]E. C. G. Sudarshan, Center for Particle Theory report Univ. of Texas—Austin, 1976 (unpublished).
[7]S. Blaha, Phys. Rev. D 10, 4268 (1974).
[8]S. Blaha, Phys. Rev. D 11, 2921 (1975).
[9]S. Blaha, Lett. Nuovo Cimento 18, 60 (1977).
[10]Cf. Ref. 2; T. W. B. Kibble, J. Math. Phys. 9, 315 (1968); Phys. Rev. 173, 1527 (1968); 174, 1882 (1968); 175, 1624 (1968);
[11]A. Salam, lecture at Center for Theoretical Studies, Miami, Florida, 1973 (unpublished).
[12]G. Mie, Ann. Phys. (Leipzig) 37, 511 (1912); 39, 1 (1912); 40, 1 (1913); H. Weyl, *Space, Time, Matter* (Dover, N.Y. 1952).
[13]H. Weyl, Z. Phys. 56, 330 (1929); T. W. B. Kibble, J. Math. Phys. 2, 212 (1961); J. Schwinger, Phys. Rev. 130, 1253 (1963); C. J. Isham, A. Salam, and J. Strathdee, Lett. Nuovo Cimento 5, 969 (1972); F. W. Hehl, P. von der Heyde, G. D. Kerlick, and J. Nester, Rev. Mod. Phys. 48, 393 (1976); and references therein.
[14]K. Schwarzschild, Göttinger Nachrichten 128, 132 (1903); H. Tetrode, Z. Phys. 10, 317 (1922); A. D. Fokker, *ibid.* 58, 386 (1929); J. Wheeler and R. P. Feynman, Rev. Mod. Phys. 17, 157 (1945); 21, 425 (1949).
[15]S. Blaha (unpublished).

Appendix F. PseudoQuantum Theory of Color Confinement

These papers are reprinted with the kind permission of Physical Review D.

S. Blaha, Phys. Rev. **D10**, 4268 (1974).

S. Blaha, Phys. Rev. D**11**, 2921 (1975).

Landau-Ginzburg theory, but $\rho \sim i\langle \phi^* \dot{\phi} - \dot{\phi}^* \phi \rangle = 0$ in the Higgs theory.

[9] M. Kalb and P. Ramond, Phys. Rev. D 9, 2273 (1974). See also E. Cremmer and J. Scherk, Nucl. Phys. B72, 117 (1974).

[10] L. N. Chang and F. Mansouri, Phys. Rev. D 5, 2235 (1972); Goto, Ref. 7; G. Goddard, J. Goldstone, C. Rebbi, and C. B. Thorn, Nucl. Phys. B56, 109 (1973).

[11] A clearcut answer to this problem seems to be lacking. See, however, Nielsen and Olesen, Ref. 1; G. 't Hooft, CERN Report No. TH-1873-CERN, 1974 (unpublished); Y. Nambu, Ref. 2.

PHYSICAL REVIEW D VOLUME 10, NUMBER 12 15 DECEMBER 1974

Towards a field theory of hadron binding*

Stephen Blaha

Laboratory of Nuclear Studies, Cornell University, Ithaca, New York 14850

(Received 17 July 1974)

A field-theoretic model for hadron binding is described in which free quarks are totally screened. Quarks interact via a dipole vector-gluon field. A second-quantization procedure for the gluon field, which reduces the field to an embodiment of a direct particle interaction, eliminates unitarity problems. A detailed description of perturbation-theory rules is given. In contrast to the results of the pseudoscalar-meson and massive-vector-meson models (without cutoff), scaling occurs in the electroproduction structure functions. Another possible model having some resemblance to the relativistic harmonic-oscillator quark model of Feynman, Kislinger, and Ravndal is also described. It is unitary and has scaling structure functions.

I. INTRODUCTION

The current understanding of hadronic structure allows two apparently contradictory statements to be made: The constituents of the hadron appear to be loosely bound, quasifree particles. The constituents of the hadron are not produced and do not occur outside of hadrons. Several attempts have been made to resolve this paradoxical situation. They may be divided into two categories: "conventional" field-theoretic approaches,[1,2] and *ad hoc* approaches which postulate manifestly non-field-theoretic structures for confinement, e.g., the "bag" model.[3] In the first approach Casher, Kogut, and Susskind[1] and Wilson[2] showed that quarks could be totally screened and not observed. However, a four-dimensional, Lorentz-invariant field-theoretic model of hadron binding with its attendant conceptual and computational advantages appears to be lacking. We shall discuss a possible candidate, the dipole gluon model, in detail. In addition, another possibility is briefly described in Appendix B which bears some comparison with the quark model of Feynman *et al.*[4] The dipole gluon model has two major qualitative features in common with the bag model[3] and the two-dimensional quantum-electrodynamic model[1]: (1) The dipole gluon field has no independent degrees of freedom; neither does a bag or the two-dimensional electromagnetic field. (2) The "Coulomb" potential between quarks is proportional to the distance between them in all three models. In a sense the bag model may be regarded as a phenomenological approximation to the dipole model, and the dipole model as a generalization of the two-dimensional model to four dimensions.

In Sec. II we describe a quantization procedure which avoids the introduction of indefinite-metric in or out states and thus leads to a unitary S matrix. In Sec. III we describe the properties of the "free" gluon Lagrangian model. In Sec. IV we describe the perturbation-theory rules of the dipole model. Section V contains a discussion of unitarity, causality, quark confinement, and scaling properties of the electroproduction structure functions. For simplicity we shall ignore all but the dipole quark interaction and do not introduce internal quark quantum numbers.

II. SECOND-QUANTIZATION PROCEDURE FOR THE GLUON FIELD

We shall not quantize the gluon field in the conventional manner for three reasons: (1) to be consistent with experiment where no such particle has been identified, (2) to avoid unitarity problems in the S matrix, and (3) to avoid infrared problems in perturbation theory. We attribute no dynamical degrees of freedom to the gluon field. Instead we regard the field as the embodiment of a direct quark-quark interaction. The gluon field can thus be removed from the Lagrangian in favor of a nonlocal current-current interaction. However, it will be of no small technical advantage to keep the gluon field in the Lagrangian. In order to do this we shall second-quantize the field following the normal prescription and then, instead of introducing a Fock space for free gluons, reduce q-number expressions in the gluon field to c-number expressions via suitable operator boundary conditions.

To illustrate this procedure we consider a scalar boson field, ϕ, with Lagrangian L. We second-quantize the in field, ϕ_{in}, with equal-time commutators

$$[\phi_{in}(x), \phi_{in}(y)] = 0 , \tag{1}$$

$$[\Pi_{in}(x), \Pi_{in}(y)] = 0 , \tag{2}$$

$$[\phi_{in}(x), \Pi_{in}(y)] = -i\delta^3(\vec{x} - \vec{y}) , \tag{3}$$

where

$$\Pi_{in}(x) = \frac{\delta L}{\delta \dot{\phi}_{in}(x)} \tag{4}$$

and L_F is the free Lagrangian part of L. The usual operator expressions and identities are established. In particular the formal expansion of the S matrix in terms of time-ordered products of in fields can be made. (We are using only in fields for convenience—our remarks apply to out fields also.)

The unequal-time commutator, $[\phi_{in}(x), \phi_{in}(y)]$, is a c-number expression which is completely determined if we require that it be consistent with the equations of motion, that it be consistent with Eqs. (1)–(3) in the limit of equal times, and that it vanish at spacelike distances. Consequently all terms with an even number of factors of $\phi(x)$ reduce to sums of c numbers times products of anticommutators $\{\phi(x), \phi(y)\}$. Terms with an odd number of factors have one factor, $\phi(x)$, times sums of c numbers times products of anticommutators. At this point we could introduce a Fock space of states to complete the reduction of q-number expressions to c number expressions. For reasons stated above we do not. In analogy to Dirac's theory[5] of Hamiltonian constraints we impose operator boundary conditions which complete the specification of the dynamics of the system. We choose

$$\Pi_{in}(x) \approx 0 \approx \phi_{in}(x) \tag{5}$$

for all x, where $\approx$ means weakly equal in the sense of Dirac, i.e., evaluate all commutators before imposing the constraints. This eliminates ϕ's degrees of freedom. The free Hamiltonian,

$$H_F = \int \Pi\dot{\phi} - L_F , \tag{6}$$

is now arbitrary to the extent that H_F may be replaced by

$$H_T = H_F + \int A\phi_{in} + \int b\Pi_{in} , \tag{7}$$

where A and b will be completely determined by requiring Eq. (5) be true for all time:

$$[\Pi_{in}, H_T] \approx 0 \tag{8}$$

and

$$[\phi_{in}, H_T] \approx 0 . \tag{9}$$

Thus

$$A = -\frac{\delta H_F}{\delta \phi_{in}} \tag{10}$$

and

$$b = -\frac{\delta H_F}{\delta \Pi_{in}} . \tag{11}$$

To see the effects of this procedure more concretely let

$$L_F = \int (\tfrac{1}{2}\partial_\mu\phi\partial^\mu\phi - \tfrac{1}{2}m^2\phi^2) ; \tag{12}$$

then (suppressing the subscript "in" for notational convenience)

$$i\Delta(x-y) = [\phi(x), \phi(y)] \tag{13}$$

$$= \int \frac{d^4k}{(2\pi)^3} \epsilon(k_0)\delta(k^2 - m^2)e^{-ik\cdot(x-y)} \tag{14}$$

and the time-ordered product becomes

$$i\bar{\Delta}(x-y) \equiv T(\phi(x)\phi(y)) = \tfrac{1}{2}i\epsilon(x_0 - y_0)\Delta(x-y) , \tag{15}$$

with $\epsilon(x) = \pm 1$ for $x \gtrless 0$. More generally, for even N

$$T(\phi(1)\phi(2)\cdots\phi(N))$$

$$\equiv \sum_{\text{permutations}} i^{N/2}\bar{\Delta}(x_1 - x_2)\bar{\Delta}(x_3 - x_4)\cdots\bar{\Delta}(x_{N-1} - x_N) , \tag{16}$$

where $\phi(i) = \phi(x_i)$. The natural correspondence to the Wick expansion

$$\langle 0|T(\psi(x_1)\cdots\psi(x_N))|0\rangle$$

$$= \sum_{\text{permutations}} i^{N/2}\Delta_F(x_1 - x_2)\cdots\Delta_F(x_{N-1} - x_N) \tag{17}$$

(where Δ_F is the Feynman propagator corresponding to the field ψ) allows us to use conventional perturbation-theory rules, except that diagrams with incoming or outgoing ϕ lines do not contribute to the S matrix and the Feynman propagator

$$\Delta_F(k) = \frac{1}{k^2 - m^2 + i\epsilon} \tag{18}$$

is to be replaced with

$$\bar{\Delta}(k) = \mathrm{P}\,\frac{1}{k^2 - m^2 + i\epsilon} = \frac{1}{2}\left(\frac{1}{k^2 - m^2 + i\epsilon} + \frac{1}{k^2 - m^2 - i\epsilon}\right) \tag{19}$$

for internal ϕ lines. In configuration space the Green's function corresponding to Eq. (19) is half the sum of the advanced and retarded Green's functions.

If we follow the above procedure in second-quantizing the electromagnetic field the resulting model quantum electrodynamics corresponds to the classical action-at-a-distance electrodynamics of Schwarzschild, Tetrode, and Fokker.[6] The fact that photon production does not occur in the model QED corresponds to the absence of radiation reaction in the classical theory. In Sec. V this will be shown to be the key to maintaining the unitarity of the S matrix in the dipole gluon model.

III. THE DIPOLE GLUON MODEL

We now consider a model[4] for hadron binding which has several major qualitative features in agreement with experimental results: large-transverse-momenta damping, scaling electroproduction structure functions, and complete screening of free quarks. The Lagrangian is

$$\mathcal{L} = -\tfrac{1}{2}F^1_{\mu\nu}F^{2\mu\nu} - \tfrac{1}{2}\lambda^2 A^2_\mu A^{2\mu} + \bar{\psi}(i\not\nabla - g\not{A}^1 - m)\psi\ , \tag{20}$$

where A^1_μ and A^2_μ are massless gluon fields, $F^i_{\mu\nu} = \partial_\nu A^i_\mu - \partial_\mu A^i_\nu$ for $i = 1, 2$, ψ is the quark field, and g is a dimensionless and λ a dimensional coupling constant. The equations of motion are

$$\partial^\mu F^1_{\mu\nu} + \lambda^2 A^2_\nu = 0\ , \tag{21}$$

$$\partial^\mu F^2_{\mu\nu} + gJ_\nu = 0\ , \tag{22}$$

$$(i\not\nabla - g\not{A}^1 - m)\psi = 0\ , \tag{23}$$

with J_μ the quark current. Equation (21) implies $\partial^\mu A^2_\mu = 0$ while A^1_μ is a gauge-invariant field. As a result we have

$$\Box\,\partial^\mu F^1_{\mu\nu} + g\lambda^2 J_\nu = 0\ . \tag{24}$$

We now consider the "free" gluon case whose Lagrangian is the first two terms on the right-hand side of Eq. (20). The canonical momentum conjugate to A^1_μ is

$$\Pi^1_\mu = F^2_{0\mu} \tag{25}$$

and that conjugate to A^2_μ is

$$\Pi^2_\mu = F^1_{0\mu}\ . \tag{26}$$

Since $\Pi^1_0 = \Pi^2_0 = 0$ we find that A^1_0 and A^2_0 are c numbers and thus $\vec{\nabla}\cdot\vec{A}^2$ is also a c number with the possible exception of the zero-frequency mode. If we choose the Coulomb gauge for A^1_μ

$$\vec{\nabla}\cdot\vec{A}^1 = 0 \tag{27}$$

then we obtain the equal-time commutation relations

$$[\Pi^a_i(x), A^b_j(y)] = i\delta^{ab}\int \frac{d^3k}{(2\pi)^3}\, e^{i\vec{k}\cdot(\vec{x}-\vec{y})}\left(\delta_{ij} - \frac{k_i k_j}{|\vec{k}|^2}\right) \tag{28}$$

for $i, j = 1, 2, 3$, in analogy to similar expressions in quantum electrodynamics. All other equal-time commutators are zero. We can define an electric field, $\vec{E}$, and magnetic field, $\vec{B}$, by

$$\vec{E} = -\vec{\nabla}A^{10} - \frac{\partial}{\partial t}\vec{A}^1 \tag{29}$$

and

$$\vec{B} = \vec{\nabla}\times\vec{A}^1\ , \tag{30}$$

which imply

$$\vec{\nabla}\times\vec{E} = \frac{\partial\vec{B}}{\partial t} \tag{31}$$

and

$$\vec{\nabla}\cdot\vec{B} = 0\ . \tag{32}$$

In the Coulomb gauge Eq. (24) can be restated as

$$\Box\vec{\nabla}\cdot\vec{E} = g\lambda^2 J^0\ , \tag{33}$$

$$\Box\left(\vec{\nabla}\times\vec{B} - \frac{\partial\vec{E}}{\partial t}\right) = g\lambda^2\vec{J}\ , \tag{34}$$

where J^μ is the quark current. Equations (29) and (33) give our analog to the differential equation for the instantaneous Coulomb potential of QED,

$$\Box\nabla^2 A^{10} = -g\lambda^2 J^0\ , \tag{35}$$

while the equivalent vector-potential differential equation is

$$\Box(\Box\vec{A}^1 + \vec{\nabla}\dot{A}^{10}) = g\lambda^2\vec{J}\ . \tag{36}$$

The free gluon unequal-time commutators may be determined from Eqs. (21), (22), (28), (35), and (36) (with the current, of course, set to zero):

$$i\Delta_{ij}^{11}(x-y)\equiv[A_i^1(x),A_j^1(y)]=-i\lambda^2(\delta_{ij}-\nabla_i\nabla_j\nabla^{-2})\left[\frac{\partial}{\partial\mu^2}\Delta(x-y,\mu)\right]_{\mu=0}, \tag{37}$$

$$i\Delta_{ij}^{12}(x-y)\equiv[A_i^1(x),A_j^2(y)]=i(\delta_{ij}-\nabla_i\nabla_j\nabla^{-2})\Delta(x-y,0), \tag{38}$$

$$i\Delta_{ij}^{22}(x-y)\equiv[A_i^2(x),A_j^2(y)]=0, \tag{39}$$

with $i,j=1,2,3$ and

$$i\Delta(x-y,\mu)=\int\frac{d^4k}{(2\pi)^3}\epsilon(k_0)\delta(k^2-\mu^2)e^{-ik\cdot(x-y)}. \tag{40}$$

The commutators, Δ^{11} and Δ^{12}, are zero at spacelike separations due to the form of Δ. They are consistent with the equal-time commutation relations in that limit and they are also consistent with the equations of motion due to the identity

$$\Box\left[\frac{\partial}{\partial\mu^2}\Delta(x-y,\mu)\right]_{\mu=0}=-\Delta(x-y,0). \tag{41}$$

Assuming that we have established all operator expressions we are now in a position to apply operator boundary conditions to the gluon field. The key quantities so far as the perturbation theory we will consider in the next section is concerned are the time-ordered propagators of the gluon field

$$T(A_i^a(x)A_j^b(y))\equiv\tfrac{1}{2}\epsilon(x_0-y_0)[A_i^a(x),A_j^b(y)] \tag{42}$$

$$=\tfrac{1}{2}i\epsilon(x_0-y_0)\Delta_{ij}^{ab}(x-y), \tag{43}$$

where we have suppressed the "in" subscript on the field operator. We can take advantage of the gauge invariance of A_μ^1 to express $T(A_\mu^1A_\nu^1)$ in the Feynman gauge,

$$T(A_\mu^1(x)A_\nu^1(y))=i\lambda^2g_{\mu\nu}\int\frac{d^4k}{(2\pi)^4}\left(\mathrm{P}\frac{1}{k^4}\right)e^{-ik\cdot(x-y)} \tag{44}$$

with

$$\mathrm{P}\frac{1}{k^4}=\frac{1}{2}\left[\frac{1}{(k^2+i\epsilon)^2}+\frac{1}{(k^2-i\epsilon)^2}\right]. \tag{45}$$

In coordinate space

$$T(A_\mu^1(x)A_\nu^1(y))=ig_{\mu\nu}\lambda^2\theta((x-y)^2)/16\pi. \tag{46}$$

The equations of motion of the dipole model display a close analogy to those of quantum electrodynamics. The main difference (with important physical consequences) is the increased degree of the differential equation for A_μ^1 vis-à-vis the corresponding QED equations. The result is a dipole propagator rather than a monopole propagator in momentum space. One could have second-quantized the dipole field in a manner which leads to dipole Feynman propagators. In that case the S matrix would not be unitary in perturbation theory.

IV. PERTURBATION THEORY

The rules for forming the integral corresponding to a Feynman diagram in the dipole model are identical with those of quantum electrodynamics[7] except that we use

$$iB_{\mu\nu}(q)=ig_{\mu\nu}\lambda^2\mathrm{P}\frac{1}{q^4} \tag{47}$$

rather than the Feynman photon propagator

$$iD_{F\mu\nu}(q)=-\frac{ig_{\mu\nu}}{q^2+i\epsilon}. \tag{48}$$

The choice of a principal-value propagator has substantial effects in perturbation theory. For example we shall show that consistency with unitarity requires no diagrams with in or out gluon lines contribute to the S matrix. In addition, there are novelties in the type of divergences in diagrams and the analytic structure of the S matrix. It also appears that conclusions based on summing only a finite number of graphs contributing to an S-matrix element may be misleading. This follows from the fact (to be shown in Sec. V) that free quarks do not exist upon summation of all orders of perturbation theory [Eq. (65)], though this is not seen in a summation to any finite order. The physical states are neutral bound states and thus it appears that the best methods of exploring the physics embodied in this model will involve Bethe-Salpeter equations[8] or eikonal summations. They are currently under study.

We now describe the modifications necessary to compute diagrams in perturbation theory. The propagator of Eq. (47) may be exponentiated through the use of the identity

$$\mathrm{P}\frac{1}{k^4}=-\tfrac{1}{2}\int_{-\infty}^{\infty}d\alpha\,\alpha\,\epsilon(\alpha)\exp(i\alpha k^2), \tag{49}$$

where $\epsilon(\alpha)=\pm1$ for $\alpha\gtrless0$. Since Feynman parameters are not necessarily positive the following identity will be useful in evaluating loop integrations:

$$\int d^4k\exp[iC(\alpha)k^2]=i\pi^2\epsilon(C)/C^2. \tag{50}$$

As a result the Feynman parameter representation of a diagram will have the form

$$I=\int_{-\infty}^{\infty}\prod_{j=1}^{p}\alpha_j\,d\alpha_j\int_0^{\infty}\frac{d\beta_1\cdots d\beta_q}{C^2}\,\epsilon(\alpha_1\alpha_2\cdots\alpha_p C)Ne^{iD/C}\,. \tag{51}$$

where α_i corresponds to an internal gluon line and β_i to an internal fermion line, N symbolizes numerator terms, and C and D are determinantal functions.[9] If we had given the gluons dipole Feynman propagators we would have obtained

$$\int_0^{\infty}\frac{\prod_{j=1}^{p}\alpha_j\,d\alpha_j\,d\beta_1 d\beta_2\cdots d\beta_q N\exp(iD/C)}{C^2}\,, \tag{52}$$

in comparison to Eq. (51). If we now scale all Feynman parameters with u and use the identity

$$\int_0^{\infty}\frac{du}{u}\,\delta\left(1-\frac{|\alpha_1+\alpha_2+\cdots+\alpha_p+\beta_1+\beta_2+\cdots+\beta_q|}{u}\right)=1 \tag{53}$$

(where $|\ |$ indicates absolute value) to introduce an integration over u in Eq. (51), we obtain

$$I=\Gamma(q+2p-2l)\int_{-\infty}^{\infty}\prod_{j=1}^{p}\alpha_j\,d\alpha_j\int_0^{\infty}\frac{d\beta_1\cdots d\beta_q\epsilon(\alpha_1\cdots\alpha_p C)\tilde{N}}{C^2(-iD/C)^{q+2p-2l}}\,\delta\left(1-\left|\sum_i\alpha_i+\sum_j\beta_j\right|\right) \tag{54}$$

where l = the number of loop integrations in the original diagram and $\tilde{N}$ is obtained from N. An example of this procedure is given in Appendix A. As an alternative to the above method one can introduce light-cone coordinates and evaluate pole terms by contour integrations with Eq. (45) specifying the location of the poles relative to the contour.

The divergences occurring in this model are somewhat novel. As one would expect, with a dipole propagator the ultraviolet divergences are restricted to some lower-order diagrams and are logarithmic in nature (see Fig. 1). The dipole propagator, because it is in principal value, does not induce infrared divergences in loop integrations. However, a third type of divergence, which may be called a light-cone divergence,[10] does occur and is connected with a divergence in a loop integration, $\int d^4k$, associated with the region where $k^2=k_0^2-k_3^2-\vec{k}_\perp^{\,2}\approx 0$ and $k_0, k_3\to\infty$. In the Feynman parameter representation of Eq. (54) the divergence will appear at the $\pm\infty$ limits of Feynman parameter integrals. The worst divergence is quadratic and associated with one-loop diagrams with one internal gluon line (Fig. 2). These divergences can be managed through the use of Pauli-Villars regularization. Some diagrams containing light-cone divergences are given in Fig. 2. It should be noted that they are necessarily one-loop diagrams. We can demonstrate this by an examination of the overall degree of light-cone divergence of a graph in the representation of Eq. (54). Let us scale all Feynman parameters in Eq. (54) with Λ and determine the leading behavior as $\Lambda\to\infty$. We find, for l = number of loops > 1,

$$\tilde{N}\sim\Lambda^0\,, \tag{55}$$

$$C\sim\Lambda^l\,, \tag{56}$$

$$D\sim\Lambda^{l+1}\,, \tag{57}$$

and as a result

$$I\sim\Lambda^{2p+q-1-2l-(q+2p-2l)}=\Lambda^{-1}\,, \tag{58}$$

or convergence of the integral as a whole. However, for one-loop diagrams ($l=1$)

$$C\sim\Lambda^0 \tag{59}$$

and consequently

$$\tilde{N}\sim\Lambda^q\,, \tag{60}$$

$$I\sim\Lambda^{3-2p}\,. \tag{61}$$

For example, the diagram of Fig. 2(a) has $p=1$ and diverges quadratically.

Light-cone divergences stem directly from the use of principal-value propagators for the gluon. As such they reflect the nontrivial nature of Wick rotation in this model and they lead to divergences in the vertex renormalization constant, the wave-function renormalization constant, and the quark self-mass which prevent this model from being a superrenormalizable theory of the conventional variety.

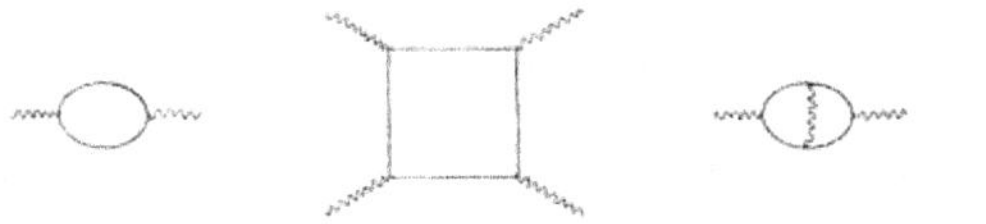

FIG. 1. Some ultraviolet-divergent diagrams.

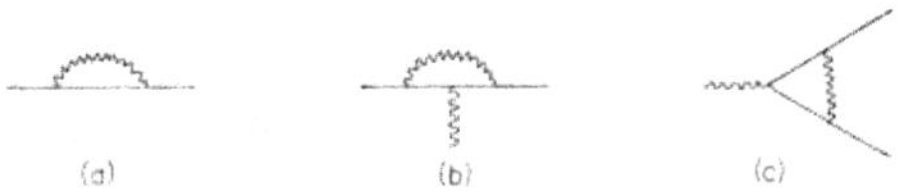

FIG. 2. Some light-cone divergent diagrams.

The Schwinger-Dyson equations for dipole electrodynamics are quite similar to those of quantum electrodynamics, with the exception of the gluon Green's functions, which we now discuss. The proper gluon self-energy, $\Pi_{\mu\nu}(q)$, which couples only to the A_μ^1 channel due to the form of our Lagrangian, satisfies

$$\Pi_{\mu\nu}(q) = i\int \frac{d^4k}{(2\pi)^4}\,\mathrm{Tr}\,\gamma_\mu S_F'(k)\Gamma_\nu(k, k+q)S_F'(k+q) \tag{62}$$

$$= (q_\mu q_\nu - g_{\mu\nu}q^2)\Pi(q^2)\,, \tag{63}$$

where S_F' is the quark propagator and Γ_ν is the proper vertex function (see Fig. 3). The gluon self-energy is related to the complete gluon propagator, $B'_{\mu\nu}$, by

$$iB'_{\mu\nu} = iB_{\mu\nu} + ig^2 iB_{\mu\lambda}\Pi^{\lambda\sigma} iB'_{\sigma\nu}\,, \tag{64}$$

with $B_{\mu\nu}$ given by Eq. (47). Using Eq. (63) we find

$$B'_{\mu\nu}(q) = \frac{\lambda^2 g_{\mu\nu}}{q^4 + g^2\lambda^2 q^2\Pi(q)} - \frac{q_\mu q_\nu g^2\lambda^4\Pi(q)}{q^8 + g^2\lambda^2 q^6\Pi(q)}\,. \tag{65}$$

The Green's function for the gluon field, A_μ^2, which is zero within the context of the free gluon Lagrangian [cf. Eq. (39)], is nonzero in the interacting theory due to vacuum-polarization effects. It is related to $B'_{\mu\nu}$ by

$$B^2_{\mu\nu}(q) = \frac{(q^4/\lambda^2)B'_{\mu\nu}(q) - g_{\mu\nu}}{\lambda^2} \tag{66}$$

$$= \frac{-g^2\Pi(q)g_{\mu\nu}}{q^2 + g^2\lambda^2\Pi(q)} \tag{67}$$

up to terms proportional to $q_\mu q_\nu$. Equation (65) will play an instrumental role in the demonstration of free-quark screening in the next section.

V. SOME GENERAL PROPERTIES

In this section we will first consider the screening mechanism for quarks and then discuss unitarity, causality, and scaling properties of the lowest-order contributions to the electroproduction structure functions.

FIG. 3. Representation of the Schwinger-Dyson equation for the gluon self-energy.

In Ref. 1 attention was drawn to the screening of free quarks due to vacuum polarization. Our mechanism is a variation of the Schwinger mechanism[11] but differs from it in an important respect: It is manifest in low order and thus not a matter of conjecture—an important consideration for insoluble field theories. Even in lowest order (Fig. 1), where $\Pi(q)$ is a constant up to a logarithmic term, we find manifest screening.

Let us consider a system containing free quarks in some bounded region. We choose to work in the Coulomb gauge. Because of Eq. (35) the total charge is proportional to

$$Q \propto \int d^3x\,\nabla^2\Box A_0^1 \tag{68}$$

$$= \int d\vec{S}\cdot\vec{\nabla}\Box A_0^1\,. \tag{69}$$

However, an examination of Eq. (65) shows that important vacuum-polarization effects occur at large distances. The potential corresponding to a static free quark located at the origin is

$$A_0^1 = \frac{-\lambda^2 g|\vec{x}|}{8\pi} \tag{70}$$

(if we ignore vacuum-polarization effects) and a finite contribution to Q would result if substituted in Eq. (69). At large distances A_0^1 is substantially modified from the expression in Eq. (70). From Eq. (65) we see that the large-distance behavior of A_0^1 is controlled by

$$\frac{1}{g^2q^2\Pi(q)}\,, \tag{71}$$

and since $\Pi(q)$ is a constant up to logarithms in lowest order and not proportional to a positive power of q^2 in any finite order of perturbation theory we find A_0^1 to be proportional to at most an inverse power of $|\vec{x}|$ at large distances. Substituting an inverse power of $|\vec{x}|$ for A_0^1 in Eq. (69) and letting the surface of integration go to infinity shows $Q=0$. Thus isolated free quarks do not exist in this model. Only neutral bound states occur.

We have chosen a propagator for the gluon which allows the S matrix to be unitarity. Our gluons are dipole ghosts, and, having indefinite metric, they would normally destroy the unitarity of the S matrix. But the quantization procedure eliminates their appearance in in or out states and their principal-value propagator precludes states containing gluons from contributing to the absorptive part of any Feynman diagram.[12] This is required if the S matrix is to be unitary. But as a result the S matrix is not analytic. The nonanalyticity is closely associated with advanced noncausal effects. Our procedure forces unitarity to be valid at the expense of noncausality. Tradeoffs of this type have recently been discussed by Coleman.[13] We return to the question of causality later.

We have verified that unitarity is maintained in perturbation theory by an explicit calculation of the lowest-order quark self-energy [Fig. 2(a)], which is

$$\Sigma(q)=\frac{-\lambda^2 g^2}{8\pi^2}\,\mathrm{P}\sum_i \frac{c_i\Lambda_i{}^2\ln(\Lambda_i{}^2/q^2)}{q^2(q^2-\Lambda_i{}^2)}\left(\frac{\Lambda_i{}^2}{q^2}\not{q}-2m\right), \tag{72}$$

where $c_1=1$, $\Lambda_1=m$, the regulator identities $\sum_i c_i=\sum_i c_i\Lambda_i{}^2=0$ hold, and P signifies $\Sigma(q^2+i\epsilon)=\Sigma(q^2-i\epsilon)$ as is demonstrated in detail in Appendix A. The fact that the singularities in Eq. (72) occur in principal value implies Σ has no absorptive part. This is to be contrasted with the corresponding quantity in QED which has an absorptive part reflecting the physically allowed decay of an off-shell electron into an electron and a photon. No similar possibility exists in our model.

We now will show that the absorptive part (in the physical region) of a Feynman diagram with one internal gluon line only receives contributions from intermediate states (obtained by appropriately "cutting" internal lines in all possible ways) which do not contain the gluon. The generalization to diagrams with many gluon lines is immediate. First we note that a principal-value propagator may be decomposed:

$$\mathrm{P}\,\frac{1}{k^2-m^2}=\frac{1}{k^2-m^2+i\epsilon}+i\pi\delta(k^2-m^2)\,. \tag{73}$$

For the sake of simplicity we shall write the integral corresponding to our hypothetical diagram as

$$I=\int d^4k\,\tilde{I}\;\mathrm{P}\,\frac{1}{k^4} \tag{74}$$

$$=\frac{\partial}{\partial\mu^2}\int d^4k\,\tilde{I}\;\mathrm{P}\,\frac{1}{k^2-\mu^2}\bigg|_{\mu^2=0}\,, \tag{75}$$

where I and $\tilde{I}$ have indices and momenta appropriate to the diagram in question and the limits we have introduced engender no infrared difficulties due to the choice of a principal-value propagator. Substituting Eq. (73) into Eq. (75) we can decompose I into three Feynman integrals (actually their derivative with respect to mass, etc.),

$$I=\frac{\partial}{\partial\mu^2}\int d^4k\left[\frac{\tilde{I}}{k^2-\mu^2+i\epsilon}+i\pi\theta(k_0)\delta(k^2-\mu^2)+i\pi\theta(-k_0)\delta(k^2-\mu^2)\right]_{\mu=0}, \tag{76}$$

in each of which only Feynman propagators are used. The last two terms correspond to opening up the loop containing the gluon. Their Feynman diagrams have in and out gluon lines of momentum k, which is summed over. Let us now restrict ourselves to the physical region[14] of our diagram so that we can take the absorptive part of I in the following way:

$$\begin{aligned}\mathrm{Abs}(I)&=i\pi\frac{\partial}{\partial\mu^2}\int d^4k[-\tilde{I}'\theta(k_0)\delta(k^2-\mu^2)+\tilde{I}'\theta(k_0)\delta(k^2-\mu^2)+\tilde{I}'\theta(-k_0)\delta(k^2-\mu^2)]+R\\&=R\,.\end{aligned} \tag{77}$$

The term in square brackets contains all contributions from intermediate states containing a gluon, while R contains the remainder of the absorptive part. The first two terms cancel, while the third term is zero in the physical region. Thus we have shown that the absorptive part receives no contributions from states containing the gluon. Consequently only states containing quarks contribute to unitarity sums for absorptive parts and diagrams containing external gluons do not contribute to the S matrix.

The principal-value gluon propagator has introduced noncausal effects into our model in the sense that the corresponding configuration-space Green's function is half-advanced and half-retarded. However, because we have maintained the commutativity of field operators at spacelike distances the principle of microscopic causality is not violated. Although advanced effects are not observed in everyday life they do not lead to internal inconsistency or paradoxes.[6] On the microscopic level, for example, within the confines of a hadron, advanced effects are not necessarily ruled out on physical grounds. From the earlier discussion of vacuum-polarization effects it is clear that noncausalities must be limited to very short distances. It thus appears that the only significant question involving causality is whether the nonanalyticity of the S matrix for low-order quark-quark scattering will be reflected in the scattering amplitudes of bound states in a manner which is in substantial disagreement with our understanding of S-matrix analyticity for physical particle scattering. The answer to this question is not known.

As an application of the dipole model we shall

study the deep-inelastic electroproduction structure functions in low-order perturbation theory. Previous calculations[15] of the structure functions in pseudoscalar-meson or neutral-vector-meson field-theoretic models (without transverse-momentum cutoffs) contained logarithmic deviations from scaling in apparent conflict with experimental results. The dipole model has strong transverse-momentum damping and as a result one obtains scaling structure functions—in fact, the only asymptotically leading contribution appears to be the diagram of Fig. 4(a). Higher-order diagrams do not scale by powers of q^2, the photon mass squared. For example the diagrams of lowest order in q^2 [Figs. 4(b)–4(d)] contributing to νW_2 are of $O(q^{-4})$. Thus the dipole model establishes a parton picture of the deep-inelastic structure functions since quarks appear to be pointlike particles in the scaling region. The choice of a principal-value propagator for the gluon has the effect of suppressing corrections to the scaling part of νW_2 which would have been of $O(q^{-2})$, such as the contribution of the diagram of Fig. 5. The absorptive part of that diagram is zero due to principal-value gluon propagator. Thus the precocious nature of scaling could be connected with the properties of the principal-value gluon propagator in electroproduction. On the other hand, the principal-value propagator will not play such an important role (at least in low order) in suppressing nonscaling contributions to the absorptive part of the amplitude associated with $e^+e^- \to$ hadrons. Thus low-order calculations are suggestive so far as scaling phenomena are concerned.

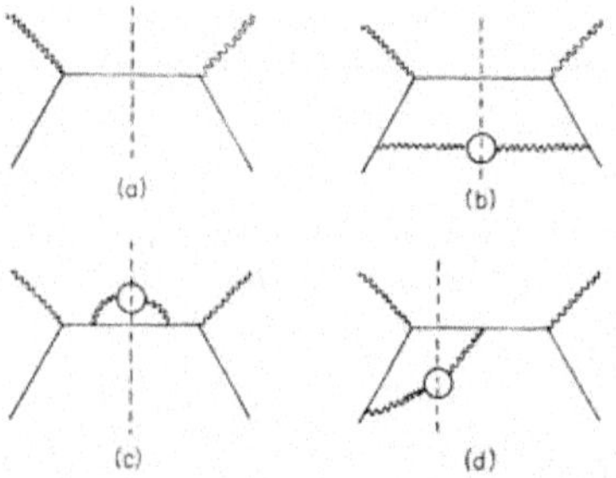

FIG. 4. Lowest-order diagrams contributing to the inelastic electroproduction structure functions. The dashed lines indicate the only contributions to the electroproduction structure functions of the absorptive part of the forward virtual Compton scattering diagram. External "wiggly" lines represent photons, while internal "wiggly" lines represent dipole gluons.

VI. CONCLUSION

The dipole electrodynamics model which we have discussed in the preceding sections is a prototype for a field theory of hadron binding. It has a number of desirable qualitative features such as quark confinement and scaling electroproduction structure functions. The physical content of this model is in the bound-states sector. This sector is currently under study using Bethe-Salpeter and eikonal techniques.

In a more realistic version of this model charge will be replaced by color in such a way that only zero-triality states are physical. The fields A_μ^1 and A_μ^2 will then become Yang-Mills fields. In that case the use of principal-value propagators appears to substantially simplify the model since closed loops of Yang-Mills fields are necessarily zero.[16]

ACKNOWLEDGMENT

I am grateful to the members of the Newman Laboratory of Nuclear Studies for helpful discussions and particularly to Dr. J. Kogut, Dr. D. K. Sinclair, and Dr. L. Susskind.

APPENDIX A

As an example of the modifications in perturbation-theory calculations resulting from principal-value propagators we evaluate the self-energy contribution of Fig. 2(a) and verify Eq. (72):

$$\Sigma(q) = \frac{ig^2\lambda^2}{(2\pi)^4} \int \frac{d^4k}{(q+k)^2 - m^2 + i\epsilon} \left(\mathrm{P}\,\frac{1}{k^4}\right) \gamma_\nu (\not{q} + \not{k} + m)\gamma^\nu \tag{A1}$$

in the Feynman gauge. Feynman parameters can be introduced, and using Eqs. (49) and (50) we obtain

$$\Sigma(q) = \frac{ig^2\lambda^2}{32\pi^2} \int_{-\infty}^{\infty} d\alpha\, \epsilon(\alpha)\alpha \int_0^{\infty} \frac{d\beta}{C^2} \left(\frac{-2\alpha}{C}\not{q} + 4m\right) \times \epsilon(C) e^{iD/C}, \tag{A2}$$

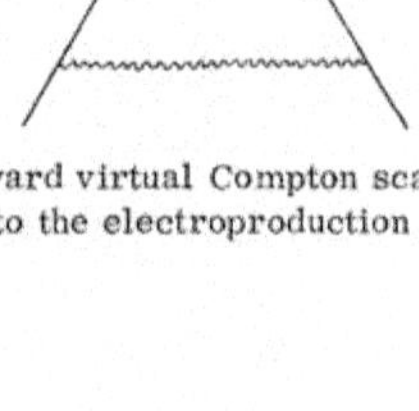

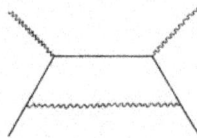

FIG. 5. A forward virtual Compton scattering diagram *not* contributing to the electroproduction structure functions.

where

$$C = \alpha + \beta \tag{A3}$$

and

$$D = \alpha\beta q^2 - \beta C m^2 . \tag{A4}$$

Scaling α and β and using Eq. (53) converts $\Sigma(q)$ to the form

$$\Sigma(q) = \frac{-g^2\lambda^2}{32\pi^2} \int_{-\infty}^{\infty} d\alpha\, \alpha\, \epsilon(\alpha) \int_0^{\infty} \frac{d\beta}{D} [-2\alpha\epsilon(C) \not{q} + 4m] \times \delta(1 - |\alpha + \beta|) . \tag{A5}$$

Of the two "points" contributing to the integral, $\alpha + \beta = \pm 1$, the contribution of the term $\alpha + \beta = -1$ can be included in the other term by an extension of the β integration domain:

$$\Sigma(q) = \frac{-g^2\lambda^2}{32\pi^2} \int_{-\infty}^{\infty} d\alpha\, \epsilon(\alpha) \alpha \int_{-\infty}^{\infty} d\beta \frac{\delta(1 - \alpha - \beta)(-2\alpha\not{q} + 4m)}{\alpha\beta q^2 - \beta m^2} . \tag{A6}$$

This may be shown by letting $\alpha \to -\alpha$ and $\beta \to -\beta$ in the term in question. Equation (A6) has divergences at $\alpha = \pm\infty$ after the β integration. These may be handled by introducing Pauli-Villars regulators of mass Λ_i satisfying

$$\sum_i c_i = 0 , \tag{A7}$$

$$\sum_i c_i \Lambda_i{}^2 = 0 , \tag{A8}$$

$$c_1 = 1 , \tag{A9}$$

$$\Lambda_1 = m . \tag{A10}$$

Equation (A6) then becomes

$$\Sigma(q) = \frac{-g^2\lambda^2}{32\pi^2} \sum_i c_i \int_{-\infty}^{\infty} \frac{d\alpha\, \alpha\, \epsilon(\alpha)(-2\alpha\not{q} + 4m)}{(1-\alpha)\{\alpha[q^2 + i\epsilon(\alpha)\delta] - \Lambda_i{}^2\}} , \tag{A11}$$

which may be shown to give Eq. (72) by elementary integrations. Apparent singularities in the denominator of the integrand of Eq. (A11) do not lead to difficulties if we take account of the $i\epsilon$'s which we have suppressed. The $i\epsilon(\alpha)\delta$ term (δ is infinitesimal) shows $\Sigma(q)$ to be in principal value $[\Sigma(q^2 + i\delta) = \Sigma(q^2 - i\delta)]$. It originates in the exponentiation of the principal-value propagator using Eq. (49).

APPENDIX B

We will briefly describe another possible model for hadron binding. Like the dipole model it is a member of a class of null-metric gluon theories with multipole Green's functions. The physical motivation for considering this model is a gross similarity to a quark model of Feynman *et al.*[4] which posited a relativistic harmonic oscillator potential, $x^\mu x_\mu$, between quarks and obtained quite successful agreement with experiment. If we neglect factors due to its vectorial nature (and also vacuum polarization effects) the interaction between quarks in our model is $x^\mu x_\mu \theta(x_\mu x^\mu)$ (note that $x_\mu x^\mu = x_0{}^2 - \vec{x}^2$). The $\theta(x^\mu x_\mu)$ factor, which is necessary for unitarity to be maintained, seems to imply that only "timelike" excitations are physical, and as a result the analysis of Feynman *et al.* cannot be directly appropriated for our use.

We shall introduce three vector-gluon fields, $A_\mu^i(x)$ $(i = 1, 2, 3)$, of which only one will interact directly with the prototype spin-$\frac{1}{2}$ quark field, $\psi(x)$:

$$\mathcal{L} = -\tfrac{1}{2}F^1_{\mu\nu}F^{3\mu\nu} + \tfrac{1}{4}F^2_{\mu\nu}F^{2\mu\nu} - \lambda^2 A^2_\mu A^{3\mu} + \bar{\psi}(i\nabla - g\not{A}^1 - m)\psi , \tag{B1}$$

with $F^i_{\mu\nu} = \partial_\nu A^i_\mu - \partial_\mu A^i_\nu$, and λ and g coupling constants. Following the canonical procedure we obtain the equations of motion

$$\partial^\mu F^1_{\mu\nu} + \lambda^2 A^2_\nu = 0 , \tag{B2}$$

$$\partial^\mu F^2_{\mu\nu} - \lambda^2 A^3_\nu = 0 , \tag{B3}$$

$$\partial^\mu F^3_{\mu\nu} + gJ_\nu = 0 , \tag{B4}$$

$$(i\not{\nabla} - g\not{A}^1 - m)\psi = 0 , \tag{B5}$$

with J^ν the quark current. The equations of motion reveal the Lagrangian to be invariant under local gauge transformations of A^1_μ and ψ while

$$\partial^\mu A^2_\mu = \partial^\mu A^3_\mu = 0 . \tag{B6}$$

Furthermore, Eqs. (B2) and (B3) imply

$$\Box F^1_{\mu\nu} - \lambda^2 F^2_{\mu\nu} = 0 , \tag{B7}$$

$$\Box F^2_{\mu\nu} + \lambda^2 F^3_{\mu\nu} = 0 , \tag{B8}$$

and as a result

$$\Box^2 \partial^\mu F^1_{\mu\nu} = g\lambda^4 J_\nu \tag{B9}$$

irrespective of the gauge choice for A^1_μ.

Following the conventional procedure we find that the canonical equal-time commutation relations in the radiation gauge $(\vec{\nabla}\cdot\vec{A}^1 = 0)$ are

$$[F^a_{0i}(x), A^b_j(y)] = ih^{ab} \int \frac{d^3k}{(2\pi)^3} e^{i\vec{k}\cdot(\vec{x}-\vec{y})} \left(\delta_{ij} - \frac{k_i k_j}{|\vec{k}|^2}\right) , \tag{B10}$$

with $h^{13} = h^{31} = -h^{22} = 1$ and all other $h^{ab} = 0$. We can now choose to quantize the theory as described in the text. Again we may use the perturbation-theory rules of QED if the photon propagator is replaced with the gluon propagator in the following manner:

$$iD_{F\mu\nu} \rightarrow iG_{\mu\nu} = i\lambda^4 g_{\mu\nu} \mathrm{P} \frac{1}{k^6} \tag{B11}$$

in the Feynman gauge, with

$$\mathrm{P} \frac{1}{k^6} = \frac{1}{2} \frac{1}{(k^2 + i\epsilon)^3} + \frac{1}{2} \frac{1}{(k^2 - i\epsilon)^3} \,. \tag{B12}$$

In coordinate space the gluon propagator is

$$T(A_\mu^1(x) A_\nu^1(y)) = i\lambda^4 g_{\mu\nu} \int d^4k \left(\mathrm{P} \frac{1}{k^6}\right) e^{-ik\cdot(x-y)} \tag{B13}$$

$$= \frac{i}{64\pi} \lambda^4 g_{\mu\nu} (x-y)^2 \theta((x-y)^2) \tag{B14}$$

in the Feynman gauge, which suggests a relationship between our model and that of Feynman, Kislinger, and Ravndal[4] as stated previously.

The discussions of unitarity, causality, and quark confinement given in the text apply to this model with only superficial changes. The light-cone divergences encountered in the dipole model are not so extreme here. For example, the overall degree of light-cone divergence for one-loop diagrams is 3-3p [where p is the number of internal gluon lines; cf. Eq. (61)] and thus the lowest-order quark self-energy (Fig. 2) is only logarithmically divergent,

$$\Sigma(q) = \frac{\lambda^4 g^2}{16\pi^2} \mathrm{P} \frac{1}{q^4} \left\{ \not{q} \left[\ln\left(\frac{\Lambda^2}{m^2}\right) - \frac{2q^2 m^2}{(q^2 - m^2)^2} + \frac{3q^4 m^2 - q^6}{(q^2 - m^2)^3} \ln\left(\frac{q^2}{m^2}\right) \right] + 2m \left[\frac{q^2(q^2 + m^2)}{(q^2 - m^2)^2} - \frac{2q^4 m^2}{(q^2 - m^2)^3} \ln\left(\frac{q^2}{m^2}\right) \right] \right\} , \tag{B15}$$

where q is the quark four-momentum, m the quark mass, Λ^2 is a regulator mass, and P signifies that all singularities are to be taken in principal value.

Finally, we would like to note again that the deep-inelastic structure functions scale in this model with leading nonscaling corrections of $O(1/q^6)$, where q is the virtual-photon four-momentum. These corrections come from the diagrams of Fig. 4.

---

*Work supported in part by the National Science Foundation.

[1]A. Casher, J. Kogut, and L. Susskind, Phys. Rev. Lett. 31, 792 (1973).

[2]K. Wilson, Phys. Rev. D 10, 2445 (1974).

[3]A. Chodos, R. L. Jaffe, K. Johnson, C. B. Thorn, and V. F. Weisskopf, Phys. Rev. D 9, 3471 (1974).

[4]R. Feynman, M. Kislinger, and F. Ravndal, Phys. Rev. D 3, 2706 (1971).

[5]P. A. M. Dirac, *Lectures on Quantum Mechanics* (Yeshiva Univ., New York, 1964).

[6]K. Schwarzschild, Göttinger Nachrichten 128, 132 (1903); H. Tetrode, Z. Phys. 10, 317 (1922); A. D. Fokker, *ibid.* 58, 386 (1929); Physica (The Hague) 9, 33 (1929); 12, 145 (1932); J. Wheeler and R. P. Feynman, Rev. Mod. Phys. 17, 157 (1945); 21, 425 (1949).

[7]J. D. Bjorken and S. D. Drell, *Relativistic Quantum Fields* (McGraw-Hill, New York, 1965), p. 382.

[8]E. Salpeter and H. Bethe, Phys. Rev. 84, 1232 (1951).

[9]R. J. Eden, P. V. Landshoff, D. I. Olive, and J. C. Polkinghorne, *The Analytic S-Matrix* (Cambridge Univ. Press, Cambridge, 1966), p. 34.

[10]Suggested by Dr. D. K. Sinclair.

[11]J. Schwinger, Phys. Rev. 128, 2425 (1962).

[12]After this work was completed Dr. K. Subbarao pointed out that Professor E. C. G. Sudarshan has considered the use of action-at-a-distance interactions to remedy unitarity problems in indefinite-metric theories: E. C. G. Sudarshan, Fields and Quanta 2, 175 (1972).

[13]S. Coleman, in *Subnuclear Phenomena,* edited by A. Zichichi (Academic, New York, 1970), Part A, p. 282.

[14]R. J. Eden, P. V. Landshoff, D. I. Olive, and J. C. Polkinghorne, *The Analytic S-Matrix* (Ref. 9), p. 112.

[15]R. Jackiw and G. Preparata, Phys. Rev. Lett. 22, 975 (1969); S. Adler and W. Tung, *ibid.* 22, 978 (1969); S. Blaha, Phys. Rev. D 3, 510 (1971).

[16]This may be proved using the representation of Eq. (54) with $q = 0$ (only principal-value propagators) and $I = 1$. If we let $\alpha_i \rightarrow -\alpha_i$ for $i = 1, 2, \ldots, p$ then we can show that $I = -I$ and thus $I = 0$. $\tilde{N}$ is arbitrary except that it can be written as a sum of terms which are homogeneous in the Feynman parameters. This condition can always be satisfied in perturbation theory.

PHYSICAL REVIEW D VOLUME 11, NUMBER 10 15 MAY 1975

Second-quantized non-Abelian field theory for hadrons with quark confinement and scaling deep-inelastic structure functions*

Stephen Blaha
Laboratory of Nuclear Studies, Cornell University, Ithaca, New York 14853
(Received 30 December 1974)

A four-dimensional second-quantized field theory with quarks bound by "colored" non-Abelian gluons is described which has the following properties: (1) the only physical particles are color singlets composed solely of quarks, (2) the deep-inelastic structure functions have Bjorken scaling, (3) gluon loops and Faddeev-Popov ghost loops are identically zero in any gauge, (4) Regge trajectories are apparently linear on a Chew-Frautschi plot, and (5) constituent motion within hadrons can be nonrelativistic.

I. INTRODUCTION

After a period of some skepticism the possibility that hadronic interactions might be understood within the framework of quantum field theory is again being seriously considered.[1] This is partly the result of the psychological climate created by the apparently successful unification of weak and electromagnetic interactions in a renormalizable field theory and partly the result of a greater appreciation of the variety of phenomena which can occur in field theories.

In this article we shall describe a field-theoretic model of hadron binding which has two major features: (1) Hadrons only occur as quark-antiquark or three-quark bound states, and (2) quarks behave as quasifree particles within hadrons. We assume that the suggestions of an internal symmetry called color[2] are correct and that the strong interaction consists of the exchange of colored Yang-Mills gluons. The nature of the interaction allows only color singlet states to occur in the gauge-invariant physical particle spectrum and consequently the first feature will be realized by choosing the color group to be SU(3). Since the (Schwinger) mechanism which produces this result is an infrared phenomenon, the second feature is not precluded and the model is essentially free in the ultraviolet region of the quark sector.

Our model is a non-Abelian version of a recently investigated Abelian field theory which had quark confinement and scaling electroproduction structure functions.[3] In that theory the free propagator of the massless gluon field embodying the quark-quark interaction was proportional to

$$\lambda^2/k^4, \tag{1}$$

where λ is a constant with the dimensions of mass and k is the gluon four-momentum. As a result the Schwinger mechanism[4] manifestly occurred, and it was shown that any charged particle was totally screened by vacuum polarization effects. In addition, explicit calculations of the deep-inelastic electroproduction structure functions in perturbation theory were in agreement with Bjorken scaling with corrections of $O(q^{-4})$, where q is the virtual photon four-momentum. These features of the Abelian model will also be shown to be true in the non-Abelian version. In addition, we shall argue that the quarks can be nonrelativistic within hadrons and that the spectrum of states has linearly rising Regge trajectories.

In spite of these salutary properties an interaction of the form of Eq. (1) could be questioned because of well-known[5] indefinite-metric difficulties which result in the violation of unitarity. While an optimist may hope that the nonappearance of colored gluons in asymptotic (color singlet) states might eliminate unitarity problems it is almost certain that the approximation techniques which will necessarily be used to find the bound states will lead to the occurrence of negative-metric states. Whether these states are "real" or artifacts of the approximation will not be clear. In view of this we suggested[3] that the gluon propagator be taken in principal value rather than as a Feynman propagator:

$$\mathrm{P}\frac{\lambda^2}{k^4} \equiv \frac{\lambda^2}{2}\left[\frac{1}{(k^2+i\epsilon)^2}+\frac{1}{(k^2-i\epsilon)^2}\right]. \tag{2}$$

As a result unitarity is maintained order by order in perturbation theory. Gluons do not appear in asymptotic states. All components of the vector-gluon propagator are "Coulombized" and the gluon field reduced to the embodiment of a direct quark interaction. There are a number of other decided advantages to principal-value propagators in the present context: (1) no color singlet states composed solely of gluons, (2) the elimination of substantial infrared divergences, (3) the suppression of corrections to Bjorken scaling in the electroproduction structure functions by a factor of q^2

vis-à-vis the corresponding Feynman-propagator result which sets the stage for precocious scaling, and (4) the elimination of closed loops of vector gluons and thus the elimination of Faddeev-Popov ghost loops.

In Sec. II we give a brief recapitulation of the Abelian model. In Sec. III we describe the canonical properties of the non-Abelian model. In Sec. IV we describe the qualitative features of the model and describe an approximation technique which appears to be naturally adopted to "solving" the theory. We shall restrict our discussion to the color binding interaction and defer the introduction of other interactions to a later work. The properties of the bound states in the non-Abelian model are currently under study and will be the subject of the next report.

II. ABELIAN MODEL

The possibility that the physical particle spectrum of a field theory consisted only of neutral states and did not include states of charged fields was first investigated in massless two-dimensional quantum electrodynamics.[6] In that case the absence of the "electron" from the gauge-invariant physical particle spectrum was directly related to the acquisition of a mass by the photon via the Schwinger mechanism. The Schwinger mechanism was manifest in the lowest-order contribution to the vacuum polarization (Fig. 1), and, taking account of the dimensionality of the coupling constant, $e \sim$ mass, could almost be considered a consequence of dimensional analysis. These vacuum polarization effects led to the total screening of the "electronic" charge, and, as a result, the "electron" was removed from the gauge-invariant physical particle spectrum. Our Abelian and non-Abelian models will display a similar pattern of events.

The Lagrangian of the Abelian model contains two gluon fields, $A^1_\mu(x)$ and $A^2_\mu(x)$, and the quark field $\psi(x)$:

$$\mathcal{L} = -\tfrac{1}{2}F^1_{\mu\nu}F^2_{\mu\nu} - \tfrac{1}{2}\lambda^2 A^2_\mu A^2_\mu + \bar\psi(i\not\nabla - g\not A^1 - m)\psi\,, \tag{3}$$

where for typographic convenience we denote the inner product of four vectors, $a\cdot b = a_\mu b_\mu = a_0 b_0 - \vec a\cdot\vec b$ throughout, λ is a constant with the dimensions of mass, g is dimensionless, and $F^i_{\mu\nu} = \partial_\nu A^i_\mu - \partial_\mu A^i_\nu$.

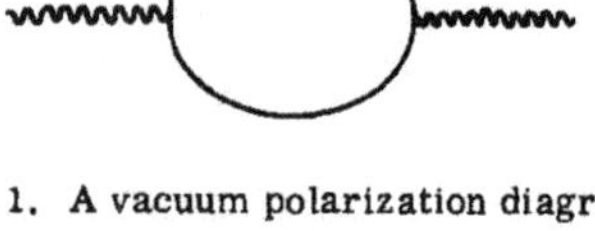

FIG. 1. A vacuum polarization diagram.

Following the canonical procedure we find the equations of motion;

$$\partial_\mu F^1_{\mu\nu} + \lambda^2 A^2_\nu = 0\,, \tag{4}$$

$$\partial_\mu F^2_{\mu\nu} + gJ_\nu = 0\,, \tag{5}$$

$$(i\not\nabla - g\not A^1 - m)\psi = 0\,, \tag{6}$$

and nonzero equal-time commutation relations [in the Coulomb gauge $\vec\nabla\cdot\vec A^1 = 0$; note $\partial_\mu A^2_\mu = 0$ by Eq. (4)]

$$[F^1_{0i}(x), A^2_j(y)] = i\Delta^{tr}_{ij}(x-y)\,, \tag{7}$$

$$[F^2_{0i}(x), A^1_j(y)] = i\Delta^{tr}_{ij}(x-y)\,, \tag{8}$$

with $i, j = 1, 2, 3$ and

$$\Delta^{tr}_{ij}(x-y) = \int \frac{d^3k}{(2\pi)^3} e^{i\vec k\cdot(\vec x - \vec y)}\left(\delta_{ij} - \frac{k_i k_j}{|\vec k|^2}\right). \tag{9}$$

It is clear from the equations of motion, Eqs. (4) and (5), tnat A^2_μ may be eliminated to obtain

$$\Box\partial_\mu F^1_{\mu\nu} + g\lambda^2 J_\nu = 0\,. \tag{10}$$

The form of the quark-gluon interaction and Eq. (10) show that only the Green's function of A^1_μ is relevant to quark-quark scattering. The perturbation theory rules of QED may be used if the photon propagator is replaced with the gluon propagator for A^1_μ:

$$iG^{11}_{\mu\nu}(k) = \frac{i\lambda^2(g_{\mu\nu} - \chi k_\mu k_\nu/k^2)}{k^4}\,, \tag{11}$$

where χ is constant, determined by the gauge choice.

In Ref. 3 we showed that choosing $G^{11}_{\mu\nu}$ to be a principal-value propagator allowed us to develop a perturbation theory which was unitary order by order:

$$G^{11}_{\mu\nu}(k^2) \equiv \tfrac{1}{2}[G^{11}_{\mu\nu}(k^2 + i\epsilon) + G^{11}_{\mu\nu}(k^2 - i\epsilon)]\,. \tag{12}$$

In addition, the equivalent of the Nambu representation of a Feynman diagram was given and some features of the perturbation theory discussed. Of particular interest was a calculation of the deep-inelastic electroproduction structure functions which scaled in the Bjorken limit. Leading corrections to scaling were of $O(q^{-4})$ as $q^2 \to \infty$ with q being the virtual photon four-momentum, and were given by the diagrams of Fig. 2(b), 2(c), and 2(d). This is to be contrasted with the logarithmic deviations from scaling found in pseudoscalar or vector meson models previously studied.[7]

The Schwinger mechanism manifestly occurred in low orders of perturbation theory. As a result quarks (and all charged objects) are removed from the gauge-invariant spectrum of physical states. The total screening of charge can be seen from the following argument.[3] Consider a spatially bounded system of charge density ρ. The total charge is

$$Q = \int d^3x\, \rho(x) \tag{13}$$

$$= \frac{-1}{g\lambda^2} \int d^3x\, \Box \nabla^2 A_0^1 \tag{14}$$

using the equations of motion in the Coulomb gauge. By Gauss's law

$$Q = \frac{-1}{g\lambda^2} \int d\vec{S} \cdot \vec{\nabla} \Box A_0^1 . \tag{15}$$

From the definition of a Green's function, we have

$$A_0^1(x) = \int d^4y\, G_{00}^{11}(x-y)\rho(y) \tag{16}$$

in the Coulomb gauge. If, for simplicity, we choose ρ to describe a static point quark charge and use the free gluon propagator [Eq. (11)], then $Q \neq 0$. However, if we take account of the effect of vacuum polarization processes (the Schwinger mechanism) we find $A_0^1(x)$ is a monotonically decreasing function of $|\vec{x}|$ for large $|\vec{x}|$ and consequently $Q = 0$ in the limit where the integration surface is taken to infinity in Eq. (15). Thus the spectrum of physical states does not include states of nonzero charge. In the next section we shall show that the proof of quark confinement is essentially the same in the non-Abelian model.

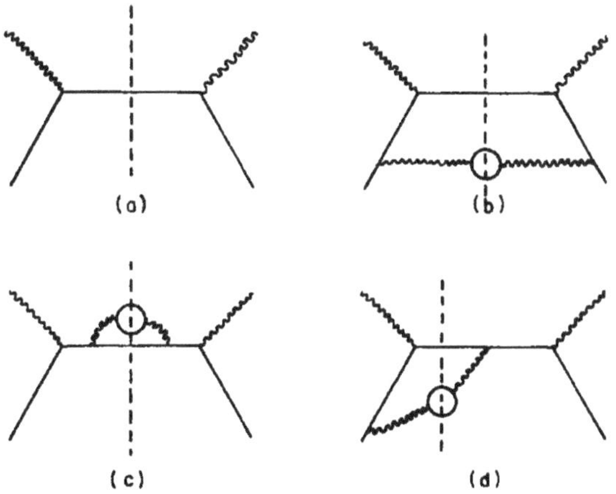

FIG. 2. Lowest-order diagrams contributing to the inelastic electroproduction structure functions. The dashed lines indicate the only contributions to the electroproduction structure functions of the absorptive part of the forward virtual Compton scattering diagram. External "wiggly" lines represent photons while internal "wiggly" lines represent gluons.

III. NON-ABELIAN MODEL

The non-Abelian model for the color sector of hadronic interactions is a direct generalization of the model of the last section.[8] There are two colored Yang-Mills fields, $A_{\mu a}^1(x)$ and $A_{\mu a}^2(x)$, which when regarded as vectors in the adjoint representation of the color group are denoted $\underline{A}_\mu^1$ and $\underline{A}_\mu^2$. The Lagrangian is

$$\begin{aligned}\mathfrak{L} = {} & \tfrac{1}{2}\underline{F}_{\mu\nu}^1 \cdot \underline{F}_{\mu\nu}^2 - \tfrac{1}{2}\underline{F}_{\mu\nu}^2 \cdot (\partial_\mu \underline{A}_\nu^1 - \partial_\nu \underline{A}_\mu^1 + g\underline{A}_\mu^1 \times \underline{A}_\nu^1) \\ & - \tfrac{1}{2}\underline{F}_{\mu\nu}^1 \cdot (\partial_\mu \underline{A}_\nu^2 - \partial_\nu \underline{A}_\mu^2 + g\underline{A}_\mu^1 \times \underline{A}_\nu^2 - g\underline{A}_\nu^1 \times \underline{A}_\mu^2) \\ & - \tfrac{1}{2}\lambda^2 \underline{A}_\mu^2 \cdot \underline{A}_\mu^2 + \bar{\psi}(i\nabla\!\!\!/ + gA\!\!\!/^1 - m)\psi \end{aligned} \tag{17}$$

$$= \mathfrak{L}_0 + \bar{\psi}(i\nabla\!\!\!/ + gA\!\!\!/^1 - m)\psi , \tag{18}$$

with ψ being the quark field.

It is invariant under the local gauge transformation

$$\psi' = S^{-1}\psi , \tag{19}$$

$$A_\mu^{1\prime} = S^{-1}A_\mu^1 S + \frac{i}{g} S^{-1}\partial_\mu S , \tag{20}$$

$$A_\mu^{2\prime} = S^{-1}A_\mu^2 S , \tag{21}$$

$$F_{\mu\nu}^{1\prime} = S^{-1}F_{\mu\nu}^1 S , \tag{22}$$

$$F_{\mu\nu}^{2\prime} = S^{-1}F_{\mu\nu}^2 S , \tag{23}$$

where S is an element in the gauge group G [which is color SU(3) in our case], and A_μ^1 is a matrix in the defining representation of G formed from

$$A_\mu^1 = \underline{A}_\mu^1 \cdot \underline{T} . \tag{24}$$

T_a is a matrix in the defining representation of G satisfying

$$[T_a, T_b] = i f_{abc} T_c , \tag{25}$$

and $\underline{T}$ is a vector formed from such matrices. We note that the homogeneity of the gauge transformation of A_μ^2 allows a mass term to occur in $\mathfrak{L}$ without breaking the gauge symmetry. We shall see that the natural gauge-fixing term to add to the Lagrangian has the form

$$-\frac{1}{\beta}\partial_\mu \underline{A}_\mu^1 \cdot \partial_\nu \underline{A}_\nu^2 . \tag{26}$$

The Euler-Lagrange equations of motion are obtained in the canonical manner:

$$(\partial_\mu + g\underline{A}_\mu^1 \times)\underline{F}_{\mu\nu}^1 - \lambda^2 \underline{A}_\nu^2 = 0 , \tag{27}$$

$$(\partial_\mu + g\underline{A}_\mu^1 \times)\underline{F}_{\mu\nu}^2 + g\underline{A}_\mu^2 \times \underline{F}_{\mu\nu}^1 + g\underline{J}_\nu = 0 , \tag{28}$$

$$\underline{F}_{\mu\nu}^1 = \partial_\mu \underline{A}_\nu^1 - \partial_\nu \underline{A}_\mu^1 + g\underline{A}_\mu^1 \times \underline{A}_\nu^1 , \tag{29}$$

$$\underline{F}_{\mu\nu}^2 = \partial_\mu \underline{A}_\nu^2 - \partial_\nu \underline{A}_\mu^2 + g\underline{A}_\mu^1 \times \underline{A}_\nu^2 - g\underline{A}_\nu^1 \times \underline{A}_\mu^2 , \tag{30}$$

$$(i\nabla\!\!\!/ + gA\!\!\!/^1 - m)\psi = 0 . \tag{31}$$

The antisymmetry of $\underline{F}_{\mu\nu}^1$ and $\underline{F}_{\mu\nu}^2$ leads to two conservation laws,

$$\partial_\nu(g\underline{A}^1_\mu\times\underline{F}^1_{\mu\nu}-\lambda^2\underline{A}^2_\nu)=0\,,\tag{32}$$

$$\partial_\nu(\underline{A}^1_\mu\times\underline{F}^2_{\mu\nu}+\underline{A}^2_\mu\times\underline{F}^1_{\mu\nu}+\underline{J}_\nu)=0\,,\tag{33}$$

which can be reexpressed as

$$(\partial_\nu+g\underline{A}^1_\nu\times)\underline{A}^2_\nu=0\tag{34}$$

and

$$(\partial_\nu+g\underline{A}^1_\nu\times)\underline{J}_\nu=0\tag{35}$$

using the equations of motion. The first of these relations acts in effect as a gauge-fixing term for $\underline{A}^2_\mu$ if a gauge is chosen for $\underline{A}^1_\mu$. The second relation has the familiar form of current-conservation equations in conventional Yang-Mills theories.

We turn now to the derivation of the perturbation-theory rules in the gluon sector. We consider the vacuum-vacuum transition amplitude in the presence of external sources[9]:

$$W(\underline{J}^1_\mu,\underline{J}^2_\mu)=\int\prod_x dA^1_\mu\, dA^2_\nu \exp\left[i\int d^4x\left(\mathcal{L}_0-\frac{1}{\beta}\partial_\mu\underline{A}^1_\mu\cdot\partial_\nu\underline{A}^2_\nu+\underline{A}^1_\mu\cdot\underline{J}^1_\mu+\underline{A}^2_\mu\cdot\underline{J}^2_\mu\right)\right].\tag{36}$$

After some functional translations we find

$$W(\underline{J}^1_\mu,\underline{J}^2_\mu)\equiv\exp\left\{-i\int d^4x\,d^4y\left[\underline{J}^1_\mu(x)\cdot G^{12}_{\mu\nu}(x-y)\cdot\underline{J}^2_\nu(y)+\tfrac{1}{2}\underline{J}^1_\mu(x)\cdot G^{11}_{\mu\nu}(x-y)\cdot\underline{J}^1_\nu(y)\right]\right\},\tag{37}$$

where we have dropped an irrelevant factor independent of $\underline{J}^1_\mu$ and $\underline{J}^2_\mu$ on the right-hand side, and

$$G^{12}_{\mu\nu ab}(x)=-\delta_{ab}\int\frac{d^4k\,e^{-ik\cdot x}}{(2\pi)^4k^2}\left[g_{\mu\nu}+(\beta-1)\frac{k_\mu k_\nu}{k^2}\right]\tag{38}$$

and

$$G^{11}_{\mu\nu ab}(x)=\frac{\lambda^2\delta_{ab}}{(2\pi)^4}\int\frac{d^4k\,e^{-ik\cdot x}}{k^4}\left[g_{\mu\nu}+(\beta^2-1)\frac{k_\mu k_\nu}{k^2}\right],\tag{39}$$

with a and b labeling color indices. The free propagators corresponding to the time-ordered products are

$$\langle TA^1_{\mu a}(x)A^1_{\nu b}(y)\rangle=i\,G^{11}_{\mu\nu ab}(x-y)\tag{40}$$

and

$$\langle TA^1_{\mu a}(x)A^2_{\nu b}(y)\rangle=i\,G^{12}_{\mu\nu ab}(x-y).\tag{41}$$

The somewhat unusual Green's functions of Eqs. (40) and (41) have their origin in the canonical equal-time commutation relations which we shall now find.

From Eqs. (27)–(30) we obtain the equations of motion

$$\partial_0\underline{A}^1_k=\underline{F}^1_{0k}+\partial_k\underline{A}^1_0+g\underline{A}^1_k\times\underline{A}^1_0\,,\tag{42}$$

$$\partial_0\underline{A}^2_k=\underline{F}^2_{0k}+\partial_k\underline{A}^2_0+g\underline{A}^2_k\times\underline{A}^1_0-g\underline{A}^2_0\times\underline{A}^1_k\,,\tag{43}$$

$$\partial_0\underline{F}^1_{0k}=(\partial_i+g\underline{A}_i\times)\underline{F}^1_{ik}-g\underline{A}^1_0\times\underline{F}^1_{0k}+\lambda^2\underline{A}^2_k\,,\tag{44}$$

$$\partial_0\underline{F}^2_{0k}=(\partial_i+g\underline{A}^1_i\times)\underline{F}^2_{ik}-g\underline{A}^1_0\times\underline{F}^2_{0k}-g\underline{A}^2_\mu\times\underline{F}^1_{\mu k}-g\underline{J}_k\,,\tag{45}$$

and equations of constraint

$$\underline{F}^1_{ik}=\partial_i\underline{A}^1_k-\partial_k\underline{A}^1_i+g\underline{A}^1_i\times\underline{A}^1_k\,,\tag{46}$$

$$\underline{F}^2_{ik}=\partial_i\underline{A}^2_k-\partial_k\underline{A}^2_i+g\underline{A}^1_i\times\underline{A}^2_k-g\underline{A}^1_k\times\underline{A}^2_i\,,\tag{47}$$

$$(\partial_i+g\underline{A}^1_i\times)\underline{F}^1_{i0}+\lambda^2\underline{A}^2_0=0\,,\tag{48}$$

$$(\partial_i+g\underline{A}^1_i\times)\underline{F}^2_{i0}+g\underline{A}^2_i\times\underline{F}^1_{i0}-g\underline{J}_0=0\,.\tag{49}$$

The Lagrangian indicates that the canonical momenta are

$$\underline{\Pi}^1_j=\underline{F}^2_{0j}\tag{50}$$

and

$$\underline{\Pi}^2_j=\underline{F}^1_{0j}\,,\tag{51}$$

for $j=1,2,3$ with $\underline{\Pi}^i_j$ conjugate to $\underline{A}^i_j$, and $\underline{A}^i_0$ having no conjugate momentum for $i=1,2$. However, the equations of constraint indicate that not all components are independent. We now find the independent components. Let us define

$$\underline{F}^a_{0i}=\underline{F}^{aT}_{0i}+\underline{F}^{aL}_{0i}\tag{52}$$

and

$$\underline{F}^{aL}_{0i}=\partial_i\underline{\phi}^a\,,\tag{53}$$

where

$$\partial_i\underline{F}^{aT}_{0i}=0\,.\tag{54}$$

Then Eq. (48) gives

$$(\partial_i+g\underline{A}^1_i\times)\partial_i\underline{\phi}^1-\lambda^2\underline{A}^2_0=-g\underline{A}^1_i\times\underline{F}^{1T}_{i0}\tag{55}$$

and Eq. (49) gives

$$(\partial_i+g\underline{A}^1_i\times)\partial_i\underline{\phi}^2+g\underline{A}^2_i\times\partial_i\underline{\phi}^1=g\underline{A}^1_i\times\underline{F}^{2T}_{i0}+g\underline{A}^2_i\times\underline{F}^{1T}_{0i}-g\underline{J}_0\,.\tag{56}$$

Rewriting Eqs. (42) and (43) after taking the divergence with respect to spatial components gives

$$(\partial_0+g\underline{A}^1_0\times)\partial_k\underline{A}^1_k=(\partial_k+g\underline{A}^1_k\times)\partial_k\underline{A}^1_0+\partial_k\partial_k\underline{\phi}^1\tag{57}$$

and

$$(\partial_0+g\underline{A}^1_0\times)\partial_k\underline{A}^2_k+g\underline{A}^2_0\times\partial_k\underline{A}^1_k=\partial_k\partial_k\underline{A}^2_0+g\underline{A}^2_k\times\partial_k\underline{A}^1_0+g\underline{A}^1_k\times\partial_k\underline{A}^2_0+\partial_k\partial_k\underline{\phi}^2\tag{58}$$

If we choose the Coulomb gauge, $\vec{\nabla}\cdot\underline{A}^1=0$, then

$$\partial_k\partial_k\underline{A}_0^1+g\underline{A}_k^1\times\partial_k\underline{A}_0^1+\partial_k\partial_k\underline{\phi}^1=0 \tag{59}$$

and

$$(\partial_0+g\underline{A}_0^1\times)\partial_k\underline{A}_k^2-\partial_k\partial_k\underline{A}_0^2-g\underline{A}_k^2\times\partial_k\underline{A}_0^1-g\underline{A}_k^1\times\partial_k\underline{A}_0^2-\partial_k\partial_k\underline{\phi}^2=0\,, \tag{60}$$

thus determining $\underline{A}_0^1$ and $\underline{A}_0^2$. Suppose we now define

$$\underline{\vec{A}}^2=\underline{\vec{A}}^{2T}+\underline{\vec{A}}^{2L}\,, \tag{61}$$

$$\underline{\vec{A}}^{2L}=\vec{\nabla}\underline{\phi}^3\,, \tag{62}$$

with

$$\vec{\nabla}\cdot\underline{\vec{A}}^{2T}=0 \tag{63}$$

Taking the divergence of Eq. (44) leads to our final equation for dependent variables

$$\lambda^2\partial_k\partial_k\underline{\phi}^3=\partial_0\partial_k\partial_k\underline{\phi}^1+g\partial_k(\underline{A}_\mu^1\times\underline{F}_{\mu k}^1)\,. \tag{64}$$

The independent dynamical variables are thus seen to be F_{0i}^{1T}, F_{0i}^{2T}, A_i^{1T}, and A_i^{2T}. Their equal-time commutation relations are

$$[F_{0ia}^{1T}(x),A_{jb}^2(y)]=i\delta_{ab}\Delta_{ij}^{tr}(x-y)\,, \tag{65}$$

$$[F_{0ia}^{2T}(x),A_{jb}^1(y)]=i\delta_{ab}\Delta_{ij}^{tr}(x-y)\,, \tag{66}$$

with $i,j=1,2,3$, Δ_{ij}^{tr} given by Eq. (9), and a and b are color indices. All other commutators of the forms $[A^1,A^1]$, $[A^2,A^2]$, $[F^1,F^1]$, $[F^2,F^2]$, $[F^1,F^2]$ are zero.

We return to our development of perturbation-theory rules. The cubic and quartic gluon vertices of our model are given by (see Fig. 3)

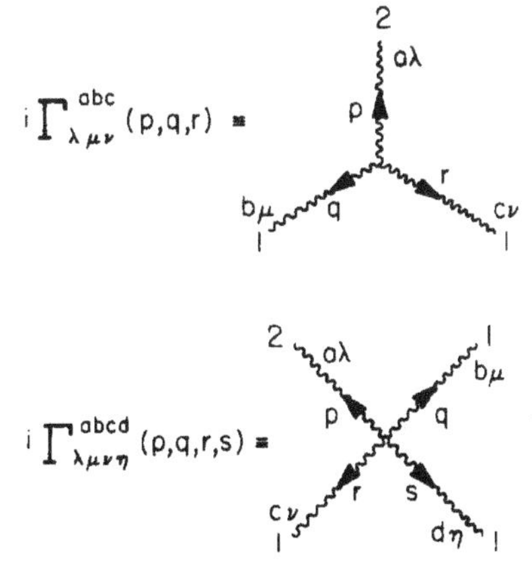

FIG. 3. Cubic and quartic vertices which are given in Eqs. (67) and (68). They introduce $1/r$ potentials in the model and may have an important effect in the baryon spectrum. The numbers 1 and 2 indicate fields $\underline{A}_\mu^1$ and $\underline{A}_\mu^2$, respectively, while p, q, r, and s are momenta, and a, b, c, and d are color indices.

$$i\Gamma^{abc}_{\lambda\mu\nu}(p,q,r)=f^{abc}[g_{\nu\lambda}(r_\mu-p_\mu)+g_{\mu\lambda}(p_\nu-q_\nu)+g_{\mu\nu}(q_\lambda-r_\lambda)]\,. \tag{67}$$

with $p+q+r=0$, and

$$i\Gamma^{abcd}_{\lambda\mu\nu\eta}(p,q,r,s)=-if^{abf}f^{cdf}(g_{\lambda\mu}g_{\nu\eta}-g_{\eta\mu}g_{\lambda\nu})-if^{acf}f^{bdf}(g_{\lambda\eta}g_{\nu\mu}-g_{\nu\lambda}g_{\mu\eta})-if^{adf}f^{bcf}(g_{\lambda\eta}g_{\nu\mu}-g_{\lambda\mu}g_{\nu\eta})\,, \tag{68}$$

with $p+q+r+s=0$.

The Faddeev-Popov ghost loops will not be relevant to our line of development so we omit their discussion. The necessity for their introduction[10] is closely related to the requirement of unitarity in Yang-Mills theories. In the present model unitarity will be necessarily violated irrespective of the ghost loops if the Green's functions [Eqs. (38) and (39)] pole ambiguities are resolved by using Feynman's $i\epsilon$ procedure. To avoid unitarity violation we have suggested an alternative procedure where the Green's function singularities are taken in principal value,

$$G_{\mu\nu ab}^{kL}(k^2)=\tfrac{1}{2}[G_{\mu\nu ab}^{kL}(k^2+i\epsilon)+G_{\mu\nu ab}^{kL}(k^2-i\epsilon)]\,, \tag{69}$$

in momentum space (cf. the Appendix). This choice has the advantage stated in the Introduction. The effects are the same as in the Abelian model[3] and may be summarized as: (1) Only states composed solely of quarks contribute to unitarity sums, (2) gluons do not appear in asymptotic states, (3) unitarity is achieved but at the price of possible advanced effects whose range is limited to hadronic dimensions and thus apparently unobservable, and (4) nonscaling corrections to Bjorken scaling in the deep-inelastic electroproduction structure functions are suppressed by a factor of q^2 vis-à-vis the corresponding result using Feynman propagators with q being the virtual photon four-momentum.

A novel feature of the use of principal-value propagators in non-Abelian models is the elimination of closed loops composed solely of gluons. If we consider a subdiagram consisting of a gluon loop with p lines, then Eq. (51) of Ref. 3 gives the Feynman parameter representation

$$I=\int_{-\infty}^{\infty}\prod_{j=1}^{p}\alpha_j\,d\alpha_j\,\frac{\epsilon(\alpha_1\alpha_2\cdots\alpha_p C)}{C^2}\,Ne^{iD/C}\,, \tag{70}$$

where C is a polynomial consisting of Feynman parameters only, while D contains scalar products of external momenta, N symbolizes appropriate numerator factors, and $\epsilon(\alpha)=\pm1$ if $\alpha\gtrless0$. Since N can be written as a sum of terms each of which is homogeneous in the Feynman parameters, we can take N to be homogeneous without loss of generality. Then scaling all parameters with u, assuming

$$N(u\alpha_1, u\alpha_2, \ldots, u\alpha_p) = u^r N(\alpha_1, \alpha_2, \ldots, \alpha_p), \quad (71)$$

with r an integer, and using

$$\int_0^\infty \frac{du}{u}\,\delta\left(1 - \frac{|\alpha_1 + \alpha_2 + \cdots + \alpha_p|}{u}\right) = 1 \quad (72)$$

we find

$$I = \Gamma(r + 2p - 2L) \times \int_{-\infty}^{\infty} \frac{\prod_{j=1}^{p} \alpha_j\, d\alpha_j\, \epsilon(\alpha_1\alpha_2\cdots\alpha_p C) N \delta\left(1 - \left|\sum_k \alpha_k\right|\right)}{C^2(-iD/C)^{r+2p-2L}}, \quad (73)$$

with L = number of loops = 1. Suppose we let $\alpha_j \to -\alpha_j$ for all j in I. Then we find $I = -I$ or

$$I = 0. \quad (74)$$

Thus any closed loop containing only principal-value propagators is zero. Since Faddeev-Popov ghosts appear only in closed loops and consistency[11] requires we use principal-value propagators for them if we use such propagators for gluons, we see that ghosts do not appear in our model. Physically we can understand this result if we remember that ghost loops were introduced to cure problems arising from contributions to unitarity sums of "opened" gluon loops.[10] In our model "opened" loops do not contribute to unitarity sums in any case so the raison d'être for ghosts is lacking.

We now derive the Ward-Takahashi-Slavnov identities using functional methods. Since we take our gluon propagators in principal value it might appear that our use of functional techniques is unjustified. We shall take the view that the functional representation of the vacuum-vacuum transition amplitude embodies the combinatorics of perturbation theory and acts as a generating function for identities, such as the Ward-Takahashi-Slavnov identities. Thus, questions of convergence of functional integrals are irrelevant—the important question is whether identities are valid in perturbation theory.

We define $W(J)$, the vacuum-vacuum transition amplitude, by

$$W(J) = \int \prod_x dA_\mu^1\, dA_\mu^2\, d\psi\, d\bar{\psi} \exp\left(i \int \bar{\mathcal{L}}\, dx\right), \quad (75)$$

with

$$\bar{\mathcal{L}} = \mathcal{L} - \frac{1}{\beta}\partial_\mu \underline{A}_\mu^1 \cdot \partial_\nu \underline{A}_\nu^2 + \underline{A}_\mu^1 \cdot \underline{J}_\mu^1 + \underline{A}_\mu^2 \cdot \underline{J}_\mu^2 + \bar{\psi}\eta + \bar{\eta}\psi,$$

with $\mathcal{L}$ given by Eq. (17). Under the infinitesimal gauge variation

$$\underline{A}_\mu^1 \to \underline{A}_\mu^1 - (\partial_\mu + g\underline{A}_\mu^1 \times)\underline{\theta}, \quad (76)$$

$$\underline{A}_\mu^2 \to \underline{A}_\mu^2 - g\underline{A}_\mu^2 \times \underline{\theta}, \quad (77)$$

$$\psi \to \psi - ig\theta\psi, \quad (78)$$

$$\bar{\psi} \to \bar{\psi} + ig\bar{\psi}\theta, \quad (79)$$

with $\theta = \underline{T} \cdot \underline{\theta}$, $\mathcal{L}$ is invariant but the remaining terms in $\bar{\mathcal{L}}$ lead to

$$\delta\bar{\mathcal{L}} \equiv \frac{1}{\beta}[(\partial_\mu + g\underline{A}_\mu^1 \times)\partial_\nu\partial_\mu \underline{A}_\nu^2 + g\underline{A}_\nu^2 \times \partial_\mu\partial_\nu \underline{A}_\mu^1] \cdot \underline{\theta} - (\partial_\mu + g\underline{A}_\mu^1 \times)\underline{J}_\mu^1 \cdot \underline{\theta} - g\underline{J}_\mu^2 \times \underline{A}_\mu^2 \cdot \underline{\theta} + ig\bar{\psi}\theta\eta - ig\bar{\eta}\theta\psi. \quad (80)$$

Since a transformation of the integration variables does not change the value of the functional integral, the variation of W with respect to θ can be taken to be zero and our equivalent of the Ward-Takahashi-Slavnov identity is

$$\left\{\frac{1}{\beta}\left[D_\nu\left(\frac{\delta}{i\delta \underline{J}_\alpha^1}\right)\partial_\nu\partial_\mu \frac{\delta}{i\delta \underline{J}_\mu^2} + g\frac{\delta}{i\delta \underline{J}_\nu^2} \times \partial_\nu\partial_\mu \frac{\delta}{i\delta \underline{J}_\mu^1}\right] + D_\mu\left(\frac{\delta}{i\delta \underline{J}_\alpha^1}\right)\underline{J}_\mu^1 - g\underline{J}_\mu^2 \times \frac{\delta}{i\delta \underline{J}_\mu^2} + g\,\underline{T}\eta\frac{\delta}{\delta\eta} - g\bar{\eta}\underline{T}\frac{\delta}{\delta\bar{\eta}}\right\} W = 0, \quad (81)$$

with

$$D_\mu\left(\frac{\delta}{i\delta \underline{J}_\alpha^1}\right) = \partial_\mu + g\frac{\delta}{i\delta \underline{J}_\mu^1} \times. \quad (82)$$

In order to investigate the structure of the gluon propagators we shall obtain the proper vertex identity equivalent to Eq. (81). We focus on the novelties of the gluon sector and neglect the quark field terms in $\mathcal{L}$ and Eq. (81). Let us define

$$W(J) = e^{iZ(J)}, \quad (83)$$

$$\underline{B}_\mu^i = -\frac{\delta Z(J)}{\delta \underline{J}_\mu^i}, \quad i = 1, 2 \quad (84)$$

$$\Gamma(B) = Z(J) + \int d^4x (\underline{J}_\mu^1 \cdot \underline{B}_\mu^1 + \underline{J}_\mu^2 \cdot \underline{B}_\mu^2), \quad (85)$$

where $\Gamma(B)$ is the generating functional of proper vertices. An immediate consequence is

$$\underline{J}^i_\mu = \frac{\delta\Gamma}{\delta \underline{B}^i_\mu}, \quad i=1,2 \tag{86}$$

and as a result Eq. (81) can be rewritten in the form

$$\frac{1}{\beta}\left[\Box\partial_\mu \underline{B}^2_\mu - g\underline{B}^1_\nu\times\partial_\nu\partial_\mu\underline{B}^2_\mu - g\underline{B}^2_\nu\times\partial_\mu\partial_\nu\underline{B}^1_\mu + g\frac{\delta}{i\delta \underline{J}^1_\nu}\times\partial_\nu\partial_\mu\underline{B}^2_\mu + g\frac{\delta}{i\delta\underline{J}^2_\nu}\times\partial_\nu\partial_\mu\underline{B}^1_\mu\right] - \partial_\mu\frac{\delta\Gamma}{\delta\underline{B}^1_\mu} + \underline{B}^1_\mu\times\frac{\delta\Gamma}{\delta\underline{B}^1_\mu} + \underline{B}^2_\mu\times\frac{\delta\Gamma}{\delta\underline{B}^2_\mu} = 0. \tag{87}$$

If we apply $\delta/\delta\underline{B}^1_\alpha$ to Eq. (87) and set $\underline{B}^i_\mu = 0$ afterwards, we find

$$-\partial_\mu \left.\frac{\delta^2\Gamma}{\delta\underline{B}^1_\alpha\delta\underline{B}^1_\mu}\right|_{B^1=B^2=0} = 0. \tag{88}$$

The second-order functional derivative of Γ is the inverse of the full propagator $G^{11\prime}_{\mu\nu ab}$ and Eq. (88) implies that the proper part of $(G^{11\prime}_{\mu\nu ab})^{-1}$ is purely transverse. We note that the "free" propagator (Eq. 39) contribution to $(G^{11\prime}_{\mu\nu ab})^{-1}$ is not one-particle irreducible and thus not constrained by Eq. (88). Therefore we find the general form

$$G^{11\prime}_{\mu\nu ab}(k) = \delta_{ab}\left(g_{\mu\nu} - \frac{k_\mu k_\nu}{k^2}\right)G^{11}(k^2) + \delta_{ab}\beta^2\lambda^2\frac{k_\mu k_\nu}{k^6}, \tag{89}$$

so that the longitudinal part of the full propagator is not renormalized.

The longitudinal part of the full propagator $G^{12\prime}_{\mu\nu ab}(k)$ is also not renormalized. This may be seen by applying $\delta/\delta\underline{B}^2_\mu$ to Eq. (87) and setting $\underline{B}^i_\mu = 0$ afterwards:

$$\frac{1}{\beta}\Box\partial_\mu\delta^4(x-y) - \partial_\nu \left.\frac{\delta^2\Gamma}{\delta\underline{B}^2_\mu\delta\underline{B}^1_\nu}\right|_{B^1=B^2=0} = 0. \tag{90}$$

This implies

$$(G^{12\prime}_{\mu\nu ab})^{-1} = \frac{\delta_{ab}(g_{\mu\nu} - k_\mu k_\nu/k^2)}{G^{12}} - \frac{k_\mu k_\nu\delta_{ab}}{\beta} \tag{91}$$

or

$$G^{12\prime}_{\mu\nu ab}(k) = \delta_{ab}\left(g_{\mu\nu} - \frac{k_\mu k_\nu}{k^2}\right)G^{12}(k) - \beta\delta_{ab}\frac{k_\mu k_\nu}{k^4}. \tag{92}$$

Having now developed the general form of the propagators we now will define the gluon vacuum polarization tensors,

$$\Pi^{11}_{\mu\nu ab}(k) = [G^{11\prime}_{\mu\nu ab}(k)]^{-1} - [G^{11}_{\mu\nu ab}(k)]^{-1}, \tag{93}$$

$$\Pi^{12}_{\mu\nu ab}(k) = [G^{12\prime}_{\mu\nu ab}(k)]^{-1} - [G^{12}_{\mu\nu ab}(k)]^{-1}, \tag{94}$$

which are transverse by our previous discussion:

$$k_\mu\Pi^{11}_{\mu\nu ab} = k_\mu\Pi^{12}_{\mu\nu ab} = 0. \tag{95}$$

Rather than write the Schwinger-Dyson equations for our polarization tensors we have given a diagrammatic representation in Fig. 4.

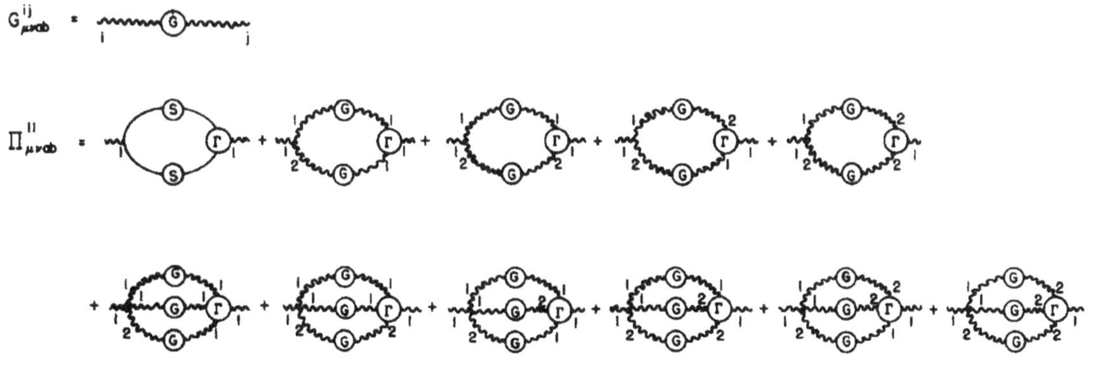

FIG. 4. Diagrammatic representation of the Schwinger-Dyson equation for the proper gluon self-energy, $\Pi^{11}_{\mu\nu ab}$. The numbers at the end of a gluon line specify whether $\underline{A}^1_\mu$ or $\underline{A}^2_\mu$ correspond to that end. The quark propagator is denoted S while Γ denotes the appropriate proper (one-particle irreducible) vertex function. A similar diagrammatic expression can be written for $\Pi^{12}_{\mu\nu ab}$.

IV. OBSERVATIONS

The Schwinger mechanism forces quark confinement to bound color singlet states in a manner which is identical to the Abelian case as described in Sec. II. In order to demonstrate that only color singlets exist in the gauge-invariant physical particle spectrum it is sufficient to show

$$\underline{Q}\psi_{\text{phys}} = 0\,, \tag{96}$$

where

$$\underline{Q} = \int d^3x\, \underline{J}_0(x) \tag{97}$$

for any physical state ψ_{phys} corresponding to a spatially localized distribution of quarks. We consider a single static quark located at the origin and choose to work in the Coulomb gauge ($\vec{\nabla}\cdot\vec{A}^1 = 0$). Then the time components of the equations of motion [Eqs. (27) and (28)] lead to (at large distance)

$$\Box\nabla^2\underline{A}_0^1 = g\lambda^2\underline{J}_0 \tag{98}$$

if we take into account the elimination of gluons' degrees of freedom through the choice of principal-value propagators and their consequent inability to act as sources. We may now repeat the arguments of Eqs. (13)–(16) for the Abelian case after noticing the occurrence of the Schwinger mechanism in the non-Abelian case which can be verified in low orders of perturbation theory for $\Pi^{11}_{\mu\nu ab}$. Thus the expectation value of the charge in the one-quark state is zero. Since the one-quark state is a charge eigenstate, we find Eq. (96) to be true in this case and more generally through the additivity of the charge operator. Thus only color singlet bound states of quarks are physical.[12]

While the infrared behavior of the theory leads to quark confinement, the ultraviolet behavior allows the quarks to appear quasifree. This is particularly noticeable when we take $\lambda^2 = 0$ in our Lagrangian and examine the corresponding perturbation theory. Taking $\lambda^2 = 0$ is equivalent to examining the short-distance behavior of the theory. The only diagrams which exist in this limit are given in Fig. 5. The quark sector of the theory is free. The only nontree structures are one-quark-loop diagrams for the scattering of gluons associated with A_μ^2 (which of course can only be generated by a hypothetical external source). (As a point of comparison we have shown in Fig. 6 the additional diagrams which would occur in the even that Feynman propagators were used—these diagrams necessarily involve gluon loops which principal-value propagators force to be zero.) The vital role of the $\lambda^2 A_\mu^2 A_\mu^2$ term in the Lagrangian in generating the interacting theory and the fact that λ^2 has the dimensions of $(\text{mass})^2$ allow a natural approximation procedure in this model. This is perhaps best seen within the context of deep-inelastic electroproduction. Just as in the Abelian case we find that the structure functions scale with leading corrections of $O(q^{-4})$, where q equals the virtual photon four-momentum. We can establish a parton picture of scattering wherein the photon is absorbed on one of the quasifree nucleon constituents [as in Fig. 2(a)] if $|q^2| \gg g^2\lambda^2$. Then leading corrections to such a picture [e.g., the diagrams of Fig. 2(b)–2(d)] will be suppressed by $(g^2\lambda^2/q^2)^2$. Thus the dimensional nature of the effective coupling constant allows a particularly simple picture to exist of the region of large spacelike virtual photon mass and the parton picture emerges as a natural approximation.

The k^{-4} form of the quark interaction also appears to have decidedly good features as far as the bound-state structure is concerned. Ignoring the numerator tensor (which does not affect our conclusions), we find the Fourier transform of the gluon propagator,

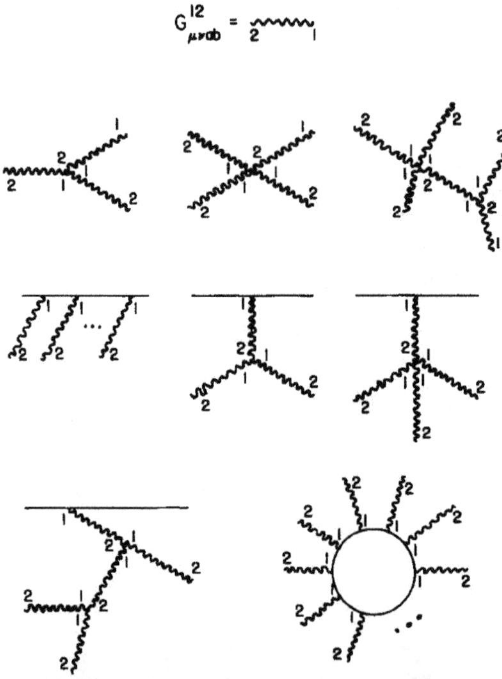

FIG. 5. Some examples of the surviving diagrams in the $\lambda^2 = 0$ limit of the non-Abelian model with principal-value gluon propagators. Except for the class of one-fermion-loop diagrams only tree diagrams exist in this limit. Note that there are no four (or more) external quark line diagrams and no two (or more) external $\underline{A}_\mu^1$ gluon "external" lines.

$$G(k) = \mathrm{P}\frac{1}{k^4}\,, \tag{99}$$

to be proportional to

$$\tilde{G}(x) = \theta(x^2)\,. \tag{100}$$

Since $\tilde{G}$ has a smooth finite limit as $x^2 \to 0$, the short-distance limit, arguments can be made[13] that low-mass bound states can occur in this model. In addition, Dalitz[14] has pointed out that the linearity of trajectories on the Chew-Frautschi plot would follow from a flat-bottomed, smooth interaction—a criterion which is met by Eq. (100). [It is interesting to note that had we used a Feynman propagator rather than principal value, then $\tilde{G}$ would have been $\ln(x^2)$ and thus the general criterion just stated would not have been met. This would appear to be another point in favor of our choice of principal-value propagators.]

Another property which is desirable in the bound-state solutions is nonrelativistic motion of the bound-state constituents.[15] Again an interaction of the form of Eq. (99) appears to realize this feature—even in the strong-binding limit. To see this we shall first take account of the Schwinger mechanism and in the spirit of Hartree-Fock theory modify the quark interaction to

$$G'(k) = \mathrm{P}\frac{1}{(k^2-\mu^2)^2}\,. \tag{101}$$

If we now take Eq. (101) to be the Green's function for the effective gluon field and calculate the "Coulomb potential" of a static, point quark source located at the origin we find

$$\varphi(r) = \frac{\varphi_0}{\mu}\, e^{-\mu r}\,, \tag{102}$$

where ϕ_0 is a constant independent of μ. In the limit $\mu \to 0$ we find

$$\varphi(r) \cong \varphi_0\left(\frac{1}{\mu} - r + \cdots\right). \tag{103}$$

The first two terms of Eq. (103) correspond to choosing Eq. (99) rather than Eq. (101) as the gluon Green's function (in the limit $\mu \to 0$). Equation (102) includes vacuum polarization effects which damp the interaction at large distances. Thus Eq. (102) imperfectly reflects the possibility that a quark-antiquark pair can separate and induce another quark-antiquark pair to be created from the vacuum so that two color singlet mesons will result (presuming it is energetically favored). At shorter distances Eq. (102) appears to be a reasonable approximation. This exponential potential was studied within the framework of the Schrödinger equation in the strong-binding limit (ϕ_0/μ large) by Greenberg.[16] He showed that the average momentum of the bound constituent in the s state satisfied

$$\frac{p}{m} \sim \left(\frac{\mu}{m}\right)^{1/3} \tag{104}$$

with m being the quark mass. Thus for μ/m small the quark motion is self-consistently nonrelativistic.

In conclusion, we have shown that a four-dimensional, Lorentz-invariant second-quantized field theory of hadron binding is possible with scaling electroproduction structure functions, only zero-triality physical particle states, and, apparently, linearly rising Regge trajectories and nonrelativistic constituents. A detailed study of the bound states is now in progress.

ACKNOWLEDGMENT

I am grateful to the members of the Newman Laboratory for interesting conversations.

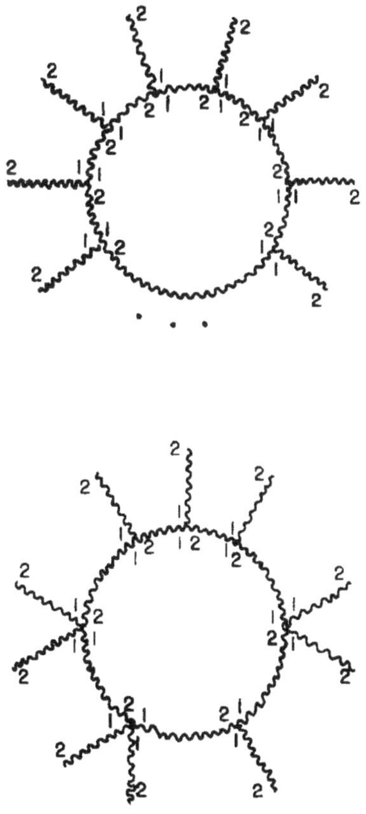

FIG. 6. Some additional diagrams which occur in the $\lambda^2 = 0$ limit of the non-Abelian model if Feynman gluon propagators are used. In addition, there will be Faddeev-Popov ghost-loop diagrams depending on the choice of gauge.

APPENDIX

In Ref. 3 semiclassical arguments based on Dirac's theory of constraints were given to introduce the use of principal-value propagators. We will now describe a second-quantized realization of those arguments for the case of a scalar Klein-Gordon field $\phi(x)$ with the Lagrangian

$$\mathcal{L} = \tfrac{1}{2}(\partial_\mu \phi)^2 - \tfrac{1}{2}m^2\phi^2 . \tag{A1}$$

The generalization to vector gluons is immediate. The canonical equal-time commutation relations are

$$[\phi, \phi] = [\dot\phi, \dot\phi] = 0 , \tag{A2}$$

$$[\dot\phi(\vec{x}, t), \phi(\vec{y}, t)] = -i\delta^3(\vec{x}-\vec{y}) . \tag{A3}$$

If we expand $\phi(x)$ in plane waves,

$$\phi(\vec{x}, t) = \sum_{\vec{k}} (A_{\vec{k}}^- e^{-ik\cdot x} + A_{\vec{k}}^+ e^{ik\cdot x}) , \tag{A4}$$

then the q-number Fourier components $A_{\vec{k}}^-$ must satisfy

$$[A_{\vec{k}}^-, A_{\vec{k}'}^-] = [A_{\vec{k}}^+, A_{\vec{k}}^+] = 0 , \tag{A5}$$

$$[A_{\vec{k}}^-, A_{\vec{k}'}^+] = \delta^3(\vec{k}' - \vec{k}) \tag{A6}$$

for consistency with Eqs. (A2) and (A3). Now the time-ordered product satisfies

$$T(\phi(x)\phi(y)) = \epsilon(x_0 - y_0)[\phi(x), \phi(y)] + \{\phi(x), \phi(y)\} , \tag{A7}$$

with $\epsilon(x_0) = \pm 1$ for $x_0 \gtrless 0$ and $\{A, B\} = AB + BA$. The first term on the right-hand side is a c number completely determined by Eqs. (A5) and (A6). If the second q-number expression were zero, then we would obtain a principal-value propagator from Eq. (A7):

$$T(\phi(x)\phi(y)) \equiv i\int \frac{d^4k}{(2\pi)^4} e^{-ik\cdot(x-y)}\, \mathrm{P}\, \frac{1}{(k^2 - m^2)} . \tag{A8}$$

We therefore require

$$\{\phi(x), \phi(y)\} = 0 , \tag{A9}$$

with the consequence

$$\{A_{\vec{k}}^-, A_{\vec{k}'}^-\} = \{A_{\vec{k}}^+, A_{\vec{k}'}^+\} = \{A_{\vec{k}}^-, A_{\vec{k}'}^+\} = 0 . \tag{A10}$$

Equations (A5), (A6), and (A10) imply

$$A_{\vec{k}}^- A_{\vec{k}'}^- = A_{\vec{k}}^+ A_{\vec{k}'}^+ = 0 , \tag{A11}$$

$$A_{\vec{k}}^- A_{\vec{k}'}^+ = \tfrac{1}{2}\delta^3(\vec{k} - \vec{k}') , \tag{A12}$$

$$A_{\vec{k}}^+ A_{\vec{k}'}^- = -\tfrac{1}{2}\delta^3(\vec{k} - \vec{k}') \tag{A13}$$

for all $\vec{k}$ and $\vec{k}'$. Thus quadratic terms in A and $A^\dagger$ are reduced to c numbers. It should further be noted that the multiplication rule is not associative.[17] In fact, the multiplication rules of the A and $A^\dagger$ operators in Eqs. (A11)–(A13) are realized by taking multiplication to be

$$UV = \tfrac{1}{2}[U, V] \tag{A14}$$

for U, V being any $A_{\vec{k}}^-$ or $A_{\vec{k}'}^+$. If we take an analogy to Lie-algebra theory seriously, where the adjoint representation of an algebra has a multiplication rule defined by commutators

$$\tilde{U} * \tilde{V} = [\tilde{U}, \tilde{V}] \tag{A15}$$

then we could call Eqs. (A11)–(A13) the adjoint representation of the Fourier components of ϕ.

The c-number nature of AA, $A^\dagger A^\dagger$, or $AA^\dagger$ can be understood physically in the following manner. Since the ϕ field has principal-value propagators it is not associated with a particle but is merely the embodiment of an interaction between other objects (which we have suppressed in our Lagrangian). Consequently an emission of a ϕ field quantum must be directly correlated with a subsequent absorption—it cannot propagate into empty space. The c-number nature of $AA^\dagger$ reflects this correlation between emission and absorption.

Finally, it should be noted that the existence of a vacuum is inconsistent with Eqs. (A11)—(A13).

---

*Work supported in part by the National Science Foundation.

[1]K. Johnson, Phys. Rev. D 6, 1101 (1972); C. M. Bender, J. E. Mandula, and G. S. Guralnik, Phys. Rev. Lett. 32, 1467 (1974); A. Chodos *et al.*, Phys. Rev. D 9, 3471 (1974); W. A. Bardeen *et al.*, *ibid.* 11, 1094 (1975); M. Creutz, *ibid.* 10, 1749 (1974); P. Vinciarelli, Nuovo Cimento Lett. 4, 905 (1972); R. Dashen, B. Hasslacher, and A. Neveu, Phys. Rev. D 10, 4114 (1974); 10, 4130 (1974); 10, 4138 (1974).

[2]Y. Nambu, in *Preludes in Theoretical Physics*, edited by A. de-Shalit, H. Feshbach, and L. Van Hove (North-Holland, Amsterdam, 1966), p. 133; H. J. Lipkin, Phys. Lett. 45B, 267 (1973).

[3]S. Blaha, Phys. Rev. D 10, 4268 (1974).

[4]J. Schwinger, Phys. Rev. 128, 2425 (1962).

[5]A. Pais and G. Uhlenbeck, Phys. Rev. 79, 145 (1950); J. Kiskis, Phys. Rev. D 11, 2178 (1975).

[6]A. Casher, J. Kogut, and L. Susskind, Phys. Rev. D 10, 732 (1974); J. Lowenstein and J. Swieca, Ann. Phys. (N.Y.) 68, 172 (1971).

[7]R. Jackiw and G. Preparata, Phys. Rev. Lett. 22, 975 (1969); S. Adler and W. Tung, *ibid.* 22, 978 (1969); S. Blaha, Phys. Rev. D 3, 510 (1971).

[8]The Lagrangian of Eq. (17) was first written by D. Sinclair as a generalization of the Abelian model of Ref. 3. An alternative non-Abelian model for quark confinement has been suggested by S. K. Kauffmann [Nucl. Phys. B87, 133 (1975)]. I am grateful to Dr. Kauffmann for sending me a copy of his paper prior to publication.

[9]E. Abers and B. W. Lee [Phys. Rep. 9C, 1 (1973)] provide a useful review of conventional Yang-Mills theories.

[10] R. P. Feynman, Acta Phys. Pol. 24, 697 (1963).

[11]B. W. Lee and J. Zinn-Justin [Phys. Rev. D 5, 3121 (1972)] point out that the $i\epsilon$ prescription in their Eq. (2.8) for the ghost loop is dictated by unitarity considerations.

[12]This does not preclude color-singlet states of the gluons from playing a role in the theory. They are not particles but can be exchanged between color-singlet quark states in scattering events. On naive dimensional grounds they should be most important in forward scattering. This leads to the possibility that the Pomeron might possibly be interpreted as a "two-gluon bound state". In the case of wide-angle scattering the predominant mechanism for large momentum transfer would appear to be constituent interchange due to the strong damping effects of k^{-4} propagators on momentum transfer.

[13]M. Böhm, H. Joos, and M. Krammer, in *Recent Developments in Mathematical Physics*, proceedings of the XII Schladming Conference (Acta Phys. Austriaca Suppl. XI), edited by P. Urban (Springer, New York, 1970), p. 3.

[14]R. H. Dalitz, a paper presented at the Topical Conference on Meson Spectroscopy, Philadelphia, 1968 (unpublished).

[15]H. J. Lipkin, Phys. Rep. 8C, 175 (1973).

[16]O. W. Greenberg, Phys. Rev. 147, 1077 (1966).

[17]Nonassociative field operators have been previously used by M. Günaydin and F. Gürsey, Phys. Rev. D 9, 3387 (1974).

REFERENCES

Akhiezer, N. I., Frink, A. H. (tr), 1962, *The Calculus of Variations* (Blaisdell Publishing, New York, 1962).

Bjorken, J. D., Drell, S. D., 1964, *Relativistic Quantum Mechanics* (McGraw-Hill, New York, 1965).

Bjorken, J. D., Drell, S. D., 1965, *Relativistic Quantum Fields* (McGraw-Hill, New York, 1965).

Blaha, S., 1995, *C++ for Professional Programming* (International Thomson Publishing, Boston, 1995).

_______, 1998, *Cosmos and Consciousness* (Pingree-Hill Publishing, Auburn, NH, 1998 and 2002).

_______, 2002, *A Finite Unified Quantum Field Theory of the Elementary Particle Standard Model and Quantum Gravity Based on New Quantum Dimensions™ & a New Paradigm in the Calculus of Variations* (Pingree-Hill Publishing, Auburn, NH, 2002).

_______, 2004, *Quantum Big Bang Cosmology: Complex Space-time General Relativity, Quantum Coordinates™ Dodecahedral Universe, Inflation, and New Spin 0, ½, 1 & 2 Tachyons & Imagyons* (Pingree-Hill Publishing, Auburn, NH, 2004).

______, *2005a, Quantum Theory of the Third Kind: A New Type of Divergence-free Quantum Field Theory Supporting a Unified Standard Model of Elementary Particles and Quantum Gravity based on a New Method in the Calculus of Variations* (Pingree-Hill Publishing, Auburn, NH, 2005).

______, 2005b, *The Metatheory of Physics Theories, and the Theory of Everything as a Quantum Computer Language* (Pingree-Hill Publishing, Auburn, NH, 2005).

______, 2005c, *The Equivalence of Elementary Particle Theories and Computer Languages: Quantum Computers, Turing Machines, Standard Model, Superstring Theory, and a Proof that Gödel's Theorem Implies Nature Must Be Quantum* (Pingree-Hill Publishing, Auburn, NH, 2005).

______, 2006a, *The Foundation of the Forces of Nature* (Pingree-Hill Publishing, Auburn, NH, 2006).

______, 2006b, *A Derivation of ElectroWeak Theory based on an Extension of Special Relativity; Black Hole Tachyons; & Tachyons of Any Spin.* (Pingree-Hill Publishing, Auburn, NH, 2006).

_______, 2007a, *Physics Beyond the Light Barrier: The Source of Parity Violation, Tachyons, and A Derivation of Standard Model Features* (Pingree-Hill Publishing, Auburn, NH, 2007).

______, 2007b, *The Origin of the Standard Model: The Genesis of Four Quark and Lepton Species, Parity Violation, the ElectroWeak Sector, Color SU(3), Three Visible Generations of Fermions, and One Generation of Dark Matter with Dark Energy* (Pingree-Hill Publishing, Auburn, NH, 2007).

______, *2008a, A Direct Derivation of the Form of the Standard Model From GL(16) (Pingree-Hill Publishing, Auburn, NH, 2008).*

______, 2008b, *A Complete Derivation of the Form of the Standard Model With a New Method to Generate Particle Masses Second Edition* (Pingree-Hill Publishing, Auburn, NH, 2008)

______, 2009, *The Algebra of Thought & Reality: The Mathematical Basis for Plato's Theory of Ideas, and Reality Extended to Include A Priori Observers and Space-Time Second Edition* (Pingree-Hill Publishing, Auburn, NH, 2009).

______, 2010a, *Operator Metaphysics: A New Metaphysics Based on a New Operator Logic and a New Quantum Operator Logic that Lead to a Mathematical Basis for Plato's Theory of Ideas and Reality* (Pingree-Hill Publishing, Auburn, NH, 2010).

______, 2010b, *The Standard Model's Form Derived from Operator Logic, Superluminal Transformations and GL(16)* (Pingree-Hill Publishing, Auburn, NH, 2010).

______, 2010c, *SuperCivilizations: Civilizations as Superorganisms* (McMann-Fisher Publishing, Auburn, NH, 2010).

______, 2011a, *21st Century Natural Philosophy Of Ultimate Physical Reality* (McMann-Fisher Publishing, Auburn, NH, 2011).

______, 2011b, *All the Universe! Faster Than Light Tachyon Quark Starships & Particle Accelerators with the LHC as a Prototype Starship Drive Scientific Edition* (Pingree-Hill Publishing, Auburn, NH, 2011).

______, 2011c, *From Asynchronous Logic to The Standard Model to Superflight to the Stars* (Blaha Research, Auburn, NH, 2011).

______, 2012a, *From Asynchronous Logic to The Standard Model to Superflight to the Stars volume 2: Superluminal CP and CPT, U(4) Complex General Relativity and The Standard Model, Complex Vierbein General Relativity, Kinetic Theory, Thermodynamics* (Blaha Research, Auburn, NH, 2012).

______, 2012b, *Standard Model Symmetries, And Four And Sixteen Dimension Complex Relativity; The Origin Of Higgs Mass Terms* (Blaha Reasearch, Auburn, NH, 2012).

______, 2013a, *Multi-Stage Space Guns, Micro-Pulse Nuclear Rockets, and Faster-Than-Light Quark-Gluon Ion Drive Starships* (Blaha Research, Auburn, NH, 2013).

______, 2013b, *The Bridge to Dark Matter; A New Sister Universe; Dark Energy; Inflatons; Quantum Big Bang; Superluminal Physics; An Extended Standard Model Based on Geometry* (Blaha Reasearch, Auburn, NH, 2013).

______, 2014a, *Universes and Megaverses: From a New Standard Model to a Physical Megaverse; The Big Bang; Our Sister Universe's Wormhole; Origin of the Cosmological Constant, Spatial Asymmetry of the Universe, and its Web of Galaxies; A Baryonic Field between Universes and Particles; Megaverse Extended Wheeler-DeWitt Equation* (Blaha Reasearch, Auburn, NH, 2014).

______, 2014b, *All the Megaverse! Starships Exploring the Endless Universes of the Cosmos Using the Baryonic Force* (Blaha Research, Auburn, NH, 2014).

______, 2014c, *All the Megaverse! II Between Megaverse Universes: Quantum Entanglement Explained by the Megaverse Coherent Baryonic Radiation Devices – PHASERs Neutron Star Megaverse Slingshot Dynamics Spiritual and UFO Events, and the Megaverse Microscopic Entry into the Megaverse* (Blaha Research, Auburn, NH, 2014).

______, 2015a, *PHYSICS IS LOGIC PAINTED ON THE VOID: Origin of Bare Masses and The Standard Model in Logic, U(4) Origin of the Generations, Normal and Dark Baryonic Forces, Dark Matter, Dark Energy, The Big Bang, Complex General Relativity, A Megaverse of Universe Particles* (Blaha Research, Auburn, NH, 2015).

______, 2015b, *PHYSICS IS LOGIC Part II: The Theory of Everything, The Megaverse Theory of Everything, U(4)⊗U(4) Grand Unified Theory (GUT), Inertial Mass = Gravitational Mass, Unified Extended Standard Model and a New Complex General Relativity with Higgs Particles, Generation Group Higgs Particles* (Blaha Research, Auburn, NH, 2015).

______, 2015c, *The Origin of Higgs ("God") Particles and the Higgs Mechanism: Physics is Logic III, Beyond Higgs – A Revamped Theory With a Local Arrow of Time, The Theory of Everything Enhanced, Why Inertial Frames are Special, Universes of the Mind* (Blaha Research, Auburn, NH, 2015).

______, 2015d, *The Origin of the Eight Coupling Constants of The Theory of Everything: U(8) Grand Unified Theory of Everything (GUTE), S^8 Coupling Constant Symmetry, Space-Time Dependent Coupling Constants, Big Bang Vacuum Coupling Constants, Physics is Logic IV* (Blaha Research, Auburn, NH, 2015).

______, 2016a, *New Types of Dark Matter, Big Bang Equipartition, and A New U(4) Symmetry in the Theory of Everything: Equipartition Principle for Fermions, Matter is 83.33% Dark, Penetrating the Veil of the Big Bang, Explicit QFT Quark Confinement and Charmonium, Physics is Logic V* (Blaha Research, Auburn, NH, 2016).

______, 2016b, *The Periodic Table of the 192 Quarks and Leptons in The Theory of Everything: The U(4) Layer Group, Physics is Logic VI* (Blaha Research, Auburn, NH, 2016).

______, 2016c, *New Boson Quantum Field Theory, Dark Matter Dynamics, Dark Matter Fermion Layer Mixing, Genesis of Higgs Particles, New Layer Higgs Masses, Higgs Coupling Constants, Non-Abelian Higgs Gauge Fields, Physics is Logic VII* (Blaha Research, Auburn, NH, 2016).

______, 2016d, *Unification of the Strong Interactions and Gravitation: Quark Confinement Linked to Modified Short-Distance Gravity; Physics is Logic VIII* (Blaha Research, Auburn, NH, 2016).

______, 2016e, *MoND: Unification of the Strong Interactions and Gravitation II, Quark Confinement Linked to Large-Scale Gravity, Physics is Logic IX* (Blaha Research, Auburn, NH, 2016).

______, 2016f, *CQ Mechanics: A Unification of Quantum & Classical Mechanics, Quantum/Semi-Classical Entanglement, Quantum/Classical Path Integrals, Quantum/Classical Chaos* (Blaha Research, Auburn, NH, 2016).

______, 2016g, *GEMS: Unified Gravity, ElectroMagnetic and Strong Interactions: Manifest Quark Confinement, A Solution for the Proton Spin Puzzle, Modified Gravity on the Galactic Scale* (Pingree Hill Publishing, Auburn, NH, 2016).

______, 2016h, *Unification of the Seven Boson Interactions based on the Riemann-Christoffel Curvature Tensor* (Pingree Hill Publishing, Auburn, NH, 2016).

______, 2017a, *Unification of the Eleven Boson Interactions based on 'Rotations of Interactions'* (Pingree Hill Publishing, Auburn, NH, 2017).

______, 2017b, *The Origin of Fermions and Bosons, and Their Unification* (Pingree Hill Publishing, Auburn, NH, 2017).

______, 2017c, *Megaverse: The Universe of Universes* (Pingree Hill Publishing, Auburn, NH, 2017).

______, 2017d, *SuperSymmetry and the Unified SuperStandard Model* (Pingree Hill Publishing, Auburn, NH, 2017).

______, 2017e, *From Qubits to the Unified SuperStandard Model with Embedded SuperStrings: A Derivation* (Pingree Hill Publishing, Auburn, NH, 2017).

______, 2017f, *The Unified SuperStandard Model in Our Universe and the Megaverse: Quarks, ...* , (Pingree Hill Publishing, Auburn, NH, 2017).

______, 2018a, *The Unified SuperStandard Model and the Megaverse SECOND EDITION A Deeper Theory based on a New Particle Functional Space that Explicates Quantum Entanglement Spookiness (Volume 1)* (Pingree Hill Publishing, Auburn, NH, 2018).

______, 2018b, *Cosmos Creation: The Unified SuperStandard Model, Volume 2, SECOND EDITION* (Pingree Hill Publishing, Auburn, NH, 2018).

______, 2018c, *God Theory (*Pingree Hill Publishing, Auburn, NH, 2018).

______, 2018d, *Immortal Eye: God Theory: Second Edition* (Pingree Hill Publishing, Auburn, NH, 2018).

______, 2018e, *Unification of God Theory and Unified SuperStandard Model THIRD EDITION* (Pingree Hill Publishing, Auburn, NH, 2018).

______, 2019a, *Calculation of: QED α = 1/137, and Other Coupling Constants of the Unified SuperStandard Theory* (Pingree Hill Publishing, Auburn, NH, 2019).

______, 2019b, *Coupling Constants of the Unified SuperStandard Theory SECOND EDITION* (Pingree Hill Publishing, Auburn, NH, 2019).

______, 2019c, *New Hybrid Quantum Big_Bang–Megaverse_Driven Universe with a Finite Big Bang and an Increasing Hubble Constant* (Pingree Hill Publishing, Auburn, NH, 2019).

______, 2019d, *The Universe, The Electron and The Vacuum* (Pingree Hill Publishing, Auburn, NH, 2019).

______, 2019e, *Quantum Big Bang – Quantum Vacuum Universes (Particles)* (Pingree Hill Publishing, Auburn, NH, 2019).

______, 2019f, *The Exact QED Calculation of the Fine Structure Constant Implies ALL 4D Universes have the Same Physics/Life Prospects* (Pingree Hill Publishing, Auburn, NH, 2019).

______, 2019g, *Unified SuperStandard Theory and the SuperUniverse Model: The Foundation of Science* (Pingree Hill Publishing, Auburn, NH, 2019).

______, 2020a, *Quaternion Unified SuperStandard Theory (The QUeST) and Megaverse Octonion SuperStandard Theory (MOST)* (Pingree Hill Publishing, Auburn, NH, 2020).

______, 2020b, *United Universes Quaternion Universe - Octonion Megaverse* (Pingree Hill Publishing, Auburn, NH, 2020).

______, 2020c, *Unified SuperStandard Theories for Quaternion Universes & The Octonion Megaverse* (Pingree Hill Publishing, Auburn, NH, 2020).

______, 2020d, *The Essence of Eternity: Quaternion & Octonion SuperStandard Theories* (Pingree Hill Publishing, Auburn, NH, 2020).

______, 2020e, *The Essence of Eternity II* (Pingree Hill Publishing, Auburn, NH, 2020).

______, 2020f, *A Very Conscious Universe* (Pingree Hill Publishing, Auburn, NH, 2020).

______, 2020g, *Hypercomplex Universe* (Pingree Hill Publishing, Auburn, NH, 2020).

______, 2020h, *Beneath the Quaternion Universe* (Pingree Hill Publishing, Auburn, NH, 2020).

______, 2020i, *Why is the Universe Real? From Quaternion & Octonion to Real Coordinates* (Pingree Hill Publishing, Auburn, NH, 2020).

______, 2020j, *The Origin of Universes: of Quaternion Unified SuperStandard Theory (QUeST); and of the Octonion Megaverse (UTMOST)* (Pingree Hill Publishing, Auburn, NH, 2020).

______, 2020k, *The Seven Spaces of Creation: Octonion Cosmology* (Pingree Hill Publishing, Auburn, NH, 2020).

______, 2020l, *From Octonion Cosmology to the Unified SuperStandard Theory of Particles* (Pingree Hill Publishing, Auburn, NH, 2020).

______, 2021a, *Pioneering the Cosmos* (Pingree Hill Publishing, Auburn, NH, 2021).

______, 2021b, *Pioneering the Cosmos II* (Pingree Hill Publishing, Auburn, NH, 2021).

______, 2021c, *Beyond Octonion Cosmology* (Pingree Hill Publishing, Auburn, NH, 2021).

______, 2021d, *Universes are Particles* (Pingree Hill Publishing, Auburn, NH, 2021).

______, 2021e, *Octonion-like dna-based life, Universe expansion is decay, Emerging New Physics* (Pingree Hill Publishing, Auburn, NH, 2021).

______, 2021f, *The Science of Creation New Quantum Field Theory of Spaces* (Pingree Hill Publishing, Auburn, NH, 2021).

______, 2021g, *Quantum Space Theory With Application to Octonion Cosmology & Possibly To Fermionic Condensed Matter* (Pingree Hill Publishing, Auburn, NH, 2021).

______, 2021h, *21st Century Natural Philosophy of Octonion Cosmology , and Predestination, Fate, and Free Will* (Pingree Hill Publishing, Auburn, NH, 2021).

Eddington, A. S., 1952, *The Mathematical Theory of Relativity* (Cambridge University Press, Cambridge, U.K., 1952).

Fant, Karl M., 2005, *Logically Determined Design: Clockless System Design With NULL Convention Logic* (John Wiley and Sons, Hoboken, NJ, 2005).

Feinberg, G. and Shapiro, R., 1980, *Life Beyond Earth: The Intelligent Earthlings Guide to Life in the Universe* (William Morrow and Company, New York, 1980).

Gelfand, I. M., Fomin, S. V., Silverman, R. A. (tr), 2000, *Calculus of Variations* (Dover Publications, Mineola, NY, 2000).

Giaquinta, M., Modica, G., Souchek, J., 1998, *Cartesian Coordinates in the Calculus of Variations* Volumes I and II (Springer-Verlag, New York, 1998).

Giaquinta, M., Hildebrandt, S., 1996, *Calculus of Variations* Volumes I and II (Springer-Verlag, New York, 1996).

Gradshteyn, I. S. and Ryzhik, I. M., 1965, *Table of Integrals, Series, and Products* (Academic Press, New York, 1965).

Heitler, W., 1954, *The Quantum Theory of Radiation* (Claendon Press, Oxford, UK, 1954).

Huang, Kerson, 1992, *Quarks, Leptons & Gauge Fields 2nd Edition* (World Scientific Publishing Company, Singapore, 1992).

Jost, J., Li-Jost, X., 1998, *Calculus of Variations* (Cambridge University Press, New York, 1998).

Kaku, Michio, 1993, *Quantum Field Theory*, (Oxford University Press, New York, 1993).

Kirk, G. S. and Raven, J. E., 1962, *The Presocratic Philosophers* (Cambridge University Press, New York, 1962).

Landau, L. D. and Lifshitz, E. M., 1987, *Fluid Mechanics 2nd Edition*, (Pergamon Press, Elmsford, NY, 1987).

Misner, C. W., Thorne, K. S., and Wheeler, J. A., 1973, *Gravitation* (W. H. Freeman, New York, 1973).

Rescher, N., 1967, *The Philosophy of Leibniz* (Prentice-Hall, Englewood Cliffs, NJ, 1967).

Rieffel, Eleanor and Polak, Wolfgang, 2014, *Quantum Computing* (MIT Press, Cambridge, MA, 2014).

Riesz, Frigyes and Sz.-Nagy, Béla, 1990, *Functional Analysis* (Dover Publications, New York, 1990).

Sagan, H., 1993, *Introduction to the Calculus of Variations* (Dover Publications, Mineola, NY, 1993).

Sakurai, J. J., 1964, *Invariance Principles and Elementary Particles* (Princeton University Press, Princeton, NJ, 1964).

Weinberg, S., 1972, *Gravitation and Cosmology* (John Wiley and Sons, New York, 1972).

Weinberg, S., 1995, *The Quantum Theory of Fields Volume I* (Cambridge University Press, New York, 1995).

INDEX

About the Author

Stephen Blaha is a well-known Physicist and Man of Letters with interests in Science, Society and civilization, the Arts, and Technology. He had an Alfred P. Sloan Foundation scholarship in college. He received his Ph.D. in Physics from Rockefeller University. He has served on the faculties of several major universities. He was also a Member of the Technical Staff at Bell Laboratories, a manager at the Boston Globe Newspaper, a Director at Wang Laboratories, and President of Blaha Software Inc. and of Janus Associates Inc. (NH).

Among other achievements he was a co-discoverer of the "r potential" for heavy quark binding developing the first (and still the only demonstrable) non-Aeolian gauge theory with an "r" potential; first suggested the existence of topological structures in superfluid He-3; first proposed Yang-Mills theories would appear in condensed matter phenomena with non-scalar order parameters; first developed a grammar-based formalism for quantum computers and applied it to elementary particle theories; first developed a new form of quantum field theory without divergences (thus solving a major 60 year old problem that enabled a unified theory of the Standard Model and Quantum Gravity without divergences to be developed); first developed a formulation of complex General Relativity based on analytic continuation from real space-time; first developed a generalized non-homogeneous Robertson-Walker metric that enabled a quantum theory of the Big Bang to be developed without singularities at t = 0; first generalized Cauchy's theorem and Gauss' theorem to complex, curved multi-dimensional spaces; received Honorable Mention in the Gravity Research Foundation Essay Competition in 1978; first developed a physically acceptable theory of faster-than-light particles; first derived a composition of extremums method in the Calculus of Variations; first quantitatively suggested that inflationary periods in the history of the universe were not needed; first proved Gödel's Theorem implies Nature must be quantum; provided a new alternative to the Higgs Mechanism, and Higgs particles, to generate masses; first showed how to resolve logical paradoxes including Gödel's Undecidability Theorem by developing Operator Logic and Quantum Operator Logic; first developed a quantitative harmonic oscillator-like model of the life cycle, and interactions, of civilizations; first showed how equations describing superorganisms also apply to civilizations. A recent book shows his theory applies successfully to the past 14 years of history and to *new* archaeological data on Andean and Mayan civilizations as well as Early Anatolian and Egyptian civilizations.

He first developed an axiomatic derivation of the form of The Standard Model from geometry – space-time properties – The Unified SuperStandard Model. It unifies all the known forces of Nature. It also has a Dark Matter sector that includes a Dark ElectroWeak sector with Dark doublets and Dark gauge interactions. It uses quantum coordinates to remove infinities that crop up in most

interacting quantum field theories and additionally to remove the infinities that appear in the Big Bang and generate inflationary growth of the universe. It shows gravity has a MOND-like form without sacrificing Newton's Laws. It relates the interactions of the MOND-like sector of gravity with the r-potential of Quark Confinement. The axioms of the theory lead to the question of their origin. We suggest in the preceding edition of this book it can be attributed to an entity with God-like properties. We explore these properties in "God Theory" and show they predict that the Cosmos exists forever although individual universes (or incarnations of our universe) "come and go." Several other important results emerge from God Theory such a functionally triune God. The Unified SuperStandard Theory has many other important parts described in the Current Edition of *The Unified SuperStandard Theory* and expanded in subsequent volumes.

Blaha has had a major impact on a succession of elementary particle theories: his Ph.D. thesis (1970), and papers, showed that quantum field theory calculations to all orders in ladder approximations could not give scaling deep inelastic electron-nucleon scattering. He later showed the eigenvalue equation for the fine structure constant α in Johnson-Baker-Willey QED had a zero at α = 1 not 1/137 by solving the Schwinger-Dyson equations to all orders in an approximation that agreed with exact results to 4th order in α thus ending interest in this theory. In 1979 at Prof. Ken Johnson's (MIT) suggestion he calculated the proton-neutron mass difference in the MIT bag model and found the result had the wrong sign reducing interest in the bag model. These results all appear in Physical Review papers. In the 2000's he repeatedly pointed out the shortcomings of SuperString theory and showed that The Standard Model's form could be derived from space-time geometry by an extension of Lorentz transformations to faster than light transformations. This deeper space-time basis greatly increases the possibility that it is part of THE fundamental theory. Recently, Blaha showed that the Weak interactions differed significantly from the Strong, electromagnetic and gravitation interactions in important respects while these interactions had similar features, and suggested that ElectroWeak theory, which is essentially a glued union of the Weak interactions and Electromagnetism, possibly modulo unknown Higgs particle features, be replaced by a unified theory of the other interactions combined with a stand-alone Weak interaction theory. Blaha also showed that, if Charmonium calculations are taken seriously, the Strong interaction coupling constant is only a factor of five larger than the electromagnetic coupling constant, and thus Strong interaction perturbation theory would make sense and yield physically meaningful results.

In graduate school (1965-71) he wrote substantial papers in elementary particles and group theory: The Inelastic E- P Structure Functions in a Gluon Model. Phys. Lett. B40:501-502,1972; Deep-Inelastic E-P Structure Functions In A Ladder Model With Spin 1/2 Nucleons, Phys.Rev. D3:510-523,1971; Continuum Contributions To The Pion Radius, Phys. Rev. 178:2167-2169,1969; Character Analysis of U(N) and SU(N), J. Math. Phys. 10, 2156 (1969); and The Calculation of the Irreducible Characters of the Symmetric Group in Terms of the

Compound Characters, (Published as Blaha's Lemma in D. E. Knuth's book: *The Art of Computer Programming Vols. 1 – 4*).
In the early 1980's Blaha was also a pioneer in the development of UNIX for financial, scientific and Internet applications: benchmarked UNIX versions showing that block size was critical for UNIX performance, developing financial modeling software, starting database benchmarking comparison studies, developing Internet-like UNIX networking (1982) and developing a hybrid shell programming technique (1982) that was a precursor to the PERL programming language. He was also the manager of the AT&T ten-year future products development database. His work helped lead to commercial UNIX on computers such as Sun Micros, IBM AIX minis, and Apple computers.
In the 1980's he pioneered the development of PC Desktop Publishing on laser printers and was nominated for three "Awards for Technical Excellence" in 1987 by PC Magazine for PC software products that he designed and developed.
Recently he has developed a theory of Megaverses – actual universes of which our universe is one – with quantum particle-like properties based on the Wheeler-DeWitt equation of Quantum Gravity. He has developed a theory of a baryonic force, which had been conjectured many years ago, and estimated the strength of the force based on discrepancies in measurements of the gravitational constant G. This force, operative in D-dimensional space, can be used to escape from our universe in "uniships" which are the equivalent of the faster-than-light starships proposed in the author's earlier books. Thus travel to other universes, as well as to other stars is possible.
Blaha also considered the complexified Wheeler-DeWitt equation and showed that its limitation to real-valued coordinates and metrics generated a Cosmological Constant in the Einstein equations.
The author has also recently written a series of books on the serious problems of the United States and their solution as well as a book on the decline of Mankind that will follow from current social and genetic trends in Mankind.

In the past twenty years Dr. Blaha has written over 80 books on a wide range of topics. Some recent major works are: *From Asynchronous Logic to The Standard Model to Superflight to the Stars, All the Universe!, SuperCivilizations: Civilizations as Superorganisms, America's Future: an Islamic Surge, ISIS, al Qaeda, World Epidemics, Ukraine, Russia-China Pact, US Leadership Crisis, The Rises and Falls of Man – Destiny – 3000 AD: New Support for a Superorganism MACRO-THEORY of CIVILIZATIONS From CURRENT WORLD TRENDS and NEW Peruvian, Pre-Mayan, Mayan, Anatolian, and Early Egyptian Data, with a Projection to 3000 AD,* and *Mankind in Decline: Genetic Disasters, Human-Animal Hybrids, Overpopulation, Pollution, Global Warming, Food and Water Shortages, Desertification, Poverty, Rising Violence, Genocide, Epidemics, Wars, Leadership Failure.*

He has taught approximately 4,000 students in undergraduate, graduate, and postgraduate corporate education courses primarily in major universities, and large companies and government agencies.

Recently he developed a quantum theory, The Unified SuperStandard Theory (UST), which describes elementary particles in detail without the difficulties of conventional quantum field theory. He found that the internal symmetries of this theory could be exactly derived from an octonion theory called QUeST. He further found that another octonion theory (UTMOST) describes the Megaverse. It can hold QUeST universes such as our own universe. It has an internal symmetry structure which is a superset of the QUeST internal symmetries.